Wireless Technology: Design, Tools and Techniques

Wireless Technology: Design, Tools and Techniques

Edited by **Dravin Rager**

Willford Press

New York

Published by Willford Press,
118-35 Queens Blvd., Suite 400,
Forest Hills, NY 11375, USA
www.willfordpress.com

Wireless Technology: Design, Tools and Techniques
Edited by Dravin Rager

International Standard Book Number: 978-1-68285-279-8 (Hardback)

Printed in China.

Contents

Preface

This book was inspired by the evolution of our times; to answer the curiosity of inquisitive minds. Many developments have occurred across the globe in the recent past which has transformed the progress in the field.

The development in wireless technology and communications has had a tremendous impact on scientific advancements and innovations. The topics included in this book such as wireless network security, modeling and propagation of channels, wireless traffic and routing, etc. provide a comprehensive insight into this discipline. As this field is emerging at a fast pace, this book will help the readers to better understand the concepts related to techniques and design of wireless technology. This book, with its detailed analyses and data, will prove immensely beneficial to professionals and students involved in this area at various levels.

This book was developed from a mere concept to drafts to chapters and finally compiled together as a complete text to benefit the readers across all nations. To ensure the quality of the content we instilled two significant steps in our procedure. The first was to appoint an editorial team that would verify the data and statistics provided in the book and also select the most appropriate and valuable contributions from the plentiful contributions we received from authors worldwide. The next step was to appoint an expert of the topic as the Editor-in-Chief, who would head the project and finally make the necessary amendments and modifications to make the text reader-friendly. I was then commissioned to examine all the material to present the topics in the most comprehensible and productive format.

I would like to take this opportunity to thank all the contributing authors who were supportive enough to contribute their time and knowledge to this project. I also wish to convey my regards to my family who have been extremely supportive during the entire project.

Editor

A security authentication method based on trust evaluation in VANETs

Ao Zhou[1*], Jinglin Li[1], Qibo Sun[1], Cunqun Fan[2], Tao Lei[1] and Fangchun Yang[1]

Abstract

Vehicular *ad hoc* networks (VANETs) have a high degree of openness. Therefore, if a new vehicle node wants to access the network, we need to validate the vehicle node carefully to ensure the security of the entire networking. There are a large number of vehicle nodes, and the strange degree among the nodes is very high in VANETs. In addition, VANETs are human-oriented networks. All vehicle nodes in the VANET have the right to decide whether to accept a new node. To attack this challenge, this paper proposes a security authentication method based on trust evaluation. The security authentication method consists of two parts: secure authentication based on direct trust evaluation and secure authentication based on indirect trust evaluation. In the direct trust evaluation, the security vector model is established based on the security behaviors of the new vehicle node. The historical security evaluation from the authority units (AU) is collected to calculate the final direct trust. In the indirect trust evaluation, the trust degree is calculated based on the recommendation trust vectors from the vehicle nodes in the network. The method employs correlation coefficient for distinguishing the malicious vehicle. Then, the recommendations from the malicious vehicle nodes are removed. The final recommended trust is gained by calculating the average recommendation trusts of remained vehicle nodes. Simulation results show the advantage of our proposed method.

Keywords: VANETs; Direct trust; Indirect trust; Correlation coefficient; Security

1 Introduction

VANETs have a high degree of openness [1]. Therefore, VANET needs face a diverse of security threats [2,3]. Firstly, through accessing the VANETs, the attacker can conduct privacy spy and obtain the moving track of the vehicles [4]. Secondly, some attackers release some false news (such as traffic accidents, road congestion, etc.) [5], and the false news can lead to chaotic traffic and accidents. What is more, the openness and highly dynamic of VANETs make the malicious attacks easy to implement and difficult to detect [6]. Due to the application characteristics and application scenarios of VANETs, these attacks can threat the information security and the property safety of users [7]. Therefore, how to accurately authenticate the new accessing vehicle node is becoming an urgently required research problem [8].

Current trust management researches are focused on message evaluation and user privacy protecting. The message evaluation can stop delivering of false message and enhance the security of VANETs. Although the trust management can be applied in VANETs to comprehensively improve the security and reliability of VANETs, these methods cannot fundamentally solve the problem. What is more, previous studies do not consider the 'suggestion' from existing nodes when validating the new accessing node. Because the VANETs are human-oriented user groups, the vehicle nodes in VANETs can subjectively judge whether to accept the new accessing node. To solve the problem and improve the accuracy of trust evaluation, this paper proposes a security authentication method based on trust evaluation. Our method does security authentication based on both direct trust evaluation and indirect trust evaluation. In the direct trust evaluation, in order to objectively determine whether the vehicle node can access the VANET, we calculate the direct trust value based on historical security behavior information. In the indirect trust evaluation, the indirect trust is calculated based on the recommendation trust from other vehicle nodes in the VANETs. Simulation results show the advantage of our method.

*Correspondence: hellozhouao@gmail.com
[1] State Key Laboratory of Networking and Switching Technology, Beijing University of Posts and Telecommunications, Beijing 100876, China
Full list of author information is available at the end of the article

The rest of this paper is organized as follows. Section 2 introduces related work. Section 3 shows our proposed security authentication method. Evaluation of the method is explained in Section 4. Section 5 concludes the paper.

2 Related work

There are plenty of trust management and encrypted authentication methods in literature. We will only review some notable work due to space limitation.

Biswas, et al. [9] proposed a safety message authentication scheme for VANETs. The scheme adopts an ID-based verification and signature mechanism. A certificate-less public key verification is offered by an ID-based technique. The message authentication is provided by a proxy signature. In this scheme, the standard ECDSA is incorporated with an ID-based proxy signature framework for the road-side unit originated messages. The transfer of signed message is specially handled to ensure the security and reliability of applications.

The work [10] pointed out that the characteristics and the security requirements of VANETs are quite different from standard *ad hoc* networks. Especially, trust management in VANETs is an urgently research problem. The paper concludes the advantage and disadvantage when adopting ordinary methods of network and standard *ad hoc* networks.

To protect the VANETs against attackers and defend VANETs against misbehavior, a threshold signature-based mechanism was proposed by the work [11]. The work also presents a privacy-preserving defense mechanism based on the threshold authentication. The paper does systematic analysis to show the strong point and the efficiency of the proposed mechanism.

The work [12] pointed out that evaluating the safety and trust level of vehicles is important to ensure the reliability of applications. The work also points out that traditional trust level is evaluated by monitoring the message generation and the behavior of the other vehicle nodes. However, the attacker can interrupt the regular communication among vehicles by creating a case of None Line of Sight. In addition, the case of None Line of Sight can prevent vehicles from monitoring other vehicle nodes. To solve the problem, the work proposes a location information-based trust evaluation model. The model can be used to evaluate the trust level of other vehicle nodes.

The emerging of VANET is to support the communication of vehicles on roads. The network allows arbitrary vehicles to broadcast traffic accident and other *ad hoc* messages. Because the attackers may release some false news, the work [13] pointed out that the concern of security and privacy needs to be taken into consideration. Therefore, all messages should be verified. However, the validation process should not reveal the real identity of vehicles. To solve the problem, a software-based solution is proposed in this paper. The method uses only two shared secret, therefore, the proposed method can satisfy the requirement. A group communication protocol is also proposed in this paper to allow communication between vehicles in the same group with a high level of security.

Certification and proof-of-work system are two basic mechanisms that have been used in security mechanisms. Palomar et al. [14] proposed a method based on the two mechanisms to provide safe communication environment and combat spam.

To guide the drivers to desired destinations, Chim et al. [4] made use of the online real-time road information collected by the vehicle nodes. When the method calculates the best route for drivers, the information source is authenticated to avoid attack. At the same time, the privacy of the drivers is protected. All nodes, including the trusted authority, cannot obtain the destination of the driver.

The release of wrong information can lead to injury to the lives of the drivers. Therefore, the Sybil attack is a serious threat in VANETs. The paper [15] proposed a Sybil attack detection algorithm to solve the problem. The algorithm is based on signature mechanism in VANETs. In the moving process, each vehicle node gathers the digital signatures at the same time; all the collected signature vectors are analyzed and compared to detect the Sybil attack.

To measure the integrity degree of security scheme in VANETs, Azogu et al. [16] proposed an asymmetric profit-loss Markov model. In the proposed asymmetric Markov model, the loss denotes the negative effect a vitiated data fragment is received by a device. Profit represents the positive effect when a vitiated data fragment is detected and disregarded. Markov chain records change of system behavior that reacts to profit and loss asymmetrically. The model adopts a black-box method when measuring the integrity level. In other words, the model does not need to know the implementation details of each security scheme. Therefore, the model is very suitable for real world applications.

There are a large number of well-performed trust evaluation methods. Nevertheless, there is lack of comprehensive security authentication method for VANETs. In addition, previous studies do not take the subjective recommendation into consideration. Different from current methods, our method can solve the problem. We will present the detail of our method in the next section.

3 Multi-level security certification of VANETs

To ensure the security of the VANETs, this paper presents a comprehensive vehicle node security authentication method. The proposed security authentication method consists of two main parts: the direct trust evaluation and the indirect trust evaluation. When a vehicle wants to access the Internet through the roadside base station, the direct trust evaluation method is adopted to validate the vehicle node. In order to objectively determine whether the vehicle node can access the VANET, the trust of the node is directly evaluated based on the historical evaluations from the authority unit (AU). When a group of vehicles form a wireless network to communicate information with each other, the indirect trust evaluation method can decide whether to accept a new vehicle node. The indirect trust evaluation mechanism of the vehicle node is established based on the recommended trust from other nodes in VANETs. The indirect trust evaluation lets the nodes within the network decide the acceptance of the new access vehicle node. We will show the detail of the trust evaluation in this section.

3.1 Security authentication based on direct trust evaluation

Vehicles need to access the VANETs to obtain the needed information. We need to validate the new vehicle nodes when they access the VANET. This section proposes a direct trust evaluation method based on historical security event record. In the interaction of AUs and vehicle nodes, the security events are recorded to the database. All AUs can access the Internet. The AUs belonging to the same organization can share the same database. We can make use of the recorded events to evaluate the new vehicle node and further determine whether the vehicle is credible. Because we try to calculate the direct trust of vehicle node based on the historical security events, we will discuss how to analyze the recorded historical events in detail.

3.1.1 *The security vector model*

The VANET can support plenty of applications. For the inherent characteristics of the vehicle, the security events of the vehicle node are very complex. There are a large number of security events. But these security events can be classified into several types, for example: physical security events, information privacy security events, information-disruptive events, et al. Therefore, security behaviors of the vehicle node can be subdivided. All security behaviors of the vehicle are classified to a security event type *VANET_Event*$_n$. We now introduce the security vector.

Definition 1. The security vector is defined by the following:

$$\text{Security\_Vector} =< \text{VE}_1, \text{VE}_2, \ldots, \text{VE}_n > \tag{1}$$

All vehicle node security events are classified based on the security vector, and each event belongs to a corresponding security component E_i. Each security component, respectively, reflects the security level of the corresponding security events.

The security vector can reflect the security level of the vehicle nodes. In VANETs environment, the security events of the vehicle nodes are very complex, and how to classify the security events is also very complex. An accurate security component partition can reflect the security of the vehicle more accurately. For example, personal security events should have a higher level compared to other security events, and information privacy security events should have a higher level compared to the interference events.

3.1.2 *Trust evaluation based on the historical security events*

In VANETs, security evaluation based on the historical events is an important part of the direct trust evaluation. If there is no insecure event of the vehicle node in the record, the vehicle node is trustworthy to some extent.

Vehicle nodes and roadside fixed access points need to exchange application data and other information with AU, so AU can evaluate the security of the vehicle node. The security event information of vehicle node is saved on the local server, and AU can evaluate the vehicle node based on the information. The evaluation data has its own specific format as follows:

$$\text{Event} = \{\text{VID, AUID, VANET\_Event, EventID, Security, Time}\}\ \text{sig}_{AU} \tag{2}$$

where VID and AUID represent the ID of the evaluated node. *VANET_Event* represents the security event component. *EventID* represents the security event ID. *Security* represents the security value of the vehicle node. *Time* represents timestamp of the system. sig_{AU} represents the private key signature of AU.

Definition 2. Vehicle historical security degree VHS:

$$P_{V_{Ei}} = \frac{1}{n}\sum_j (\text{Event}_j.\text{Security}),\ \text{Event}_j.\text{VANET\_Event is Ei} \tag{3}$$

where $Security_i$ represents the security degree evaluated by AU. n represents the number of events belonging to E_i. The events long before may have little relevance to current security. Therefore, only recent events are considered. The 'period of validity' of the events can be decided by the AU owner. P_v can represent the average security degree of a vehicle node evaluated by AU.

The time complexity of the direct trust evaluation method is $O(n)$. n is the number of recent history events.

3.2 Security authentication based on indirect trust evaluation

When a group of vehicles form a wireless network to communicate information with each other, the vehicle nodes in the network have the right to determine whether to accept the new vehicle node. The vehicle network is largely similar to the interpersonal network. In interpersonal network, the acceptance of a new node mainly depends on the trust value and the recommendation of other individuals. Therefore, the trust value of the vehicle node depends on the recommendation trust from other nodes in the network. However, some selfish vehicle nodes in the VANET may maliciously deny the new vehicle node. The condition is unfavorable for the VANET. We need to distinguish the malicious nodes before calculating the indirect trust value. Based on above analysis, we propose an indirect trust evaluation method. In the indirect trust evaluation, we make use of the correlation coefficient of the recommendation trust value to distinguish the malicious nodes. All the recommendation trust values from the malicious nodes are removed. We do not take the suggestion of malicious nodes into consideration when calculating the indirect trust value.

3.2.1 *The recommendation trust vector model*

The security behaviors of vehicle nodes are diverseness and complexity. The security behavior of vehicles generally include: safety drive, information security, information authenticity, information accuracy, etc. All security behaviors can be quantized. The recommendation node scores the access node in each security behavior. All the scores are stored in a vector. The vector is called recommendation trust vector in our paper.

Definition 3. The recommendation trust vector is defined by the following:

$$\text{Trust\_Vector}_n = \langle P_{S1}^{XB}(t), P_{S2}^{XB}(t), \ldots, P_{Sm}^{XB}(t) \rangle \tag{4}$$

It represents recommendation trust vector from node X to node B. X denotes the Xth recommendation node, and B denotes the node to be evaluated. S_i denotes the ith type of security event. $P_{Si}^{XB}(t)$ is called the trust component. The trust component denotes the score for a type of security behavior. The score from a vehicle node may change with time. A vehicle node may give a different score for the same access node at different time. Therefore, we give each score a timestamp. t represents the specific timestamp when node X gives the score.

In VANET, the score from different recommendation nodes is quite different and denotes the subjective will of each recommendation node. Some vehicle nodes in the VANET may maliciously give a low score to the vehicle node. However, the maliciousness can be detected by analyzing each recommendation trust component P_{Si} carefully. In this way, we can evaluate the trust of the vehicle node more accurately. The detail will be discussed in the next section.

3.2.2 *Trust evaluation based on the recommendation trust value*

The score from recommendation vehicle nodes may not consist with the real behavior of the new accessing node. VANET is an open network and faces many risks. Some inner and outer factors may cause some node to become selfish and malicious. In general, we can divide the recommendation node into two types: general recommendation node and malicious recommendation node. The inconsistency is mainly caused by the malicious recommendation node. To eliminate the influence of malicious recommendation nodes, we propose a malicious recommendation node detection method. Firstly, the method calculates the average recommendation trust vector. The average recommendation trust vector can be obtained by calculating the mean score of each trust component. Then, by analyzing the correlation coefficient between average recommendation trust vector and recommendation trust vector of each recommendation node, we can detect the malicious recommendation node. The recommendation nodes that have a relatively larger deviation are the malicious recommendation nodes. The average recommendation trust vector can be calculated by the following:

$$\text{Trust\_Vector}_{\text{average}} = \langle P_{S1}^{B}(t), P_{S2}^{B}(t), \ldots, P_{Sm}^{B}(t) \rangle \tag{5}$$

where $P_{Sm}^{B}(t) = \frac{1}{n} \sum_{x \in \text{network}} \left[P_{Sm}^{xB}(t) \right]$, x denotes the xth recommendation node, and n represents the number of recommendation nodes.

All existing vehicle nodes in the VANET should send their recommendation trust vectors. A vehicle node can send a blank vector to give up the right. After receiving the recommendation trust vector from all vehicle nodes in the VANET, we need to compare each recommendation trust vector with the average recommendation trust vector. In this way, we can distinguish malicious recommendation nodes. Linear interpolation method is a basic method to solve the problem. However, this method is obviously not able to accurately distinguish the malicious recommendation nodes. Therefore, we will adopt a correlation coefficient based method to solve the problem. Firstly, we will introduce the definition of the correlation coefficient:

Definition 4. (Correlation coefficient) The correlation coefficient ρ is an index which represents the degree of correlation between variables. The correlation coefficient is larger than -1 and smaller than 1. When the value of

$|\rho|$ is large, the error Q is small; and the linear correlation degree between variables is high. When the value of $|\rho|$ is small, the error Q is large; and the linear correlation degree between variables is low. When there are two sample functions X and Y, the cross-correlation of X and Y can be calculated by the following:

$$\rho_{xy} = \frac{\sum_{i=1}^{N}(X_i - \bar{X})(Y_i - \bar{Y})}{\sqrt{\sum_{i=1}^{N}(X_i - \bar{X})^2}\sqrt{\sum_{i=1}^{N}(Y_i - \bar{Y})^2}} \tag{6}$$

To express our work more concisely, we employ $T_V$instead of *Trust_Vector* in the rest of our paper. The correlation coefficient between average recommendation trust vector and a single vehicle node recommendation trust vector can be calculated by the following:

$$\mathrm{Cov}(T\_V_a, T\_V_n) = E(T\_V_a, T\_V_n) - E(T\_V_a)E(T\_V_n) \tag{7}$$

$$\rho_{xy} = \frac{\mathrm{Cov}(T\_V_a, T\_V_n)}{\sqrt{D(T\_V_a)}\sqrt{D(T\_V_n)}} \tag{8}$$

To distinguish the malicious nodes, we will do a quantitative comparison based on the correlation coefficient. We set two thresholds to divide the range of $|\rho|$ into three intervals. Then, we determine whether the vehicle node is in a trusted state. Suppose the two thresholds are $\Delta\rho_1$ and $\Delta\rho_2$. The three intervals are $(0, 1 - \Delta\rho_1 - \Delta\rho_2]$, $(1 - \Delta\rho_1 - \Delta\rho_2, 1 - \Delta\rho_1]$, and $(1 - \Delta\rho_1, 1)$.

- When the correlation coefficient ρ_{ns} of T_V_a and T_V_n belongs to $(1 - \Delta\rho_1, 1)$, T_V_a and T_V_n are close to linear correlation. Therefore, we can consider that the node recommendation trust consists with the average recommendation trust. We think the recommendation trust of the node is reasonable. The recommendation node is not a malicious node.
- When the correlation coefficient ρ_{ns} of T_V_a and T_V_n belongs to $(1 - \Delta\rho_1 - \Delta\rho_2, 1 - \Delta\rho_1]$, the linear relationship between T_V_a and T_V_n is not so obvious. Therefore, the reasonableness of the recommended trust value needs a further discussion. We introduce a norm method to discuss reasonableness of the recommendation trust. The single recommendation trust vector and average recommendation trust vector are normalized by using the following equations:

$$\|T\_V_n\| = \sqrt{T\_V_n, T\_V_n} = \sqrt{P_{S1}^{XB}(t)^2 + P_{S2}^{XB}(t)^2, \ldots, P_{Sm}^{XB}(t)^2} \tag{9}$$

$$\|T\_V_a\| = \sqrt{T\_V_a, T\_V_a} = \sqrt{P_{S1}^{B}(t)^2 + P_{S2}^{B}(t)^2, \ldots, P_{Sm}^{B}(t)^2} \tag{10}$$

The deviation between $\|T\_V_n\|$ and $\|T\_V_a\|$ can be calculated by the following:

$$\Delta\|T\_V\| = |\|T\_V_n\| - \|T\_V_a\|| \tag{11}$$

We set the minimum deviation of normalized trust vector as $\|T\_V\|'_{\min}$. When $\Delta\|T\_V\| > \|T\_V\|'_{\min}$, we think the recommendation node is not trusted. Otherwise, the deviation is small. We think that the recommendation node is in a temporary trusted state.

- When the correlation coefficient ρ_{ns} of T_V_a and T_V_n belongs to $(0, 1 - \Delta\rho_1 - \Delta\rho_2]$, T_V_a and T_V_n do not show a linear relationship. Therefore, we can consider that the deviation is very large. We think the recommendation trust value of the node is unreasonable. The recommendation node is a malicious node.

The value of $\Delta\rho_1$ and $\Delta\rho_2$ can be adjusted according to the actual situation. We will discuss the problem in the experimental section.

3.2.3 *The recommendation trust*

After distinguishing the malicious nodes, we can obtain the unreasonable recommendation trust vector. The unreasonable recommendation trust vectors should be abandoned. When calculating the average recommendation trust value, we just use the reasonable recommendation trust vectors. The formula is as follows:

$$\mathrm{Trust\_Vector}_B = \left\langle P_{S1}^{B}(t), P_{S2}^{B}(t), \ldots, P_{Sj}^{B}(t)\right\rangle \tag{12}$$

where *Trust_Vector*$_B$ is the average recommendation trust vector from other vehicle nodes. The trust value of the target node can be evaluated according to the above equation. Then, whether to accept the access node can be determined based on the vector.

The time complexity of the average recommendation trust vector calculation is O(m*n). n is the number of recommendation nodes, and m is the number of security event type. The time complexity of the correlation coefficient calculation is O(m*n). The time complexity of the malicious node detection is $O(n)$. The time complexity of the final recommendation trust calculation is O(m*n). Therefore, the time complexity of the indirect trust evaluation method is O(m*n).

4 Experimental results

We proposed a security authentication method based on trust evaluation. The method is composed of two parts: Firstly, direct trust evaluation is presented based on historical security event record. Secondly, the vehicle nodes in the network can determine whether to accept the new vehicle node. To verify the effectiveness of our method, we implement our method in Matlab and conduct experiments on it. The following sections first outline the experimental setting. We then discuss the results and show the advantage of our method.

4.1 Experimental setup

The physical machine configurations of our experiments are as follows: CPU core is i3-2310M 2.10 GHz, RAM is 2 GB, and operating system is Windows 7. The method is implemented in Matlab 7.1.

There are 80 target nodes when we do the direct trust evaluation simulation experiment. Then, all the security events are divided into five levels in the security settings, and the security weights are as follows: 0.1, 0.15, 0.2, 0.25, 0.3. The security degree of target nodes is calculated as follow:

$$\text{Security}_i = 0.1V_{E1} + 0.15V_{E2} + 0.2V_{E3} + 0.25V_{E4} + 0.3V_{E5} \quad (13)$$

There are 20 recommendation vehicle nodes when we do the indirect trust evaluation simulation experiment. All the security events are divided into five levels, and their security weights are as follows: 0.1, 0.15, 0.2, 0.3, 0.35. The security degree of target nodes can be calculate by the following:

$$\text{Trust}_B = 0.1P^B_{S1}(t) + 0.15P^B_{S2}(t) + 0.2P^B_{S3}(t) + 0.3P^B_{S4}(t) + 0.35P^B_{S5}(t) \quad (14)$$

4.2 Experimental results

To verify the accuracy of the proposed method, we discuss the result from two aspects. We first study the results on direct trust evaluation experiment. Then, we discuss the results on indirect trust evaluation experiment.

4.2.1 Direct trust evaluation

Figure 1 shows the security degree distribution of the 80 nodes, and Figure 2 shows the number of nodes that can be trusted in after the minimum accepted security degree has been selected.

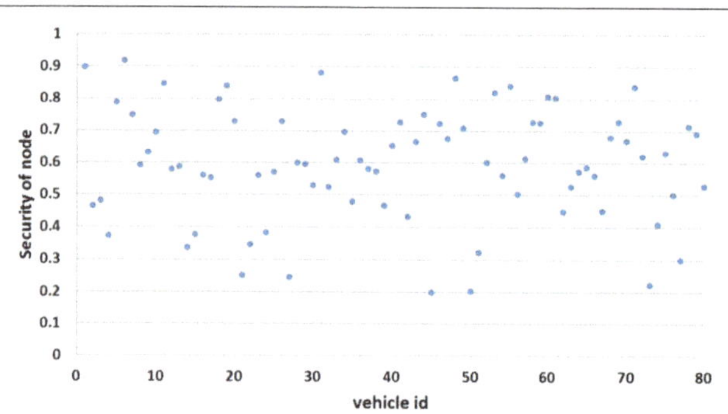

Figure 1 Distribution of security degree.

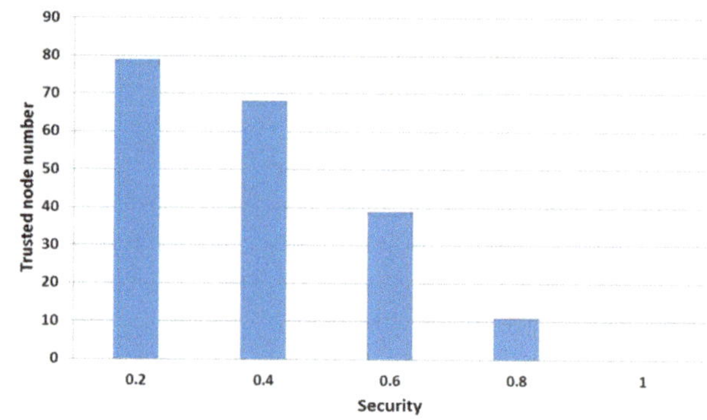

Figure 2 Trusted node number.

As shown in Figure 1, on various security degrees, how to distinguish the security of nodes is relatively obvious. As shown in Figure 2, the security degree of most nodes is greater than 0.5. When the minimum accepted security degree is 0.2, there are 79 nodes than can meet the security demands. When the minimum accepted security degree is 0.4, there are 68 nodes that can meet the security demands. When the minimum accepted security degree is 0.6, there are 39 nodes that can meet the security demands. When the minimum accepted security degree is 0.8, there are 11 nodes that can meet the security demands.

After calculating the security degree of the nodes, you can decide the appropriate minimum accepted security degree based on the actual situation.

4.2.2 Indirect trust evaluation

To study the performance of our indirect trust evaluation method (ITE), we compare ITE with an indirect trust evaluation method without malicious nodes detecting (WMD). Different from ITE, WMD does not detect malicious nodes and remove the malicious recommendation trust vectors.

When we evaluate the node by using ITE, all existing vehicle nodes in the VANET should send their recommendation trust vectors. The correlation coefficient values of all recommendation trust vectors are shown in Figure 3.

As shown in Figure 3, there are four nodes whose recommendation trust vector correlation coefficient is smaller than 0.6. There is a high chance that the four nodes are malicious nodes and may reject new nodes viciously. Among 20 recommendation nodes, there are 16 recommendation nodes whose recommendation trust vector correlation coefficient is larger than 0.6. It indicates that trust vectors from these 16 recommendation nodes consist with the overall distribution of trust vector. Therefore, there is a high chance that these recommendation nodes are not malicious. Therefore, we set $\Delta\rho_1 = 0.4$ and

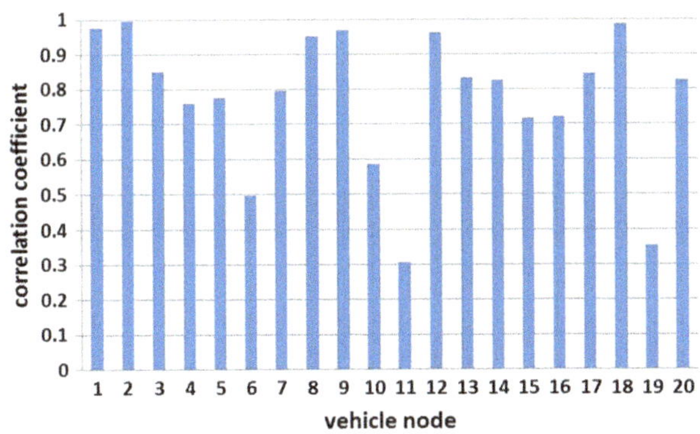

Figure 3 The correlation coefficient of trust vector.

$\Delta\rho_2 = 0.2$. The larger $\Delta\rho_1$ and $\Delta\rho_2$ are, the lesser recommendation nodes that can meet the demands. In the real environment, the value of $\Delta\rho_1$ and $\Delta\rho_2$ can be adjusted according to the actual situation.

Figure 4 shows the trust degree of the ten target nodes. We set $\Delta\rho_1 = 0.4$ and $\Delta\rho_2 = 0.2$. As shown in Figure 4, when employing ITE, the trust degree of the targets nodes is higher than WMD. That is because LTE employs correlation coefficient for distinguishing the malicious recommendation trust vector. LTE abandons the unreasonable recommendation trust first when calculating the final trust degree. As shown in Figure 4, there are nine nodes whose trust degree is greater than 0.6, there are seven nodes whose trust degree is greater than 0.7, and there are six nodes whose trust degree is greater than 0.8. A higher trust degree means more nodes in the network trust the new node. Therefore, there is higher chance that the node is more trustworthy than other nodes. As we can see from the above discussion, our method can accurately evaluate the vehicle nodes.

5 Conclusions

Because of the openness of VANETs, the network needs to face many security risks. If a new vehicle node wants to access the VANET, we need to validate the new vehicle node to improve the security of the VANET. We propose a security authentication method based on trust evaluation. Firstly, when a vehicle wants to access the Internet through the roadside base station, we evaluate the access node by employing the direct trust evaluation. Based on the historical security event record, the direct security degree of the new vehicle node is determined. When a group of vehicles form a wireless network to communicate information with each other, we adopt the indirect trust evaluation mechanism to evaluate the new vehicle node. All vehicle nodes in the network can determine whether to accept the new vehicle node. Each node sends a vector to show its recommendation trust value. Based on the correlation coefficient, we distinguish the malicious vehicle nodes and remove all recommendation trust value from the malicious vehicle nodes. Then, the indirect trust value is calculated by averaging all the remaining recommendation trust values. Simulation results show our method can accurately validate the vehicle node.

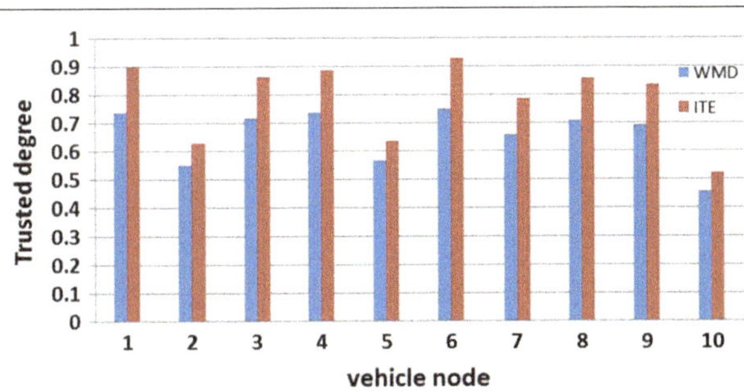

Figure 4 Trust degree.

Competing interests
The authors declare that they have no competing interests.

Acknowledgements
The work presented in this study is supported by NSFC (61202435, 61472047); Beijing Natural Science Foundation (4132048).

Author details
[1]State Key Laboratory of Networking and Switching Technology, Beijing University of Posts and Telecommunications, Beijing 100876, China. [2]National Satellite Meteorological Centre, China Meteorological Administration, Beijing 100876, China.

References

1. H Hartenstein, K Laberteaux, *VANET: Vehicular Applications and Inter-networking Technologies*. (Wiley Online Library, Hoboken, 2010)
2. A Buchenscheit, F Schaub, F Kargl, M Weber, in *Vehicular Networking Conference (VNC), 2009 IEEE*. A VANET-based emergency vehicle warning system (IEEE Piscataway, 2009), pp. 1–8
3. S Wang, C Fan, C-H Hsu, Q Sun, F Yang, A vertical handoff method via self-selection decision tree for internet of vehicles. IEEE Syst. J. **pp**(99), 1–10 (2014)
4. T Chim, S Yiu, L Hui, V Li, Vspn: VANET-based secure and privacy-preserving navigation. Compt. IEEE Trans. **63**(2), 510–524 (2012)
5. S Xi, X-M Li, in *Wireless Communications, Networking and Mobile Computing, 2008. WiCOM'08. 4th International Conference On*. Study of the feasibility of VANET and its routing protocols (IEEE, Piscataway, 2008), pp. 1–4
6. A Mahajan, N Potnis, K Gopalan, A Wang, in *Proceedings of the 10th ACM Symposium on Modeling, Analysis, and Simulation of Wireless and Mobile Systems*. Modeling VANET deployment in urban settings (ACM, New York, 2007), pp. 151–158
7. K Sampigethaya, M Li, L Huang, R Poovendran, Amoeba: robust location privacy scheme for vanet. Sel. Areas Commun. IEEE J. **25**(8), 1569–1589 (2007)
8. K Sampigethaya, L Huang, M Li, R Poovendran, K Matsuura, K Sezaki, Caravan: Providing location privacy for vanet. Technical report, DTIC Document (2005)
9. S Biswas, J Misic, V Misic, in *Distributed Computing Systems Workshops (ICDCSW), 2011 31st International Conference On*. ID-based safety message authentication for security and trust in vehicular networks (IEEE, Piscataway, 2011), pp. 323–331
10. P Wex, J Breuer, A Held, T Leinmuller, L Delgrossi, in *Vehicular Technology Conference, 2008. VTC Spring 2008. IEEE*. Trust issues for vehicular ad hoc networks (IEEE, Piscataway, 2008), pp. 2800–2804

11. J Sun, C Zhang, Y Zhang, Y Fang, An identity-based security system for user privacy in vehicular ad hoc networks. Parallel Distributed Syst. IEEE Trans. **21**(9), 1227–1239 (2010)
12. O Abumansoor, A Boukerche, in *Global Telecommunications Conference (GLOBECOM 2011), 2011 IEEE*. Towards a secure trust model for vehicular ad hoc networks services (IEEE, Piscataway, 2011), pp. 1–5
13. TW Chim, S-M Yiu, LC Hui, VO Li, Specs: Secure and privacy enhancing communications schemes for vanets. Ad Hoc Networks. **9**(2), 189–203 (2011)
14. E Palomar, JM de Fuentes, AI González-Tablas, A Alcaide, Hindering false event dissemination in vanets with proof-of-work mechanisms. Transportation Res. Part C: Emerging Technol. **23**, 85–97 (2012)
15. C Chen, W Han, X Wang, Sybil attack detection based on signature vectors in vanets. Int. J. Crit. Comput.-Based Syst. **2**(1), 25–37 (2011)
16. IK Azogu, MT Ferreira, H Liu, in *Global Communications Conference (GLOBECOM), 2012 IEEE*. A security metric for vanet content delivery (IEEE, Piscataway, 2012), pp. 991–996

2

Efficiency analysis of the dynamic traffic control for an urban highway

Ali Sercan Kesten[1*] and Murat Ergün[2]

Abstract

In this study, dynamic traffic control strategies, namely dynamic ramp metering and dynamic speed limit control, have been examined through microscopic traffic simulation based on site measurements. In this context, the traffic flow data at a particular highway intersection have been analyzed to determine the pattern of the traffic. Then, the traffic model has been built in a traffic micro-simulation software and calibrated with the field data. The foci of the study are to measure the efficiency of the dynamic traffic control strategies and to compare it with the uncontrolled case considering various performance indicators such as total travel time, average delay time per vehicle, and average number of stops per vehicle. For the dynamic ramp metering strategies, the ALINEA (Asservissement Lineaire d'Entree Autoroutiere - French for Linear Utilization for Highway Entrances) control algorithm is implemented with different fixed-time cycle lengths. It has been observed that various ramp metering implementations decreased the average delay time per vehicle up to 30%. The dynamic speed limit control strategies are set according to the occupancy rates that are measured at the bottleneck downstream. The examined speed limit control strategies decreased the average delay time per vehicle to around 7%. The results also revealed that the implemented dynamic traffic control strategies help alleviate congestion by increasing the capacity of the bottleneck section.

Keywords: Intelligent transportation systems; Dynamic ramp metering; Dynamic speed limit control

1 Introduction

Traffic congestion and management have been an important problem for traffic engineers. Traffic congestion along the freeway sections is expected to increase in upcoming years due to the continuous increasing demand of mobility [1]. Almost half of the congestion experienced in the modern world happens virtually every day which can be also defined as 'recurring.' This is a type of congestion where there are simply more vehicles than roadway capacity. The other half of congestion is caused by temporary disruptions that take away part of the roadway from use, namely 'nonrecurring' congestion [2]. There are many ways that congestion can occur along the highway. One of the most important reasons of congestion is the bottleneck sections which can be formed by either a lane drop or a temporary occupation of a lane due to an accident, incident, or roadwork. The disruption of traffic flow can be observed at the vicinity of bottlenecks caused by merging maneuvers. These merging maneuvers tend to create shock wave effects that cause long delays and slow traffic.

Congestion has a direct effect on travel speed and brings out safety concerns [3,4]. The continuously increasing traffic congestion problem has led to the application of various control strategies. Basically, these are formed by controlling the number of vehicles entering the freeway and/or by changing the speed limit of a designated section along the freeway. Advanced urban traffic networks including both urban roads and freeways utilize control strategies like signal control, ramp metering, variable message signs, and route guidance [5].

This paper investigates the performance of dynamic traffic control strategies as an alternative congestion management tool especially for the freeway recurring traffic congestion. An intercontinental highway section was selected as the study field that suffers from recurring traffic congestion caused by lane drop and a ramp merge just before the toll booth stretch of FSM Bridge in Istanbul.

* Correspondence: sercan.kesten@isikun.edu.tr
[1]Division of Civil Engineering, Department of Engineering, Işık University, Istanbul, Turkey
Full list of author information is available at the end of the article

The remainder of this paper is organized as follows: In Section 2, the ramp metering control, speed limit control models, and the model predictive control will be presented. Following this section, the simulation study will be introduced emphasizing the simulation study area, data, and simulation procedures. The simulation results will be given in Section 4, and the discussion and the conclusions are drawn in the corresponding sections of this paper.

2 Dynamic traffic control strategies

In order to control traffic in an optimal state, various approaches are proposed by traffic engineers since the 1970s [6]. The increasing demand of automobile use and therefore the use of highways cannot be managed by increasing the supply respectively. It is obvious that resources are limited and it is not feasible to have a capacity which is required for only a limited time. Intelligent traffic control systems are one of the prominent alternatives to road expansion and should be the first option to consider. The models examined in this study are widely used traffic control strategies which are proven to improve traffic conditions by increasing the capacity utilization, namely ramp metering control, variable speed limit control, and model predictive control.

2.1 Ramp metering control

The use of traffic signals on ramps to control the merging on freeways is called ramp metering. Ramp meters are installed to control the rate of vehicles moving into to the mainline traffic; thus, they prevent the critical volume of a freeway in order to control the demand and, moreover, break the platoon of vehicles entering the freeway upstream of the signal to decrease the weaving phenomenon at the merge area.

Ramp metering is a well-known strategy in freeway traffic control. In fact, various techniques of ramp control were used in the late 1950s and through 1970s in Japan and USA [7]. By the early 1990s, the technological advancement both in computing and measurement techniques make more sophisticated ramp metering systems possible to analyze and implement.

Ramp metering strategies could be categorized according to their level of scope ramp metering which are

- local ramp metering (isolated) and
- system-wide (coordinated) ramp metering.

In isolated or local ramp metering, the control system deals with an isolated highway section rather than the whole network. Most of the local ramp metering strategies can be categorized into four types [8]:

- *Demand capacity control.* In this type of control system, metering rate is determined by the upstream volume and downstream capacity. The difference between the upstream volume and downstream capacity determines the metering rate for the next cycle.
- *Upstream occupancy control.* In this strategy, real-time occupancy upstream of the on-ramp is used in order to determine the metering rate for the next cycle.
- *Gap acceptance control.* For gap acceptance control, occupancy measurements from upstream of the ramp are measured to determine the metering rate.
- *Closed-loop local control strategies.* For closed-loop control, system output is fed back and the input is modified with respect to the output. Here the aim is to set the output value at a desired level.

System-wide ramp metering is a control system which considers a network with various sequential or coordinated ramps. Improving traffic flow at a highway intersection may not be sufficient for a relaxation in the whole network. It may also be harmful to the other consecutive segments.

That is why the whole network should be under consideration while optimizing the traffic flow. For that purpose, system-wide (coordinated) ramp metering control strategies are taken into account. Examples of system-wide ramp metering control strategies are FLOW [9], Zone (Minnesota) algorithm [10], Stratified Zone (new Minnesota) algorithm [11], Helper [12], METALINE [13], and SWARM [14].

2.2 Variable speed limit control

Variable speed limit systems consist of variable message signs placed on gantries along the freeway and connected to a traffic control center. The variable message signs, rather than traditional static signs, are used to display the regulatory or advisory speed limit, enabling freeway system controllers to dynamically intervene to the corresponding traffic conditions. In general, variable speed limit control is implemented to homogenize traffic flow, improve safety, and reduce driver stress. Many variable speed limit control strategies have been put into action in USA, UK, the Netherlands, Germany, Australia, Austria, Japan, and Turkey [4].

There are several recent studies investigating the impact of variable speed limit on safety and traffic flow [15]. Much of the focus of VSL system evaluation studies has been on safety [16]. There appears to be even less evidence to suggest that speed control strategy increases traffic flow efficiency. Papageorgiou et al. [17] utilized the data of a German motorway to investigate the impact on aggregate traffic flow behavior and efficiency.

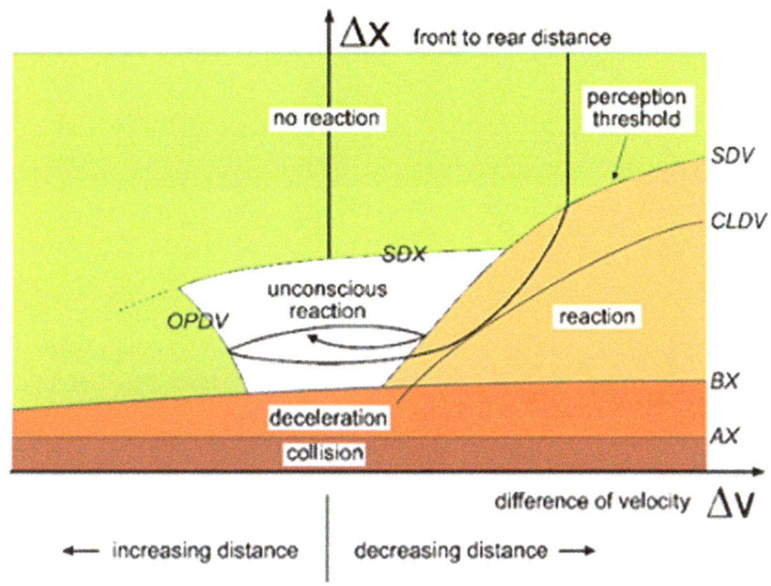

Figure 1 Wiedemann's car-following model [22].

2.3 Model predictive control

Model predictive control (MPC) models predict the change in the dependent variables of the modeled system that will be caused by changes in the independent variables. MPC is an optimal control method applied in a rolling horizon framework. Optimal control has been successfully applied by several researchers in traffic control over the years [9,18].

Either optimal control or MPC has the advantage of having the controller generate control decisions that are optimal according to a controller-supplied objective function. However, MPC offers some important advantages over conventional optimal control.

First, optimal control has an open-loop structure, which means that the disturbances (in our case, the traffic demands) have to be completely and exactly known before the simulation and that the traffic model has to be very accurate to ensure sufficient precision for the whole simulation. MPC operates in a closed-loop structure, which means that the traffic state and the current demands are regularly fed back to the controller, and the controller can take disturbances into account and correct for prediction errors resulting from model mismatch.

Second, adaptivity is easily implemented in MPC because the prediction model can be changed or replaced during operation. This may be necessary when traffic behavior significantly changes (e.g., in case of incidents, changing weather conditions, and lane closures for maintenance).

Third, for MPC, a shorter prediction horizon is usually sufficient, which reduces complexity and makes the real-

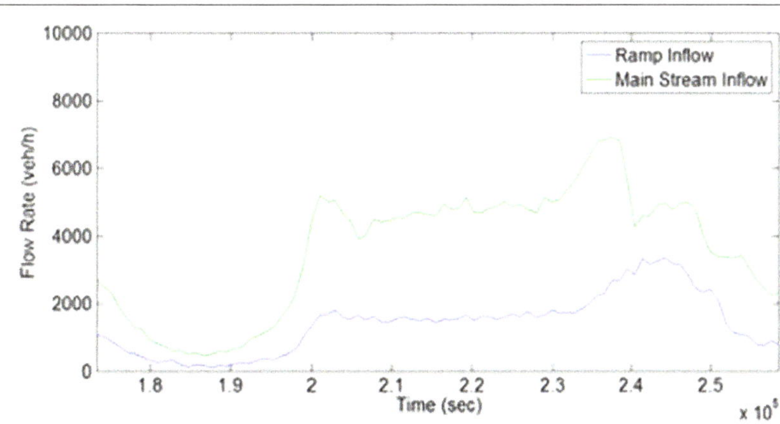

Figure 2 Flow data for 8 July 2010 (Thursday).

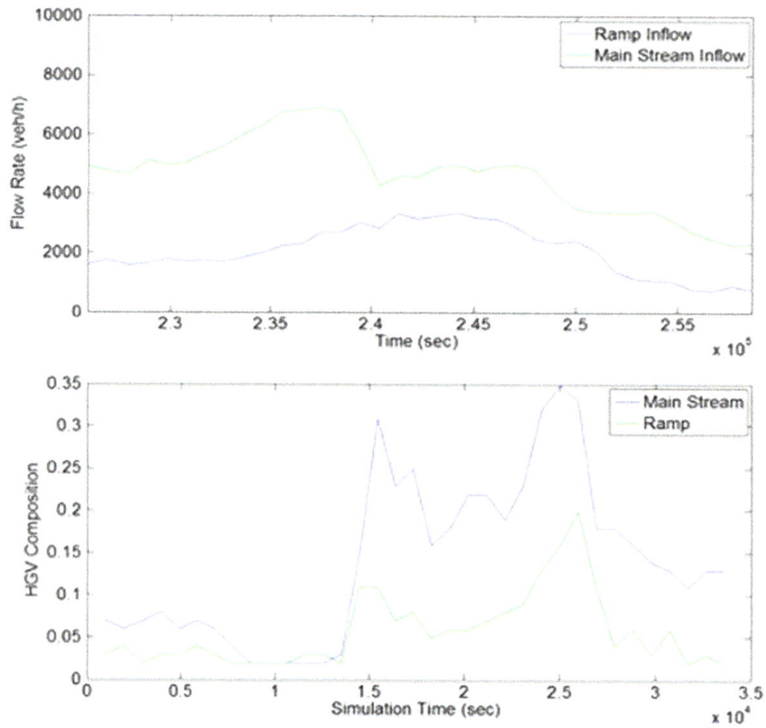

Figure 3 **Flow data (up) HGV composition for 8 July 2010 (Thursday).**

time application of MPC feasible [9]. In this study, a linear regression model is used to predict the occupancies of the next time frame.

3 Microscopic traffic simulation

There are three main classifications of traffic models according to the approach used in the analysis. One of them is the classification with respect to the given input. If traffic is invariant over time, this type of simulation models the steady-state average traffic conditions and is stated as static; if traffic changes over time, this type models the variant nature of traffic and is stated as dynamic.

Another classification of traffic models is defined by their statistical point of view. If the outputs of the traffic simulations are different by their executions, i.e., if there is randomness at the modeling process, it is called stochastic modeling of traffic. If the simulations demonstrate the same output in every execution, it is called deterministic modeling of traffic. The last and the most important classification of traffic models is the level of

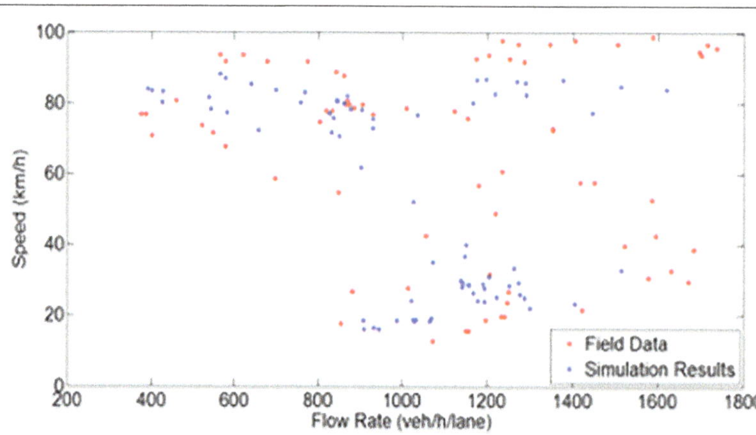

Figure 4 **Speed-flow calibration/verification results for the entire stretch.**

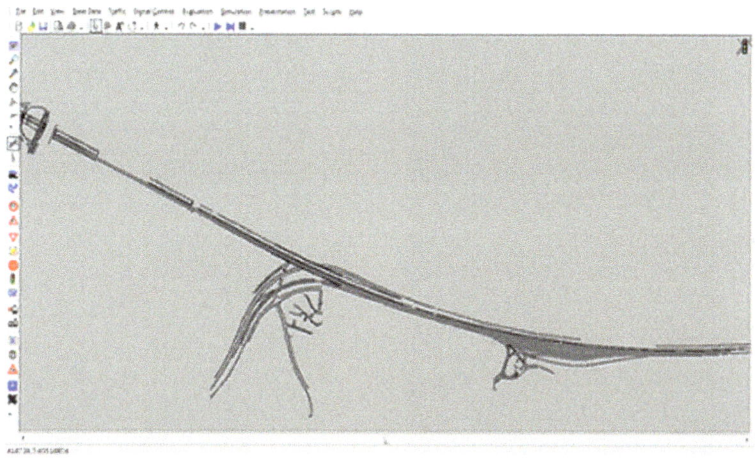

Figure 5 Study network generated in PTV VISSIM.

detail in terms of traffic flow, which are microscopic, mesoscopic, and macroscopic simulations.

3.1 Simulation model

In this study, the microscopic traffic simulations are performed by using PTV VISSIM simulation software. PTV VISSIM is a microscopic simulation program based on the driving behavior model of Wiedemann proposed in 1974 [19] and in 1999 [20]. PTV VISSIM is developed by PTV AG (Planung Transport Verkehr Aktiengesellschaft) and written in C++ considering guidelines of object-oriented programming (OOP) [21,22].

PTV VISSIM enables not only monitoring of the actual state of traffic but also simulation and evaluation of the macroscopic traffic variables as in real life. In many cases, the precautions, suggestions, or changes are done with regard to the simulation results. PTV VISSIM is also able to demonstrate the multimodal structure of the area such as public transit busses, heavy rail, or pedestrians, which is defined by the user. There are two main driving behaviors which are logically based on longitudinal and lateral movements, since all the movements on the road surface can be represented as a superposition of those two main movements.

The Wiedemann car-following models used for longitudinal movements in the PTV VISSIM simulation program are mainly based on the psychophysical driving behavior of the drivers.

Four main phases of driving are defined as free driving, approaching, following, and braking. These states are represented in Figure 1.

Lateral movements are mainly considered in lane changing movements in PTV VISSIM. There are two main lane change types considered - which lane is changing from the fast lane to the slow lane and from the slow lane to the fast lane - which are highly considered in [22].

3.2 Study field and data set

Levent ramp of Istanbul Outer Beltway (O-2) was determined as the study field. The data were collected by remote traffic microwave sensors (RTMS) between 6 July 2010 (Tuesday) and 13 July 2010 (Tuesday) which includes regular days (non-holidays, regular weather conditions, etc.).

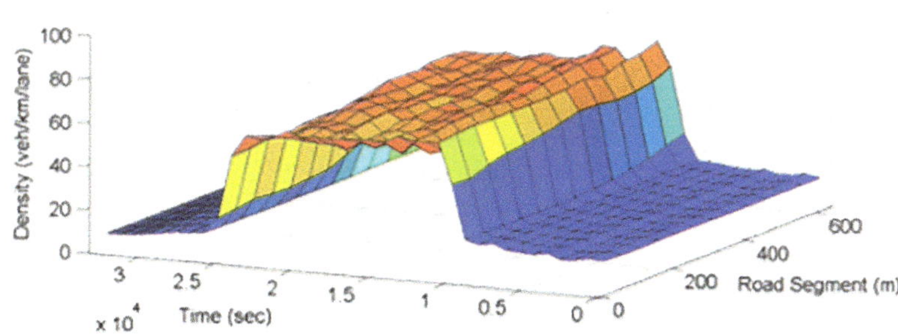

Figure 6 No-control density profile.

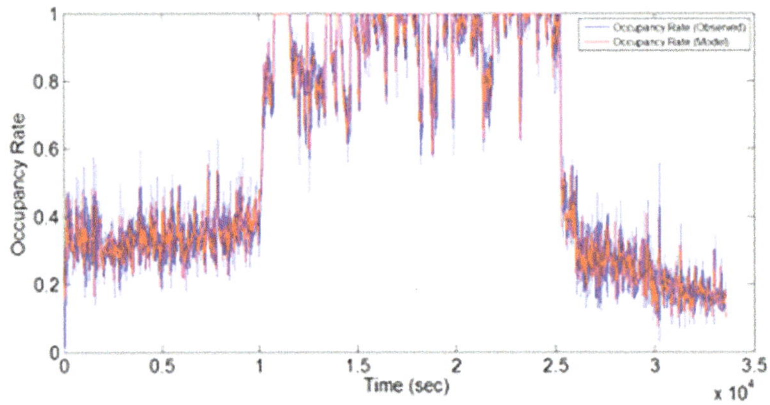

Figure 7 **Occupancy rate prediction model.**

For vehicle input scenario, 8 July 2010 (Thursday) 14:45:28 to 23:48:56 is selected where afternoon peak, low inflow, moderate inflow, and high inflow for both main stream and ramp are seen. Flow data for the whole day (8 July 2010), for the selected hours and heavy goods vehicle (HGV) percentage in traffic composition change, are shown in Figures 2 and 3, respectively.

3.3 Calibration, verification and implementation of proposed strategies

A speed-volume diagram is plotted for calibration in order to check and validate the accuracy of the speed-volume diagram between the simulation results and field observations as proposed in [21]. These diagrams are very useful for calibration because they contain information about a broad range of traffic situations. In particular, they show how the traffic flow behaves around capacity. This is the reason why these speed-flow diagrams are often used for comparing simulation results with real-world data. Another purpose for looking at the speed-volume scatter plots is to see whether the desired volumes can pass through an intersection with the desired speeds or not. It is found that the simulation is valid and accurate by looking at the speed-volume diagram. The results are presented in Figure 4.

The linearized local-feedback control algorithm ALINEA (Asservissement Lineaire d'Entree Autoroutiere - French for Linear Utilization for Highway Entrances), proposed by Papageorgiou et al. in 1991 [1,17], is used in this study. This strategy aims to keep the occupancy rate at the level of the desired occupancy rate which is generally taken as 0.30, as proposed by the authors. The main step of this strategy is given as follows:

$$r(t+\Delta t) = r(t) + (K_r(\hat{o} - o(t+\Delta t))$$

where $r(t + \Delta t)$ is called the metering rate to release to the downstream at the next cycle, $r(t)$ is the metering rate which is released at the current cycle, K_r is a regulator parameter which is taken generally as 70 by the authors and the applicators of this algorithm, $\widehat{o}$ is the desired occupancy rate, and $o(t + \Delta t)$ is the predicted occupancy for the next cycle. There are also other closed-loop local control strategies such as fuzzy logic control [12].

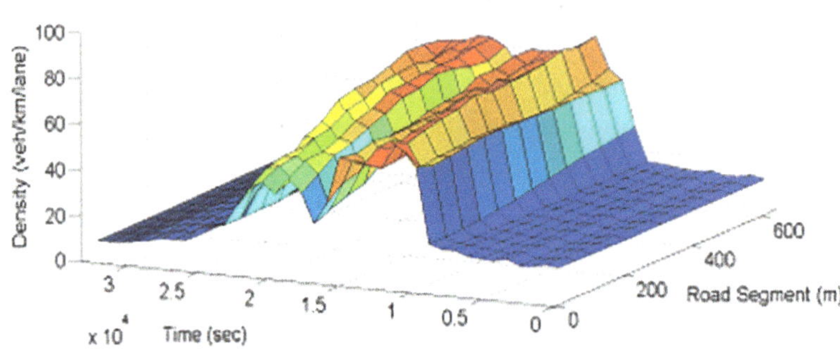

Figure 8 **Density profile for ramp metering approach (ALINEA 1).**

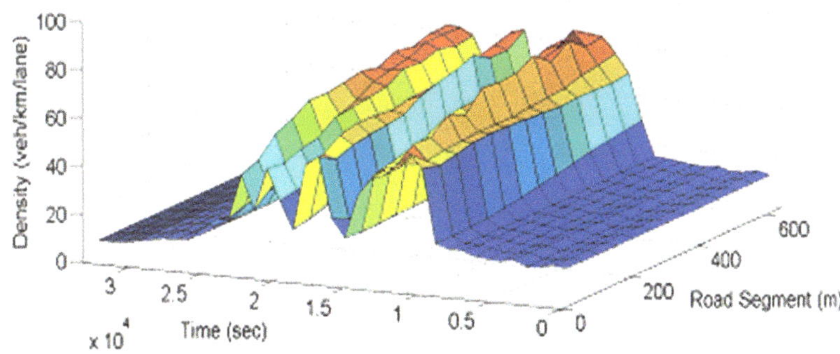

Figure 9 Density profile for ramp metering approach (ALINEA 2).

For the study field and data described, the network is generated, and the vehicle inputs (traffic volumes and vehicle composition) are entered as proposed. The simulation interface is shown in Figure 5. Random seeds are selected in a range varying from 5 to 95 with five increments, and the arithmetic average is taken into account while interpreting the results. The downstream link is segmented into 50-m multiple segments, and from the link evaluation option, density is collected in order to plot the density profile along the downstream link. The result is demonstrated in Figure 6 for the no-control case.

For the evaluation process to be used in dynamic traffic flow control approaches, data collection points are positioned at the 50th meter of the downstream link as proposed in [14].

Occupancy rates are measured with a time interval of 20 s since one of the ramp metering approach and all the variable speed limit approaches have the cycle time of 20 s. For the obtained results, a regression model is proposed in order to predict the traffic and to take precautions. The regression model is as follows:

$$o(t + \Delta t) = c_0 o(t) + c_1 o(t-\Delta t) + c_2 o(t-2\Delta t)$$

where $o(t + \Delta t)$ is the next occupancy rate which is predicted to be measured in the next cycle; $o(t)$ is the current occupancy rate which is measured in the same time interval; $o(t)$, $o(t - \Delta t)$, and $o(t - 2\Delta t)$ are the occupancy rates which have been measured in the past first, second, and third time intervals, respectively; and ci's are normalized constants to be determined. After the regression analysis, c_0 = 0:29, c_1 = 0:02, and c_2 = 0:69 are taken, and R^2 is found as 0:96; thus, this model can be considered as highly representative. The occupancy rate prediction model is shown in Figure 7.

In order to apply the ALINEA algorithm to the Vehicle Actuated Programming (VAP) module for PTV VISSIM, in the simulation, a 5-m detector is placed at the 50th meter of the bottleneck downstream. The green times are calculated as

$$\text{Green time}\,(t + \Delta t) = \frac{r(t + \Delta t)}{r_{\text{sat}}} \times \text{Cycle time}$$

for each cycle where r_{sat} is the saturated flow and taken as 2,000 vehicles/h.

The two versions of the ALINEA approach are coded in VAP. In the first version (ALINEA 1), cycle time t is taken as 20 s and the minimum green time is taken as 10 s. The density profile for the first approach is demonstrated in Figure 8.

In the second version (ALINEA 2), cycle time t is taken as 15 s and the minimum green time is taken as 5 s. The density profile for the second approach is demonstrated in Figure 9.

Instead of taking speed or volume as a control variable, the occupancy measure is proposed as an optimal control variable for the dynamic speed limit control strategy. The algorithm includes the time-dependent occupancies and variable speed limit intervals as control variables and parameters. Table 1 shows the dynamic speed limit control

Table 1 Dynamic speed limit control strategies

Dynamic speed limit control strategies	Occupancy (%)	Speed (km/h)
DSLC 1	15	120
	20	110
	30	90
	35	80
DSLC 2	10	120
	15	110
	20	100
	25	90
	30	80
	35	70

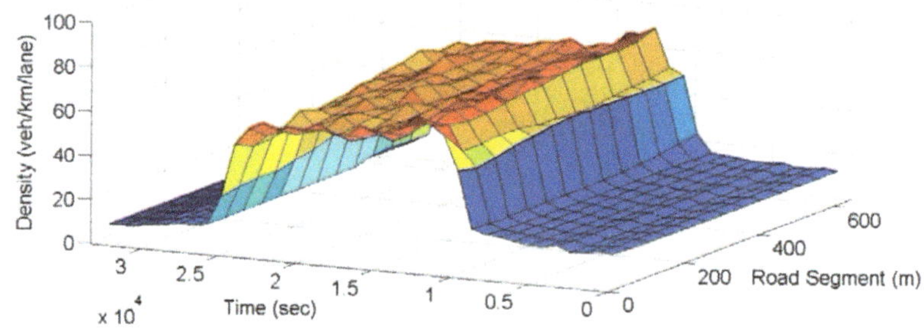

Figure 10 Density profile for dynamic speed limit control 1 (DSLC 1).

strategies employed in this study. The dynamic speed limit control strategies take the occupancy measurements at the downstream section of the ramp bottleneck, which is the most critical section of the stretch that is located just before the entrance of the FSM Bridge where six-lane highways narrow down to four-lane highways. The density profiled for the speed limit control strategies is demonstrated in Figures 10 and 11.

4 Dynamic traffic control simulation results

The performance measures are determined as the average delay time per vehicle, average number of stops, average speed, average stopped delay per vehicle, total delay time, number of stops, total stopped delay, and total travel time.

The first dynamic ramp metering approach, ALINEA 1, managed to decrease the average delay time per vehicle by 32%, average number of stops by 21%, average stopped delay per vehicle by 34%, total delay time by 32%, number of stops by 21%, total stopped delay by 34%, and total travel time by 20%. This approach also increased average speed of vehicles by 24%. The second dynamic ramp metering approach, ALINEA 2, accomplished to reduce the average delay time per vehicle by 29%, average number of stops by 21%, average stopped delay per vehicle by 35%, total delay time by 30%, number of stops by 22%, total stopped delay by 35%, and total travel time by 18%. This approach also increased the average speed of vehicles by 21%.

The simulation results also show that the first dynamic speed limit control strategy (DSLC 1) decreased the average delay time per vehicle by 7%, average speed of vehicles by 3%, average stopped delay per vehicle by 6%, total delay time by 8%, number of stops by 1%, and total stopped delay by 7%. However, the total travel time could not be significantly increased with this control strategy. For the second dynamic speed limit control strategy (DSLC 2), it decreased the average delay time per vehicle by 7%, average speed of vehicles by 6%, average stopped delay per vehicle by 7%, total delay time by 8%, number of stops by 2%, and total stopped delay by 9%. The total travel time remained the same for this control strategy as well.

5 Conclusions

Two main dynamic traffic control strategies are proposed and examined for the traffic congestion management for Istanbul Outer Beltway (O-2). For the ramp metering strategy, different versions of the ALINEA control algorithm are employed. The simulations are run multiple times in order to check for the consistency of

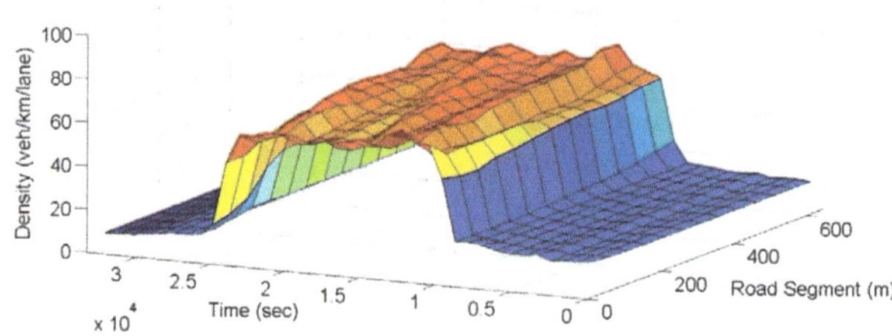

Figure 11 Density profile for dynamic speed limit control 2 (DSLC 2).

the random seeds which is also proposed in the PTV VISSIM manual [22].

The results indicated that especially the total delays can be significantly diminished by the examined control strategies. The best results are obtained for total travel time in the ramp metering approach whose cycle time is equal to 15 s and minimum green time to 5 s.

Vehicle-to-vehicle (V2V) and vehicle-to-infrastructure (V2I) communications are not included within the context of this study. However, it is obvious that communication technologies are increasingly advancing and the effects of communication on traffic control performance are considered as a future work besides the investigation of coordinative dynamic ramp metering approaches, dynamic lane closure, or various variable speed limit approaches.

Competing interests
The authors declare that they have no competing interests.

Acknowledgements
The first author would like to thank PTV Japan for providing the license key of PTV VISSIM. The Ministry of Education, Culture, Sports, Science & Technology of Japan supported the PhD research of the first author.

Author details
[1]Division of Civil Engineering, Department of Engineering, Işık University, Istanbul, Turkey. [2]Transportation Engineering Study Group, Department of Civil Engineering, Istanbul Technical University, Istanbul, Turkey.

References

1. M Papageorgiou, H Hadj-Salem, JM Blosseville, ALINEA: a local feedback control law for on-ramp metering. Transp Res Rec **1320**, 58–64 (1991)
2. CF Daganzo, *Fundamentals of Transportation and Traffic Operations* (Pergamon, Great Yarmouth, UK, 1997)
3. GB Whitham, *Linear and Nonlinear Waves* (Wiley, New York, 1974)
4. AS Kesten, M Ergün, T Yai, Traffic congestion management on Bosporus Bridge by considering speed limit control, in *The 10th International Congress on Advances in Civil Engineering ACE, Ankara, Turkey*, 2012
5. NH Gartner, CJ Messer, AK Rathi, *Revised monograph on traffic flow theory: a state-of-the-art report. Special Report by the Transportation Research Board of the National Research Council*, 2005
6. TRB, *Highway Capacity Manual (HCM 2010)* (Transportation Research Board, National Research Council, Washington, DC, 2010)
7. A Akbas, M Ergün, Freeway traffic flow control by lumped parameter system approach. Indian J Engineer Mater Sci **13**, 322–332 (2006)
8. M Papageorgiou, C Diakaki, V Dinopoulou, A Kotsialos, Y Wang, Review of road traffic control strategies. Proc IEEE **91**(12), 2043–2067 (2003)
9. A Hegyi, B De Schutter, H Hellendoorn, Model predictive control for optimal coordination of ramp metering and variable speed limits. Transportation Res Part C Emerging Technol **13**(3), 185–209 (2005)
10. NA Chaudhary, Z Tian, CJ Messer, C-L Chu, *Ramp metering algorithms and approaches for Texas. Texas Transportation Institute Technical Report*, 2004
11. P Kachroo, K Ozbay, *Feedback Ramp Metering Intelligent Transportation Systems* (Kluwer Academic/Plenum Publishers, New York, 2004)
12. C Taylor, D Meldrum, D McCormick, *Evaluation of a Fuzzy Logic Ramp Metering Algorithm* (Washington State Transportation Center, ., 2000)
13. LN Jacobson, KC Henry, O Mehyar, *Real-Time Metering Algorithm for Centralized Control* (Transportation Research Record, ., 1989), p. 1232
14. R Lau, *Ramp Metering by Zone - the Minnesota Algorithm* (Minnesota Dept. of Transportation, USA, 1997)
15. W Xin, PG Michalopoulos, J Hourdakis, D Lau, Minnesota's new ramp control strategy: design overview and preliminary assessment. Transportation Res Record J Transportation Res Board **1867**(1), 69–79 (2004)
16. LJ Corcoran, GA Hickman, *"Freeway Ramp Metering Effects in Denver", in Institute of Transportation Engineers Meeting*, 1989
17. M Papageorgiou, H Hadj-Salem, F Middelham, ALINEA local ramp metering: summary of field results. Transportation Res Record J Transportation Res Board **1603**(1), 90–98 (1997)
18. G Paesani, J Kerr, P Perovich, F Khosravi, System Wide Adaptive Ramp Metering (SWARM), in *Merging the Transportation and Communications Revolutions* (Abstracts for ITS America Seventh Annual Meeting and Exposition, Washington, DC, 1997)
19. R Wiedemann, *"Simulation des Straßenverkehrsflusses," Schriftenreihe des Instituts fur Verkehrswesen der Universit¨ at Karlsruhe, vol. Heft 8*, 1974
20. R Wiedemann, *Modeling of RTI-elements on multi-lane roads, in Advanced Telematics in Road Transport* (European Community, DG XIII, Brussels, Belgium, 1991)
21. M Fellendorf, P Vortisch, *Microscopic Traffic Flow Simulator VISSIM. Fundamentals of Traffic Simulation* (Springer Verlag, New York, USA, 2010), pp. 63–93
22. PTV AG, *VISSIM 5.40 User Manual*, Karlsruhe, Germany, 2011.

3

A reservation-based call admission control scheme and system modeling in 4G vehicular networks

Hassan Halabian[1], Perumalraja Rengaraju[2*], Chung-Horng Lung[1] and Ioannis Lambadaris[1]

Abstract

In 4G cellular networks, call admission control (CAC) has a direct impact on quality of service (QoS) for individual connections and overall system efficiency. Reservation-based CAC schemes have been previously proposed for cellular networks where a certain amount of system bandwidth is reserved for high-priority calls, e.g., hand-off calls and real-time new calls. Traditional reservation-based schemes are not efficient for 4G vehicular networks, as the reserved bandwidth may not be utilized effectively in low hand-off rates. We propose a channel borrowing approach in which new best effort (BE) calls can borrow the reserved bandwidth for high-priority calls. Later, if a hand-off call arrives and all the channels are busy, it will pre-empt the service of a borrower BE call if there exists any. The pre-empted BE calls are kept in a queue and resume their service whenever a channel becomes available. The analytical model for this scheme is a mixed loss-queueing system for which it is difficult to calculate call blocking probability (CBP) and call dropping probability (CDP). Our focus in this paper is on the system modeling and performance evaluation of the proposed scheme. We present two system models that approximate the operation of the proposed scheme. For these models, we derive the CBP and CDP analytically. It is shown that our analytical results are very close to the ones obtained from simulations. Furthermore, it is observed that our channel borrowing approach decreases the CBP considerably while increases the CDP slightly over a large range of hand-off rates.

Keywords: Reservation-based call admission control; Channel borrowing; CBP and CDP

1 Introduction

Initially, vehicular networks are introduced to provide communications for safety applications such as collision avoidance, hard braking warnings, accident reporting, intersection announcements, etc. Safety applications often require fast message exchanges but do not use much bandwidth where vehicles are enabled to communicate with one another, i.e., vehicle-to-vehicle or vehicle-to-roadside communications. On the other hand, evolution of vehicular networks is aimed to support non-safety multimedia applications which require high-speed Internet for mobile users. Fortunately, the telecommunication industry landscape for cellular networks is growing rapidly from 2G to 4G to accommodate the increasing usage of multimedia applications and users' mobility. In 4G, Worldwide Interoperability for Microwave Access (WiMAX) and Long-Term Evolution (LTE) are two emerging broadband wireless technologies aimed to provide high-speed Internet of 100 Mbps at a vehicular speed of up to 350 km/h [1,2]. Recently, research on the performance evaluation and improvement of 4G/5G vehicular networks has been under the focus of researchers in the filed of vehicular networking [3,4].

In 4G vehicular networks, call admission control (CAC) is considered as one of the radio resource management (RRM) functionalities and has a direct impact on quality of service (QoS) for individual connections and overall system efficiency. The goal of the CAC mechanism is to regulate the admission of new users, while controlling the quality of current connections without any call drops. In the traditional mobile networks, e.g., cellular and vehicular networks, CAC schemes have been designed to handle the voice traffic. On the other hand, design of practical and efficient CAC schemes for 4G vehicular networks is a

*Correspondence: rpr@vcet.ac.in
[1] Department of Systems and Computer Engineering, Carleton University, 1125 Colonel By Dr, Ottawa K1S 5B6, Canada
Full list of author information is available at the end of the article

challenging task due to the heterogeneous nature of multimedia traffic, user mobility, etc. However, CAC design is left open for innovation in 4G wireless network standards, such as WiMAX and LTE.

In the design of CAC schemes, the most common QoS parameters for performance evaluation are call blocking probability (CBP) and call dropping probability (CDP). Call blocking means denying new calls due to insufficient bandwidth in the network or the QoS requirements. Call dropping means dropping an existing call during a hand-off process due to users' mobility (vehicular or pedestrian). Forced termination of a call in progress is more frustrating than blocking a new call. As a result, hand-off calls are treated differently by being given a higher priority over new calls in cellular/vehicular networks. Particularly, we may either reserve certain amount of channels from the total available channels in a cell for hand-off calls or dynamically allocate channels for an individual cell, based on the time-varying status of vehicular traffic. The amount of channel reservation for hand-off calls is mostly based on users' mobility pattern, i.e., using vehicular traffic modeling that aggregates variables such as traffic density, mean speed, etc. [5]. In various research efforts on mobile cellular networks, the hand-off call arrival rate can be derived from the mobility models [6-9]. CAC designs based on mobility and traffic models will be *depended* on the assumptions made in modeling of the *users' mobility and traffic pattern*. In vehicular networks, due to the fast movement of vehicles, the variation range of hand-off rates is large. A good CAC design for 4G vehicular networks should be robust enough to support a vast range of hand-off rates. This motivates us to propose a CAC scheme which is robust enough to support a vast range of hand-off rates. We will propose a CAC scheme that *irrespective of the traffic model and mobility pattern* can improve the CBP considerably while not affecting CDP over a large range of hand-off rates. This motivates application of the proposed scheme for 4G vehicular networks.

The main goal of using conventional reservation-based CAC schemes in the cellular and vehicular networks is to reduce CDP. Although bandwidth reservation for hand-off calls decreases CDP, it also increases CBP in 4G wireless networks due to the inefficient channel utilization. By using reservation, a portion of the bandwidth is always reserved for hand-off calls. Hence, new calls may arrive in the system and are blocked, while reserved channels still exist in the system but are unused. On the other hand, newly originated real-time (high priority) calls should be treated differently to satisfy the high-priority customers in 4G wireless networks. Therefore, we propose the idea of channel borrowing for BE calls which carry non-delay-sensitive traffic. In our approach, new BE calls can borrow channels from the reserved channels for high-priority (hand-off) calls. Later on, if a hand-off call arrives and there is no channel available in the system, it pre-empts the service of the borrower BE call. The pre-empted BE call is kept inside a queue and resumes its service as soon as a channel becomes free.

It is observed that the model associated to the channel borrowing scheme is a mixed loss-queuing system [10]. To the best of our knowledge, no research has adopted the mathematical model for the channel borrowing scheme with the mixed loss-queuing system. However, it is difficult to mathematically analyze the mixed loss-queuing system. Therefore, the objective of the paper is to mitigate the issue of complicated analysis of the mixed loss-queuing system. To achieve the objective, we propose two system approximations for the mixed loss-queuing model and introduce the Markov chain model associated with each system approximation. In Section 4, we will observe that to solve the first system approximation, we need to solve a system of non-linear equations, whereas for the second approximation, we only need to solve a system of linear equations (which is much simpler and more efficient to solve). We will show through numerical analysis and simulations that our system approximations and the proposed channel borrowing scheme perform closely in terms of CBP and CDP. We also show that the proposed channel borrowing scheme considerably decreases CBP compared with the new call bounding scheme [11] while the amount of increase in CDP is minor. Note that the hand-off dynamics in a cellular network is mostly dependent on users' mobility pattern and traffic modeling that can be represented by aggregate variables such as traffic density, mean speed, etc. [5]. In this paper, our focus would be on *modeling and performance analysis* of a reservation-based CAC scheme in which we assume that the hand-off calls arrive based on a Poisson arrival process. Users' mobility and traffic modeling of the incoming calls are out of the scope of this paper and can be studied in a future work.

The rest of the paper is organized as follows. In Section 2, we review the traditional reservation-based CAC in cellular and vehicular networks. Section 3 presents the system model for conventional bandwidth reservation (new call bounding) scheme and the proposed bandwidth reservation with channel borrowing scheme. In Section 4, we study the performance of the proposed channel borrowing scheme analytically and introduce two system approximations for our channel borrowing scheme. Section 5 illustrates the numerical analysis where we show the effectiveness of the channel borrowing idea. Section 6 presents the conclusions of the paper.

2 Literature review

There has been a great body of research on CAC design reported in the literature [1,5,6,12-18]. In this section, we highlight a few important papers that mainly focus on reservation-based schemes in cellular and vehicular networks. The *guard channel* policy was proposed by Hong and Rappaport in [6], in which a certain amount of channels is reserved permanently for hand-off calls only, while the rest of the channels is shared by both new calls and hand-off calls. Given the total number of channels in the system is C, we reserve T channels ($T < C$) for hand-off calls. Hence, if the total number of calls in the system exceeds $C - T$, we do not admit new calls anymore. In this scheme, hand-off calls are given higher priority over new calls to reduce the CDP. The authors in [12] proposed fractional guard channel scheme in which the new calls are admitted with a certain probability. Such a probabilistic call admission prevents the system approaching congestion. In [11], the authors showed that when the service time of new calls and hand-off calls is not identically distributed, the traditional one-dimensional Markov chain cannot be used for performance analysis. In this case, two-dimensional Markov chain models must be invoked. Moreover, they proposed *new call bounding* scheme in which the total number of new calls in the system is restricted to a fixed threshold. Hence, if the total number of channels in the system is C, this scheme reserves T channels for hand-off calls and admits a new call if the total number of new calls in the system is less than $C - T$. In all the reservation-based schemes, a critical parameter is the number of reserved channels for hand-off calls since it provides a tradeoff between minimizing CBP and minimizing CDP. To deal with this problem, several dynamic reservation-based schemes were proposed in the literature, e.g., in [1]. Dynamic reservation-based schemes are not easy to analyze, and simulations are used for their performance evaluation. Researchers have also considered multiple threshold reservation schemes in [13,19]. In multiple threshold CAC design, the complexity is increased for the selection of thresholds and the system performance is unpredictable.

The CAC design for vehicular networks is discussed in [5,15,16], where the vehicular environment and movement has a major impact on the system design. Hence, the macroscopic vehicular modeling (CMVM) was introduced in CAC design in [5] where the macroscopic modeling describes vehicular traffic with aggregate variables such as traffic density, mean speed, etc. In [15], both fixed and dynamic channel assignments are discussed for CAC design. Fixed channel assignment requires extensive frequency planning and maintains a constant frequency reuse distance for all cells implementation. Hence, neighboring cells are allocated with mutually disjoint radio channels to cater for mobile host communications in the cell-specific region. On the other hand, dynamic channel assignment enables the fixed channel allocation methods to maintain a pre-set range of channels allocated to a cell and then varies based on the users' mobility due to vehicular movement. The CAC proposed in [16] uses fixed channel reuse technique, channel borrowing process, and the speed of the vehicle to reserve the channel for hand-off calls to improve quality of service (QoS) in VANETs. The authors in [14] proposed the idea of channel borrowing for the originating new calls along with the idea of channel adaptation for reserved channels. However, the model in [14] does not distinguish between different classes of new calls. Moreover, there is no notion of call pre-emption by hand-off calls, i.e., if a hand-off call arrives and all the reserved channels are borrowed, the hand-off call will be dropped. This will increase the CDP for hand-off calls. In the proposed scheme in our paper, hand-off calls can pre-empt the service of a borrower BE call. This keeps CDP about the same level while increases the channel utilization efficiency and decreases CBP. The work in [14] does not consider any mathematical modeling, and the effectiveness of the proposed scheme is shown through simulations.

In vehicular networks, the hand-off rate varies fast due to the speed of vehicles. A good CAC design for 4G vehicular networks should be robust enough to *support a vast range of hand-off rates*. Most of the research on handling hand-off in vehicular networks are trying to use the mobility/traffic pattern to enhance the call admission process [5,15,16,20-22]. In our work, we propose the channel borrowing idea which improves the channel utilization of conventional channel borrowing CAC schemes and therefore results in improved CBP while does not affect CDP. In Section 5, it will be shown that by applying our CAC scheme, the amount of improvement in CBP is considerable for a vast range of hand-off rates which motivates its application in 4G vehicular networks. Moreover, we propose a mathematical system modeling and provide an analytical performance evaluation framework for the proposed channel borrowing technique.

3 System model for the conventional and the proposed reservation-based schemes

Different classes of traffic have been defined in 4G wireless mobile networks. For example, WiMAX standard supports five different traffic classes and LTE defines nine traffic classes. A new call belongs to one of the traffic classes defined in the network. Other than the new call arrivals, there exist hand-off calls which are generated in a cell due to the mobility of users into the cell. Note that the hand-off calls have the same traffic classification. Assuming a wireless network with N traffic classes, the Markov modeling of the system will end up to a Markov chain with $2 \times N$-dimensional state space which makes it unfeasible to analyze mathematically. To show the effectiveness of our proposed reservation-based CAC scheme and to provide an analytical performance evaluation for it, we consider a 4G wireless network with two main priority traffic classes. In our model, we assume that the high-priority class contains only the hand-off calls that is coming from other cells in the cellular network. We may also categorize the real-time new calls in the current cell into this class. The low-priority class contains the new calls generated in the current cell. We assume that the new calls are subcategorized to non-real-time (nRT) new calls and best effort (BE) new calls. Examples of nRT and BE calls are non-real-time multimedia traffic (e.g., Youtube video) and web traffic, respectively. Example of real-time traffic is live broadcasting.

It is assumed that the arrival processes of hand-off, nRT, and BE calls are Poisson distributed with parameters λ_H, λ_{nRT}, and λ_{BE}, respectively. We further assume that the service processes of the hand-off, nRT, and BE calls are exponentially distributed with parameters μ_H, μ_{nRT}, and μ_{BE}, respectively. The system bandwidth is channelized, and the number of channels (bandwidth units) in the system is C. Note that C is not the network capacity in terms of amount of served traffic (which is dependent on the users' wireless channel model and interference). In this model, C denotes the number of physical network resources in a cell which should be allocated to the arriving calls, e.g., number of physical resource blocks in an LTE network.

We build our channel borrowing idea on the *new call bounding scheme* proposed in [11], and henceforth, we call this CAC as the conventional scheme. We first review the new call bounding scheme and then explain how we incorporate our channel borrowing approach in this scheme.

3.1 Review of new call bounding scheme

In the conventional scheme (i.e., new call bounding scheme introduced in [11]), a new call is blocked if the total number of admitted new calls in the system is more than $C - T$ or there is no more channel available in the system. T denotes the number of channels reserved for high-priority traffic. Note that $T < C$. A hand-off call is blocked only if there is no channel available in the system to serve it. By defining the state of the system as the pair (n_1, n_2) where n_1 and n_2 are the number of hand-off and new calls in the system, respectively, the authors in [11] derived a two-dimensional Markov chain for this system (Figure two in [11]). Since this Markov chain is *time-reversible*, one may easily write the local balancing equations for it and derive the closed-form formulas for CBP of new calls and CDP of hand-off calls (refer to [11], Equations (1) and (2)). Global balance equations also may be used to solve this Markov chain. This is involved in solving a system of linear equations.

3.2 Channel borrowing in reservation-based schemes

In this section, we introduce our channel borrowing approach in the new call bounding scheme. The idea of channel borrowing can be implemented in any reservation-based CAC algorithm. Before we proceed to introduce our CAC scheme, we will clarify our motivations by mentioning the following points:

- Among the new call arrivals, BE calls are the ones that are flexible in terms of QoS requirements. BE traffic is usually dedicated to non-critical services such as web traffic or file transfers. Therefore, the service of a BE call can be delayed or can be disrupted.
- In reservation-based CAC approaches, in cases where the hand-off arrival rate is relatively low and the arrival rate for BE calls is relatively high, the channels statically reserved for hand-off calls may be wasted since we idle the reserved channels and keep them for hand-off calls arriving in future. Hence, the channel utilization in such schemes is not efficient.

By considering the inefficiency of bandwidth utilization in reservation-based schemes and the flexibility of the service of BE calls, we propose the following channel borrowing scheme: We consider a reserved bandwidth of T channels for hand-off calls. We admit an nRT call only if there exists a free channel in the system, and the total number of new calls (nRT and BE) does not exceed $C - T$ (similar to what we do in the new call bounding scheme). However, we allow the BE calls to use all the channels in the system (upon the availability). When admitting a new BE call even if we violate the reservation of hand-off calls, we still admit it if there exists any free channel in the system. Hence, a new BE call can borrow a channel from the reserved channels of hand-off calls. In this case, if in the near future a hand-off call arrives and there is no available channel in the system to serve it while the total number of hand-off calls in the system is less than T calls, the hand-off arrival can pre-empt the service of the borrower BE call. In this case, the BE call returns the borrowed channel. Nevertheless, we do not drop the service of the pre-empted BE call; instead, we will keep the pre-empted BE call in a queue and will resume its service as soon as a channel becomes available. We denote the size of this queue at time t by $X(t)$. At the arrival time of a BE call if $X(t) > 0$, we will block the new BE arrival.

If we denote the number of hand-off calls, nRT calls, and BE calls in the system by n_1, n_2, and n_3, respectively, we have the following properties for the dropping of hand-off calls and blocking of nRT calls and BE calls.

- A hand-off arrival will be dropped if $n_1 + n_2 + n_3 = C$ and $n_1 \geq T$.
- If $n_1 + n_2 + n_3 = C$ and $n_1 < T$, a hand-off arrival will be admitted by pre-empting a BE call.

- An nRT arrival will be blocked if $n_1 + n_2 + n_3 = C$ or $n_2 + n_3 \geq C - T$.
- A BE arrival will be blocked if $X(t) > 0$ or if $X(t) = 0$ and $n_1 + n_2 + n_3 = C$.

4 Performance analysis of the proposed channel borrowing scheme

In the proposed channel borrowing CAC scheme described above, there is no queue for the hand-off and nRT calls while we keep the pre-empted BE calls in a queue. This CAC system model is a mixed loss-queueing system [10], and therefore, it is not easy to analyze mathematically. In the following section, we will introduce two system approximations for the channel borrowing scheme described in Section 3.2 by using two pure loss systems. We can analyze the approximation systems mathematically by solving the global balance equations for each system. We will show that these two approximation systems perform very closely with respect to the the original mixed loss-queueing system in terms of CBP and CDP.

4.1 System approximation 1

Let $p(n_1, n_2, n_3)$ denote the steady-state probability of being at state (n_1, n_2, n_3). It can be verified that in the proposed channel borrowing CAC scheme, the input arrival rate to the queue keeping the pre-empted BE calls is equal to:

$$\lambda_H \sum_{\substack{n_1+n_2+n_3=C \\ n_1<T}} p(n_1, n_2, n_3), \tag{1}$$

where $\sum_{\substack{n_1+n_2+n_3=C \\ n_1<T}} p(n_1, n_2, n_3)$ is the probability of the system being in the state that all the channels are utilized while the number of hand-off calls is less than T. In these states, if a hand-off call arrives, it will request a channel borrowed by a BE call previously. Since the hand-off calls arrive with rate λ_H, the total rate of pre-empted BE calls is given by (1).

In the first system approximation, we remove the queue in the system and instead modify the arrival rate of the BE calls to:

$$\lambda_{BE}^{(new)} = \lambda_{BE} + \lambda_H \sum_{\substack{n_1+n_2+n_3=C \\ n_1<T}} p(n_1, n_2, n_3). \tag{2}$$

As it is observed, the arrival rate $\lambda_{BE}^{(new)}$ depends on the state probabilities and the arrival rate of the hand-off calls. Such a dependency is expected since there is a loop in the system which reenters the pre-empted BE calls into the system. By applying such an approximation, we will get rid of the queue in the system and obtain a pure loss system with a modified BE arrival rate (which is dependent on the system state probabilities). We can derive the three-dimensional Markov chain for this approximation with the state space:

$$\mathcal{S} = \{(n_1, n_2, n_3) \mid n_1, n_3 \geq 0, 0 \leq n_2 \leq C - T, n_1 + n_2 + n_3 \leq C\}. \tag{3}$$

We can check that the transition probabilities for the Markov chain associated to this system approximation are the following:

$$q(n_1, n_2, n_3; n_1, n_2, n_3 - 1) = n_3\mu_{BE}$$
$$(0 \leq n_1 < C, n_3 > 0, 0 \leq n_2 \leq C - T, n_1 + n_2 + n_3 \leq C)$$
$$q(n_1, n_2, n_3; n_1, n_2, n_3 + 1) = \lambda_{BE}^{(new)}$$
$$(0 \leq n_1 < C, 0 \leq n_2 \leq C - T, n_1 + n_2 + n_3 < C)$$
$$q(n_1, n_2, n_3; n_1, n_2 - 1, n_3) = n_2\mu_{nRT}$$
$$(0 \leq n_1 < C, 0 < n_2 \leq C - T, n_1 + n_2 + n_3 \leq C)$$
$$q(n_1, n_2, n_3; n_1, n_2 + 1, n_3) = \lambda_{nRT}$$
$$(0 \leq n_1 < C, 0 \leq n_2 + n_3 < C - T, n_1 + n_2 + n_3 < C)$$
$$q(n_1, n_2, n_3; n_1 - 1, n_2, n_3) = n_1\mu_H$$
$$(0 < n_1 \leq C, 0 \leq n_2 \leq C - T, n_1 + n_2 + n_3 \leq C)$$
$$q(n_1, n_2, n_3; n_1 + 1, n_2, n_3) = \lambda_H$$
$$(0 \leq n_1 < C, 0 \leq n_2 \leq C - T, n_1 + n_2 + n_3 < C)$$
$$q(n_1, n_2, n_3; n_1 + 1, n_2, n_3 - 1) = \lambda_H$$
$$(0 \leq n_1 < T, 0 \leq n_2 \leq C - T, n_1 + n_2 + n_3 = C)$$

It is hard to illustrate the three-dimensional Markov chain on paper. If we assume that the nRT class does not exist (or $\lambda_{nRT} = 0$), we will obtain a two-dimensional Markov chain since we will have only two classes of BE and hand-off calls. Note that the idea of channel borrowing is not affected by this simplification and can be still applied without nRT class, since in channel borrowing scheme, a BE call is borrowing a channel from the reserved channels for hand-off calls. Figure 1 depicts the two-dimensional Markov chain associated to this system approximation. The states shown in black are the states that also appear in the conventional new call bounding scheme. The states shown in red are the states that do not exist in the new call bounding scheme, but they appear in the proposed channel borrowing scheme. In this figure, the arrival rate of BE calls is denoted by λ'_{BE}. For the conventional new call bounding scheme, we have $\lambda'_{BE} = \lambda_{BE}$. For the system approximation we introduced in this subsection, we have $\lambda'_{BE} = \lambda_{BE}^{(new)}$.

We use global balance equations and solve the Markov chain numerically. In this system approximation, instead of constant λ_{BE} value, we have to use the new BE arrival rate $\lambda_{BE}^{(new)}$ in the global balance equations. Since $\lambda_{BE}^{(new)}$ is a function of the system states, the global balance equations will result into a system of non-linear equations that should be solved numerically using non-linear solvers.

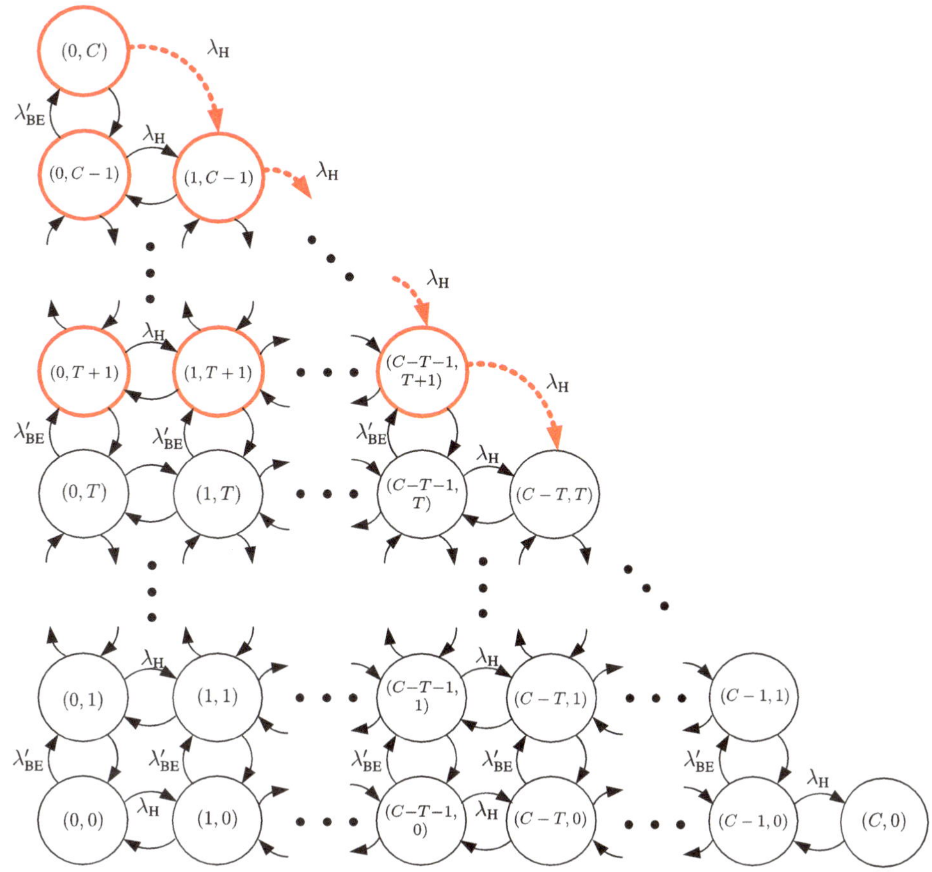

Figure 1 Markov chain associated to the new call bounding scheme, system approximations 1 and 2. The black states are the ones appear in the Markov chain of the new call bounding scheme with $\lambda'_{\text{BE}} = \lambda_{\text{BE}}$. The black and red states appear in the system approximations 1 and 2. For the system approximation 1, we have $\lambda'_{\text{BE}} = \lambda^{(\text{new})}_{\text{BE}}$, and for system approximation 2, we have $\lambda'_{\text{BE}} = \lambda_{\text{BE}}$.

To clarify this fact, consider the global balance equation associated to state $(0, C)$ in Figure 1.

$$p(0, C)(\lambda_{\text{H}} + C\mu_{\text{H}}) = p(0, C-1)\lambda^{(\text{new})}_{\text{BE}}$$
$$= \lambda_{\text{BE}} p(0, C-1) + \lambda_{\text{H}} \sum_{\substack{n_1+n_3=C \\ n_1<T}} p(0, C-1)p(n_1, n_3)$$

As it is observed, in this equation, we have multiplicative terms $p(0, C-1)p(n_1, n_3)$ which makes it a non-linear equation. As C increases, the size of such a non-linear system of equations increases rapidly. For example, for a system with 16 channels, we obtain a system of non-linear equations with more than 800 variables which may become very time-consuming and inefficient to solve. Therefore, we propose the following system approximation presented in the following subsection which will result in a system of linear equations.

4.2 System approximation 2

In the second system approximation, we assume that $\lambda'_{\text{BE}} = \lambda_{\text{BE}}$. This is equivalent to assume that the pre-empted BE calls are dropped without being returned to the system inside a queue. Clearly, the accuracy of this approximation is dependent on the rate of pre-emption which itself is dependent on the input arrival rates. Such an approximation will result in a system of *linear* global balance equations. A large system of linear equations can be efficiently solved in a reasonable amount of time. We will show that the first and the second system approximations will result in very close CBP and CDP values. Therefore, for system performance evaluation, we propose to use the second approximation especially for systems with large number of channels (e.g., in realistic wireless systems).

5 Performance evaluation

In this section, we will evaluate the performance of the proposed channel borrowing scheme and compare it with that of the conventional new call bounding scheme in terms of CBP and CDP. We also validate our system approximations by comparing the derived CDP and CBP values obtained from global balance equations and the CDP and CBP values obtained from simulation of the original mixed loss-queueing model.

First, we show that the two system approximations perform closely in terms of CBP and CDP. To show this, we have derived the CBP and CDP associated to the system approximations 1 and 2 by solving the global balance equations numerically. The CBP and CDP values for these approximations are calculated then as follows:

$$\mathrm{CBP} = \sum_{(n_1,n_2,n_3):n_2+n_3\geq C-T} \frac{\lambda_{\mathrm{nRT}}}{\lambda_{\mathrm{nRT}}+\lambda'_{\mathrm{BE}}} p(n_1,n_2,n_3) + \sum_{\substack{(n_1,n_2,n_3):n_2+n_3\geq C-T \\ n_1+n_2+n_3=C}} \frac{\lambda'_{\mathrm{BE}}}{\lambda_{\mathrm{nRT}}+\lambda'_{\mathrm{BE}}} p(n_1,n_2,n_3) + \sum_{\substack{(n_1,n_2,n_3):n_2+n_3<C-T \\ n_1+n_2+n_3=C}} p(n_1,n_2,n_3) \quad (4)$$

$$\mathrm{CDP} = \sum_{\substack{(n_1,n_2,n_3):n_2+n_3<C-T \\ n_1+n_2+n_3=C}} p(n_1,n_2,n_3) \quad (5)$$

The CBP for system approximation 1 is calculated if in (4) we put $\lambda'_{\mathrm{BE}} = \lambda^{(\mathrm{new})}_{\mathrm{BE}}$ and is calculated for system approximation 2 if we put $\lambda'_{\mathrm{BE}} = \lambda_{\mathrm{BE}}$. To compare the analytical results for system approximations 1 and 2, we considered a system with $C = 12$ and $T = 6$ and solved the global balance equations using Matlab. Solving the global balance equations for system approximation 1 with more than 12 channels takes a considerable amount of time, and therefore, we did the numerical analysis only for $C = 12$. The service rates of hand-off, BE, and nRT traffic are given by $\mu_{\mathrm{H}} = \mu_{\mathrm{BE}} = \mu_{\mathrm{nRT}} = 1$. We also assume that the arrival rate of nRT traffic is $\lambda_{\mathrm{nRT}} = 0.5$. We plot the CBP and CDP for $\lambda_{\mathrm{H}} = 0.5 \times i$ and $\lambda_{\mathrm{BE}} = 0.5 \times i$ where $i = 1, 2, \ldots, 10$. The parameters of the numerical analysis for comparing the performance of approximations 1 and 2 are presented in Table 1. Figure 2a,b shows the results in 3D graphs. As we observe in the figures, the two system approximations will result in very close CBP and CDP values. The CBP and CDP values of the system approximation 2 are slightly lower than that of the system approximation 1, since in system approximation 2, we will drop all the pre-empted BE calls, and therefore, the BE load in the system approximation 2 is slightly smaller than that of the system approximation 1. As a result, the system approximation 2 is suggested for the performance evaluation for large networks (with large amount of channels).

As explained earlier, we used system approximations 1 and 2 to derive approximate values of CDP and CBP for the original mixed loss-queueing model which implements the proposed bandwidth borrowing scheme. It is due to the fact that analytical derivation of the CDP and CBP of the original mixed loss-queueing model is unfeasible. Therefore, we invoked simulations to derive those values. The simulations and the analytical study are performed both in Matlab. To show how precise are our approximations and to compare the analytical results with the simulation results for the proposed bandwidth borrowing scheme and then with the conventional scheme, a system with $C = 16$ and $T = 8$ is considered. In comparison, the CDP and CBP values are calculated for the system approximation 2 by solving the system of global balance (linear) equations. We also perform simulation of the original mixed loss-queueing model to derive the actual CDP and CBP values of the proposed channel borrowing scheme. The CDP and CBP values of the conventional scheme are also calculated based on the results in [11]. The CBP and CDP values are calculated for $\lambda_{\mathrm{H}} = 4, 7, 10$ and $\lambda_{\mathrm{BE}} = i$ where $i = 1, 2, \ldots, 10$. We present the results for two values of $\lambda_{\mathrm{nRT}} = 10$ and $\lambda_{\mathrm{nRT}} = 1$ in Figures 3 and 4, respectively. Parameter setting for the analysis of this part are summarized in Table 2.

Table 1 Parameter setting in numerical analysis for comparing the performance of approximations 1 and 2

Parameter	Setting
Number of channels (C)	12
Reserved channels (T)	4
$\mu_{\mathrm{H}}, \mu_{\mathrm{BE}}, \mu_{\mathrm{nRT}}$	1
λ_{nRT}	0.5
λ_{H}	$0.5 \times i, i = 1, 2, \ldots, 10$
λ_{BE}	$0.5 \times i, i = 1, 2, \ldots, 10$

From the CBP performance results, it is evident that the channel borrowing scheme results in a considerable decrease in CBP compared with the conventional bandwidth reservation scheme. Similarly, the CDP in the channel borrowing scheme is slightly lower than that of the conventional bandwidth reservation scheme. Conversely, the channel borrowing scheme admits more BE calls for the unused reserved channel and then keeps the pre-empted ones in a queue to resume their service once the channel is available. Therefore, more BE calls are using the unreserved channels, and less hand-off calls have the opportunity to use the unreserved channels. This makes the probability of the call dropping of hand-off calls increases. However, the amount of increase for CDP in the proposed channel borrowing scheme is very small compared to the gain obtained in terms of the CBP decrease. For example, for $\lambda_{\mathrm{H}} = 4$, $\lambda_{\mathrm{nRT}} = 10$, and $\lambda_{\mathrm{BE}} = 10$, the CBP of the proposed scheme is 0.11 less than that

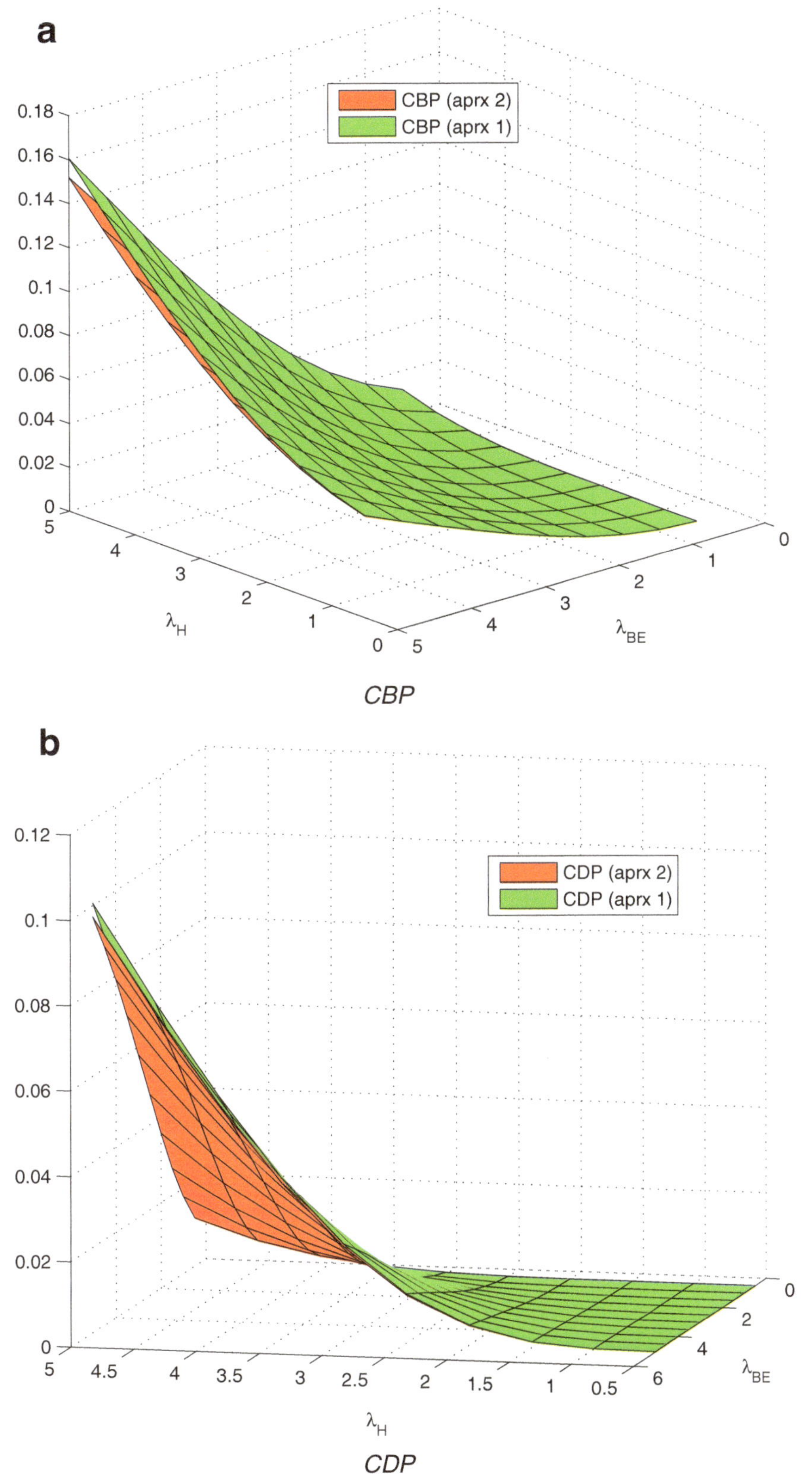

Figure 2 CBP and CDP performance of system approximations 1 and 2 for a system with $C = 12$, $T = 6$, and $\lambda_{nRT} = 0.5$. **(a)** CBP and **(b)** CDP.

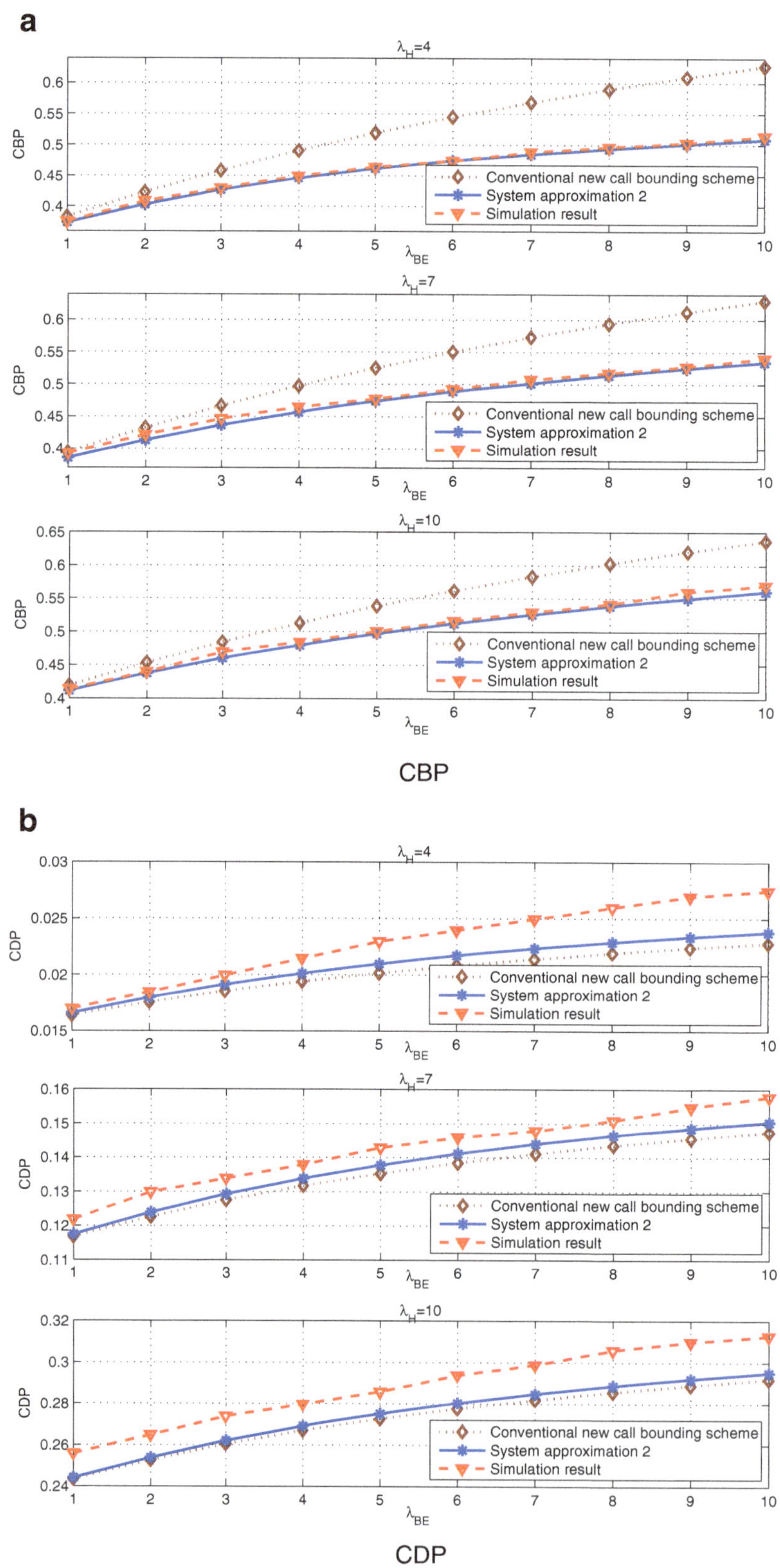

Figure 3 CBP and CDP performance of the system for simulation, approximation 2, and the conventional scheme with $C = 16$, $T = 8$, and $\lambda_{nRT} = 10$. **(a)** CBP and **(b)** CDP.

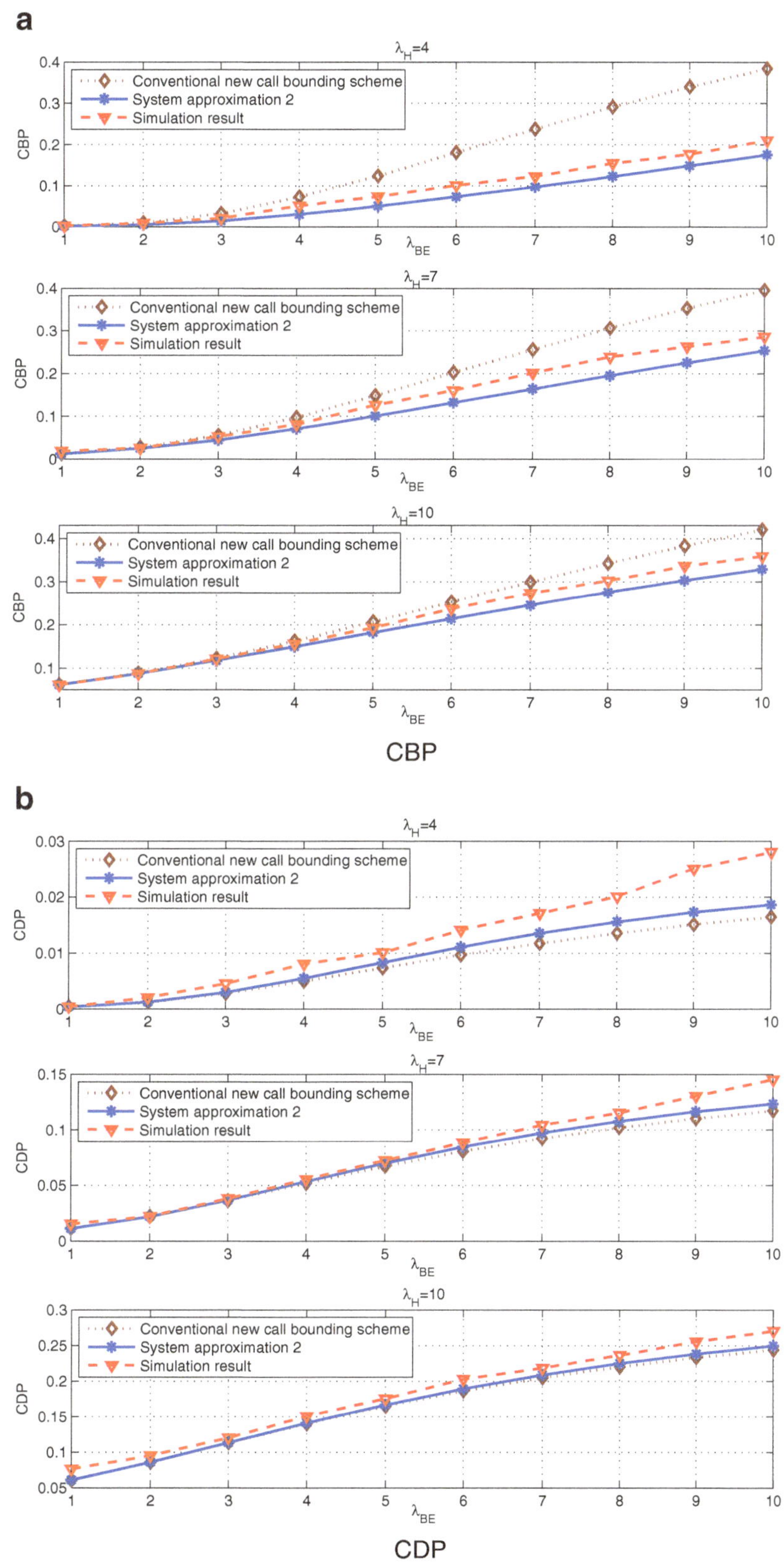

Figure 4 CBP and CDP performance of the system for simulation, approximation 2, and the conventional scheme with $C = 16$, $T = 8$, and $\lambda_{nRT} = 1$. **(a)** CBP and **(b)** CDP.

Table 2 Parameter setting for comparing the performance of simulation and approximations 2 and the conventional scheme

Parameter	Setting
Number of channels (C)	16
Reserved channels (T)	8
$\mu_H, \mu_{BE}, \mu_{nRT}$	1
λ_{nRT}	1, 10
λ_H	4, 7, 10
λ_{BE}	$i = 1, 2, \ldots, 10$

of the conventional scheme, while the CDP of the conventional scheme is 0.004 less than that of the channel borrowing scheme. Another example is the case $\lambda_H = 4$, $\lambda_{nRT} = 1$, and $\lambda_{BE} = 10$ where CBP of the proposed borrowing scheme is 0.175 less than that of the conventional scheme while the difference of CDP values in these two schemes is 0.012. Also, the simulation results show that the system approximation 2 performs very closely to the original mixed loss-queuing system. Thus, we propose to use this approximation for system performance evaluation of channel borrowing scheme in reservation-based CAC schemes.

Since solving the set of non-linear equations for a system with more than 12 servers is very time-consuming, we used the system approximation 2 (the one with linear equations) for our analytical study. Therefore, when we compare the analytical results with the simulation results (simulation of the borrowing idea in a mixed loss-queueing model), we only considered system approximation 2. Note that system approximation 2 is a more relaxed system approximation (with respect to approximation 1) for the mixed loss-queueing model since in the second approximation, we assume that the pre-empted BE calls are dropped. By comparing the CDP and CBP derived from simulation of the original mixed loss-queueing model with those of the system approximation 2 and the conventional reservation-based CAC scheme, we conclude the following:

- The proposed channel borrowing idea outperforms the conventional reservation-based CAC scheme. This is concluded by comparing the simulation results of the proposed scheme with the conventional scheme.
- The system approximation 2 can provide reasonably close CDP and CBP values for the mixed loss-queueing model.

System approximation 1 is expected to even perform better than the approximation 2. However, due to its computational overhead, it is not desirable.

Finally, we observe that the CBP improvement is observable for a vast range of hand-off rates (4, 7, and 10) which implies that the proposed CAC scheme is robust enough to support high and low hand-off rates which motivates the application of the proposed scheme in 4G vehicular networks.

6 Conclusions

In this paper, we introduced the idea of channel borrowing for reservation-based CAC schemes. The lack of efficiency in bandwidth utilization is the main motivation to use this idea in reservation-based CAC schemes. In our approach, BE calls are able to borrow channels from the reserved channels for hand-off calls. A hand-off arrival call can pre-empt the service of a borrower BE call, and the pre-empted BE calls are stored in a queue to resume their service in future. Moreover, we modeled the channel borrowing scheme using a mixed loss-queuing system and introduced two system approximations for the proposed borrowing scheme to simplify the mathematical analysis. By using simulations and numerical analysis, we showed that the two approximations result in very close CBP and CDP values with respect to the actual CBP and CDP values. Furthermore, it was shown that channel borrowing decreases CBP considerably while it only increases CDP slightly. In our study, the number of reserved channels for high-priority calls, i.e., T, is assumed to be fixed. Optimization of T with respect to the number of users in the cell and the arrival rate of high-priority calls can be consider as a future work in this area of research.

Competing interests

The authors declare that they have no competing interests.

Author details

[1]Department of Systems and Computer Engineering, Carleton University, 1125 Colonel By Dr, Ottawa K1S 5B6, Canada. [2]Department of Information Technology, Velammal College of Engineering and Technology, Viraganoor, Madurai 625 009, India.

References

1. D Niyato, E Hossain, Call admission control for qos provisioning in 4G wireless networks: issues and approaches. IEEE J. Network. **19**(5), 5–11 (2005)
2. M-S Pan, Chen Lin T-M W-T, An enhanced handover scheme for mobile relays in LTE-A high-speed rail networks. IEEE Trans. Vehicular Technol. **64**(2), 743–756 (2014)
3. G Araniti, C Campolo, M Condoluci, A Iera, LTE for vehicular networking: a survey. IEEE Wireless Commun. Mag. **51**(5), 148–157 (2013)
4. ZH Mir, F Filali, LTE and IEEE 802.11p for vehicular networking: a performance evaluation. EURASIP J Wireless Commun. Networking. **89**, 1–15 (2014)
5. T Kwon, Y Choi, in *Proc. of 48th IEEE Vehicular Technology Conference*. Call admission control in mobile cellular network based on macroscopic modeling of vehicular traffic (Ottawa, Canada, 1998)
6. D Hong, S Rappaport, Traffic model and performance analysis for cellular mobile radio telephone systems with prioritized and nonprioritized handoff procedures. IEEE Trans. Vehicular Technol. **35**(3), 77–92 (1986)

7. SA El-Dolil, W-C Wong, R Steele, Teletraffic performance of highway microcells with overlay macrocell. IEEE J. Select. Areas Commun. **7**(1), 71–78 (1989)
8. R Steele, M Nofal, Teletraffic performance of microcellular personal communication networks. IEE Proc. I, Commun., Speech Vision. **139**(4), 448–461 (1992)
9. H Xie, S Kuek, in *Proc. IEEE VTC '93*. Priority handoff analysis (Secaucus, NJ, USA, 1993)
10. AM Lee, *Applied Queueing Theory*, (New York, Macmillan, 1966)
11. Y Fang, Y Zhang, Call admission control schemes and performance analysis in wireless mobile networks. IEEE Trans. Vehicular Technol. **51**(2), 371–382 (2002)
12. R Ramjee, D Towsley, R Nagarajan, On optimal call admission control in cellular networks. Wireless Networks. **3**(1), 29–41 (1997)
13. PM Tsang, KF Tsang, JC-K Wong, HY Tung, in *Proc. IET Int'l Conf. on Wireless Mobile and Computing*. Call admission control for long term evolution (Shanghai, China, 2011)
14. S Alagu, T Meyyappan, Efficient utilization of channels using dynamic guard channel allocation with channel borrowing strategy in handoffs, Jun 2012. [Online]. Available: http://arxiv.org/abs/1206.3375
15. BR Adnan Muhtar, JMH Elmirghani, in *Proc. of 7th Int'l Conf. on Wireless And Optical Communications Networks*. Performance of fixed and dynamic channel allocation schemes in a motorway environment (Colombo, Sri Lanka, 2010)
16. V Saritha, VM Viswanatham, Approach for channel reservation and allocation to improve quality of service in vehicular communications. Int'l J IET Networks. **3**(2), 150–159 (2014)
17. SB Rejeba, N Nasser, S Tabbane, A novel resource allocation scheme for LTE network in the presence of mobility. J. Network Comput. Appl., Elsevier. **46**, 352–361 (2014)
18. QA Zeng, DP Agrawal, *Handoff in Wireless Mobile Networks in Handbook of Wireless Networks and Mobile Computing*. (John Wiley & Sons, Inc, Hoboken, NJ, 2002). ch 1
19. S Yi, Yilin S, S Jinglin, D Eryk, in *Proc. IEEE Int'l Conf. on Telecom. and Malaysia Int'l Conf. on Comm*. Research on bandwidth reservation in IEEE 802.16 (wimax) networks (Kuala Lampur, Malaysia, 2007)
20. L Banda, M Mzyece, Maximisation of correct handover probability and data throughput in vehicular networks. J. Machine to Machine Commun. **1**(2), 123–144 (2014)
21. GH Mohimani, F Ashtiani, A Javanmard, M Hamdi, Mobility modeling, spatial traffic distribution, and probability of connectivity for sparse and dense vehicular ad hoc networks. IEEE Trans. Vehicular Technol. **58**(4), 1998–2007 (2009)
22. P Chowdhury, IS Misra, in *Proc. of 2nd, Int'l. Conf. on Radio and Antenna Days of the Indian Ocean*. An improved call admission control mechanism with prioritized handoff queuing scheme for BWA networks (Republic of Mauritius, 2014)

4

Enabling smartphone-based HD video chats by cooperative transmissions in CRNs

Tao Jing[1], Xuewei Cui[1], Wei Cheng[2*], Shixiang Zhu[1], Yan Huo[1] and Xiuzhen Cheng[3]

Abstract

Smartphones have been equipped with the cameras that can shoot HD videos, and the video chat apps such as Skype are becoming popular. We can, therefore, intuitively predict the trend that users are expecting to enjoy HD video chats via utilizing their smartphones. Most of the current Internet services, however, cannot support the live HD video transmissions because of their low uplink rate. In order to overcome this limit, we propose to offload the uplink transmissions to cooperative users via cognitive radio networks. Specifically, we first divide the video stream into several substreams according to the H.264/SVC standard and the cooperative users' uplink rates. Then, the cooperative users are selected by employing our proposed optimal multiple stopping method. Finally, the substreams are assigned to the selected cooperative users by a 0-1 Knapsack-based allocation algorithm. The simulation results demonstrate that our proposed scheme can successfully support 720P HD video chats.

Keywords: Video chat; Cooperative relay selection; Optimal multiple stopping theory

1 Introduction

High-definition cameras are currently available on popular smartphones. These cameras have been physically ready to support HD video shootings (such as 720P and 1,080P). According to [1], 720P and 1,080P videos require the transmission speeds of 6 and 12 Mbps, respectively. However, most of the popular Internet services are mainly optimized for downlink transmission. The uplink transmission rate supported by most popular Internet access networks is only 2 Mbps, and the typical upload speed on 3G/4G approximately ranges from 0.45 to 1.93 Mbps [2]. As a result, a single user's Internet upload speed is generally not enough to support live HD video transmission. Therefore, an open problem is how could users enjoy the HD video chats without upgrading their Internet services, which may be too expensive and/or not necessary for other applications.

In order to enable the smartphone-based HD video chat, we propose to utilize cooperative users to help with the uplink transmission in cognitive radio networks (CRNs). The challenge of this task is how to efficiently select appropriate CR relays for cooperative uplink transmission as the *real-time* requirement of video chat is very strong. In other works, the sender does not have enough time to look at all the candidate relays to select the best subset for the transmission. The sender has to do quick decisions on relay selection while guaranteeing the HD video transmission with a reasonable low cost. Therefore, our objective is to design an efficient relay selection method, which can support HD video chat with the lowest cost, and a load assignment method, which can optimally distribute the video data among the selected relays. Specifically, the major work and the contributions can be summarized as the following:

- Overall, we propose a scheme to support smartphone-based HD video chat without upgrading the user's internet service by utilizing the cooperative CR users' unlink resource. Simulations verify that our proposed scheme can successfully support 720P HD video chat.
- Intuitively, the selected relays can use more time to forward packets if the relay selection time can be reduced. According to the requirement of HD video transmission rate and the cooperative CR users' uplink rate, given a number of time slots, we analyze the relationship between the relay selection time and

*Correspondence: wcheng3@vcu.edu
[2]Department of Computer Science, Virginia Commonwealth University, 401 West Main Street, 23284 Richmond, VA, USA
Full list of author information is available at the end of the article

the relay forwarding time so that the requirement can be satisfied.

- To the best of our knowledge, this is the first work to formulate an optimal multiple stopping model to solve the problem of relay selection for supporting HD video chat via cooperative transmission. We derive the multiple optimal stopping rules by jointly considering the instantaneous reward (of selected relays) and the expected sum reward (of the unobserved candidate relays). The proposed selection method can select the relays, whose instantaneous reward is at least the same as the expected sum reward.
- Extensive simulations have been conducted to investigate the impact of the parameters on the performance of the proposed scheme.

The rest of this paper is organized as follows. The most related work is summarized in Section 2. The overview of H.264/SVC, the network model, and the adopted relaying framework are illustrated in Section 3. Section 4 introduces our optimal stopping policy-based cooperative relay selection scheme and the packets assignment algorithm. The results of the performance evaluation are reported in Section 5. The paper is concluded in Section 6.

2 Related work

In recent years, researchers have shown a great interest in HD video technology. Jansen, 2011 [3] studies video-conferencing system for home. Lu, et al. 2010 [4] studies and compares mechanisms and performance of the existing video conference systems. Mirta, et al. 2010 [5] introduces a HD video broadcasting scheme by using scalable video coding so that the devices under various network environments can obtain the video with different resolutions. All of these work are studied under ideal network situations. However, for smartphone users, their uplink rate is generally not enough for HD transmission.

Several current video coding standards such as H.264/AVC, Dirac, AVS China, and VC-1 are introduced in [6]. Schwarz, et al. 2012 [7] provides the comparison of the coding efficiency for these video coding standards. Schwarz, et al. 2007 [8] gives a detailed overview of the scalable video coding extension of the H.264/AVC standard. In our work, we consider H.264/SVC as the coding standard utilized by HD video chat app.

An overview of existing cooperative relaying selection schemes is provided in [9]. They require channel-related information from all the candidate relay nodes, which is inefficient when the number of candidate relays is large or the time for relay selection is limited [10]. For example, channel state information and SNR are required by the relay selection approaches proposed in [11,12], respectively. Moreover, [12] needs to compute the SNR thresholds for all candidate relays.

Jing, et al. 2013 [13] studies the relay selection problem in cognitive networks by applying optimal stopping theory. This approach does not look at the information from all the candidate relay nodes as it scans the candidate secondary user (SU) relays one by one and stops when a suitable relay is identified. As a result, this relay selection process is efficient for single relay selection. As a comparison, our work is the first one to apply optimal multiple stopping theory for multiple relay selection in cognitive radio networks, to our best knowledge. We design an optimal multiple stopping rule to find out the relays with good performance and within a short observation time.

3 Preliminaries

3.1 Overview of H.264/SVC

H.264/SVC standard, an extension of the H.264/AVC standard, is generally applied in HD video stream transmission. The speeds for 720P and 1,080P HD videos are 6 and 12 Mbps, respectively, according to [5]. An SVC video bit stream is essentially constructed with a base layer sub-stream and more than one scalable enhancement layer sub-stream referring to different video layers. If part of the scalable stream is lost, the rest of the sub-stream forms a new valid bit stream with a certain bit-rate and reduced quality level which can be decoded by the target decoder. Hence, SVC can be applied in lossy transmission environments where graceful degradation of rate, format, and power adaptation exists. It is believed that H.264/SVC allows 20% packet loss in the network without affecting the quality [14].

Compared to the bit stream derived by dropping packets, a sub-stream can tolerate a lower temporal resolution (lower frame rate), lower spatial resolution (smaller screen), or lower quality video signal, which are named by these three modalities of scalability: temporal scalability, spatial scalability, and quality scalability.

3.2 System model

We consider a simple time-slotted cooperative transmission model depicted in Figure 1 which consists of a pair of primary users and a number of secondary users. To implement a HD video chat, a primary user (PU) transmitter, denoted by P_t, transmits its HD video stream to a primary user receiver, denoted by P_r, with the assistance of multiple cooperative users, who are the secondary users (SUs). The n secondary users, represented by S_i, i=1, 2, $\cdots$, n, have the ability to help transmit packets for the primary users, which are called *candidate relays.* When P_t needs to transmit packets to P_r, m free secondary users, which have favorable channel condition, can be selected as relay nodes by the PU transmitter. The m

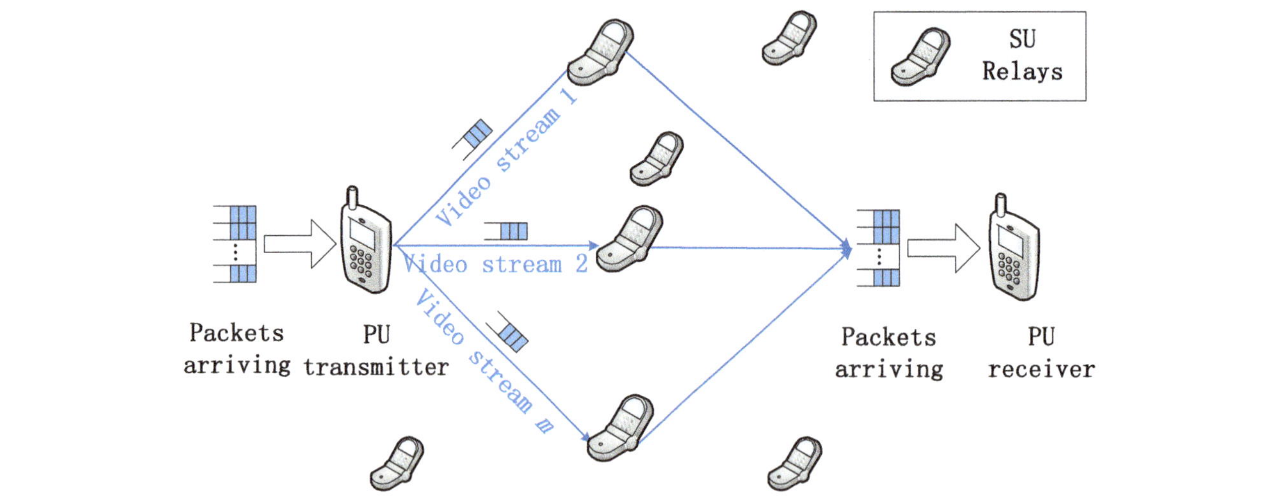

Figure 1 Cooperative transmission model.

secondary users finally selected by the PU transmitter are called *cooperative relays*.

It is assumed that the multiple cooperative relay selection is performed at each time slot. The duration of a time slot is $T = \frac{D_{hd}}{R_{hd}}$, where D_{hd} represents the amount of the packets to be transmitted for HD video and R_{hd} represents the required transmission rate for HD video.

As illustrated in Figure 2, each time slot T is partitioned into three components T_{se}, T_{sr}, and T_{rd}. Let τ be the time needed for observing a candidate relay. We assume that τ is identical for different SUs and for different time slots. Denoted by $S = \{s_1, s_2, \cdots, s_n\}$ an observation order/sequence, which is a permutation of the SU candidate relays index set $\{1, 2, \cdots, n\}$. At the beginning of a time slot, P_t starts to observe the SU candidate relay nodes sequentially according to the observation sequence. If the reward of the kth observation satisfies a specific criterion, P_t stops at the kth SU candidate relay node and then continues to observe candidate relay nodes for the following stops. The whole observation ends after P_t stops m times.

T_{se} represents the time of selecting relays in each time slot. After the cooperative relay selection process, P_t transmits packets to relay node i in T_i. The total time for transmission between PU transmitter and the relay nodes is denoted by T_{sr}, which is the sum of $T_1, T_2, \cdots, T_m$. Then, the m cooperative relays forward the video stream packets to P_r simultaneously in T_{rd}.

From Figure 2, we can see that:

$$T = T_{se} + T_{sr} + T_{rd} \tag{1}$$

T and T_{sr} are given by:

$$T = \frac{D_{hd}}{R_{hd}}, \; T_{sr} = \frac{D_{hd}}{R_{sr}} \tag{2}$$

where R_{sr} represents the transmittion rate between the PU transmitter and the cooperative relay node. Then we can get:

$$T_{se} + T_{rd} = T - T_{sr} = \frac{D_{hd}}{R_{hd}} - \frac{D_{hd}}{R_{sr}} \tag{3}$$

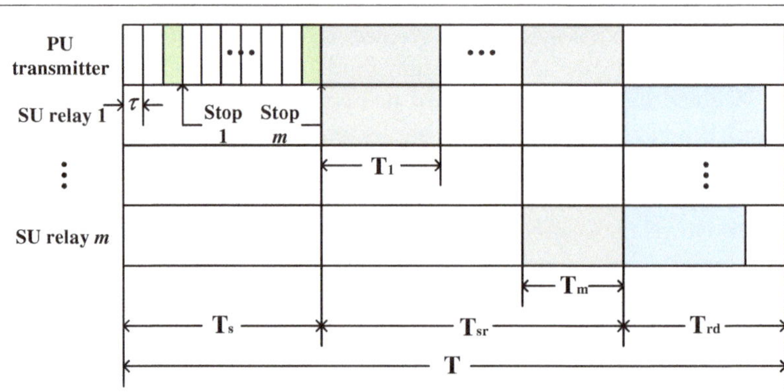

Figure 2 The time slot structure.

During T_{rd}, the cooperative relays forward the video packets to the PU receiver simultaneously. When there are m relays forwarding packets, the estimated value of T_{rd} is given by:

$$T^*_{\text{rd}} = \frac{D_{\text{hd}}}{mR^*_{\text{rd}}} \tag{4}$$

where R^*_{rd} denotes the expected rate of a relay node in a flat Rayleigh fading channel. Then, we can get the estimated observation time and the estimated size of observation sequence n below:

$$T^*_{\text{se}} = \frac{D_{\text{hd}}}{R_{\text{hd}}} - \frac{D_{\text{hd}}}{R_{\text{sr}}} - \frac{D_{\text{hd}}}{mR^*_{\text{rd}}} \tag{5}$$

$$n = \frac{T^*_{\text{se}}}{\tau} \tag{6}$$

Obviously, the size of observation sequence should be larger than the number of cooperative relays. Since the condition that observation time T_{se} is larger than zero should be satisfied, we estimate the value of m with a variable parameter K as:

$$m^* = \min\left\{m|n > m + K, T^*_{\text{se}} > 0\right\} \tag{7}$$

There are two types of relaying techniques: decode and forward (DF) and amplify and forward (AF) [15,16]. In DF, a relay node first decodes the encoded data from the source node and then recodes before forwarding it to the destination node [17]. Different from the DF technique, a relay node in AF simply amplifies the signal of the received packets and then delivers them [18,19]. In this paper, we adopt AF to illustrate our designs.

This cooperative video stream transmission scheme can be divided into three steps: (i) the video source selects a group of cooperative users that have the independent Internet access and are willing to help; (ii) after encoding the video in H.264-SVC, the video source distributes the video frames to the cooperative users via CR channels; and (iii) the cooperative users forward the video frames to the destination via their own Internet access networks, then, the video destination can get the HD video after decoding it successfully. Undoubtedly, the whole process is transparent to the destination. The procedure of HD video transmission is depicted in Figure 3.

4 Optimal multiple stopping policy

4.1 Problem formulation

In this paper, we formulate the cooperative transmission problem as multiple stopping problem in CRNs. In order to select m relays for cooperative transmissions, the PU transmitter observes multiple SU candidate relays one by one in an observation sequence. After contacting with a candidate relay, the PU decides whether to make a stop to select the current relay as a cooperative relay. The observation comes to the end if the PU finishes selecting m relays. When PU makes a decision, it considers the reward in the sum case. In other words, we assume the PU has selected l relays before observing ith relay, if PU selects the current relay, we can get a sum reward denoted by $y_i + V_i^{m-l-1}$, where y_i represents the instantaneous reward of the ith relay and V_i^{m-l-1} represents an expected reward of the $m - l - 1$ relays to be selected from the remaining candidate relays; if PU does not select the current relay, we can get an expected sum reward of $m - l$ relays to be selected from the remaining candidate relays V_i^{m-l}. If the sum reward with selecting the current relay is larger than the one without selecting the current relay, the PU makes a stop and vice versa. Therefore, the multiple relay selection problem can be further formulated as a sequential decision problem which can be solved by the optimal multiple stopping theory.

We first introduce a definition of multiple stopping theory in the sum case. We, then, formulate our cooperative relay selection problem as an optimal multiple stopping problem in the sum case.

Definition 1. *A multiple stopping rule problem in the sum case is defined by two parts:*

- *A sequence of random variables, $X_1, X_2, \cdots$, which have a known joint distribution*
- *A sequence of sum reward function $z_0^l = y_0 + V_0^{m-l}$, $z_1^l(x_1) = y_1(x_1) + V_1^{m-l}, \cdots, z_j^l(x_1, x_2, \cdots, x_j) = y_j(x_1, x_2, \cdots, x_j) + V_j^{m-l}$, where $y_j(x_1, x_2, \cdots, x_j)$ represents the instantaneous reward of jth relay, V_j^{m-l} represents the expected sum reward of $m - l$ relays to be selected in the remaining candidate relays.*

The objective is to find out the variables in the sequence such that the sum reward function is maximized.

In order to further investigate the channel quality in our cooperative relay selection problem, we assume that the underlying channel is a flat Rayleigh fading channel [20], in which the instantaneous signal-to-interference-plus-noise ratio (SNR) is received by the destination with

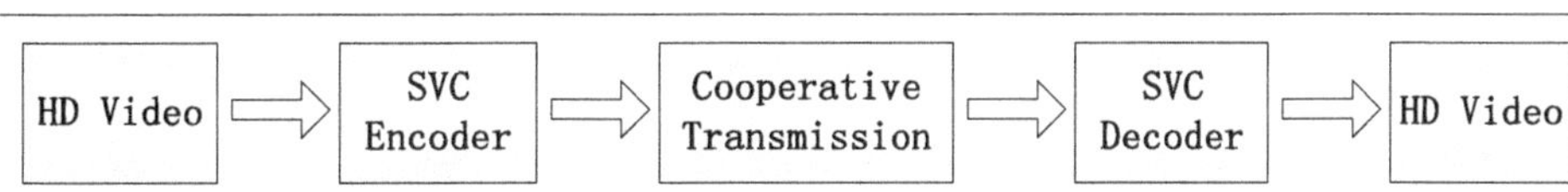

Figure 3 The procedure of HD video transmission.

an exponential distribution having a probability density function (PDF) $f(\gamma) = \frac{1}{\overline{\gamma}}e^{-\frac{\gamma}{\overline{\gamma}}}$, where $\overline{\gamma}$ denotes the average SNR in the channel model. Then, we can model the Rayleigh fading channel as a finite state Markov chain (FSMC) as proposed in [6]. In the FSMC, we partition the SNR into U intervals and then divide SNR into a finite-state space. Thus, the SNR thresholds are denoted by $\Upsilon = \{\gamma_1 = 0, \gamma_2, \cdots, \gamma_U, \gamma_{U+1} = \infty\}$. If an instantaneous SNR Γ is in $[\gamma_u, \gamma_{u+1})$, the channel of the SU candidate relay is said to be in state s_u. When the PU pair observes the channel of the candidate relay, the probability of the SU being in state s_u for the channel can be given by:

$$q_u = \int_{\gamma_u}^{\gamma_{u+1}} f(\gamma)d\gamma = e^{-\frac{\gamma_u}{\overline{\gamma}}} - e^{-\frac{\gamma_{u+1}}{\overline{\gamma}}}, u = 1, \cdots, U \tag{8}$$

In our cooperative transmission problem, the achievable transmission rate is viewed as a metric for the channel quality in wireless communications. Let r_k denote the achievable transmission rate between the PU pair and the SU candidate relay node k. According to the Shannon's theorem [21], r_k is calculated as follows:

$$r_k = W \log(1 + \gamma_k) \tag{9}$$

where W denotes the bandwidth of the spectrum. Thus, the corresponding date rate can be denoted as $R = \{r_1, r_2, \cdots, r_U\}$. We can also model the transmission rate as a discrete random variable, which has a same distribution as the channel state:

$$Pr\{R = r_u\} = q_u, u = 1, 2, \cdots, U \tag{10}$$

The PU pair acquires the achievable transmission rate of the channel between itself and the SU candidate relay by executing the observation in relay selection. The process of observation is similar to the request-to-send (RTS)/clear-to-send (CTS) access mechanism designed for the 802.11 technique [22,23]. At each observation step, the PU transmitter sends a RTS frame to the candidate relay. Upon receiving of a RTS frame, the candidate relay returns a CTS frame, which includes the information for calculating the achievable rate. We define that $X_k = R_k$ is the valid transmission rate of the kth observation step. Then, the distribution of X_k can be calculated as follows:

$$\begin{aligned} &p_u = Pr\{X_k = x_u = r_u\} = q_u \\ &\text{for } 1 \le u \le U, 1 \le k \le n \end{aligned} \tag{11}$$

Then, we derive the instantaneous reward function denoted by Y_k based on the valid transmission rate and two scaling factors. We denote $e(k, l)$ as the scaling function if the PU pair stops at the kth observed candidate relay node with l selected relays:

$$e(k, l) = 1 - \frac{k(m - l)}{M} \tag{12}$$

where m also means the stopping times for the whole observation and l also means the times we have stopped. M is a variable parameter. From (12), we can see that $e(k, l)$ is an increasing function of k and a decreasing function of l. In other words, a larger number of SU candidate relay nodes, which the PU transmitter has observed, means the lower efficiency of the cooperative relay selection process. More relays, which have been selected by the PU transmitter, means the less number of stops remains. As a result, the efficiency of the cooperative relay selection process becomes higher. Accordingly, the reward after the kth observation attempt with l stops is given by:

$$Y_k = e_k X_k \tag{13}$$

After deriving the reward function, we summarize the optimal multiple stopping problem in cooperative relay selection as follows: the PU pair receives the reward Y_k after the kth observations, then the PU transmitter makes a decision on whether to make a stop at the current candidate relay. The PU transmitter finally stops the observation till m cooperative relays are all selected. Note that no recall is allowed since the channel quality is changing rapidly in cognitive radio networks due to the complicated conditions such as the mobility of the users.

4.2 Optimal multiple stopping rule

In this subsection, we solve the multiple stopping problems in the sum case by deriving the optimality equations. We also prove that the proposed rule is better than the random stopping rule, in which the PU stops m times randomly.

We denote F_i as the probability distribution function of the instantaneous reward Y_i after the ith observation. For the finite payoff reward sequence $Y_1, \cdots, Y_n$ that are independent and identically distributed as depicted in Section 4.1, the optimality equations can be stated as follows for the m-stopping problems. In case $m = 1$, a well-known recursive solution of stopping problem is defined by:

$$V_n := -\infty \tag{14}$$

$$V_i := E[Y_{i+1} \vee V_{i+1}|F_i] \tag{15}$$

where V_i^m represents the expected sum reward of continue observing for m stopping times after ith observation and $\vee$ denotes the maximum. We need a version of this classical result for stopping times larger than a given stopping time C.

Proposition 1. *Recursive solution of one-stopping problems.*

(a) For any time point $0 \le k \le n - 1$*, the stopping time*

$$T(k) := min\{k < i \le n : Y_i > V_i\} \tag{16}$$

is optimal in the sense that for any stopping time we have

$$E\left[Y_{T(k)}|F_k\right] = V_k \geq E\left[Y_T|F_k\right] \tag{17}$$

(b) For any stopping time C, the stopping time

$$T(C) := min\{C < i \leq n : Y_i > V_i\} \tag{18}$$

is optimal in the sense that for any stopping time T with $C < T$ conditionally on $\{C < n\}$ and $C = T$ conditionally on $C = n$, we have

$$E\left[Y_{T(C)}|F_C\right] = V_C \geq E\left[Y_T|F_C\right] \tag{19}$$

For m-stopping problems, the following variant of Proposition 1 will also be needed. Let C be a stopping time, let B be measurable with the same probability distribution with Y_i and h be measurable with $Eh(Y_i, B)^+ < \infty$. We can also define recursively for b:

$$V_n(b) := h(Y_n, b) \tag{20}$$

$$V_i(b) := E\left[h(Y_{i+1}, b) \vee V_{i+1}(b)|F_i\right] \tag{21}$$

Then the stopping time:

$$T(C, B) := \min\{C < i \leq n : h(Y_i, B) > V_i(B_i)\} \tag{22}$$

is optimal in the sense that for any further stopping time T with $C < T$ conditionally on $C < n$ and $C = T$ conditionally on $C = n$, we have:

$$E\left[h(Y_{T(C,B),B})|F_C\right] = V_C(B_C) \geq E\left[h(Y_T, B)|F_C\right] \tag{23}$$

The idea of solving m-stopping problem is based on the one for solving the one-stopping problem. The lth stopping time T_l should be i, if the total expected reward by taking i as a stop is larger than the one, where i will not be a stop. We define the thresholds V^m_{n-m+1}, V^0_i and V^m_i by:

$$V^m_{n-m+1} := -\infty \tag{24}$$

$$V^0_i := 0 \tag{25}$$

$$V^m_i := E\left[\left(Y_{i+1} + V^{m-l}_{i+1}\right) \vee V^m_{i+1}|F_i\right] \tag{26}$$

From (26), we know that V^{m-1}_{i+1} and V^m_{i+1} can be calculated by backwards induction. We compute V^m_i as follows:

$$\begin{aligned} V^m_i &= E\left[\left(Y_{i+1} + V^{m-1}_{i+1}\right) \vee V^m_{i+1}|F_i\right] \\ &= E\left\{\left[e(k,l)X_{i+1} + V^{m-1}_{i+1} \vee V^m_{i+1}\right]|F_i\right\} \\ &= \sum_\alpha \left[e(k,l)x_\alpha + V^{m-1}_{i+1}\right]p_\alpha + \sum_\beta V^m_{i+1}p_\beta \end{aligned} \tag{27}$$

where $\alpha \in \left\{k \mid e(k,l)x_k + V^{m-1}_{i+1} > V^m_{i+1}, k = 0, 1, 2, \cdots, U\right\}$, $\beta \in \left\{k \mid e(k,l)x_k + V^{m-1}_{i+1} < V^m_{i+1}, k = 0, 1, 2, \cdots, U\right\}$, subject to:

$$\begin{cases} 0 \leq \alpha \leq U \\ 0 \leq \beta \leq U \\ \alpha + \beta = U \end{cases} \tag{28}$$

Then, we can get the sum reward function:

$$z^l_i(x_1, x_2, \cdots, x_i) = y_i(x_1, x_2, \cdots, x_i) + V^{m-l}_i \tag{29}$$

where $y_i(x_1, x_2, \cdots, x_i)$ denotes the instantaneous reward after the ith observation. It is optimal to stop if $z^l_i(x_1, x_2, \cdots, x_i)$ is larger than V^{m-l+1}_i.

The related threshold stopping times are defined recursively for $1 \leq m \leq n$, $0 \leq k \leq n - m$ by:

$$T^m_1(k) := \min\left\{k < i \leq n - m + 1|z^1_i(x_1, x_2, \cdots, x_i) > V^m_i\right\} \tag{30}$$

$$T^m_l(k) := \min\left\{T^m_{l-1}(k) < i \leq n - m + l\,|z^l_i(x_1, x_2, \cdots, x_i) > V^{m-l+1}_i\right\}, 2 \leq m \leq n. \tag{31}$$

As in the one-stopping problem, it holds that $\{T^m_i(k)\}$ are optimal stopping times.

Proposition 2. *For multiple stopping problem in the sum case, $\{T^m_i(k)\}$ are optimal m-stopping times in the sense that for all stopping times $k < T_1 < \cdots < T_m \leq n$ it holds that:*

$$E\left[\sum_{l=1}^m Y_{T^m_l(k)}|F\right] \geq E\left[\sum_{l=1}^m Y_{T_l}|F\right] \tag{32}$$

Proof 1. *This proof is by induction in m. As depicted above, $\{Y_i\}$ are independent and identically distributed. In the case $m = 1$, we can get:*

$$\begin{aligned} E\left[Y_{T^1_1(k)}|F\right] &= V^1_k \\ &= E\left[\left(Y_{k+1} + V^0_{k+1}\right) \vee V^1_k|F\right] \\ &= E\left\{Y_{k+1} \vee \cdots E\left[Y_{n-1} \vee E(Y_n)\right]|F\right\} \\ &\geq (n-k) \cdot \frac{1}{n-k} E\left[Y_i|F\right] = E\left[Y_{T_1}|F\right] \end{aligned} \tag{33}$$

For the induction step $m \rightarrow m + 1$, by induction hypothesis we obtain:

$$\begin{aligned} &E\left[Y_{T_1} + \cdots + Y_{T_{m+1}}|F\right] \\ &= E\left[Y_{T_1}|F\left[+E\right]\left[Y_{T_2} + \cdots + Y_{T_{m+1}}|Y_{T_1}\right]|F\right] \\ &\leq E\left[Y_{T^{m+1}_1(k)}|F\right] + E\left[Y_{T^m_1(T_1)} + \cdots + Y_{T^m_m(T_1)}|F\right] \\ &= E\left[\sum_{l=1}^{m+1} Y_{T^{m+1}_l(k)}|F\right] \end{aligned} \tag{34}$$

This is maximized by choosing T_1 to be the optimal one-stopping time of the process $(Y_i + V^m_i)$. This *optimal stopping time is given from the case* $m = 1$ *by* $T_1 = T_1^{m+1}$. *In consequence, the optimal stopping times for* $1 \leq l \leq m$ *are obtained by* $T_l^m(T_1^{m+1}) = T_{l+1}^{m+1}$.

Therefore, we propose the optimal multiple stopping rule in Algorithm 1. The PU observes an SU candidate relay node according to the observation sequence S and obtains an instantaneous reward y_i after the ith observation (line 4). Then, the PU compares the sum reward z_i^l with the expected sum reward V_i^{m-l+1}. If $z_i^l \leq V_i^{m-l+1}$, the PU continues to observe the next SU candidate relay. Otherwise, the PU pair makes a stop at the ith relay. When m cooperative relays are all selected, PU finishes the whole observation. Note that if the PU observes the $(n-m+l)$th candidate relay with having selected l cooperative relays, it has to select the rest of SU nodes as the cooperative relays for supporting the transmissions regardless of the value of the sum reward.

Algorithm 1 The optimal multiple stopping rule.

1: Construct the observation sequence $S = \{s_1, s_2, \cdots, s_n\}$;
2: Decide the value of m according to the (7);
3: Let S_a and l denote the set and the number of cooperative relay nodes that have been selected, $S_a \leftarrow \emptyset, l \leftarrow 0$;
4: **for** $i \leftarrow 1$ to $n - m$ **do**
5:   **if** $m - l = n - i + 1$ **then**
6:     Select $S_a \bigcup \{s_i, s_{i+1}, \cdots, s_n\}$ as the output;
7:     Break;
8:   **else**
9:     Compute the instantaneous reward y_i given by (13) after obtaining the achievable transmission rate r_i by observing the ith relay;
10:     Compute the sum reward z_i^l given by (29).
11:     Compute the expected sum reward V_i^{m-l+1} given by (27);
12:     **if** $z_i^l < V_i^{m-l+1}$ **then**
13:       Continue;
14:     **else**
15:       Stop at the current step and select the ith SU node as one of the cooperative relays, $S_a \leftarrow S_a \bigcup s_i, l \leftarrow l + 1$;
16:       **if** $l < m$ **then**
17:         Continue;
18:       **else**
19:         Break;
20:       **end if**
21:     **end if**
22:   **end if**
23: **end for**

4.3 Video frames allocation algorithm

Since different cooperative relay nodes may have varied transmission rates, we target on maximizing the amount of transmission on each relay under certain rate limits. In this subsection, we intend to find an optimal video frames allocation algorithm by formulating it as 0-1 Knapsack problem.

The 0-1 Knapsack problem [24] is a subproblem of Knapsack problem, which is a problem in combinatorial optimization. We first give an brief definition of 0-1 Knapsack problem, which is employed in our allocation model.

Definition 2. *The 0-1 Knapsack problem is defined as follows: given a set of items, each with a weigh and a value, determine the number (0 or 1) for each item to include in a collection, so that the total weight is less than or equal to a given limit, and the total value is maximized.*

The PU video source encoder divides the video stream into a base sub-stream and M enhancement streams. Let $D = \{D_1, D_2, \cdots, D_M\}$ denote the set of divided packets, where D_k $(k = 1, 2, \cdots, M)$ represents the amount of a specific video packet. We allocate the packets to the selected cooperative relays one by one according to the descending order of their transmission rates. For the ith cooperative relay s_i, we denote set $D^i = \left\{D_1^i, D_2^i, \cdots, D_{M_i}^i\right\}$ as the set of the amount of packets that have not been allocated.

We formulate the video frames allocated problem as follows. The PU selects m cooperative relays with the rate $\{r_1, r_2, \cdots, r_m\}$ for forwarding packets to the PU receiver. Let A_i, $i = 1, 2, \cdots, m$, denote relay s_i'a available amount of transmissions, which is defined as the amount of packets that can be forwarded by s_i. In T_{rd}, the amount of packets, which relay s_i can forward, is $r_i \cdot T_{\text{rd}}$. Therefore, we obtain $A_i = r_i \cdot T_{\text{rd}}$. Then, we define an available transmission amount of vector A as $A = \{A_1, A_2, \cdots, A_m\}$.

In our allocation model, we take $D^i = \left\{D_1^i, D_2^i, \cdots, D_{M_i}^i\right\}$ as the weigh and the value for items of base layer and enhancement layers. We regard set $A = \{A_1, A_2, \cdots, A_m\}$ as the given limits depicted in Definition 2.

In 0-1 Knapsack problem, the weigh and the limit should be integers, we, therefore, construct a corresponding integer sets $D^{i*} = \left\{D_1^{i*}, D_2^{i*} \cdots, D_{M_i}^{i*}\right\}$ and $A^* = \left\{A_1^*, A_2^*, \cdots, A_m^*\right\}$. Where $D_k^{i*} = \left\{D_k^i\right\} + 1$ and $A_i^* = [A_i]$. Here, $[\cdot]$ represents the integral function. Note that after the allocation for a relay, the set D^{i*} changes regularly.

The objective of our video frames allocation is to determine the subset $Q_i \subseteq \{1, 2, \cdots, M_i\}$ such that:

$$\max \sum_{t \in Q_i} D_t^{i*}, \tag{35}$$

$$\text{subject to} \sum_{t \in Q_i} D_t^{i*} \leq A_i^*. \tag{36}$$

For $1 \leq k \leq M_i, 0 \leq r \leq A_i^*$, we set:

$$G(0,r) = 0, \; G(k,0) = 0 \tag{37}$$

$$\text{if } D_k^{i*} > r, \; G(k,r) = G(k-1,r) \tag{38}$$

$$\text{if } D_k^{i*} \leq r,$$
$$G(k,r) = \max\left\{D_k^{i*} + G\left(k-1, r - D_k^{i*}\right), G(k-1,r)\right\} \tag{39}$$

From (38) and (39), we can see that there are only two choices for $\{D_1^{i*}, D_2^{i*} \cdots, D_{M_i}^{i*}\}$ to compute $G(k,r)$.

1. Leave D_k^{i*}: the maximal amount that relay i can forward from $\left\{D_1^{i*}, D_2^{i*} \cdots, D_{M_i}^{i*}\right\}$ with limit r is $G(k-1,r)$.
2. Take D_k^{i*} (only possible if $D_k^{i*} \leq r$): we gain D_k^{i*} of transmission amount and the maximal amount that the relay can transmit from $\left\{D_1^{i*}, D_2^{i*} \cdots, D_{M_i}^{i*}\right\}$ with rate limit $r - D_k^{i*}$ is $G\left(k-1, r - D_k^{i*}\right)$. Totally, we get $D_k^{i*} + G(k-1)$.

To allocate packets for relay i, we first do the bottom-up computation using iteration by computing the table using (38) and (39) row by row as shown in Figure 4 .

The set Q_i is derived by the process depicted in Algorithm 2, where $T_{x,y}$ represents the element in row x column y of the table. The video frames allocation algorithm which is based on the Knapsack problem is depicted in Algorithm 3.

G(k,r)	r=0	1	2	3	···	A_i^*	
k=0	0	0	0	0	···	0	Bottom
1	→						↓
2	→						↓
⋮	→						↓
Mi	→						Up

Figure 4 Bottom-up computation table. Figure 4 has been moved to this new location, and the citation for Figure 4 has been added in the above paragraph.

Algorithm 2 Deriving set Q_i.

1: $Q_i \leftarrow \varnothing$;
2: $R \leftarrow A_i^*, K \leftarrow M_i$;
3: **for** $t^* \leftarrow 1$ to $K-1$ **do**
4:   **if** $T_{K-t^*,R} < T_{K,R}$ **then**
5:     $Q_i \leftarrow Q_i \bigcup (K - t^* + 1), K \leftarrow (K - t^*)$ and $R \leftarrow (R - D_{K-t^*+1}^{i*})$;
6:     **if** $K < 1$ or $R < 1$ **then**
7:       Break;
8:     **else**
9:       Return to line 2;
10:     **end if**
11:   **else**
12:     Continue;
13:   **end if**
14: **end for**

Algorithm 3 Video stream allocation.

1: Construct the not-increasing sequence of the rate limit $A = \{A_1, A_2, \cdots, A_m\}$ and evolve it to a not-increasing integer sequence $A^* = \{A_1^*, A_2^*, \cdots, A_m^*\}$.
2: Construct the set of weigh and value $D^i = \left\{D_1^i, D_2^i \cdots, D_{M_i}^i\right\}$ and its integer set $D^{i*} = \left\{D_1^{i*}, D_2^{i*} \cdots, D_{M_i}^{i*}\right\}$.
3: **for** $i \leftarrow 1$ to m **do**
4:   **if** $D^{i*} \neq \emptyset$ **then**
5:     **for** $r \leftarrow 0$ to A_i^* **do**
6:       $G(0,r) \leftarrow 0$;
7:     **end for**
8:     **for** $k \leftarrow 1$ to M_i **do**
9:       $G(k,0) \leftarrow 0$;
10:     **end for**
11:     Compute the bottom-up computation table;
12:     Derive set Q_i;
13:   **else**
14:     break;
15:   **end if**
16: **end for**

5 Performance evaluation

In this section, we evaluate the performance of our cooperative HD video transmission scheme by simulations. The video stream is divided into five parts: one base layer substream and four pieces of enhancement layer substream. It is assumed that the transmission rate r_k of the kth SU candidate relay node does not change within one slot. We divide the finite-state space of SNR received by the receiver into $U = 30$ intervals. The bandwidth W is set to be 1 MHz. The rate of PU source is set to be the average rate of channel. The controlling factor M is set to be 10^6. We set the transmission rate of 720P HD video as 6 Mbps. Each result reported in this section is the average of 100 instances.

To provide a deep insight into the feasibility of our cooperative HD video transmission scheme, we first make a comparison between our cooperative relay selection scheme and the random cooperative relay selection scheme. Then, we investigate the performance of our video frames allocation algorithm by comparing it with the random video frames allocation scheme. Moreover, we study the impact of the average SNR $\overline{\gamma}$ and the observation duration τ on the transmission performance.

5.1 The comparison with random schemes

In this subsection, we compare our cooperative relay selection scheme and video frames allocation algorithm with the corresponding random methods, respectively. We set the average SNR $\overline{\gamma}$ to be 30 dB, the time duration τ to be 3μs.

We change the transmission time T from 0.4 to 0.85 ms and investigate the available cooperative transmission rate and the realistic cooperative transmission rate of schemes. Figure 5 reports the performance comparison of our cooperative relay selection scheme and the random relay selection scheme. We can see that selecting cooperative relays via optimal multiple stopping rule can achieve higher available cooperative transmission rate than selecting relays randomly. With the increasing of HD video transmission time, the available cooperative transmission rate of our scheme increases linearly, while the available cooperative transmission rate of random cooperative relay selection is distributed irregularly. From the equations in Section 4.1, we can see that with a specific SNR $\overline{\gamma}$, the number of candidate relays increases with T, which is explained in Section 5.2 in detail. So we have greater chance to find cooperative relays with good performance.

Then, we compare the proposed video frames assignment algorithm with the random allocation scheme. In this part, all the relay nodes are selected by our cooperative relay selection scheme. As Figure 6 shows, the realistic cooperative rate in the situation where applying the proposed video frames allocation is almost as same as the transmission rate for 720P HD video, which implies that our scheme can support HD video transmissions. We can also see that the realistic rate in the situation, where the packets are assigned randomly, is always lower than the transmission rate required by 720P HD video.

5.2 The impact of the parameter $\overline{\gamma}$

In this subsection, we investigate the impact of the parameter $\overline{\gamma}$ on the number of nodes, available cooperative transmission rate, and realistic cooperative transmission rate under the situation, where the average SNR $\overline{\gamma}$ changes regularly. We set the observation duration τ to be 3μs. The transmission time T is set to be 0.5, 0.6, and 0.7ms. We set $K = 10$ in (7) when $\overline{\gamma} \geq$ 22dB and $K = 0$ when $\overline{\gamma} <$ 22dB, respectively.

The relationship between $\overline{\gamma}$ with the the number of candidate relays and cooperative relays can be seen from Figure 7. We can see that the lager the average SNR $\overline{\gamma}$ is, the less cooperative relays we need. According to the

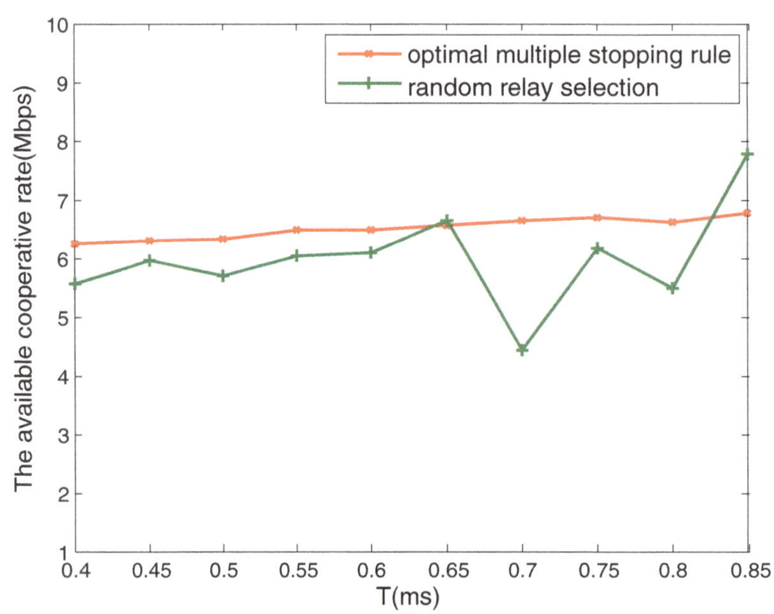

Figure 5 Our cooperative relay selection scheme vs. the random cooperative relay selection scheme.

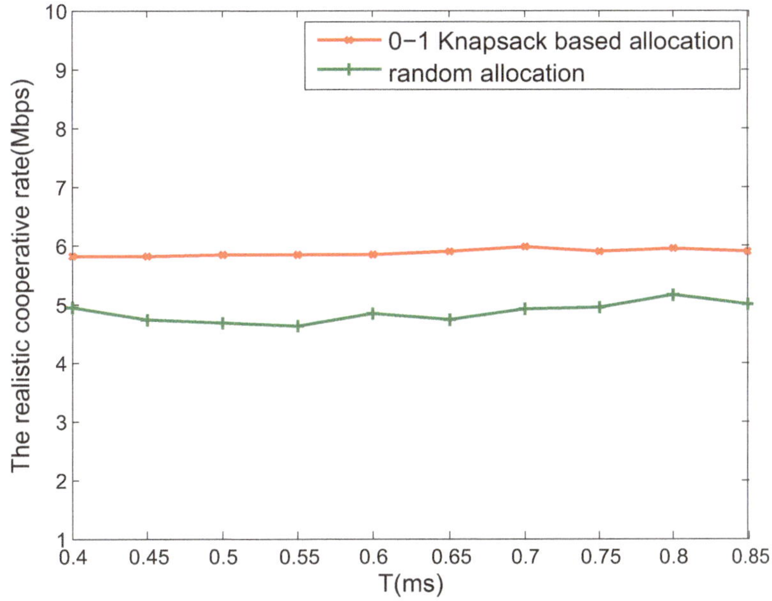

Figure 6 **Assignment based on 0-1 Knapsack problem vs. random assignment.**

analysis in Section 4.1, we can know that the observation time is decided by the number of cooperative relays and average SNR $\overline{\gamma}$, but the number of cooperative relays has a larger impact on the observation time for all candidate relays. Therefore, when the number of cooperative relays decreases one, the number of candidate relays that is in proportion to observation time reduces greatly; when the number of cooperative relays remains unchanged, the number of candidate relays impacted by $\overline{\gamma}$ increases linearity.

Note that when $\overline{\gamma}$ is below 22 dB, the available transmission rate of candidate relays is low. As a result, we need to select more cooperative relays for cooperative transmissions. However, the number of candidate relays is less than the number of cooperative relays with the parameter K in (7) is set to be 0. Therefore, our

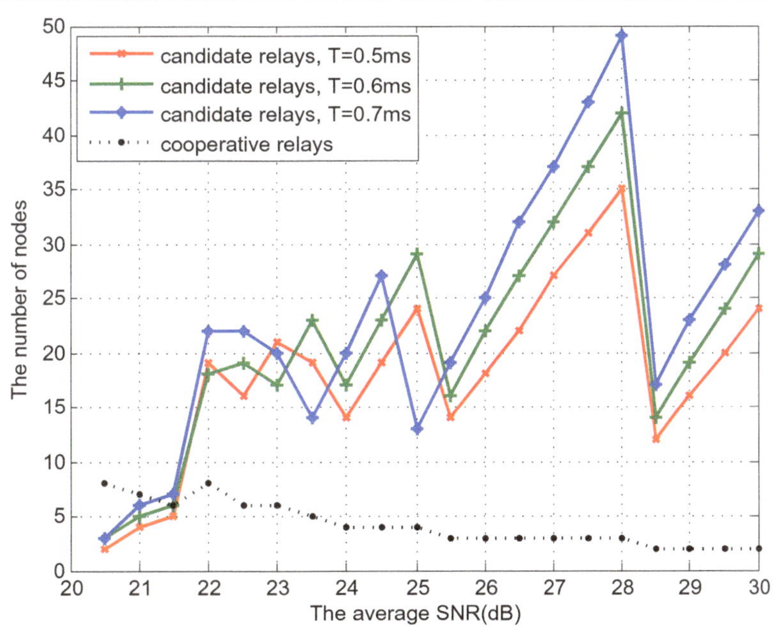

Figure 7 **The number of relays vs. the parameter $\overline{\gamma}$.**

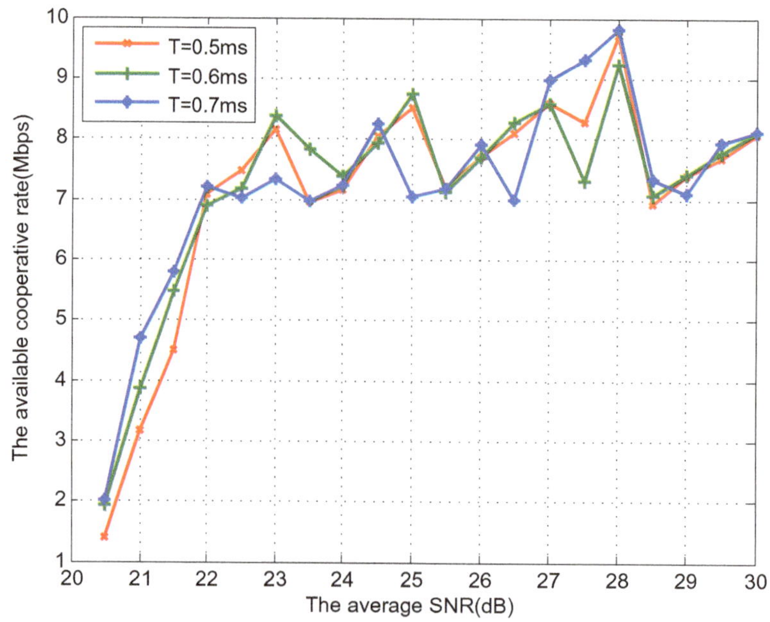

Figure 8 The available transmission amount vs. the parameter $\overline{\gamma}$.

proposed scheme is infeasible when the average SNR $\overline{\gamma}$ is low.

Form Figure 8, we can see that the available cooperative transmission rate increases with rising average SNR. We know from Figure 7 that the higher average SNR results in the larger number of candidate relays when the number of cooperative relays remains unchanged. With the increasing number of candidate relays, the PU transmitter has higher probability to choice relays with better performance. Therefore, the available cooperative transmission rate increases. Similar results can be obtained in Figure 9. With the increasing of $\overline{\gamma}$, the realistic cooperative transmission rate increases at the beginning, then stays in a stable state that can satisfy the transmission rate for supporting 720P HD video.

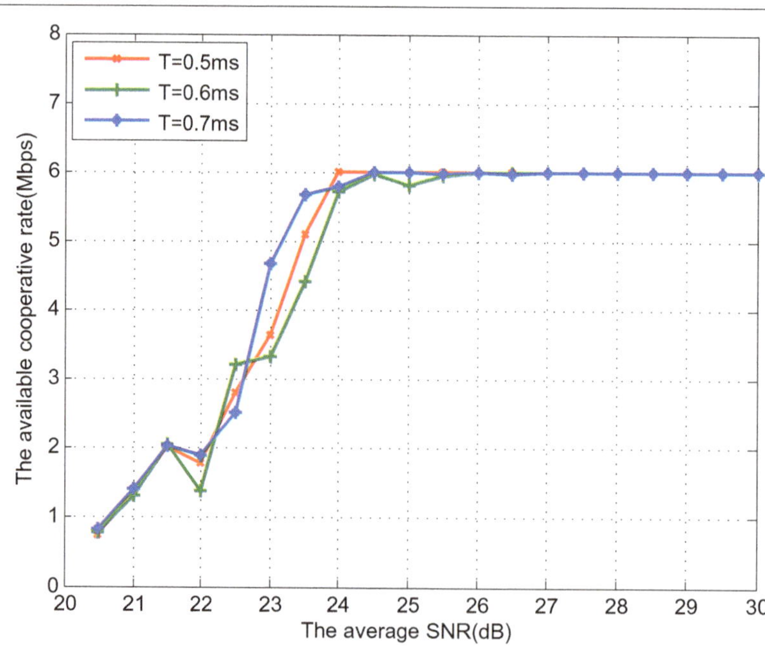

Figure 9 The realistic transmission amount vs. the parameter $\overline{\gamma}$.

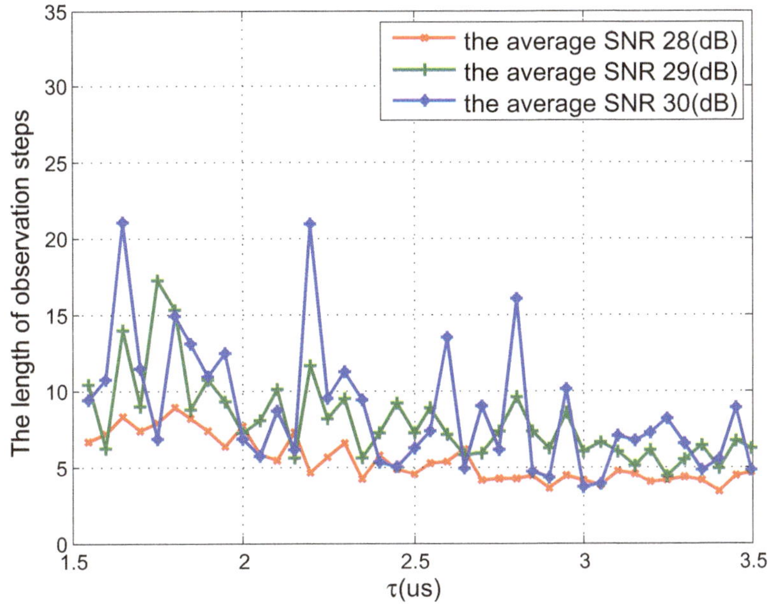

Figure 10 The length of observation steps vs. time for one observation.

5.3 The impact of observation duration τ

In this subsection, we investigate the impact of observation duration τ on the length of observation steps and available cooperative transmission rate. We set the average SNR $\overline{\gamma}$ to be 28, 29, and 30 dB.

From Figure 10, we can see that the length of observation steps decreases slowly with the observation duration τ increases regularly. In other words, a larger τ results in a smaller number of observation steps. This is because the value of τ represents the cost of observing one SU candidate relay, and the PU needs to stop observing as soon as possible to avoid generating a large cost. On the contrary, when the value of τ is small, the cost for observation is low, and the PU tends to observe more candidate relays to find better cooperative relays. We also can see that different $\overline{\gamma}$ impacts the length of observation. A larger $\overline{\gamma}$ results

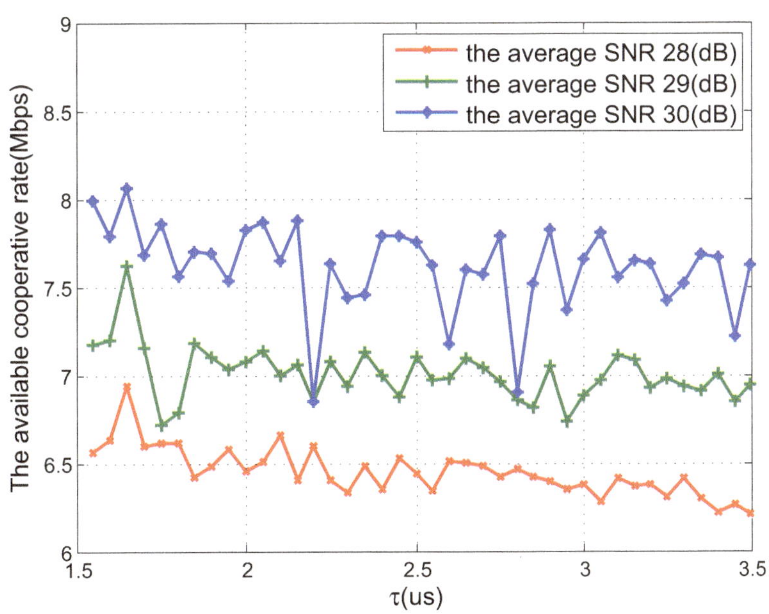

Figure 11 The available transmission amount vs. Time for one observation.

in a larger the number of candidate relays according to Figure 7.

We can see from Figure 11 that the available cooperative transmission rate declines slowly with the increasing of the time duration for each observation τ. This relationship can be obtained from the analysis in Section 4.2. When τ increases, the expected sum reward decreases. The decrease of the performance of selected relays, therefore, leads to a lower available cooperative transmission rate. On the other hand, the cost increases with the τ's increasing. Therefore, the available cooperative transmission rate declines. Similarly, different $\overline{\gamma}$ impacts the available transmission rate. This observation is consistent with the analysis in Section 5.2 for Figure 7.

6 Conclusions

In this paper, we propose an optimal multiple stopping scheme to solve the problem of multiple cooperative relay selection for supporting smartphone-based HD video chat in CRNs. In the proposed scheme, the video source selects the candidate CR relays by following a proposed optimal multiple stopping rule to guarantee the success of HD video chat. The proposed scheme has been verified to be able to support 720P video chat via simulations. In the future, we will implement the proposed scheme on android smartphones.

Competing interests
The authors declare that they have no competing interests.

Acknowledgements
The authors would like to thank the support from the National Natural Science Foundation of China (Grant No. 61272505, 61172074 and 61371069) and the National Science Foundation of the US (CNS-1265311).

Author details
[1]School of Electronics and Information Engineering, Beijing Jiaotong University, No.3 Shangyuancun, 100044 Beijing, China. [2]Department of Computer Science, Virginia Commonwealth University, 401 West Main Street, 23284 Richmond, VA, USA. [3]Department of Computer Science, The George Washington University, 801 22nd Street NW, 20052 Washington DC, USA.

References

1. B Gonzalez, All about Internet speed requirements for Hulu, Netflix, and Vudu Movie Viewing. http://hometheater.about.com/od/internethometheater2/a/Internet-Speed-for-Netflix-Vudu.htm
2. M Sullivan, 3G/4G performance map: data speeds for AT & T Sprint, T-Mobile, and Verizon. http://www.techhive.com/article/254888/3g_4g_performance_map_data_speeds_for_atandt_sprint_t_mobile_and_verizon.html
3. J Jansen, P Cesar, DC Bulterman, T Stevens, I Kegel, J Issing, Enabling composition-based video-conferencing for the home. Multimedia IEEE Trans. **13**(5), 869–881 (2011)
4. Y Lu, Y Zhao, F Kuipers, P Van Mieghem, in *NETWORKING 2010*. Measurement study of multi-party video conferencing (Springer-Verlag Berlin, Heidelberg, 2010), pp. 96–108
5. S Mirta, T Schierl, T Wiegand, P Inigo, C LeGuern, C Moreau, L Guarnieri, J Tronc, in *Advanced Satellite Multimedia Systems Conference (asma) and the 11th Signal Processing for Space Communications Workshop (spsc)*. HD video broadcasting using scalable video coding combined with dvb-s2 variable coding and modulation (Cagliari, Italy, 2010), pp. 114–121
6. KR Rao, Do Nyeon Kim, in *System Theory (SSST), 42nd Southeastern Symposium On*. Current video coding standards: H. 264/avc, dirac, avs china and vc-1 (IEEE Tyler, TX, USA, 2010), pp. 1–8
7. H Schwarz, G Sullivan, T Tan, T Wiegand, in *Circuits Syst. Video Technol. IEEE Trans*. Comparison of the coding efficiency of video coding standards-including high efficiency video coding (HEVC), vol. 22, (2012), pp. 1669–1684
8. H Schwarz, D Marpe, T Wiegand, Overview of the scalable video coding extension of the h.264/avc standard. Circuits Syst. Video Technol. IEEE Trans. **17**(9), 1103–1120 (2007)
9. W Elmenreich, N Marchenko, H Adam, C Hofbauer, G Brandner, C Bettstetter, M Huemer, Building blocks of cooperative relaying in wireless systems. Elektrotechnik und Informationstechnik. **125**, 353–359 (2008)
10. Z Cai, S Ji, J He, L Wei, AG Bourgeois, Distributed and asynchronous data collection in cognitive radio networks with fairness consideration. Parallel Distributed Syst. IEEE Trans. **25**(8), 2020–2029 (2014)
11. Y Jing, H Jafarkhani, Single and multiple relay selection schemes and their achievable diversity orders. Wireless Commun. IEEE Trans. **8**(3), 1414–1423 (2009)
12. G Amarasuriya, M Ardakani, C Tellambura, in *Wireless Communications and Networking Conference (WCNC)*. Adaptive multiple relay selection scheme for cooperative wireless networks (IEEE Sydney, Australia, 2010), pp. 1–6
13. T Jing, S Zhu, H Li, X Cheng, Y Huo, in *IEEE INFOCOM Mini-Conference*. Cooperative relay selection in cognitive radio networks (Turin, Italy, 2013)
14. N Stratton, Reality of H.264 SVC. http://www.robotics.net/2013/02/04/reality-of-h-264-svc/
15. W Su, AK Sadek, KR Liu, Cooperative communication protocols in wireless networks: performance analysis and optimum power allocation. Wireless Pers. Commun. **44**(2), 181–217 (2008)
16. J Li, S Li, F Zhao, R Du, Co-channel interference modeling in cognitive wireless networks. Commun. IEEE Trans. **62**(9), 3114–3128 (2014)
17. S Li, G Bi, Time domain averaging and correlation-based improved spectrum sensing method for cognitive radio. EURASIP J. Adv. Signal Process. **46**(1), 520–532 (2014)
18. Y Chen, C-X Wang, H Xiao, D Yuan, Novel partial selection schemes for AF relaying in Nakagami-fading channels. Vehicular Technol. IEEE Trans. **60**(7), 3497–3503 (2011)
19. H Xiao, S Ouyang, Power control game in multisource multirelay cooperative communication systems with a quality-of-service constraint. Int. Trans. Syst. IEEE Trans. **16**(1), 41–50 (2015)
20. Z Cai, J He, AG Bourgeois, S Ji, in *2012 IEEE 32nd International Conference On Distributed Computing Systems (ICDCS), Macau, China*. Optimal distributed data collection for asynchronous cognitive radio networks, (2012), pp. 245–254
21. Z Cai, Y Duan, AG Bourgeois, Delay efficient opportunistic routing in asynchronous multi-channel cognitive radio networks. J. Combinatorial Optimization, 1–21 (2013)
22. G Bianchi, Performance analysis of the IEEE 802.11 distributed coordination function. Selected Areas Commun. IEEE J. **18**(3), 535–547 (2000)
23. Y Duan, Z Cai, G Liu, Opportunistic channel-hopping based effective rendezvous establishment in cognitive radio networks. Wireless Algorithms Syst. Appl. **7405**, 324–336 (2012)
24. M Hristakeva, D Shrestha, in *The Midwest Instruction and Computing Symposium*. Different approaches to solve the 0/1 knapsack problem (Eau Claire, WI, USA, 2005)

Heterogeneous spectrum sensing: challenges and methodologies

Wei Liu[1*], Mikolaj Chwalisz[2], Carolina Fortuna[1], Eli De Poorter[1], Jan Hauer[2], Daan Pareit[1], Lieven Hollevoet[3] and Ingrid Moerman[1]

Abstract

Distributed sensing is commonly used to obtain accurate spectral information over a large area. More and more heterogeneous devices are being incorporated in distributed sensing with the aim of obtaining more flexible sensing performance at lower cost. Although the concept of combining the strengths of various sensing devices is promising, the question of how to compare and combine the heterogeneous sensing results in a meaningful way is still open. To this end, this paper proposes a set of methodologies that are derived from several spectrum sensing experiments using heterogeneous sensing solutions. Each of the solutions offers different radio frequency front-end flexibility, sensing speed and accuracy and varies in the way the samples are processed and stored. The proposed methodologies cover four fundamental aspects in heterogeneous sensing: (i) storing experiment descriptions and heterogeneous results in a common data format; (ii) coping with different measurement resolutions (in time or frequency domain); (iii) calibrating devices under strictly controlled conditions and (iv) processing techniques to efficiently analyse the obtained results. We believe that this paper provides an important first step towards a standardized and systematic approach of heterogeneous sensing solutions.

Keywords: Cognitive radio; Heterogeneous spectrum sensing; Methodology; Experimentation

1 Introduction

It has been shown that a licensed spectrum is often underused in time and space. As a concrete example, a 6-day measurement campaign in multiple cities of Europe indicates that the UHF TV bands are less than 62.2% utilized [1]. In order to improve spectrum usage, the research community proposed the concept of cognitive radio (CR) [2,3], which offers a mechanism for unlicensed users (i.e. secondary users) to use the licensed spectrum in the absence of the primary users. There are two fundamentally opposite approaches to determine whether the primary user is active. In the first approach, the secondary user senses the spectrum [4] to decide whether it is available or not; while in the second approach, a geo-location database built upon information of primary transmitters and radio propagation models is accessed in order to locate spectrum opportunities [5]. The first approach tends to be less accurate while the second has difficulties adapting to dynamic scenarios. More recently, also hybrid geo-location databases have been considered to improve accuracy [6].

While initially the Federal Communications Commission (FCC) only allowed opportunistic access to licensed spectrum via the geo-location database approach, in a recent revision, requirements for devices relying solely on spectrum sensing have been included in the FCC regulation [7]. The new regulation states that sensing devices must demonstrate a very high degree of confidence to avoid interfering with the incumbents and they must meet the minimum sensitivity requirements for several types of primary signal (i.e. −114 dBm for analog and digital TV signal and −107 dBm for low power auxiliary).

Currently, mainly due to cost considerations, these constraints are hard to satisfy for commercial devices. Fortunately, the lack of individual sensitivity can be compensated by using cooperative sensing with multiple sensing devices [8]. These sensing devices can be installed as part of the infrastructure for the sole purpose of spectrum sensing [9], or existing user devices can be involved

*Correspondence: wei.liu@intec.ugent.be
[1] University Ghent - iMinds, Gaston Crommenlaan 8, 9050 Ghent, Belgium
Full list of author information is available at the end of the article

through a crowdsourcing approach [10]. In both cases, the devices involved in the collaborative spectrum sensing are likely to be heterogeneous, thus performing *heterogeneous spectrum sensing*. Unlike spectrum sensing using homogeneous devices, heterogeneous sensing is more challenging due to the fact that it involves high- to low-end sensing devices and their corresponding approaches to process or store the results. More particularly, heterogeneous spectrum sensing experiments are faced by the following challenges:

1.1 Storage format

Spectrum sensing devices may record data in different formats. Each type of device has its own way of logging data, which could be plain text, or different types of binary formats. Apart from the format of actual measurements, the lack of common configuration and well-defined meta information makes it hard to initialize the measurements and interpret the results. Therefore, we need a uniform and well-structured storage mechanism for heterogeneous sensing devices.

1.2 Measurement resolution

In addition, measurements from heterogeneous devices may have different resolutions in time and frequency domains. Consider the following example: two devices (A and B) are both monitoring the same range of spectrum; device A measures the power spectrum with a 1-MHz resolution bandwidth updating each second, while device B measures with a resolution bandwidth of 10 KHz at the rate of two times per second. Obviously, it is not possible to directly compare or combine the results from A and B. When dealing with the output from heterogeneous devices, the situation in the above example often occurs. Thus, we need a method to obtain a common resolution before a meaningful comparison or combination of the heterogeneous data sources can be made.

1.3 Calibration

Some devices, particularly the low-end ones, may be uncalibrated. Simple wireless devices typically provide the spectrum information using the existing channel assessment module for the purpose of wireless medium access control (MAC). For MAC purposes, the output of the channel assessment module is used within the device, thus it only needs to be relatively correct, rather than absolutely. However, for the purpose of heterogeneous spectrum sensing, it is important that sensing results are produced against a common reference; hence, there is a need for calibration. Though most high-end devices are calibrated individually by their manufactures, a uniform calibration process is still desired in case there are noticeable differences in the factory calibration process or even individual differences among the same type of devices. For instance, prior to experiments, the authors in [11] calibrate the internal noise level of multiple radio receivers of the same type in a shielded environment. However, this approach cannot be used to calibrate heterogeneous devices, as internal noise level is one of the primary heterogeneities among devices. In conclusion, there is a need for a uniform calibration mechanism for heterogeneous spectrum sensing.

1.4 Processing methods

Due to the use of different devices and complex experiment scenarios (i.e. distributed measurements, multiple iterations), heterogeneous spectrum sensing usually generates a large amount of data. Efficient processing methods are crucial to reach an objective conclusion in a reasonable amount of time. Although for certain performance metrics, well-accepted processing approaches exist; most of the time, it is not feasible to pursue a uniform processing mechanism for all experiments. A more pragmatic way is to simply make use of existing methods when applicable. Thus, we believe there is a need to share experiences related to processing heterogeneous data.

In this paper, we identify the challenges and discuss the methodologies of heterogeneous spectrum sensing regarding the following aspects: storage format, measurement resolution, calibration and processing methods. First, we propose a common data format for uniformly defining experiments and storing the results. Then we provide a set of methodologies regarding measurement resolution, device calibration and data processing, which are implemented, validated and evaluated on reference scenarios. Thus, the main contributions of this paper are the common data format, the methodologies, the validation and the evaluation.

The remainder of this paper is structured as follows: first, the related work is discussed in Section 2; after that, Section 3 describes the proposed methodologies regarding the aforementioned four challenges, while the performance of these methodologies are verified with concrete implementations and real-life experiments in Section 4. Finally, we conclude this paper in Section 5.

2 Related work

This section gives an overview of related work for heterogeneous spectrum sensing. The following aspects are discussed: (i) the storage format, (ii) the measurement resolution, (iii) calibration methods and (iv) processing mechanisms.

2.1 Storage format

The IEEE 1900.6 standard [12] defines spectrum sensing-related parameters and data structures, and as such may be considered as a guideline of data storage for spectrum sensing. Since this is also the main interest of our work, the terminologies available in this standard have been considered and used whenever applicable. In comparison, the IETF PAWS standard [13] is limited to the communication among TV white space devices and databases, thus it is less relevant to the low level challenges in heterogeneous spectrum sensing.

Apart from existing standards, authors in [14,15] use heterogeneous spectrum sensing to construct a radio environment map (REM), which includes a dedicated spectrum data server (SDS) to collect heterogeneous sensing data. After the data collection phase, the server fusion (SF) interface is used to communicate with other processing units. It is seen that the architecture of splitting storage and processing units performs well in real experiments and can be easily adapted as part of existing LTE network [16]. Compared to [14,15], this work has a more general perspective for sensing experimentation. We aim to provide tools for common data storage that focuses more on low level spectrum data and not specifically tailored to any application purposes.

2.2 Measurement resolution

Measurement resolution is an important indication of the sensing capabilities of a device. Advanced sensing devices usually have more flexibilities for configuring which resolution to use for the measurements.

A straightforward way to achieve a common resolution is to use common configurations at the measurement time (e.g. choose a resolution bandwidth that is available on all devices) [17]. When no common setting is available among the considered devices, the authors in [17] determine the settings for each device by performance (e.g. choose a sweeping pattern or detector type after a number of initial trial measurements). While this approach is practical and reasonable, it is not clear how experimenters can achieve data with common resolutions for further processing when no common settings could be found.

The authors of [18] use experimental measurements to illustrate how the choices of measurement resolutions can influence the detection performance. However, the methodologies discussed in [18] do not take heterogeneous devices into account.

Therefore, this paper proposes a post-processing approach to derive data with common resolutions regardless of the settings chosen at the measurement time.

2.3 Calibration

As described in the previous section, the internal noise level used in [11] is not an ideal metric for calibrating heterogeneous sensing devices, because it is expected to be different. The authors of [8] propose to use satellite band signals for cooperative sensing devices to identify shadow effects. Since satellite signals have relatively constant strength over a wide area, devices that receive weaker satellite signals have a higher probability to be shadowed by obstacles. Though this solution is not directly related to calibration, the idea of using satellite band signals as reference could be used to calibrate devices in outdoor measurements. In general, for the calibration of heterogeneous devices, a known and constant power level is needed as input reference [14,15]. This is also the main principle of calibration followed by our previous work [19]. Apart from a stable reference signal, calibration experiments need to be strictly controlled. In this paper, we share our experience of using coaxial-cable-based experiments to achieve accurate calibration solutions.

2.4 Processing methods

Unlike the previous challenges, the processing mechanisms are very experiment specific. For energy detection, the magnitude of the complex samples is calculated to represent the received signal strength. Alternatively, the magnitude of the output from the fast Fourier transform (FFT) could be used instead of time domain samples [14,15]. For the purpose of detection performance, a receiver operating characteristic (ROC) plot is often used to observe the device's sensitivity under various input conditions. The work in [14,15] goes one step further by using the inverse distance weighting (IDT) technique to interpolate discrete energy measurements into a REM. Though it is not feasible to have a uniform processing approach, we believe there is a need to share processing experience in order to improve the efficiency.

3 Methodologies for realizing heterogeneous sensing

This section describes the methodologies for realizing heterogeneous sensing regarding the previously identified challenges.

Conceptually, we propose the following workflow for heterogeneous spectrum sensing (see Figure 1). The ini-

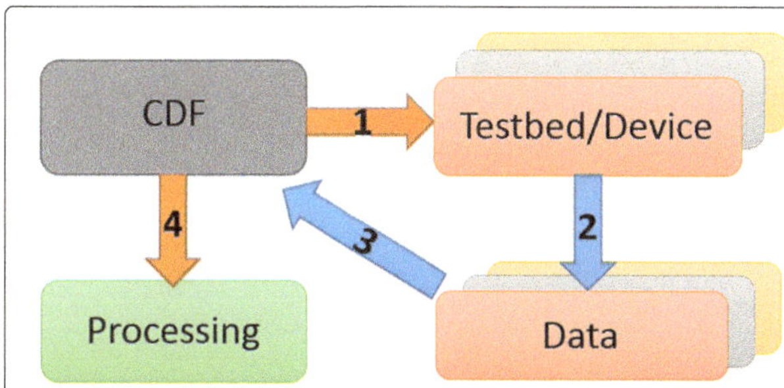

Figure 1 Conceptual workflow towards heterogeneous spectrum sensing. (i) Uniform CDF experiment descriptions are translated into device-specific configuration scripts. **(ii)** The collected results are converted into a common data format. **(iii)** The results are further processed based on the common data formant.

tial phase for performing sensing experiments consists of configuring the heterogeneous devices, sending them the instructions for starting the sensing and collecting the data. This involves creating a series of device-specific scripts. As one of the contributions of this paper, we propose a uniform way of providing the configuration to the devices and storing the data from the devices. We use a common data format (CDF) for experiment description and data storage that is device independent and machine readable. Spectrum sensing descriptions and settings are defined using this common data format as depicted in Figure 1. In the first step, these uniform descriptions are then converted into device (or testbed/infrastructure)-specific configurations and control scripts. In the second step, the results of the experiments are transformed into a common representation format. As the third and final step, the resulting uniformly described data is further processed by a set of tools to align the resolution, achieve calibration and compute spectrum occupancy-related metrics.

It should be noted that, the information needed in the calibration phase comes from separate experiments, which will be denoted as *calibration experiments* in the remaining part of the paper. Calibration experiments are not necessarily part of every experiment iteration, but it should be at least performed once before the real sensing measurements start.

3.1 Common data format

The proposed common data format has been developed to ease spectrum sensing experimentation across devices and testbeds and contains three main parts. The first part refers to the description of the experiment abstract, the second part refers to the spectrum sensing experiment, thus the so-called meta-data. The third part focuses on the actual traces resulted from the experiment.

The experiment description provides a detailed description of the experiment, such as how it was performed and what kind of data was collected. From the top level, the description contains the following fields: *experiment abstract*, *meta-information* and *experiment iteration(s)*. Below each field (except for experiment abstract), some sub fields are defined, as shown in Figure 2.

3.1.1 Experiment abstract

Experiment abstract is a high level description of the experiment, providing a basic idea of the experiment motivation, as well as the expected output. It is possible to relate to other experiments by adding relevant information. For instance, when experiment B is a scaled extension of experiment A, the following sentence 'repetition of experiment A on a larger scale' can be noted in the abstract of experiment B. In addition, we provide means to link to related documentations, such as publications that are based on a given experiment.

3.1.2 Meta-information

Meta-information is the information required for describing, understanding and evaluating the experiment. All experimental details except the data itself should be described in this field. The most important items are the description of involved devices, physical setup of the experiment, the selected signal type and frequency, as well as the description of the measurements.

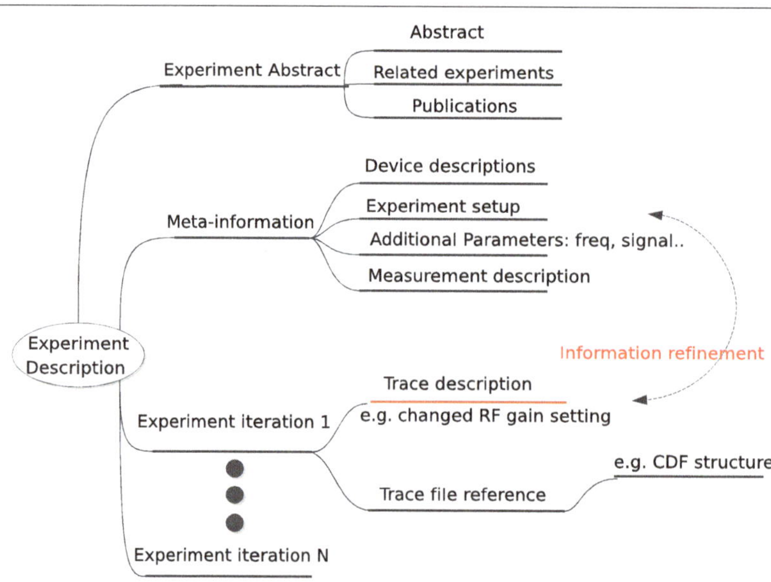

Figure 2 The architecture of the CDF. (i) experiment abstract, **(ii)** meta information and **(iii)** experiment iterations containing data traces.

The description of the involved devices is critical to reproduce the experiment. It should not only be limited to textual description but also provide references to the relevant data sheets. Moreover, we recommend to include information of related software and, if necessary, the operating system. The bottom line is that the collected information must suffice to repeat the experiment from scratch, starting from finding the same devices to setting up the identical software environment.

The physical experiment setup mainly refers to the description of how devices are positioned and connected. Ideally, there should be a location map to indicate the topology of the devices. Wireless experiments are sensitive to environmental factors, such as if an experiment is conducted indoor or outdoor, or if an experiment is conducted under a static or rather dynamic environment. Thus, we recommend to document this information in the meta-information as well.

Furthermore, the operating frequency and the characteristics of the used signals are noted as additional parameters. This creates a convenient way of indexing the existing experiments, e.g. one can easily find all sensing experiments in the TV white space. Thus, it allows experimenters to reuse past experiments more efficiently.

Finally, the measurement description contains a common description of the recorded data of all the *experiment iterations*, allowing experimenters to understand and process the data more smoothly. It specifies the configuration used by each device (e.g. gain settings, sample frequency) and the collected data types (e.g. frequency, signal power, time stamp). In addition, each data type is associated with a measurement unit (e.g. Hz, dBm, μs). For more information related to defining measurement units, readers are referred to the IEEE 1900.6 [12] standard.

3.1.3 Experiment iteration(s)

Experiment iteration provides information that is related to the execution of a particular experiment round. There are two sub fields in each experiment iteration: the trace description and the trace file reference. The trace description is similar to the description in the *meta-information* but may extend or refine the *meta-information* partially if necessary, as shown by the red line in Figure 2. For instance, if a set of measurements is used to compare the influence of different radio frequency (RF) front-end gain settings, trace description is an ideal place to indicate what gain setting is used in each experiment iteration. This way, different settings among experiment iterations can be highlighted without the need of describing the entire experiment setup over and again.

The trace file reference is a 'pointer' towards the measurement data, which indicates where the measurement trace is physically stored.

A reference implementation of the CDF architecture is presented in Section 4.1.

3.2 Measurement resolution

Typically, one spectrum sensing trace cannot be directly compared to another, due to the differences in frequency and/or time domain. To overcome the heterogeneous *frequency resolution*, the easiest and most straightforward approach is to integrate the power spectral density (PSD) in a certain frequency interval and use the integrated power as the metric for comparison. This also implies that the selected interval for integration needs to be wider than the largest resolution bandwidth among all the sensing solutions.

There are different approaches to overcome the differences in *time resolution*. The easiest way is to apply averaging on the traces obtained in the same time duration. Alternatively, instead of using averaging, one can apply max-hold filtering, so the combined trace contains every transient signal that ever appeared in the observation period. By using integration in the frequency domain and averaging or max-holding in the time domain, a common metric is derived from various raw spectra. This is referred to as the common metric in the remainder of the paper. We provide a reference implementation of this processing scheme in the CDF toolbox (`pw_integration` function).

3.3 Calibration

Calibration of heterogeneous devices essentially means comparing the received power of each device to its corresponding input signal strength. The calibration process consists of four steps. First, a set of reference signals has to be selected. Second, the path loss between the signal source and the devices under calibration must be strictly controlled. Third, a suitable metric for performing the calibration has to be identified and fourth, the offset between the reference signal and the signals received by the devices has to be computed.

For the first step, it is generally advisable to use a set of diversified input signals (i.e. different bandwidth and signal strength) so that the calibration experiment is general enough to cope with different types of input. Also the generated signal needs to be continuous so that the recorded signal has a constant amplitude. This ensures that the sensing performance in terms of timing does not affect the performance in terms of power accuracy. The produced signal strength needs to be tuned within the dynamic range of all devices. If the signal is too strong, it may saturate the device under calibration; on the other hand, when it is too weak, the signal might be buried by noise. Both situations should be avoided. Ideally, a high-end signal generator should be used as the signal source to meet the above constraints.

For the second step, the most at-hand method is to use a coaxial cable for controlling the path loss between the signal source and the sensing devices. An alternative way would be to use an anechoic chamber where the path loss is not affected by the multi-path effect.

In the third step, the received signal strength needs to be calculated from the power spectral density, which comes down to performing integration over the interval where the signal is transmitted in the frequency domain. If the integration interval is not the same as the signal bandwidth, the obtained metric will rely partially on the device's noise floor instead of solely on the input signal, thus, not qualified for power calibration.

Finally, in the fourth step, the power offset is then computed according to Equation 1, where the transmit power is denoted as P_{tx}, the received power is denoted as P_{rx}, and the total attenuation caused by coaxial cables and splitters is denoted as P_{atten}:

$$P_{offset} = P_{tx} - P_{atten} - P_{rx} \tag{1}$$

In Equation 1, P_{offset} accounts for the combined heterogeneity of the RF front-end, analog-to-digital converter (ADC) and the processing unit. However, it does not include the influence of the antenna, as the antenna is replaced by the coaxial cable connections. For devices using different types of antenna, the power offset needs to be readjusted with the antenna gain.

If the relative position of the transmitter and receiver is known, the influence of the radiation pattern should also be taken into account. For omnidirectional antenna, the radiation pattern changes with the elevation angle between the transmitter and receiver; while for directional antenna, the radiation pattern varies with both horizontal and vertical angles [20]. When the relative position of transmitter and receiver is unknown, it is necessary to rotate the directional antenna several times to cover the 360° [18].

Sometimes, P_{offset} varies with the input signal strength and the settings of the sensing device (i.e. gain settings). For instance, it is mentioned in [15] that the RFX2400 daughter-board of USRP does not have a linear input and output (IO) relationship. In this case, more measurements need to be performed to cope with different input signal strength and sensing configurations.

3.4 Processing

Sensitivity and accuracy are two important metrics to compare spectrum sensing devices. For heterogeneous sensitivity analysis, experimenters tend to form a mainstream processing style, which is discussed in the first part of this section. As for power accuracy, generally, a high-end device (i.e. spectrum analyser) is used as benchmarker in various measurements. However, when it comes to large scale heterogeneous measurements, this approach becomes very tedious. Thus, there is a need to process data in a more elegant approach, which is what we discuss in the second part of this section.

3.4.1 Heterogeneous sensitivity analysis

The sensitivity of a sensing device is reflected by its noise floor. Unlike power accuracy, sensitivity cannot be evaluated by the common metric derived in Section 3.2. This is because the noise floor is affected by the resolution bandwidth, thus the integrated power metric will always be higher than the original noise floor.

The most straightforward way is to observe the mean and variance value of the spectrum trace when no signal is present. Alternatively, we can also use the receiver operating characteristic. The ROC is obtained by expressing the probability of detection (P_d) as a function of the probability of false alarm (P_f). Some papers utilize the probability of missed detection (P_m) which is simply given by $1 - P_d$.

Despite of the heterogeneity in power spectra, ROC can be obtained via a common approach:

- Record spectrum traces when no signal is present.
- Vary P_f from 0% to 100% in small steps and determine a detection threshold for each P_f based on the previously recorded trace.
- Apply a signal at the input of the sensing device and record spectrum trace again.
- Compute P_d or P_m for all the detection thresholds determined in the second step from the trace recorded in the previous step

The advantage of ROC analysis is that it is device independent, as for a given false alarm, each device can have its own threshold. The only constraint is that the detection threshold should be calculated in an uniform approach for all devices. This is why it is commonly applied in the heterogeneous sensitivity studies [14,15,19].

As for the method to obtain detection threshold, there are many optimized variants [21-23]. As an example, the constant false alarm (CFA) approach [11] is described in Equation 2, where σ_n denotes the variance of the noise samples, N denotes the number of spectrum samples, P_f denotes the target false alarm and λ denotes the calculated detection threshold, respectively.

$$\lambda = \sigma_n^2 (1 + \frac{Q^{-1}(P_f)}{\sqrt{N/2}}) \tag{2}$$

3.4.2 Heterogeneous accuracy analysis

As stated previously, distributed heterogeneous measurements usually generate a large amount of data, which needs more efficient processing mechanisms. When processing a large dataset, the basic approach is to look at how the data is distributed, which can be achieved by computation of several statistics (i.e. mean, variance). However, to gain more insights of the data (i.e. discover a common behaviour, or a group of data that displays similarity within the entire set), then more advanced techniques, such as correlation, various linear regression algorithms need to be involved. Essentially, we recommend to use the basic techniques of data mining for analysing large scale heterogeneous sensing experiments, among which four most relevant techniques are exemplified as follows:

- Dependency modelling - the establishment of relationships between variables. This could be that the detection probability depends on the target signal strength or the distance between the transmitter and the sensing device.
- Outlier detection - the identification of the unusual spectrum records, which could be caused by malfunctioning devices or other unknown interferences.
- Regression - is a statistical way to explore the relationship among variables which models the data with the least error.
- Clustering - is the task of discovering groups and structures in the data that are in one way or another 'similar.' In case of spectrum sensing, this could be that a group of sensing devices are shadowed by a common obstacle.

The outlier detection is a rather basic step, which can be achieved by many statistical tools or simply manual observations. The procedure of 'Clustering' and 'Regression' are addressed with a concrete experiment in Section 4.4. For the dependency modelling, we find that the path loss model (the relationship between received signal strength and distance) is generally applicable for dependency modelling in the case of distributed sensing measurements. More particularly, the well-known log-distance path loss model can be expressed by two parameters - the path loss coefficient exponent α and the path loss offset β:

$$PL(d) = 20 \times \alpha \times \log_{10}(d) + \beta, \tag{3}$$

where d is the distance between the transmitter and the receiver. When using the logarithmic distance as the argument, the equation reduces to a simple linear expression. Hence, various approaches, such as least square regression, can be used to estimate α and β.

The role of the path loss model is essentially a way to extract new parameters out of the raw data sets. It is a tool to correlate data from distributed locations. Although deriving the path loss model is not always easy or feasible, the basic idea of correlating data to extract new parameters is generally applicable and highly valuable in our experience.

4 Reference implementation and experimentation

This section first presents the implementation of the common data format in Section 4.1 and then illustrates how to apply the methodologies defined in the previous section with real-life experiments. More specifically, Section 4.2 describes a calibration experiment, which uses the method proposed in Section 3.2 to overcome the different measurement resolutions and then obtains the power offset following the mechanisms proposed in Section 3.3. Sections 4.3 and 4.4 present two experiments that use processing techniques discussed in Section 3.4 to evaluate two fundamental sensing performance metrics (sensitivity and accuracy), respectively.

4.1 Common data format implementation

The reference implementation of the CDF architecture consists of three parts: the CDF experiment description, the CDF data structure for common storage and the CDF toolbox for additional functionalities such as conversion between formats and result analysis.

4.1.1 The CDF experiment description

The CDF experiment described in Section 3.1 can be easily translated into modern markup languages such as XML and JSON. We made a design choice to use XML because (i) it can be read and processed by a large set of programming languages, and (ii) it is internally used by OMF - a testbed cOntrol and Management Framework [24] which is widely adopted by many modern wireless testbed facilities [25,26]. The goal here is to ensure that the CDF experiment description can be easily translated to testbed/device-specific implementations. Additionally, we provide an XML schema [27] to validate the semantic correctness of an experiment description.

4.1.2 The CDF data structure

The CDF data structure is one of the formats that the trace file field in the CDF experiment description can reference to, it is meant to be a starting point where users can easily load data from different devices. The content of the CDF data structure is illustrated below:

```
p             = common data format structure
p.Name        = Unique identifier of the sensing device
p.Location    = Location of the sensing device (m) e.g. [x,y,z]
p.CentreFreq  = Array of centre frequencies corresponding to
                             the columns of the power matrix (Hz)
p.BW          = Bandwidth around each centre frequency (Hz)
p.Tstart      = Start time of the measurement in datestr format
                 e.g. '24-Jan-2003 11:58:15'
p.SampleTime  = Array of timestamps relative to Tstart (s)
                 corresponding to rows of the power matrix (Hz)
p.Power       = Matrix containing power measurements (dBm)
                row contains all frequencies for one timestamp
```

For FFT-based sensing devices, defining both 'BW' and 'CentreFreq' is unnecessary. However, for pure sweeping-based sensing devices, the resolution bandwidth depends on the width of the band pass filter at the RF front-end, which is not necessarily the same as the distance between the consecutive RF centre frequencies. Therefore, both fields are included in the data structure so that it is suitable to store results from all types of sensing devices.

The field 'Location' is included in both the *CDF experiment description* and the *CDF data structure*, because it not only is important as 'meta-information' but also needed in various of calculations. Note that the device location in 'meta-information' is a description of the general experiment topology, while in the CDF data structure, it has to be expressed incoordinates.

The remaining fields of the CDF data structure are self-explaining, hence, omitted from further explanations.

4.1.3 The CDF toolbox

The CDF toolbox is a set of functions implemented in Matlab to support the usage of the CDF data structure. We choose Matlab as the programming environment because it is powerful in matrix processing and widely used among research and academic institutions. The most often used functions are listed below:

- The `create_structure` function extracts information from the input spectrum trace, and store the data in the CDF data structure.
- The `pw_integration` computes the power of a certain frequency band by integrating the PSD over the corresponding interval. It takes three input parameters - the spectrum trace in the CDF data structure, the begin and the end of the frequency interval, and produces the integrated power as output.
- Several plotting and analysis tools are implemented based on the fields of the CDF data structure, which makes the CDF more attractive from practical point of view.

Sample scripts of the CDF toolbox are made available online [27]. The scripts are currently only developed for devices used in the CREW[a] project. However, they can serve as valuable examples for other devices, given the fact that the existing examples already cover large varieties of sensing devices and data formats. Although it is not jet a full-fledged toolbox for sensing analysis, we believe it is one step towards the support of heterogeneous devices.

4.2 Calibration and resolution

This section first gives an overview of the devices involved in the experimental evaluations of the entire Section 4 and then describes a calibration experiment using the common metric derived in Section 3.2 and the instructions given in Section 3.3.

4.2.1 Overview of sensing devices

Telosb [28] is a sensor node developed at UC Berkeley. It is widely used by wireless sensor network community. The platform uses the IEEE 802.15.4-compliant CC2420 transceiver [29], which operates in the 2.4 GHz ISM band. The sensing application is built above TinyOS [30]. In our experiments, the device sweeps over the target spectrum in steps of 2 MHz and measures RF energy in each step. A single received signal strength indication (RSSI) is collected at every RF centre frequency. The RSSI is transferred to the host computer via USB connection in real-time. It takes around 2 ms to sweep over the entire 2.4 GHz ISM band. The collected RSSI value and its timestamp are stored in a comma-separated value (CSV) file. TelosB has less flexibility towards spectrum sensing applications, both in processing algorithms and RF functionalities; however, it has the lowest price as well.

The *metaGeek Wi-Spy 2.4x* and [31] *Airmagnet* [32] are both low-cost spectrum analysing devices, and they both use USB dongles as the RF front-end. The sensing mechanism of Wi-Spy resembles TelosB in the sense that it also uses a narrow-band RF receiver to scan across the band of interest in tiny steps. The step width ranges from 50 KHz to over 600 KHz, depending on the width of the frequency span. This essentially determines the resolution bandwidth of the spectrum trace. In our experiments, the Wi-Spy is used jointly with the Kismet Spectools [33] in

Linux environment, instead of the standard 'Chanalyzer' software. By doing so, the power spectrum trace can be stored in a non-proprietary format, which is more convenient for further processing. Unlike Wi-Spy, Airmagnet relies on FFT-based sensing algorithm. The radio of Airmagnet has a 20-MHz instantaneous bandwidth. It performs sweeping in steps of 20 MHz to cover a bandwidth that is wider than the instantaneous span. For 2.4 GHz ISM band, it has a fixed span of 83 MHz with 156 KHz resolution bandwidth. The PSD of Airmagnet can be stored with either CSV format or other proprietary formats.

USRP [34] is a relatively low-cost SDR platform that consists of two parts - a fixed motherboard and a removable daughterboard. The motherboard contains ADC and digital-to-analog converter (DAC), a field programmable gate array (FPGA) for digital down sampling and an interface connected to a host computer. The daughterboard provides RF front-end functionalities. There are many third-party software platforms, such as GNU Radio [35] and Iris SDR platform [36], which can communicate with the USRP. Thus, spectrum sensing applications can be implemented in many ways.

The imec sensing engine [37] is an integrated sensing device based on a custom design that targets for low-power and hand-held devices. Hence, it is powered and configured over a single USB connection. Similar to USRP, it has a separate PCB for the RF front-end functionality. The imec sensing engine has a very wide RF frequency range (from 100 MHz up to 6 GHz) and a programmable instantaneous bandwidth between 1 MHz and 40 MHz. Additionally, it uses a dedicated integrated circuit (IC) for signal processing instead of using the host computer. There are several pre-defined modes in the IC, including sensing based on FFT and sensing based on fast sweeping over a set of consecutive RF frequencies. The host application of imec sensing engine is written in C; therefore, the storage format is also flexible.

The overview of the devices are summarized in Table 1.

4.2.2 Calibration experiment

For the first step, the reference signals are defined as continuous OFDM signals with two different bandwidths (22 MHz and 5 MHz), transmitted on three different channels (Wi-Fi channel 1,6,11 for the 22 MHz signal and Zigbee channel 11,16,26 for the 5 MHz signal), with three input signal strength (−60 dBm, −70 dBm, −80 dBm). The 22 MHz and 5 MHz bandwidth are selected to emulate Wi-Fi and Zigbee signals, respectively. A Rohde & Schwarz signal generator is used as the signal source.

The signal generator is connected to the sensing devices with coaxial cables and splitters as shown in Figure 3. The idle terminals are properly connected to terminators with matching impedance (50Ω). The calibration experiment consists of two simultaneous operations: continuous RF signal is produced by the signal generator, and at the same time, all devices record the sensing data to cover the same frequency span (2.4 GHz ISM band) for the same amount of time (1 min). This process is repeated for all the predefined reference signals, which means 18 iterations in total (2 types of bandwidth, 3 input levels, 3 channels).

After the recording, the raw spectra traces are converted in to the CDF data structure. Then the received power within the transmitted signal bandwidth is calculated from the raw PSD. More particularly, the `create_structure` and the `pw_integration` functions within the CDF toolbox are used to perform these operations. As the input signal has a constant amplitude, and the devices are configured to sense for the same amount of time, averaging the P_{rx} over the entire sensing duration is the most logical way to compute the common metric. Finally the power offset of each device against each reference signal is calculated according to Equation 1.

The minimum, maximum and average value of the measured offsets are plotted in Figure 4. The results indicate that the USRP solution has the largest offset, most likely due to the fact that it is a general research platform and the output of the customized software is not strictly calibrated. Airmagnet and Wi-Spy are both commercial USB spectrum devices, but their offsets are the opposite of each other. Additionally, Airmagnet has much bigger variations during the measurements (seen by the difference between the maximum and minimum offset values). We believe that this is most likely caused by the difference in sensing approach - Airmagnet uses FFT-based processing approach while Wi-Spy relies on pure narrow band sweeping (see Table 1). As this paper does not focus on

Table 1 Overview of the sensing solutions

Device name	Resolution bandwidth	Frequency span	Processing	Sweeping	Cost
TelosB	2 MHz	2.4 GHz ISM band	RSSI	Narrow band	Medium
Wi-Spy2.4x	50 to 600 KHz	2.4 GHz ISM band	RSSI	Narrow band	Medium
Airmagnet	155 KHz	2.4 and 5 GHz ISM bands	FFT	In 20 MHz	Low
USRP	48 KHz*	RF board dependent	FFT	In 20 MHz*	Medium
imecse	78 KHz*	100 MHz to 6 GHz	FFT	In 20 MHz*	High

*Indicates that the entry is configurable and only a typical value is entered.

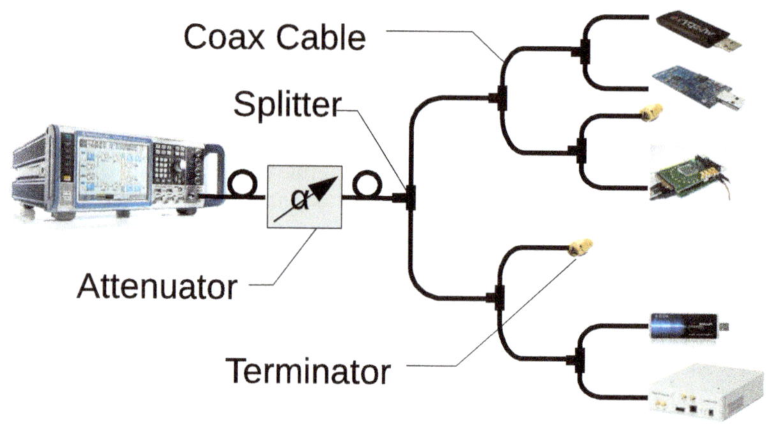

Figure 3 Replace wireless medium with coaxial cable and splitters.

the performance of sensing devices, the exact cause of the different offsets is not examined. However, the fact that heterogeneous devices have very different offsets confirms the need for calibration.

Finally, we would like to examine the calibration result by looking at the spectra of different devices with the same resolution bandwidth. To do so, we first subtract the mean P_{offset} from the collected raw spectra, and then divide the entire 2.4 MHz ISM band into a set of 2 MHz wide consecutive intervals (2 MHz is the largest resolution bandwidth among the considered devices, see Table 1) and perform integration over each of these intervals using the CDF toolbox. This operation essentially brings all spectra to the same frequency resolution. For comparison, the original raw spectra and the resulting spectra of the measurement obtained under one type of reference signal (22 MHz OFDM on Wi-Fi channel 1 with −60 dBm input power) are shown in plots (a) and (b) of Figure 5, respectively. Plot (a) shows that devices with larger resolution bandwidth have higher level of PSD than devices with finer resolution bandwidth. This behaviour can be best illustrated when comparing the raw spectra of TelosB and imec sensing engine. Compared to plot (a), the spectra in plot (b) are much smoother due to the coarser resolution bandwidth and closer to each other within the interval where the signal is present, which is the expected behaviour after the calibration and resolution conversion. At the same time, we notice that there is still big differences between the spectra where the signal is not present. This difference is no longer linked to the influence of resolution bandwidth, but solely reflects the internal noise level of devices. Finally, the envelope of the 22 MHz OFDM signal is slightly shifted to the right for the case of TelosB, this is because the resolution bandwidth of TelosB is not fine enough to resolve the exact boundary of the OFDM signal in the frequency domain.

4.3 Reference experiment for heterogeneous sensitivity analysis

This section describes an experiment for sensitivity analysis following the instructions in Section 3.4.1. The experiment setup involves an MS3700A signal generator, placed 9 m away from the rest of the sensing devices. This experiment does not include the Airmagnet device due to

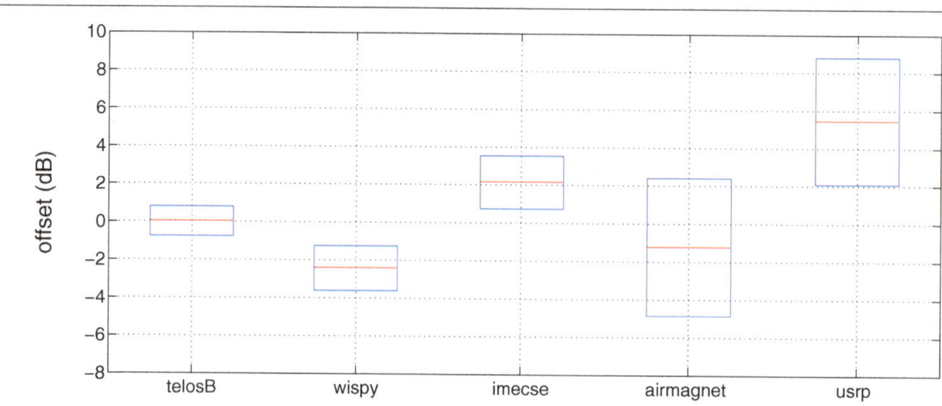

Figure 4 Offset vs device.

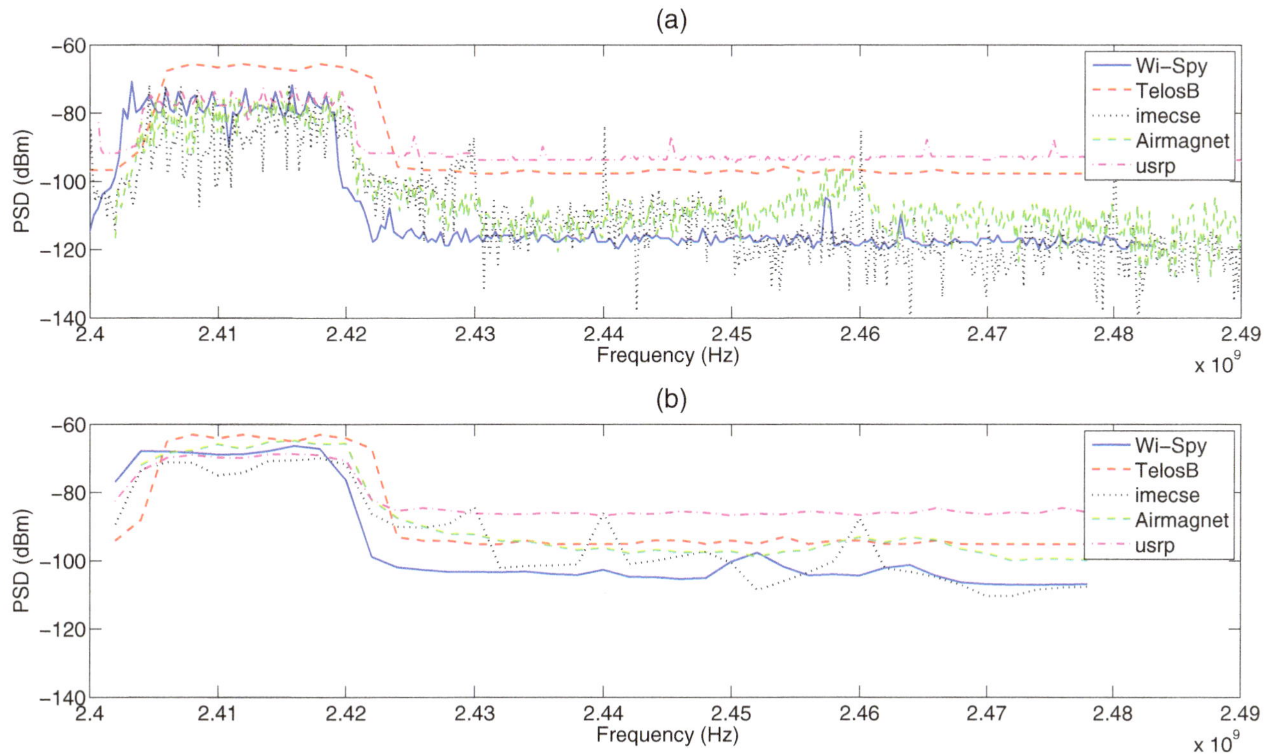

Figure 5 Raw spectra **(a)** vs calibrated spectra with common frequency resolution **(b)**.

practical limitations at the time of this measurement. All devices are placed on the same horizontal level, with no obstacles in between.

First, samples are recorded when devices are shielded away from external signals. Then recordings are made when an 8 MHz wide OFDM signal is transmitted with various signal strength. The raw traces are converted to the CDF data structure via the `create_structure` function in the CDF toolbox. In the next step, the detection thresholds for a set of P_f are calculated according to Equation 2, and finally the corresponding P_d values are computed.

The ROC plot obtained under −4 dBm signal strength is displayed in Figure 6. It shows that imec sensing

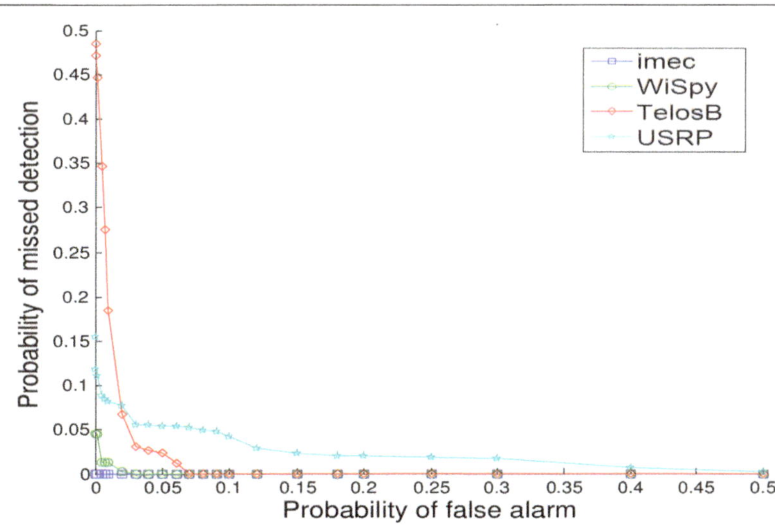

Figure 6 Example ROC plot obtained with heterogeneous devices.

engine and Wi-Spy have better sensitivity than USRP and TelosB. The lack of sensitivity for TelosB is clearly caused by the limitation of its large resolution bandwidth. While for USRP, the low sensitivity is due to insufficient amplification applied at the time of the experiment, which can be resolved in more recent sensing implementations [38].

4.4 Reference experiment for heterogeneous accuracy analysis

This section uses an experiment to illustrate how the processing mechanisms described in Section 3.4.2 can be used in power accuracy analysis. The experimental setup is shown in Figure 7, where 23 measurement locations and one transmitter's location are chosen within an indoor cafeteria. During the measurement, the signal generator transmits a constant 20 MHz wide OFDM signal on Wi-Fi channel 8 (2.447 GHz), with 3 dBm transmit power. Each of the aforementioned sensing devices is configured to record the spectrum at all locations for a duration of minimum 30 s.

As stated in 3.4.2, the simple logarithmic path loss model is used for *dependency modelling*. Based on Equation 2, path loss model is characterized by two parameters: the path loss exponent α and offset β. Therefore, the dependency modelling is realized by estimating the two parameters with the distributed power measurements.

Only one measurement from Wi-Spy device is identified as *outlier* due to the fact that its output was abnormal, the rest of the measurements are considered valid.

Two types of *regression* techniques are applied for the estimation: the least square regression and the robust regression. The least square regression attributes an equal weight to all input data so that the resulting α and β give the minimum mean squared error over the input dataset. On the other hand, the robust regression iteratively attributes weights to different ranges of the dataset, so that the impact of potential outliers is minimized.

Finally, the measurement locations are grouped into different *clusters* so that the locations which has a line-of-sight (LOS) topology, with respect to the transmitter, are separated from those that are in a none line-of-sight (NLOS) condition (shadowed by the coffee machine). The resulting path loss models estimated with the least

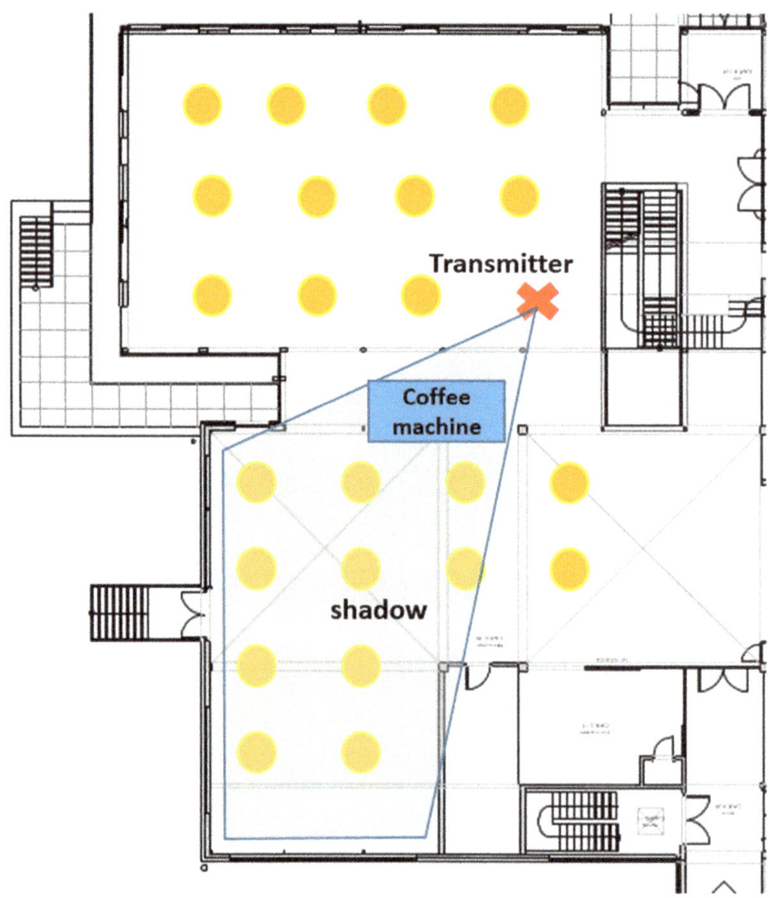

Figure 7 Experiment setup and 'Cluster' of locations adapted from [39].

square regression for both location clusters are shown in Figure 8. We observe that compared to LOS model, NLOS has a smaller slope but higher offset. The high offset in NLOS estimation is caused by the shadow effect. For the same reason, around shadow, the increment of path loss caused by distance can be compensated by the decreasing amount of shadowing; hence, the path loss exponent appears to be smaller than the LOS estimation. This analysis gives extra insight on the impact of shadowing in the power accuracy measurement, thanks to the 'cluster' of devices. More technical details about this experiment and its result analysis are presented in [39].

5 Conclusions

In this paper, we identify and address several challenges in heterogeneous spectrum sensing. First, we provide a *common data format (CDF)* (consisting of data structure and toolbox) to configure sensing devices and store measurement results in a uniform approach. We show that the use of CDF can effectively reduce the experiment overhead, however its implementation requires device-specific scripts.

Second, we apply aggregation techniques to raw spectra in both frequency and time domain (through CDF toolbox) to overcome heterogeneous measurement resolution. We show that this technique can be used to compare heterogeneous spectra conveniently, however, it cannot be applied in sensitivity analysis, since sensitivity and resolution are related.

Third, we propose the use of a strict calibration process: replace the wireless medium by coaxial cables, use high-end signal generator as reference, derive the power offsets among devices and perform calibration. We validate experimentally that this approach is highly reliable and repeatable. The drawback is that the usage of coaxial cable leaves the antenna out of the calibration system. This can be resolved by adjusting the measured power offset with the offsets among antenna gains.

Finally, we share our experience of analysing two fundamental heterogeneous sensing metrics. More particularly, a well-accepted common procedure is applied to heterogeneous devices to achieve fair sensitivity analysis; basic data-mining techniques are used to extract new parameters concerning distributed power accuracy analysis. It can be seen that these techniques greatly improve the processing efficiency and trigger profound understanding of the measurements. Though processing mechanisms generally vary with experiment details, sharing these to the community will lead to a quicker harmonisation of approaches.

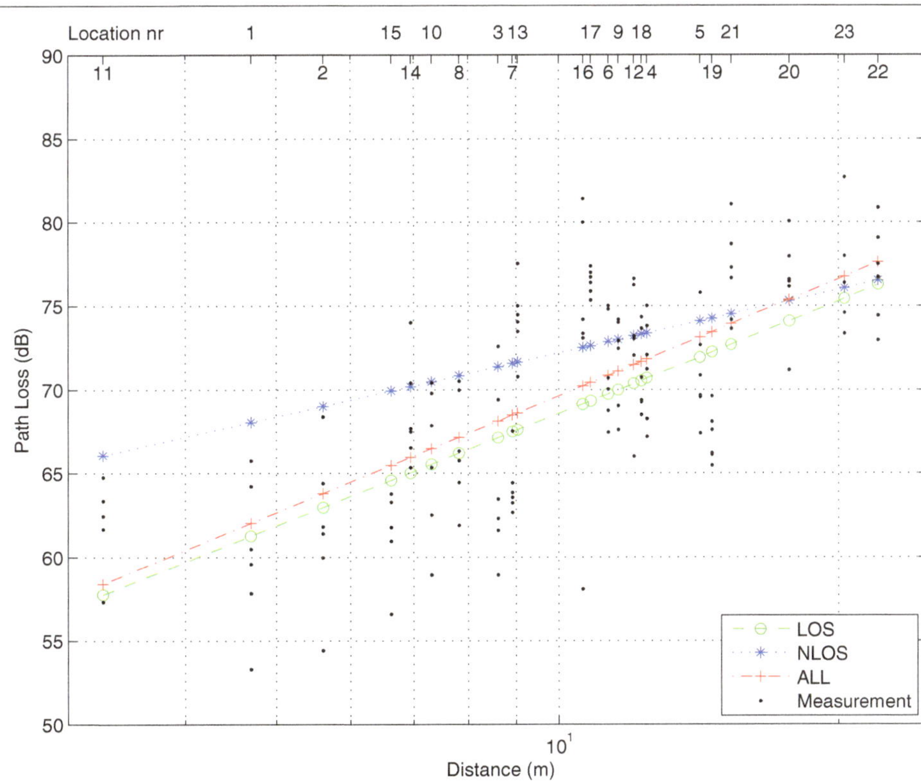

Figure 8 The path loss models of LOS and NLOS clusters.

In the future, before extending the CDF to support more devices, the device-specific implementation can be simplified and validated via standard procedures. Also more advanced aggregation techniques could be explored in order to meet the needs of different analysis and the influence of the antenna could be studied more in depth to improve the calibration accuracy.

Endnote

[a]Cognitive Radio Experimental World (CREW) is a project in European Unions Seventh Framework Programme FP7/2007-2013.

Competing interests

The authors declare that they have no competing interests.

Acknowledgements

The research leading to these results has received funding from the European Union's Seventh Framework Programme FP7/2007-2013 under grant agreement n 258301 (CREW project).

Author details

[1]University Ghent - iMinds, Gaston Crommenlaan 8, 9050 Ghent, Belgium. [2]Telecommunication Networks Group (TKN)/TU Berlin, Sekretariat FT 5, Einsteinufer 25, 10587 Berlin, Germany. [3]imec, Kapeldreef 75, Leuven, 3001, Belgium, 3000 Leuven, Belgium.

References

1. V Valenta, R Marsalek, G Baudoin, M Villegas, M Suarez, F Robert, in *Cognitive Radio Oriented Wireless Networks & Communications (CROWNCOM), 2010 Proceedings of the Fifth International Conference On*. Survey on spectrum utilization in europe: measurements, analyses and observations (IEEE, Cannes, France, 2010), pp. 1–5
2. S Haykin, Cognitive radio: brain-empowered wireless communications. Sel. Areas Commun. IEEE J. **23**(2), 201–220 (2005)
3. J Walko, Cognitive radio. IEEE Rev. **51**(5), 34–37 (2005)
4. T Yucek, H Arslan, A survey of spectrum sensing algorithms for cognitive radio applications. Commun. Surveys Tutorials, IEEE. **11**(1), 116–130 (2009)
5. HR Karimi, in *New Frontiers in Dynamic Spectrum Access Networks (DySPAN), 2011 IEEE Symposium On*. Geolocation databases for white space devices in the uhf tv bands: Specification of maximum permitted emission levels (IEEE, Aachen, Germany, 2011), pp. 443–454
6. J Zander, LK Rasmussen, KW Sung, P Mahonen, M Petrova, R Jantti, J Kronander, On the scalability of cognitive radio: assessing the commercial viability of secondary spectrum access. Wireless Commun. IEEE. **20**(2), 28–36 (2013)
7. Federal Communications Commission — Rules and Regulations for Title 47. http://www.fcc.gov/encyclopedia/rules-regulations-title-47. Last accessed on 02/21/2015.
8. R Tandra, SM Mishra, A Sahai, What is a spectrum hole and what does it take to recognize one? Proc. IEEE. **97**(5), 824–848 (2009)
9. CF Tomaz Solc, M Mohorcic, in *Cognitive Radio and Networking in Heterogeneous Wireless Networks*. Low-cost testbed development and its applications in cognitive radio prototyping (Springer, 2015)
10. A Nika, Z Zhang, X Zhou, BY Zhao, H Zheng, in *Proceedings of the 1st ACM Workshop on Hot Topics in Wireless*. Towards commoditized real-time spectrum monitoring (ACM, Los Angeles, California, USA, 2014), pp. 25–30
11. D Cabric, A Tkachenko, RW Brodersen, in *Proceedings of the First International Workshop on Technology and Policy for Accessing Spectrum*. Experimental study of spectrum sensing based on energy detection and network cooperation (ACM, Boston, Massachusetts, USA, 2006), p. 12
12. M Murroni, RV Prasad, P Marques, B Bochow, D Noguet, C Sun, K Moessner, H Harada, Ieee 1900.6: spectrum sensing interfaces and data structures for dynamic spectrum access and other advanced radio communication systems standard: technical aspects and future outlook. Commun. Mag. IEEE. **49**(12), 118–127 (2011)
13. J Caufield, Protocol to Query a White Space Database. http://tools.ietf.org/html/draft-caufield-paws-protocol-for-tvws-01
14. D Denkovski, V Rakovic, M Pavloski, K Chomu, V Atanasovski, L Gavrilovska, in *Wireless Communications and Networking Conference (WCNC), 2012 IEEE*. Integration of heterogeneous spectrum sensing devices towards accurate rem construction (IEEE, Paris, France, 2012), pp. 798–802
15. V Atanasovski, J van de Beek, A Dejonghe, D Denkovski, L Gavrilovska, S Grimoud, P Mahonen, M Pavloski, V Rakovic, J Riihijarvi, B Sayrac, in *New Frontiers in Dynamic Spectrum Access Networks (DySPAN), 2011 IEEE Symposium On*. Constructing radio environment maps with heterogeneous spectrum sensors (IEEE, Aachen, Germany, 2011), pp. 660–661
16. J Van De Beek, T Cai, S Grimoud, B Sayrac, P Mahonen, J Nasreddine, J Riihijarvi, How a layered rem architecture brings cognition to today's mobile networks. Wireless Commun. IEEE. **19**(4), 17–24 (2012)
17. TF TRL, FC TRL, SS TRL, W Workpackage, Flexible and spectrum aware radio access through measurements and modelling in cognitive radio systems. Evaluation. **5**(2) (2009)
18. M López-Benítez, F Casadevall, Methodological aspects of spectrum occupancy evaluation in the context of cognitive radio. Eur. Trans. Telecommunications. **21**(8), 680–693 (2010)
19. D Finn, JC Tallon, LA DaSilva, P Van Wesemael, S Pollin, W Liu, S Bouckaert, J Vanhie-Van Gerwen, N Michailow, J Hauer, D Wilkomm, C Heller, in *Proceedings of the 6th ACM International Workshop on Wireless Network Testbeds, Experimental Evaluation and Characterization*. Experimental assessment of tradeoffs among spectrumsensing platforms (ACM, New York, USA, 2011), pp. 67–74
20. WL Stutzman, WA Davis, *Antenna Theory*. (Wiley Online Library, 1998)
21. Y Liu, C Zeng, H Wang, G Wei, in *Advanced Computer Control (ICACC), 2010 2nd International Conference On*. Energy detection threshold optimization for cooperative spectrum sensing, vol. 4 (IEEE, Shenyang, Liaoning, China, 2010), pp. 566–570
22. W Zhang, RK Mallik, K Letaief, in *Communications, 2008. ICC'08. IEEE International Conference On*. Cooperative spectrum sensing optimization in cognitive radio networks (IEEE, Beijing, China, 2008), pp. 3411–3415
23. N Wang, Y Gao, X Zhang, Adaptive spectrum sensing algorithm under different primary user utilizations. **17**(9), 1838–1841 (2013)
24. T Rakotoarivelo, M Ott, G Jourjon, I Seskar, Omf: a control and management framework for networking testbeds. ACM SIGOPS Operating Syst. Rev. **43**(4), 54–59 (2010)
25. S Bouckaert, P Becue, B Vermeulen, B Jooris, I Moerman, P Demeester, in *Testbeds and Research Infrastructure. Development of Networks and Communities*. Federating wired and wireless test facilities through emulab and omf: the ilab. t use case (Springer, thessaloniki, Greece, 2012), pp. 305–320
26. V Handziski, A Köpke, A Willig, A Wolisz, in *Proceedings of the 2nd International Workshop on Multi-hop Ad Hoc Networks: from Theory to Reality*. Twist: a scalable and reconfigurable testbed for wireless indoor experiments with sensor networks (ACM, Florence, Italy, 2006), pp. 63–70
27. CREW Project. http://www.crew-project.eu/repository/. Last accessed on 02/16/2015.
28. Ultra low power IEEE 802.15.4 compliant wireless sensor module. http://www.eecs.harvard.edu/~konrad/projects/shimmer/references/tmote-sky-datasheet.pdf. Last accessed on 02/16/2015.
29. 2.4 GHz IEEE 802.15.4 ZigBee-ready RF Transceiver. http://www.ti.com/lit/ds/symlink/cc2420.pdf. Last accessed on 02/16/2015.
30. P Levis, S Madden, J Polastre, R Szewczyk, K Whitehouse, A Woo, D Gay, J Hill, M Welsh, E Brewer, D Culler, in *Ambient Intelligence*. Tinyos: An operating system for sensor networks (Springer, 2005), pp. 115–148
31. How Wi-Spy Works. http://www.metageek.net/blog/2011/01/how-wi-spy-works. Last accessed on 02/16/2015.
32. Airmagnet Survey. http://airmagnet.flukenetworks.com/assets/datasheets/AirMagnet_Survey_Datasheet.pdf. Last accessed on 02/16/2015.
33. J Bock, M Lynn, *Hacking Exposed Wireless*. (McGraw-Hill, Inc., 2007)
34. Ettus Research. http://www.ettus.com/. Last accessed on 02/16/2015.
35. E Blossom, Gnu radio: tools for exploring the radio frequency spectrum. Linux J. **2004**(122), 4 (2004)

36. PD Sutton, J Lotze, H Lahlou, SA Fahmy, KE Nolan, B Ozgul, TW Rondeau, J Noguera, LE Doyle, Iris: an architecture for cognitive radio networking testbeds. Commun. Mag. IEEE. **48**(9), 114–122 (2010)
37. S Pollin, L Hollevoet, P Van Wesemael, M Desmet, A Bourdoux, E Lopez, F Naessens, P Raghavan, V Derudder, S Dupont, A Dejonghe, in *New Frontiers in Dynamic Spectrum Access Networks (DySPAN), 2011 IEEE Symposium On*. An integrated reconfigurable engine for multi-purpose sensing up to 6 ghz (IEEE, Aachen, Germany, 2011), pp. 656–657
38. W Liu, D Pareit, E De Poorter, I Moerman, Advanced spectrum sensing with parallel processing based on software-defined radio. EURASIP J Wireless Commun. Netw. **2013**(1), 1–15 (2013)
39. P Van Wesemael, W Liu, M Chwalisz, J Tallon, D Finn, Z Padrah, S Pollin, S Bouckaert, I Moerman, D Willkomm, in *Future Network & Mobile Summit (FutureNetw), 2012*. Robust distributed sensing with heterogeneous devices (IEEE, 2012), pp. 1–9

6

Measurements of ultra wide band in-vehicle channel - statistical description and TOA positioning feasibility study

Jiri Blumenstein[1*], Ales Prokes[1], Tomas Mikulasek[1], Roman Marsalek[1], Thomas Zemen[2] and Christoph Mecklenbräuker[1,3]

Abstract

This paper reports on a real-world wireless channel measurement campaign for in-vehicle scenarios in the UWB frequency range of 3 to 11 GHz. The effects of antenna placement in the vehicle's passenger compartment as well as the effects due to the presence of passengers are studied. The measurements have been carried out in the frequency domain, and the corresponding channel impulse responses (CIRs) have been estimated by inverse Fourier transform. The influence of a specific band group selection within the whole UWB range is also given. Statistical analysis of the measured channel transfer functions gives a description of the wireless channel statistics in the form of a generalized extreme value process. The corresponding parameter sets are estimated and documented for all permutations of antenna placement and occupancy patterns inside the vehicle's passenger compartment. Further, we have carried out a feasibility study of an in-vehicle UWB-based localization system based on the TOA. The positioning performance is evaluated in terms of average error and standard deviation.

Keywords: UWB; In-vehicle environment; Channel model; Positioning; TOA

1 Introduction

The onboard electrical power distribution, communication, and networking functionalities are realized by cable bundles in today's vehicles. We observe a trend towards increasing numbers of sensors, actuators, control units, and infotainment systems in cars and trucks. As a direct result, the weight of the wiring in all types of vehicles increases. Moreover, their flexible installation and reliability represent a challenging and costly task [1]. The weight of the wiring becomes even more serious when the vehicles are powered fully electrically.

In [2,3], the authors conclude that ultra wide bandwith (UWB) technology with its favorable radio environment characteristics for indoor areas such as low transmit power and robustness against multipath fading could be extrapolated even for the vehicular passenger compartment. Naturally, attempts to replace cable bundle start up with in-vehicle radio channel measurements were performed by authors in [4-9] and by channel modeling in [3,10], and a clustering approach for intra-bus channel modeling is studied in [11,12]. Attempts to build a prototype of an UWB-based wireless sensor network within a vehicle, both in the passenger and the engine compartments, are published in [13,14]. In [15], the topic of wireless in-vehicle communication links based on LTE is discussed while reckoning with specific in-vehicle impulse noise. In [16], the UWB channel inside a vehicle is studied from a spatial stationarity point of view. The necessity of detailed knowledge of the channel characteristics is of highest importance for the proper physical layer design of any wireless communication system.

Together with this motivation to substitute at least part of the vehicle's cable bundles by wireless links, a wireless localization service within the vehicle is desirable. Future applications of such a localization service include remote keyless entry and ignition systems, advanced child passenger safety, and beamsteering for in-vehicle high-speed Internet access.

*Correspondence: blumenstein@feec.vutbr.cz
[1]Department of Radio Electronics, Brno University of Technology, Technicka 12, 612 00, Brno, Czech Republic
Full list of author information is available at the end of the article

In [14,17], a localization service utilizing UWB is studied and it is concluded that in general, thanks to the high time resolution of UWB impulses, the time of arrival (TOA) technique is capable of providing sufficient spatial resolution for a variety of applications. Although the TOA works reliably in environments with LOS, it can be used with some restriction also in NLOS scenarios. In the multipath environment, the important component for ranging based on the TOA technique is the direct ray, which propagates from the transmitter towards the receiver. When the beam penetrates some obstacles whose attenuation does not avoid the beam detection, the TOA technique is applicable. Note that, for example, in the US, the frequency range of 1.99 to 10.6 GHz is deregulated for communications and wall-penetrating radars, enabling looking into or through non-metallic materials [14,18]. Thus, the presumption is that even the harsh in-vehicle ambiance with OLOS propagation may provide sufficient positioning accuracy.

In [4,6,7,19], the path-loss, seat material, and occupancy influences are presented for the frequency range of 3 to 8 GHz. Since the positioning service deployment is not seen as the aim of [4,6,7,19], the placement of transmit antennas is inappropriate from that point of view. Thus, resulting parameters could differ from parameters obtained by measurement campaigns which take the positioning into account in the first place.

1.1 Contribution of the paper

While taking into account the influence of the occupancy, antenna placement, and the influence of a specific frequency band group selection, in this paper, we address the following:

- Intra-vehicle channel measurement and statistical evaluation via GEV. This allows a reproducibility of the measured results for 90 selected wireless links within a passenger car compartment.
- Statistical analysis and the in-vehicle positioning in the UWB range of 3 to 11 GHz. The aim of the article is to give a general overview of the achievable accuracy of ranging regardless of LOS and NLOS scenarios.

The paper is organized as follows. In Section 2, we provide an overview of our measurement site including a hardware description. In Section 3, we present our channel measurement tools including our conical monopole antenna design [20] and we define the sought channel parameters. In Section 4, the feasibility of the positioning service deployment within a vehicle compartment is assessed, while the conclusion in Section 5 sums up the paper.

2 Measurement setup

2.1 Measurement bandwidth and dynamic range

The scheme of the measurement setup is shown in Figure 1. The complex CIRs (below introduced by (1)) corresponds to the s_{41}, s_{42}, and s_{43} scattering parameters which are measured in the frequency domain for two different frequency bands, 3 to 11 GHz (entire UWB band with a bandwidth of $B = 8$ GHz) and 3.3168 to 4.752 GHz (first band group with a bandwidth of $B = 1.58$ GHz), utilizing a four-port vector network analyzer Agilent Technologies E5071C (VNA; Agilent Technologies Inc., Santa Clara, CA, USA) (Figure 2). The spatial placement of the receiving (RX) and transmitting (TX) antennas inside the vehicle is depicted in Figure 3.

The dynamic range of the measurement setup is higher than 90 dB ($P_{\text{outVNA}} = 5$ dBm, IF bandwidth = 100 Hz). The chosen frequency step of 10 MHz results in 801 frequency points in the case of the entire UWB band and 159 frequency points in the case of the first band group.

In order to avoid a degradation of the measured phase accuracy due to movements of the RX antenna, phase-stable coaxial cables were used and included in the calibration process. The measurement is carried out in the Skoda Octavia 1.8 TSI car.

2.2 Antenna placement

As depicted in Figure 3, the RX antenna is placed at various locations inside the car compartment (on all seats and in the boot) and the TX antennas are placed on the left and right sides of the dashboard, top corners of the windshield, and at the rear part of the ceiling.

The channel measurements are carried out for both LOS and NLOS scenarios. NLOS is caused by the backrest of the seats, the dashboard, and/or persons sitting inside the vehicle.

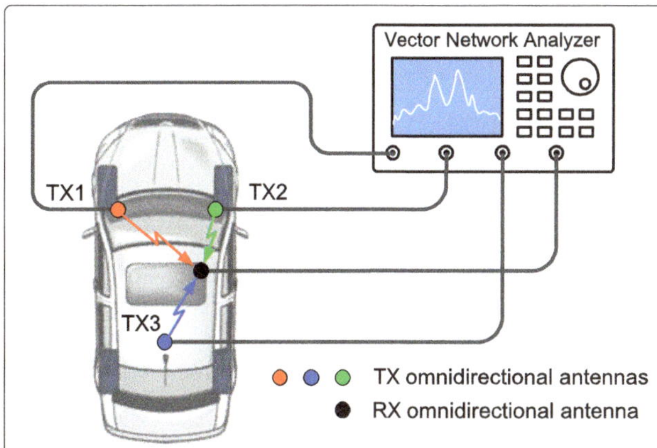

Figure 1 Measurement setup containing the vector network analyzer Agilent Technologies E5071C and the car Skoda Octavia 1.8 TSI.

Figure 2 Images of the conical monopole antenna, measurement position, and four-port VNA. [left] Detail of the conical monopole antenna mounted on the front windshield. [middle] One measurement position inside the vehicle. [right] Four-port VNA connected with antennas inside the measured vehicle.

Since the radiation pattern of the conical monopole antenna [20] is very close to the omnidirectional radiation pattern, we were able to capture a maximal number of multipath components (reflected waves).

In Figure 4, the conical monopole has an omnidirectional H-plane radiation pattern which is invariant in the frequency band of interest. Due to a variable gain in the lower half E-plane radiation pattern (elevation angle from 90° to −90°), the antennas were placed in the car compartment so that the upper half E-plane radiation pattern (almost constant) was used. It means that when the antenna was placed at the cabin ceiling, it was situated bottom up. Thus, the LOS and NLOS are minimally affected by the radiation pattern; however, with the reflected waves arriving from the TX antenna or incident on the RX antenna, the lower elevation angle might be affected by the non-ideal radiation pattern of the antennas.

3 Channel parameters

The CIR describes the wireless channel. We utilize an inverse discrete Fourier transform of the windowed scattering parameter series, expressed as:

$$h^{\alpha}(n) = \sum_{k=0}^{N-1} w(k) s_{\zeta}^{\alpha}(k) e^{jkn2\pi/N}, \tag{1}$$

where $s_{\zeta}^{\alpha}(k)$ corresponds to the kth measured scattering parameter (as described in Section 2) and $w(k)$ represents the Blackman window. Parameter α denotes the spatial positions of the transmit and the receive antenna in the measured vehicle and $\zeta \in \{41, 42, 43\}$. For practicality in the following statistical processing, we arbitrarily merge indices α and ζ into one measurement number $\alpha \in \{1, \ldots, 90\}$. Hence, in the following, it is not possible to assign the specific measured data to the actual spatial positions.

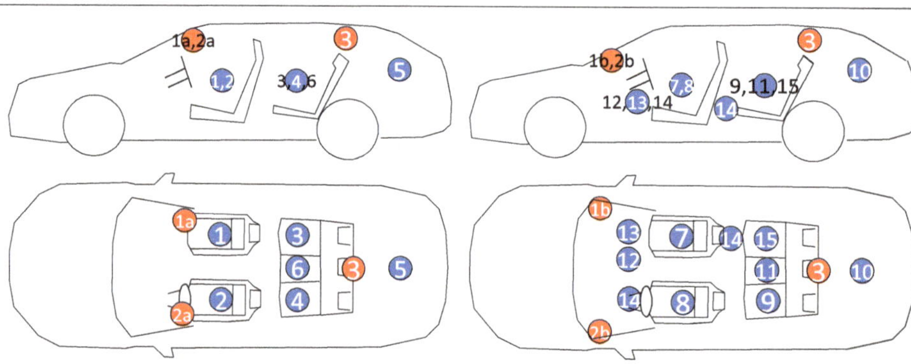

Figure 3 The positions of transmitting (red) and receiving (blue) antennas. We employ two possible receive antenna placement patterns. As seen on the left part, the antennas 1a and 2a occupy the left and right top corners of the windshield, while on the right part, the antennas 1b and 2b are positioned in the lower corners. Please note that all measurements have been measured for various passenger layouts. We have considered (1) empty vehicle and (2) driver and two to three passengers.

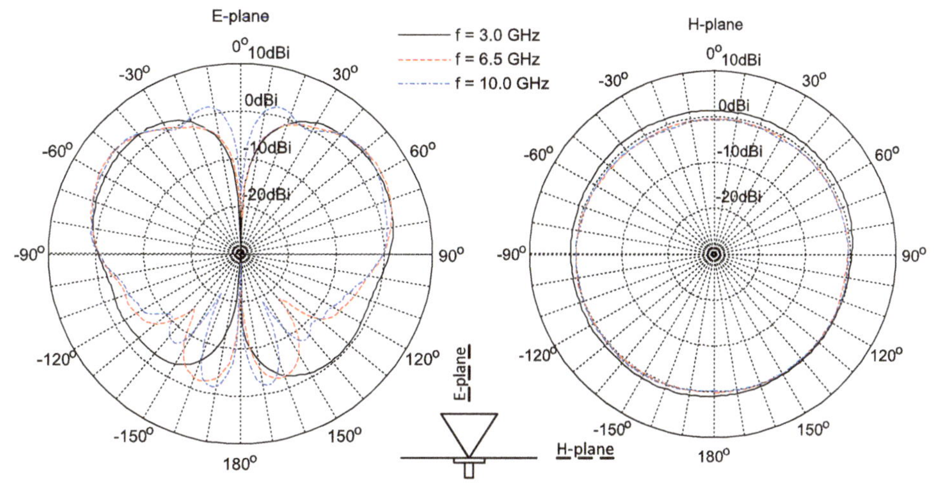

Figure 4 Measured gain pattern of the conical monopole antenna [20].

The number of measured frequency points $N = 801$ for the entire UWB or $N = 159$ for the first band group. Since the in-vehicle channel is assumed to be time invariant, we performed one repetition of the scattering parameter measurement.

The relationship between discrete time delay n and continuous time delay τ is given by:

$$\tau_n = n\frac{1}{B}, \tag{2}$$

where $1/B$ stands for the time resolution (see Equation 7).

For a statistical characterization of the UWB channel, we use the MDP which is defined as the magnitude of complex CIR:

$$A(\tau) = |h(\tau)|. \tag{3}$$

3.1 Statistical description of the received signal

3.1.1 Independent identically distributed (IID) phase

In this chapter, the phase statistics of the measured CIRs are presented. As visible in Figure 5 [right], according to the ecdf evaluated for each measured CIR, the phase $\Phi^\alpha(\tau)$ is uniformly distributed.

Moreover, utilizing the Pearson correlation coefficient $\rho_{\alpha,\beta}$, we evaluate the mutual dependence between phases for all measured positions denoted as α and β. The Pearson correlation coefficient is given as:

$$\rho_{\alpha,\beta} = \frac{\mathrm{E}\left[\left(\Phi^\alpha(\tau) - \xi^\alpha\right)\left(\Phi^\beta(\tau) - \xi^\beta\right)\right]}{\wp^\alpha \wp^\beta}, \tag{4}$$

where $\wp^\alpha$ denotes the standard deviation and ξ^α the mean of $\Phi^\alpha(\tau)$. The evaluation of the Pearson correlation coefficient is visible in Figure 5 [left] showing uncorrelated behavior of $\Phi^\alpha(\tau)$. The operator $\mathrm{E}[\cdot]$ denotes the expected value.

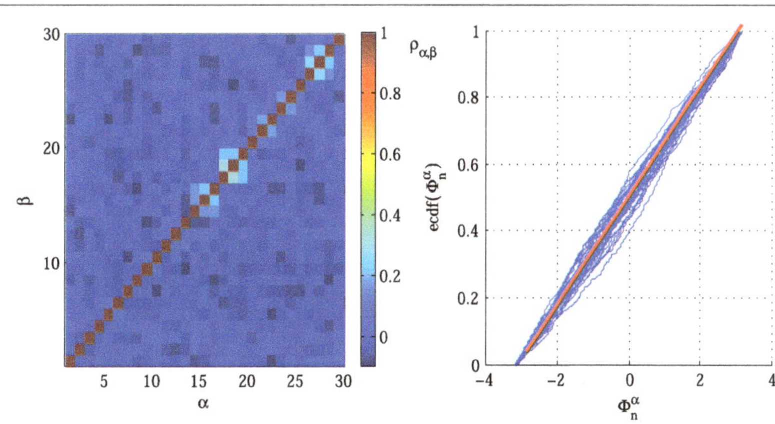

Figure 5 Pearson correlation coefficient and ecdf curves. [left] The Pearson correlation coefficient $\rho_{\alpha,\beta}$ evaluation of the measured $\Phi^\alpha(\tau)$. [right] The ecdf curves of Φ^α_n. The closer the blue ecdf curve to the red line, the closer the probability distribution of $\Phi^\alpha(\tau)$ to the uniform distribution.

According to the results presented in Figure 5, we conclude that $\Phi^{\alpha}(\tau)$ is iid uniformly distributed with respect to the measurement number α.

3.1.2 Statistics of the received signal magnitudes and GEV

Utilizing MLEs [21], we have found a statistical model of received signal magnitudes. As seen in Figures 6 and 7, the received signal magnitudes can be approximated using the GEV distribution [22] with the PDF given by:

$$f(x \mid 0, \mu, \sigma) = \frac{1}{\sigma}\exp\{-z - \exp(-z)\}, \text{ where } z = \frac{x-\mu}{\sigma}, \tag{5}$$

with μ being the location parameter and σ the distribution scale parameter. Equation 5 represents the GEV type I distribution, also known as log-Weibull distribution, where the shape parameter defined in the regular GEV is set to zero. This approach is justified in Section 3.1.3.

In order to capture the statistical nature of the environment, we have performed 90 measurements permuting both the TX and RX antenna placements as well as the in-car seat occupancy. In Figure 6, we can see the CDF curves for all permutations of the antenna placement and occupancy, while each curve is fitted by a GEV type I random process obtained by the MLE fitting.

3.1.3 GEV parameters as a random process

According to the observations of resulting GEV parameters, we approximate the corresponding μ, k, and σ parameters with the statistical model obtained by MLE. Figures 8, 9, and 10 compare the CDFs of the measured μ, k, and σ parameters with random processes of corresponding distributions.

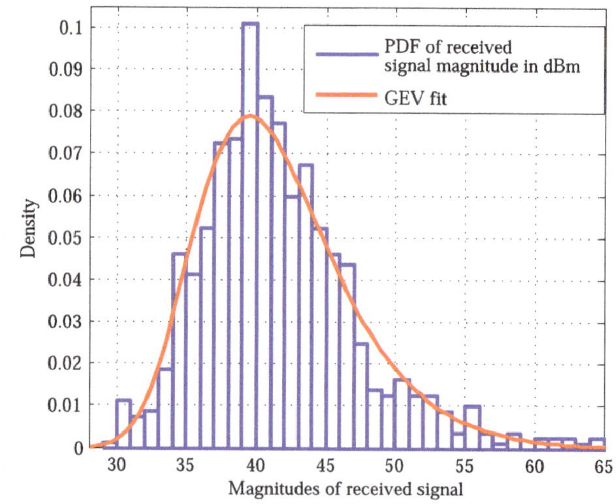

Figure 7 PDF of the received signal magnitudes in dBm for one measurement setup. The measured PDF is fitted with the GEV type I procedure.

The location parameter μ follows the lognormal distribution given as:

$$f(x \mid \eta, \nu) = \frac{1}{x\eta\sqrt{2\pi}}\exp\left(\frac{-(\ln x - \nu)^2}{2\eta^2}\right), \tag{6}$$

where ν is the mean and η represents the standard deviation. The extracted shape parameter k is of significantly low values; therefore, our choice of the GEV type I (also known as log-Weibull) characterized by $k = 0$ is appropriate (see Equation 5). The scale parameter σ is normally distributed.

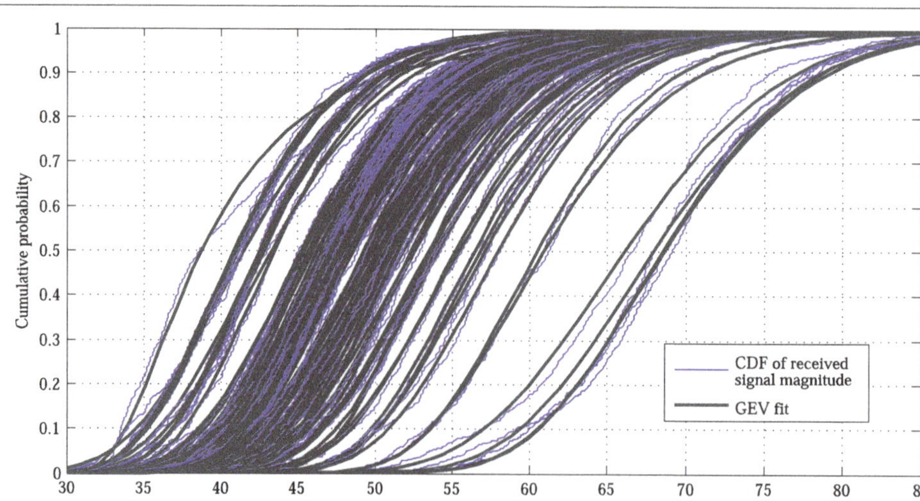

Figure 6 The CDF of received signal magnitude in dBm for all permutations of antenna placement and occupancy. Measured curves are fitted with the GEV type I procedure.

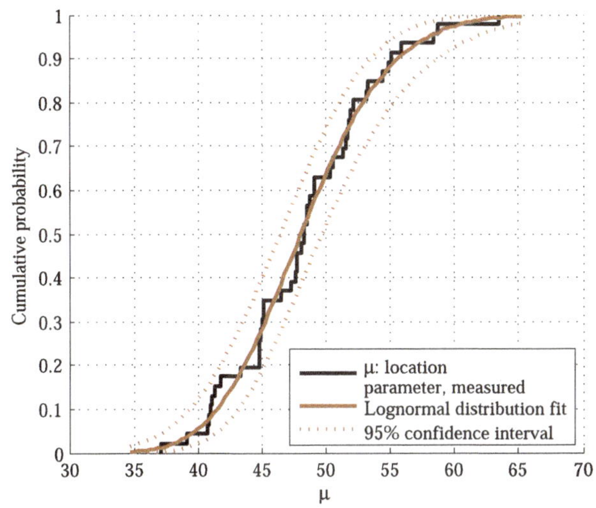

Figure 8 The CDF plot of location parameter μ fitted with lognormal distribution (with 95% confidence interval).

A summarized overview, including the specific values of μ and η, is given in Table 1. Thanks to a high number of performed measurements, the tabulated values represent typical data for an in-vehicle channel which also applies to vehicles of similar size, seat configurations, and materials utilized for its manufacture.

A correlation between the derived parameters μ and η ($k = 0$) exhibits a very weak positive correlation value of 0.35 with a p value below 6×10^{-4}. Thus, to recreate the received signal magnitudes, one can arbitrarily choose the parameters μ and η according to Table 1.

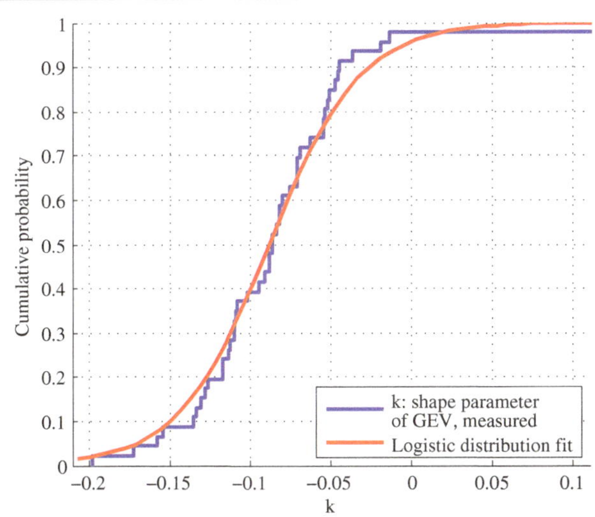

Figure 9 The CDF plot of shape parameter k fitted with a logistic distribution (with 95% confidence interval). Since the values of the k parameter are very small, in the following we set $k = 0$; thus, the GEV becomes a GEV type I (or log-Weibull) distribution.

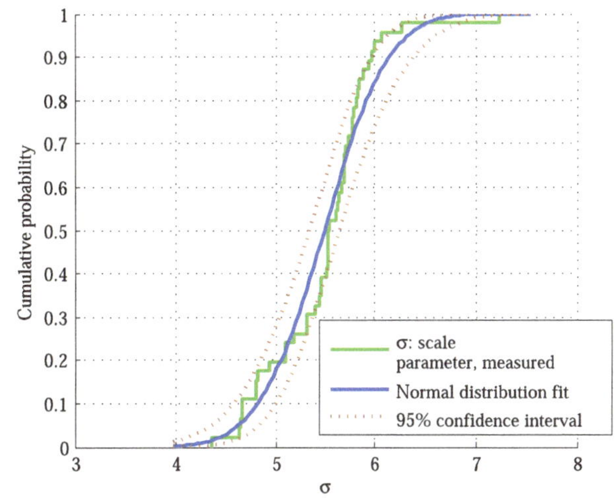

Figure 10 The CDF plot of scale parameter σ fitted with a normal distribution (with 95% confidence interval).

Due to a high flexibility of the GEV fit, which is given by three input parameters as opposed to usual two parameters, the MLE metric recommends the GEV distribution. On other hand, authors in [23] claim that there is no theoretical explanation for encountering this distribution type. We may, however, add that the GEV contains the accepted log-Weibull distribution as a special case for $k = 0$.

4 Localization

One of the often discussed UWB applications is precise ranging and localization especially when the TOA technique is used. As mentioned above, this is because the large UWB bandwidth allows excellent time resolution (see Equation 7) and MPC separation. Because we had measured the channel transfer function for many different antenna positions, we wanted to get some insight into attainable ranging accuracy. Our estimation of the distance results from the CIR calculated from the complex transfer function. This approach gives some limitations compared to a direct channel sounding in the time domain where some advanced techniques such as the matched filtering of the known Gaussian pulses or a well-correlated binary sequences can be used [24].

We calculated the antenna distance using the TOA technique based on the detection of the first ray transmitted

Table 1 Summarization of GEV type I parameters characterizing in-vehicle environment for 90 permutations of antenna placement and car seat occupancy

	μ	σ	k
Distribution type	Lognormal	Normal	Logistic
Mean ν	48.37	5.48	−0.08
Variance η	31.05	0.26	0.002

from a particular antenna. The proposed threshold-based search algorithm compares individual signal samples of the CIR with a certain threshold in order to identify the amplitude peak corresponding to the first MPC. This approach allows to calculate distance also in the NLOS scenario because the first ray may not be the strongest ray. However, penetration of the obstacles can cause some measurement accuracy degradation (see below). The aim of this chapter is to give a basic idea about accessible average error and standard deviation of the measured distances for the entire UWB band and for the first band group and also for the empty and occupied car. For more information about the measurements and the distance calculation, see [25]. Because all the particular measurements were done for three TX antennas and one RX antenna, we also calculated RX antenna position using the 2D localization technique in order to assess whether it corresponds at least roughly to reality, i.e., whether it is possible for example to reliably detect a device on a particular seat. Note that most of the above mentioned application does not need an accurate localization but only rough estimation of the device position.

4.1 Basic system parameter calculation

The frequency band B determines the time resolution of measurement by [24]:

$$T_r = \frac{1}{B} = \frac{1}{F_U - F_L}, \tag{7}$$

where F_U is the upper frequency and F_L is the lower frequency of the band. The propagation distance resolution is then

$$D_r = T_r c, \tag{8}$$

where $c = 2.998 \times 10^8$ m/s is the speed of the light. The maximum measurable propagation distance depends on the number of measured frequency values N_M inside the frequency band, i.e., on the frequency step f_s according to

$$D_{\max} = \frac{c}{B} N_M = \frac{c}{f_s}. \tag{9}$$

It is obvious from the equations above that narrowing the bandwidth decreases the distance resolution and the reduction in the measured frequency points shortens the measurable propagation distance.

4.2 Ranging and localization of the receiving antenna

As mentioned above, the main aim of this section is calculation of the average error and standard deviation of the ranging and verification of the RX antenna position (Table 2). The processing of the measured data consists in the following steps:

- Calculation of the CIR
- Detection of the first incident ray
- Calculation of the RX - TX1 to TX3 distances and ranging errors
- RX antenna localization

The CIR was calculated using the IFFT in combination with a Blackman window (see Figure 11) applied to all 801 frequency response points. Before error statistics calculation, we tested a few windows (rectangular, Hann, Hamming, flattop, Blackman, and Kaiser-Bessel). Although some windows (e.g., rectangular) are generally recommended for the applications where good component separation is required, these windows could be inapplicable in our case as they may produce large side lobes that cross the threshold and cause incorrect first ray detection. Experimentally, we found that the best results giving the distances closest to reality are given by the Blackman window. The second best results can be then obtained using the Hann window.

The threshold for the first ray detection is generally determined by the noise floor. Its value is equal to the level of the peaks of noise, i.e., to the maximum amplitude of the CIR where the multipath component amplitudes are below noise level. It is obvious that the proposed algorithm works reliably in both LOS and NLOS scenarios, but it fails in some NLOS cases when the first (direct) ray is strongly attenuated and drowned in noise.

The distance of RX and TX antennas is given by the formula $D_A = cT_D$, where T_D is the first detected ray arrival time. The error statistics were calculated separately for the empty and occupied car. It was experimentally discovered that the two or three passengers sitting in the car compartment cause very similar results, and therefore, these cases were joined into one set of results. For the RX antenna localization, the trilateration technique [24] was applied. Using the three calculated distances, this technique allows 2D localization.

Calculation of the average error and standard deviation of the measured distances is summarized in Table 3 (for

Table 2 The parameters used for the ranging

	Bandwidth [GHz]	Freq. step [MHz]	Time resolution [ns]	Distance resolution [cm]	Max. propag. distance [m]	Max. measurement time [ns]
UWB	8	10	0.125	3.750	30	100
First band group	1.58		0.633	18.987		

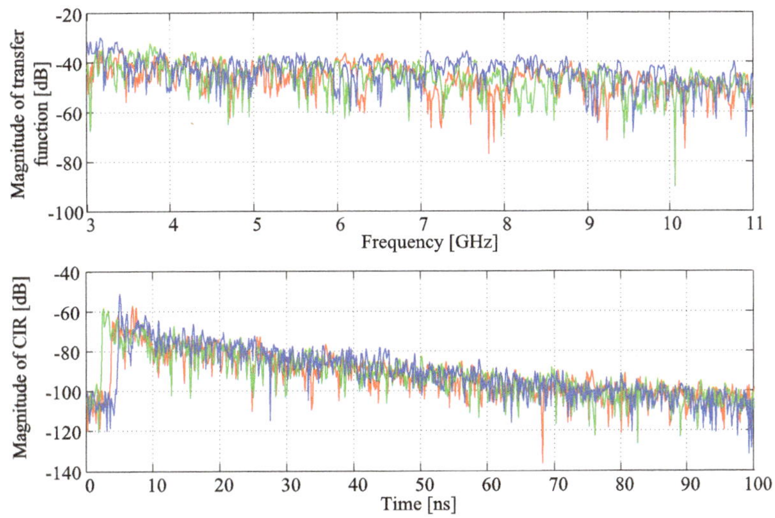

Figure 11 Magnitudes of channel transfer functions and channel impulse responses (RX antenna was placed on right rear seat).

the entire UWB) and Table 4 (for the first band group). The time intervals used for the noise peak detection were 0 to 1.25 ns (before receiving of the first MPC) and 80 to 100 ns (where the MPC can be neglected). These time intervals correspond to the following distances: 0 to 37.5 cm (minimum distance of RX-TX antennas in all scenarios is 50 cm) and 24 to 30 m. The reference antenna distances were measured by a ruler. We compared 15 × 3 distances without passengers and 15 × 3 distances with two or three passenger sitting on the seats surrounding the RX antenna. An example of peak detection for the empty car is shown in Figure 12 (upper part for the entire UWB and lower part for the first band group), while Figure 13 depicts the 2D localization result also for the UWB and first band group.

4.3 Positioning results and sources of error

It is obvious that the rough distance resolution in the case of the first band group measurement causes markedly higher average error and standard deviation compared to the measurement of the entire UWB band. The calculated distances exhibit noticeable positive bias caused by a few phenomena:

- *Existence of difference between the calibration plane and phase center of the antenna.* The coaxial interfaces of the antennas (line between the connector and phase center of the antenna) were not included when the VNA was calibrated. They were applied only during channel measurement and increased the total antenna distance.
- *Inaccurate reference measurement.* Distance measured between the antennas by the ruler was performed between the centers of the top of cones which are not identical to the phase centers of antennas. In many cases, the measured distance were slightly shorter (when the TX antenna was upside down with regard to RX antenna) .
- *Time lag in the first ray detection.* The first ray (peak) detection above the threshold exhibits random delay in the interval 0 to D_r due to the discrete nature of the CIR time axis. Received ray cannot be generally detected in advance.

Table 3 Average error and standard deviation of the measured distances for the first band group

	TX1	TX2	TX3	Total
Average error without passengers [cm]	6.76	6.30	5.75	6.27
Average error with two or three passengers [cm]	11.83	10.37	7.62	9.94
Standard deviation without passengers [cm]	7.49	6.86	2.10	5.87
Standard deviation with two or three passengers [cm]	11.13	9.28	8.95	9.80

Table 4 Average error and standard deviation of the measured distances for the first band group

	TX1	TX2	TX3	Total
Average error without passengers [cm]	25.85	21.04	14.3	20.39
Average error with two or three passengers [cm]	34.73	23.50	10.82	23.02
Standard deviation without passengers [cm]	20.76	13.88	8.74	15.67
Standard deviation with two or three passengers [cm]	20.13	12.63	7.88	17.26

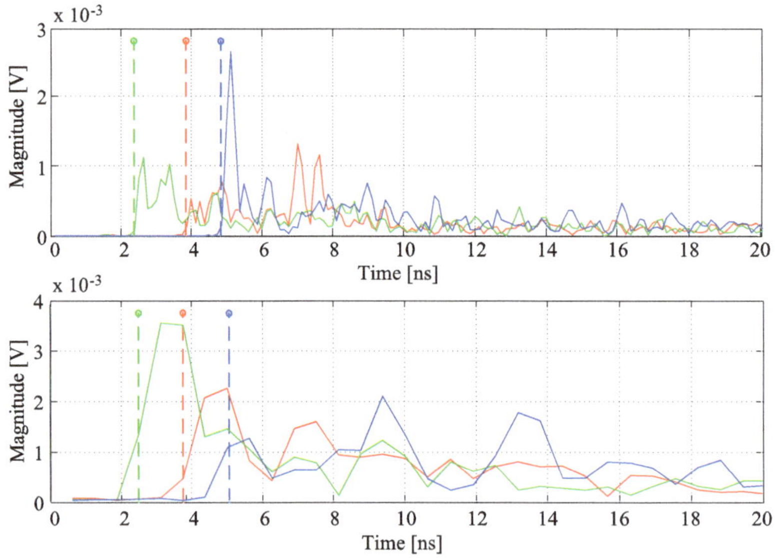

Figure 12 First peak detection of CIRs (for entire UWB) [upper] for empty and [lower] occupied vehicle.

- *Incorrect MPC component detection.* Large attenuation of some obstacles in the car may avoid correct detection of the direct ray. In this case, the other reflected MPC which travels on a longer path is regarded as the first ray.
- *Lower wave propagation velocity in media.* The velocity of an electromagnetic wave penetrating an obstacle is less than that in free space, and it depends on the obstacle material constants.

The first phenomenon is systematic and can be subtracted (it is about 2 cm together for two antennas). The two last phenomena occur only in the NLOS scenario. In the last case, the velocity in some material can be calculated according the formula $v_p = c/\sqrt{\varepsilon_r \mu_r}$, where v_p is the velocity of propagation in m/s, μ_r is the material relative permeability, and ε_r is the relative permittivity. It is easy to find that when, for example, the wave passes the 10-cm-thick plastic obstacle ($\varepsilon_r = 2$ to 3, $\mu_r = 1$ [26]), the propagation time delays are in the interval 0.138 to 0.244 ns which results in the distance bias from 4.1 to 7.3 cm.

5 Conclusions

We performed an extensive UWB measurement campaign for the vehicular passenger compartment. The measured

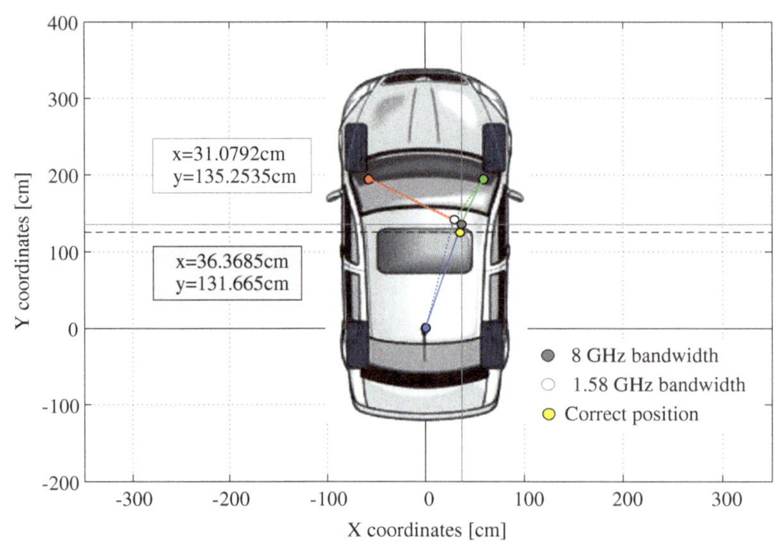

Figure 13 Localization of the RX antenna using TOA (RX antenna was placed on front passenger seat).

channel impulse responses are modeled using the GEV distribution; its parameters are estimated using a MLE. As a result, our statistical description of the received amplitude and phase distribution in the in-vehicle environment fits almost perfectly to the empirical measurement results. We showed that the measured phase is uniformly distributed with iid behavior.

Based on the measurement data, a feasibility study on the use of UWB-based positioning inside the vehicle was conducted. We could show that the accuracy of the transmitter location could be obtained with a standard deviation smaller than 10 cm for the full UWB bandwidth. The standard deviation was smaller than 16 cm for the first UWB band group only. The influence of the antenna position on the localization accuracy was lower than the effect of the occupancy level of the car.

Competing interests
The authors declare that they have no competing interests.

Acknowledgements
This work was supported by the Czech Science Foundation Project No. 13-38735S Research into wireless channels for intra-vehicle communication and positioning. Research described in this paper was financed by Czech Ministry of Education in frame of National Sustainability Program under grant LO1401. For research, infrastructure of the SIX Center No. CZ.1.05/2.1.00/03.0072 was used. The cooperation in the COST IC1004 action was supported by the MEYS of the Czech Republic Project No. LD12006 (CEEC).

Author details
[1]Department of Radio Electronics, Brno University of Technology, Technicka 12, 612 00, Brno, Czech Republic. [2]AIT Austrian Institute of Technology GmbH, Donau-City-Straße 1, 1220 Vienna, Austria. [3]Institute of Telecommunications, Vienna University of Technology, Gußhausstraße 25/E 389, 1040 Vienna, Austria.

References
1. G Leen, D Heffernan, Expanding automotive electronic systems. Computer. **35**(1), 88–93 (2002)
2. M Win, R Scholtz, Characterization of ultra-wide bandwidth wireless indoor channels: a communication-theoretic view. Selected Areas Commun. IEEE J. **20**(9), 1613–1627 (2002)
3. R-M Cramer, R Scholtz, M Win, Evaluation of an ultra-wide-band propagation channel. Antennas Propag. IEEE Trans. **50**(5), 561–570 (2002)
4. M Schack, J Jemai, R Piesiewicz, R Geise, I Schmidt, T Kurner, in *IEEE Vehicular Technology Conference, 2008*. Measurements and analysis of an in-car UWB channel (VTC Spring 2008 Singapore, 11–14 May 2008), pp. 459–463
5. T Kobayashi, in *2006 IEEE Ninth International Symposium on Spread Spectrum Techniques and Applications, Manaus-Amazon*. Measurements and characterization of ultra wideband propagation channels in a passenger-car compartment, (28–31 Aug 2006), pp. 228–232
6. T Tsuboi, J Yamada, N Yamauchi, in *7th International Conference on ITS Telecommunications, 2007. ITST '07*. UWB radio propagation inside vehicle environments (Sophia Antipolis, 6–8 June 2007), pp. 1–5
7. M Schack, R Geise, I Schmidt, R Piesiewiczk, T Kurner, in *3rd European Conference on Antennas and Propagation, 2009. EuCAP 2009*. UWB channel measurements inside different car types (Berlin, 23–27 March 2009), pp. 640–644
8. A Moghimi, H-M Tsai, C Saraydar, O Tonguz, Characterizing intra-car wireless channels. Vehic. Technol. IEEE Trans. **58**(9), 5299–5305 (2009)
9. J Blumenstein, T Mikulasek, R Marsalek, A Prokes, T Zemen, C Mecklenbrauker, in *2014 IEEE 80th Vehicular Technology Conference (VTC Fall)*. In-vehicle mm-wave channel model and measurement (Vancouver, BC, 14–17 Sept 2014), pp. 1–5
10. PC Richardson, W Xiang, W Stark, Modeling of ultra-wideband channels within vehicles. Selected Areas Commun. IEEE J. **24**(4), 906–912 (2006)
11. L Liu, Y Wang, N Zhang, Y Zhang, in *2010 12th IEEE International Conference on Communication Technology (ICCT)*. UWB channel measurement and modeling for the intra-vehicle environments (Nanjing, 11–14 Nov 2010), pp. 381–384
12. B Li, Z Zhou, D Li, S Zhai, Efficient cluster identification for measured ultra-wideband channel impulse response in vehicle cabin. Prog. Electromagnetics Res. **117**, 121–147 (2011)
13. J Li, T Talty, in *Military Communications Conference, 2006. MILCOM 2006*. Channel characterization for ultra-wideband intra-vehicle sensor networks (Washington, DC, 23–25 Oct 2006), pp. 1–5
14. R Thoma, O Hirsch, J Sachs, R Zetik, in *The Second European Conference on Antennas and Propagation, 2007. EuCAP 2007*. UWB sensor networks for position location and imaging of objects and environments (Edinburgh, 11–16 Nov 2007), pp. 1–9
15. J Blumenstein, R Marsalek, A Prokes, C Mecklenbrauker, in *Multiple Access Communcations*. Impulse noise mitigation for OFDM by time-frequency spreading, vol. 8310 (ser. Lecture Notes in Computer Science. Springer International Publishing Vilnius, Lithuania, 16–17 Dec 2013), pp. 8–20
16. J Blumenstein, T Mikulasek, R Marsalek, A Chandra, A Prokes, T Zemen, C Mecklenbrauker, in *IEEE Vehicular Networking Conference (VNC)*. In-vehicle UWB channel measurement, model and spatial stationarity (Paderborn, 3–5 Dec 2014), pp. 77–80
17. R Fontana, E Richley, J Barney, in *2003 IEEE Conference on Ultra Wideband Systems and Technologies*. Commercialization of an ultra wideband precision asset location system (Reston, VA, USA, 16–19 Nov 2003), pp. 369–373
18. R-R Lao, J-H Tarng, C Hsiao, in *The 57th IEEE Semiannual Vehicular Technology Conference, 2003. VTC 2003-Spring*. Transmission coefficients measurement of building materials for UWB systems in 3-10 GHz, vol. 1 (Jeju, Korea, 22–25 April 2003), pp. 11–14
19. Q Liang, A Audu, H Khani, H Nie, W Xiang, Z Chen, in *2013 IEEE Radio and Wireless Symposium (RWS)*. Measurement and analysis of intra-vehicle UWB channels (Austin, TX, USA, 20–23 Jan 2013), pp. 166–168
20. JD Kraus, *McGraw-Hill Education*, (New York, 1988)
21. RA Fisher, in *Mathematical Proceedings of the Cambridge Philosophical Society*. Theory of Statistical Estimation, (1925), pp. 700–725
22. S Kotz, S Nadarajah, *Extreme Value Distributions*, (World Scientific, Washington, D.C, 2000)
23. H Hashemi, The indoor radio propagation channel. Vehicular Technology, IEEE Transactions on. **81**(7), 594–606 (2002)
24. Z Sahinoglu, S Gezici, I Guvenc, vol. 2. (Cambridge University Press, Cambridge, 2008)
25. J Vychodil, J Blumenstein, T Mikulasek, A Prokes, V Derbek, in *International Conference on Connected Vehicles & Expo 2014 ICCVE 3rd*. Measurement of in-vehicle channel feasibility of ranging in UWB and MMW band (Vienna, 3–7 Nov 2014)
26. DR Lide, *CRC Handbook of Chemistry and Physics*. (CRC Press, 2001)

7

Blind identification of code word length for non-binary error-correcting codes in noisy transmission

Yasamine Zrelli[1,2†], Roland Gautier[1,2*†], Eric Rannou[1,3†], Mélanie Marazin[1,2†] and Emanuel Radoi[1,2†]

Abstract

In cognitive radio context, the parameters of coding schemes are unknown at the receiver. The design of an intelligent receiver is then essential to blindly identify these parameters from the received data. The blind identification of code word length has already been extensively studied in the case of binary error-correcting codes. Here, we are interested in non-binary codes where a noisy transmission environment is considered. To deal with the blind identification problem of code word length, we propose a technique based on the Gauss-Jordan elimination in GF(q) (Galois field), with $q = 2^m$, where m is the number of bits per symbol. This proposed technique is based on the information provided by the arithmetic mean of the number of zeros in each column of these matrices. The robustness of our technique is studied for different code parameters and over different Galois fields.

Keywords: Cognitive radio; Blind identification; Non-binary error-correcting codes; Galois field

Introduction

Error-correcting codes are frequently used in modern digital transmission systems in order to improve the communication quality. These codes are designed to achieve a good immunity against channel impairments by introducing redundancy in the informative data. Due to the complexity of both encoding and especially decoding procedures, the majority of research and practical implementations of real-time embedded systems were often restricted to encoders manipulating binary data, i.e., elements of the Galois field GF(2). Over the last decade, low-density parity check (LDPC) codes and turbo codes over GF(2) have attracted considerable interest of many researchers due to their excellent error correction capability. They have been generalized to finite fields GF(q) [1,2], where $q = 2^m$, and are among the most widely used error-correcting codes in wireless communication standards. It has been shown in [1] that non-binary LDPC codes perform generally better than binary LDPC codes and turbo codes. However, the major drawback of these codes is their decoding complexity for a large Galois field order q [3,4]. Low complexity decoding algorithms have recently been proposed [5,6], thus allowing the use of non-binary LDPC codes in practical implementations.

Our main research interests are focused on non-binary error-correcting codes in order to blindly identify their parameters. This topic is a part of a non-cooperative context like a military interception or cognitive radio applications. In this case, the receiver has no knowledge about the parameters used to encode the information at the transmitter. The solution is to design an intelligent receiver which is able to blindly identify the encoder parameters from the only knowledge of the received data stream. This blind identification function of the receiver permits to increase the data rate transmission, since it will be unnecessary to transmit supplementary information about the encoder parameters with the useful data. Such intelligent receiver is able to adapt automatically itself to the development of new high-performance coding schemes and the fast evolution of new communication standards without equipment change. In this work, we are only interested

*Correspondence: roland.gautier@univ-brest.fr
†Equal contributors
[1]Université Européenne de Bretagne, 5 Boulevard Laënnec, 35000 Rennes, France
[2]Université de Brest; CNRS, UMR 6285 Lab-STICC, 6 avenue Victor Le Gorgeu, 29238 Brest, France
Full list of author information is available at the end of the article

in blindly identifying the code word length of linear non-binary block codes. In the case of the interception, this parameter can not be transmitted. Likewise, if we want to change the encoder or get out of the list of possible choice of encoders, the code word length is not transmitted.

In this context, the published research results have been restricted so far to the blind recognition of the code word length of binary codes. To the best of our knowledge, this paper introduces, for the first time, an approach to blindly identify the code word length of non-binary codes in noisy conditions. In this work, the aim is to blindly identify the code word length from the only knowledge of received data. The authors in [7] proposed a technique of identification of non-binary LDPC parameters, but the identification is not blind because it is based on using a predefined candidate set of encoders which is known by both the transmitter and the receiver. Furthermore, this technique only works with LDPC codes unlike our proposed technique, which is general and suitable for all block codes. In our paper, the proposed blind identification technique is based on a generalization of an existing method used for binary codes. The principle of this generalization will be explained in this paper without specifying in details its detection performances. So, we present here state-of-the-art techniques to identify the code word length of binary linear block codes. The idea of these techniques is to find a basis of a dual code composed of parity check relations. For this purpose, an approach based on finding code words of small Hamming weight [8,9] was improved by Valembois [10] by using statistical hypothesis tests and recently by Cluzeau [11,12] and Côte [13]. A second approach based on linear algebra theory was introduced in [14] for noiseless channel. This approach permits to recover the length of code words by studying behaviors of the rank of matrices composed of received bits. However, the rank criterion was exploited without providing an algebraic and theoretical justification of such behavior. In [15], the use of this criterion was justified. In [16], the rank criterion approach was generalized to convolutional codes over GF(q), where $q > 2$, assuming a noiseless transmission, but it was shown that this generalized technique can be also performed to non-binary linear block codes. In noisy transmissions, a technique based on the Gauss elimination in GF(2) was applied in [17-19] to matrices composed of noisy received bits in order to find the number of almost dependent columns permitting the identification of the code word length in the case of binary error-correcting codes. Indeed, an almost dependent column of a matrix composed of noisy received symbols corresponds to a column which may be a linear combination of some preceding columns without the presence of erroneous symbols and which leads to a column that contains more zero elements after the Gauss elimination.

Compared to previous works, we demonstrate here that it is possible to generalize the blind identification technique proposed in [17,18] to non-binary block codes provided that the Galois field parameters (the cardinality and the primitive polynomial) are known by the receiver. To identify the primitive polynomial, an algorithm of identification was proposed in [20]. To achieve our purpose, it is necessary to identify the number of almost dependent columns in the matrices composed of noisy symbols of GF(q) by studying the probability of detection of these columns, denoted as P_i. In fact, the computation of P_i is essential in order to determine an optimal detection threshold. Assuming a transmission over q-ary symmetric channel with an error probability p_e, the techniques based on finding a base of a dual code [18,19] for binary codes require the knowledge of p_e, where a hard decision demodulation is considered. For this reason, we propose here an approach which is more robust because it allows us the blind identification of the code word length of non-binary and binary block codes without using the error probability p_e. This approach is based on analyzing behaviors of the arithmetic mean of the number of zeros in the columns of the matrices constructed by the Gauss elimination in GF(q). In this paper, the proposed method is a general method that should be applied to all non-binary block codes even though most examples of codes given here are non-binary LDPC codes. For this reason, the properties of LDPC codes are not exploited by our method.

This paper is organized as follows. In the 'Technical background' section, we present the encoding process of non-binary error-correcting codes. Then, the principle of the blind identification of code parameters in the noiseless case is described. The channel model used in this study is also defined and justified in this section. In the 'Blind identification of code word length in the noisy case' section, the blind identification method of the code word length in noisy environment is described. A comparison in terms of error probability and detection performances is shown in the 'Analysis and performances' section. Finally, some conclusions are drawn in the 'Conclusions' section and planned future work is pointed out.

Technical background

Non-binary error-correcting codes

The use of an efficient coding system in the transmitter as error-correcting codes is essential in order to fight disturbances present on the transmission channel. For a long time, cyclic codes such as BCH codes [21,22] and Reed-Solomon codes [23] have been the most commonly used as codes based on finite fields since they are characterized by large minimum distances for a hard decision decoding. The non-binary LDPC codes described by a sparse parity check matrix with elements in GF(q) have been developed by Davey and MacKay in 1998 [1]. Significant works on the design and the decoding complexity reduction of these codes have shown that they have a great potential to replace Reed-Solomon codes in some applications of communication, such as space communications [24], and storage systems [25,26]. In this paper, we focus on the blind identification of code word length for the non-binary block codes, but this proposed method can also be applied to convolutional codes and concatenated codes.

Let us present the encoding process of these codes over GF(q). Actually, the principle of a transmission chain is to send digital information from a source to one or more receivers. The information yielded by the source is binary data $\{0, 1\} = \text{GF}(2)$. Each block of m information bits are combined to generate a symbol of GF(q). Then, the generated non-binary information, denoted as $\mathbf{d}$, is encoded by one of the block codes over GF(q) listed above. For most block error-correcting codes, a code word, denoted as $\mathbf{c}$, composed of non-binary symbols is obtained by the multiplication of the information $\mathbf{d}$ and a non-binary generator matrix $\mathbf{G}$:

$$\mathbf{c} = \mathbf{d} \cdot \mathbf{G} \tag{1}$$

In the case of LDPC codes, the encoding process needs the use of the parity check matrix, which is always sparse compared to the other codes.

In most of the standards, such as long-term evolution (LTE) standard [27], the encoding is performed in a systematic form in order to facilitate the decoding process without degrading performances of the error correction. For this reason, in the case of block codes, the required parameters to perform the decoding operation are the number of inputs, denoted as k, the code word length, denoted as n, and a parity check matrix, denoted as $\mathbf{H}$. Indeed, the matrix $\mathbf{H}$ will be used by the decoder to detect or/and to correct the errors. The recovered information will be the first k symbols of the recovered code word due to the systematic form used in the encoding. Our aim in this research work is to blindly identify the parameter n from non-binary received symbols which are affected by noisy transmissions. In the noiseless context, we have already demonstrated in [16] that we can identify this parameter with the only knowledge of the received data, provided that the Galois field parameters are known. The principle of blind identification of the code parameter n in the noiseless case is recalled in the following subsection.

Principle of blind identification method of code word length in the noiseless case

In this part, we assume that the channel introduces no error. In [16], we have adapted the method proposed in [28] to identify the parameters of convolutional codes over GF(q), where $q = 2^m$. We have shown that our method for the noiseless case can be applied to block codes. This method reshapes row-wise the received symbols, denoted as $\mathbf{r}$, under a matrix form, denoted as $\mathbf{R}_l$, of size $(M \times l)$. Indeed, $\mathbf{R}_l$ is filled by received symbols from the top left corner to the bottom right as illustrated in Figure 1.

The number of columns l varies between 1 and $l_{\max}$ and the number of rows M which depends on l is given by the integer part $\lfloor \frac{L}{l} \rfloor$ where L is the length of a received symbol stream. Then, the rank over GF(q) is calculated for each matrix $\mathbf{R}_l$. When all matrices $\mathbf{R}_l$ have full rank, it is impossible to detect the existence of a code. Nevertheless, the redundancy introduced by the code leads to rank deficiencies in some matrices $\mathbf{R}_l$. Henceforth, the rank behaviors of $\mathbf{R}_l$ allow us to detect the code and to identify its parameters, in particular the code word length. As demonstrated in [15] and studied in [16], there are two possible rank behaviors according to the number of columns l. If l is a multiple of n (i.e., $l = \alpha \cdot n$, $\alpha \in \mathbb{N}$), the ranks of the matrices $\mathbf{R}_l$ are proportional to the code rate k/n (i.e., $\text{rank}(\mathbf{R}_l) = l \cdot k/n$). Otherwise (i.e., $l \neq \alpha \cdot n$), $\mathbf{R}_l$ have full rank (i.e., $\text{rank}(\mathbf{R}_l) = l$). Thus, the value of the rank deficiency depends on code parameters (k and n). Indeed, only two consecutive rank deficiencies are necessary to determine all code parameters. The code word length n can be determined by the difference between two values of l corresponding to two consecutive rank deficiencies of $\mathbf{R}_l$. As shown in [16], the rank method gives good results in a noiseless environment. A theoretical and algebraic study of the behavior of the rank criterion, as well as particular cases which can occur for specific parameters of

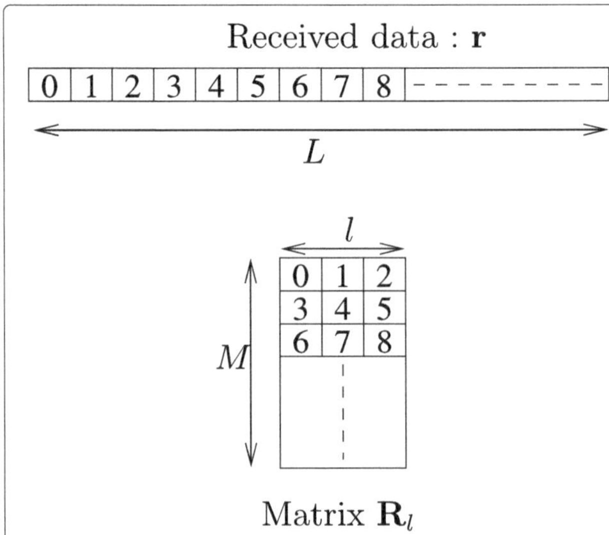

Figure 1 Example of a matrix $\mathbf{R}_l$. Example of a matrix construction from the received symbols.

codes, were presented in [15]. It was demonstrated that most matrices $\mathbf{R}_l$ have full rank when l is not a multiple of n, except for some particular cases which depend on codes (generator matrix). In a noisy environment, the rank method can not be used, since all the matrices $\mathbf{R}_l$ have full rank in this case.

Non-binary channel

In order to evaluate our blind identification algorithm, we assume that the encoded sequences are transmitted through a q-ary (non-binary, for $q = 2^m > 2$) symmetric channel (QSC) which is the simplest channel. However, our proposed algorithms can work for every type of channel provided that the error probability p_e computed at the output of the demodulator is known. Indeed, we consider that the blocks of the transmission chain, the modulator, the transmission channel, and the demodulator can be modeled by a non-binary channel, where a hard decision demodulation is considered. In a cognitive radio context, a multipath fading channel is used. This realistic channel leads to burst errors which can be corrected by using an interleaver and error-correcting codes. In this context, the errors at the output of a deinterleaver at the receiver side can be modeled by a QSC when a decoding process with hard decision will be used. The problem of a blind identification of the interleaver period, as well as a blind synchronization with the interleaver blocks was handled in [14,18].

Let us define the q-ary symmetric channel which is the generalization of the binary symmetric channel (BSC). In fact, it is a discrete memoryless channel with an error probability p_e and composed of non-binary inputs and non-binary outputs belonging to the GF(q), where $q = 2^m$. The symbols at the input of the channel are independent and distributed uniformly with a probability equal to $1/q$. A symbol $\delta \in \mathrm{GF}(q)$ at the channel input is received incorrectly with a probability $p_e/(q-1)$ [29]. In other words, it is replaced at the receiver by a different symbol β of GF(q). The probability of correctly receiving a symbol is equal to $1 - p_e$. The QSC channel is characterized by the conditional probabilities:

$$\begin{aligned} p(\tilde{r}_i = \beta | r_i = \delta) &= \frac{p_e}{q-1}, \ \delta \neq \beta \\ p(\tilde{r}_i = \delta | r_i = \delta) &= 1 - p_e \end{aligned} \tag{2}$$

where the transmitted symbol is denoted r_i, i.e., $r_i = c_i$, for $i \in \{1, \cdots, L\}$, and the noisy received symbol is denoted $\tilde{r}_i$ such that $\tilde{r}_i = r_i + e_i$ with e_i the transmission error introduced in the symbol r_i. An example of a non-binary symmetric channel for $q = 2^2$ is depicted in Figure 2.

In the following section, we present the blind identification method of the parameter n in a noisy framework.

Blind identification of code word length in the noisy case

In this part, we present the implementation method which allows us to identify the code word length of a non-binary code in a noisy environment. This method is based on the concept of finding the rank-deficient matrices among $\tilde{\mathbf{R}}_l, \forall l \in \{1, \ldots, l_{\max}\}$, corresponding to matrices having at least one almost dependent column. Indeed, the matrices $\tilde{\mathbf{R}}_l$ are reshaped in the same way as $\mathbf{R}_l$ using the noisy received symbols $\tilde{r}_i$. In [19], a method devoted to determine these matrices in the case of binary codes was presented. However, this method requires the knowledge of the error probability p_e. In order to avoid this constraint, we propose a method based on using the arithmetic mean criterion in order to detect the rank-deficient matrices which have some almost dependent columns without the need of the error probability p_e.

Principle

In a noiseless case, the rank criterion is used to find the maximum number of linearly independent columns in the matrices $\mathbf{R}_l$. This allows us to derive the number of linearly dependent columns in $\mathbf{R}_l$ (columns which are linear combinations of other columns). The finite-field Gauss

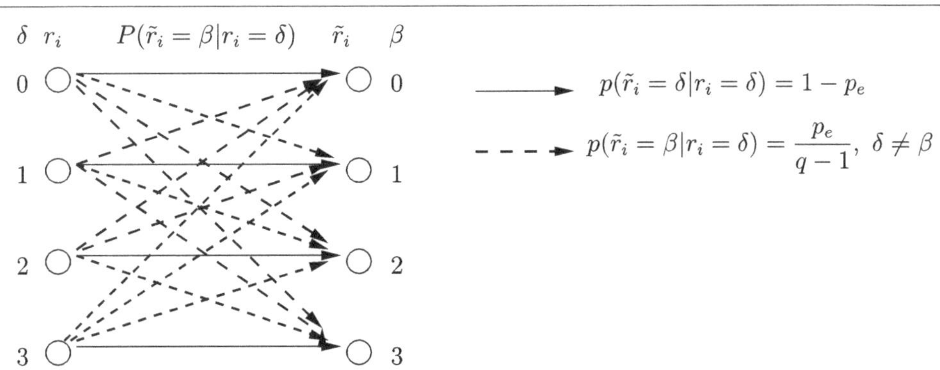

Figure 2 Non-binary symmetric channel. An example of a non-binary symmetric channel for $q = 2^2$.

elimination method [30] has to be used to eliminate those linear-dependent columns to zero. In noisy transmissions, all matrices $\tilde{\mathbf{R}}_l$ have full rank. A matrix $\tilde{\mathbf{R}}_l$ can be expressed according to $\mathbf{R}_l$ by:

$$\tilde{\mathbf{R}}_l = \mathbf{R}_l + \mathbf{E}_l \tag{3}$$

where $\mathbf{E}_l$ is the error matrix of size $(M \times l)$ constructed in the same way as $\mathbf{R}_l$ using the errors induced by the channel. Therefore, the dependence of the columns is disturbed by the presence of errors in some received symbols. In such context, the authors in [17,18] proposed to look for the number of almost dependent columns in the matrices composed of noisy received bits by using the Gauss elimination over GF(2). Inspired by this idea, it is sufficient, in the case of non-binary error correcting codes, to apply the finite-field Gauss elimination in GF(q) to $\tilde{\mathbf{R}}_l$ in order to obtain a new matrix $\tilde{\mathbf{T}}_l$ of size $(M \times l)$. This algorithm gives also at output a matrix of size $(l \times l)$, denoted $\tilde{\mathbf{A}}_l$, that describes the combination operations performed to the columns of the matrix $\tilde{\mathbf{R}}_l$ in order to obtain the transformation matrix $\tilde{\mathbf{T}}_l$. A recall of the finite-field Gauss elimination over GF(q) is presented in Algorithm 1. To describe this algorithm, we denote $\mathbf{I}_l$ the identity matrix of size $(l \times l)$, $\mathbf{x}_i^{(l)}$ the i-th column of a given matrix $\mathbf{X}_l$ and $x_i^{(l)}(j)$ a coefficient of a matrix $\mathbf{X}_l$ placed in the i-th column and in the j-th row.

Algorithm 1 The finite-field Gauss elimination over GF(q).

Require: $\tilde{\mathbf{R}}_l$
Ensure: $\tilde{\mathbf{T}}_l$ and $\tilde{\mathbf{A}}_l$
  Initialization: $\tilde{\mathbf{T}}_l \leftarrow \tilde{\mathbf{R}}_l$ and $\tilde{\mathbf{A}}_l \leftarrow \mathbf{I}_l$
  for $i = 1$ to l **do**
    if the i-th element of the i-th column $\tilde{t}_i^{(l)}(i) = 0$ **then**
      Permute the i-th column $\tilde{\mathbf{t}}_i^{(l)}$ with the first column $\tilde{\mathbf{t}}_{i'}^{(l)}$ $(i' > i)$ that has a non-zero on its i-th element
      Permute the column $\tilde{\mathbf{a}}_i^{(l)}$ with $\tilde{\mathbf{a}}_{i'}^{(l)}$
    end if
    Multiply the columns $\tilde{\mathbf{t}}_i^{(l)}$ and $\tilde{\mathbf{a}}_i^{(l)}$ by $v = 1/\tilde{t}_i^{(l)}(i)$ in order to have $\tilde{t}_i^{(l)}(i) = 1$
    for $j = i + 1$ to l **do**
      Let $b = \tilde{t}_j^{(l)}(i)$. Apply the following operation to the columns $\tilde{\mathbf{t}}_i^{(l)}$ and $\tilde{\mathbf{t}}_j^{(l)}$ in order to have $\tilde{t}_j^{(l)}(i) = 0$:

$$\tilde{\mathbf{t}}_j^{(l)} = \tilde{\mathbf{t}}_j^{(l)} - b \cdot \tilde{\mathbf{t}}_i^{(l)}$$

      Apply the following operations to the columns of the matrix $\mathbf{A}_l$:

$$\tilde{\mathbf{a}}_j^{(l)} = \tilde{\mathbf{a}}_j^{(l)} - b \cdot \tilde{\mathbf{a}}_i^{(l)}$$

    end for
  end for

By means of this algorithm, the linear-dependent columns in the matrix will be eliminated to zeros. The whole matrix is considered in our proposed method instead of only the lower part of the matrix $\tilde{\mathbf{R}}_l$ as mentioned in [17]. It would be more accurate than assuming that errors do not occur in the upper part of the matrix, but it is not the real case.

We can note that the finite-field Gauss elimination over GF(q) can be defined by a linear application given by:

$$\tilde{\mathbf{R}}_l \cdot \tilde{\mathbf{A}}_l = \tilde{\mathbf{T}}_l \tag{4}$$

In noiseless transmissions, the number of dependent columns in $\mathbf{R}_l$, for $l = \alpha \cdot n$, $\alpha \in \mathbb{N}$, corresponds to the number of the zero columns in the matrix $\mathbf{T}_l$ which is the result of the transformation of $\mathbf{R}_l$ by the finite-field Gauss elimination in GF(q) ($\mathbf{R}_l \cdot \mathbf{A}_l = \mathbf{T}_l$). The matrix form of $\mathbf{T}_l$ is described in Figure 3.

In fact, the dimension identification of a vector space generated by a code $\mathcal{C}$ is equivalent to finding the

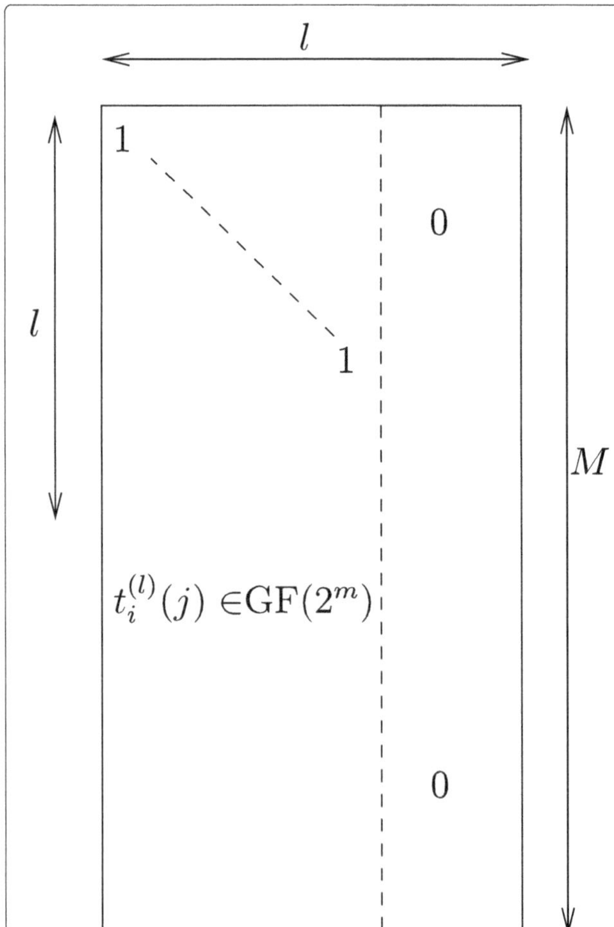

Figure 3 Matrix form of $\mathbf{T}_l$. An example of a matrix form of $\mathbf{T}_l$ for $l = \alpha \cdot n$.

dimension of a vector space generated by its dual code $\mathcal{C}^{\perp}$. For any vector $\mathbf{h}$ belonging to $\mathcal{C}^{\perp}$ and for any code word $\mathbf{r}$ of $\mathcal{C}$, the relation between both is defined by $\mathbf{r} \cdot \mathbf{h}^T = 0$. In noiseless conditions, the matrix $\mathbf{R}_n$, for $l = n$, which is composed of M code words of length n, should satisfy:

$$\mathbf{R}_n \cdot \mathbf{h}^T = \mathbf{0} \tag{5}$$

We can note that $\mathbf{h}$ belongs to the kernel of $\mathbf{R}_n$, denoted as $\ker(\mathbf{R}_n)$. So, we have $\mathcal{C}^{\perp} \subset \ker(\mathbf{R}_n)$. Since the dependent columns in $\mathbf{R}_l$ multiplied by the columns $\mathbf{a}_i^{(l)}$ permit to have the zero columns in the matrix $\mathbf{T}_l$, the corresponding columns $\mathbf{a}_i^{(n)}$ will belong to $\ker(\mathbf{R}_n)$ in which the dual code $\mathcal{C}^{\perp}$ is contained. Therefore, finding the dependent columns in $\mathbf{R}_l$ is equivalent to finding the columns $\mathbf{a}_i^{(l)}$ which belong to the dual code $\mathcal{C}^{\perp}$.

Due to the presence of errors induced by the channel in $\tilde{\mathbf{R}}_l$, for $l = \alpha \cdot n$, the columns of $\tilde{\mathbf{T}}_l$ corresponding to the almost dependent columns in $\tilde{\mathbf{R}}_l$ will contain some non-zero symbols. Assuming that the first l rows and the pivots of the matrix $\tilde{\mathbf{T}}_l$ do not contain transmission errors, using (3) and (4) allows us to write the matrix $\tilde{\mathbf{T}}_l$ as:

$$\tilde{\mathbf{T}}_l = \mathbf{T}_l + \mathbf{E}_l \cdot \mathbf{A}_l \tag{6}$$

In this case, a vector $\mathbf{h}$ is a parity check relation (i.e., $\mathbf{h} \in \mathcal{C}^{\perp}$) with high probability if the relation $\tilde{\mathbf{R}}_l \cdot \mathbf{h}^T$ has a low Hamming weight [11]. However, the opposite is not necessarily true. We can conclude that $\tilde{\mathbf{a}}_i^{(l)}$ belongs to $\mathcal{C}^{\perp}$ if the corresponding $\tilde{\mathbf{t}}_i^{(l)} = \tilde{\mathbf{R}}_l \cdot \tilde{\mathbf{a}}_i^{(l)}$ has a small Hamming weight. In GF(q), the Hamming weight of a vector is the number of non-zero elements in this vector. So, our aim is to determine the columns $\tilde{\mathbf{t}}_i^{(l)}$ which have a high number of zeros. The idea is to study the number of zeros in the columns of the $\tilde{\mathbf{T}}_l$ in order to detect the almost dependent columns in $\tilde{\mathbf{R}}_l$.

Behaviors of the number of zeros in the columns of $\tilde{\mathbf{T}}_l$

Let $B_l(i)$ be the number of zeros in the i-th column of $\tilde{\mathbf{T}}_l$, $\tilde{\mathbf{t}}_i^{(l)}$. Hence, the variable $B_l(i)$ has two behaviors depending on whether the column $\tilde{\mathbf{a}}_i^{(l)}$ belongs to the dual code $\mathcal{C}^{\perp}$ or not. This variable will be studied as a function of $\tilde{\mathbf{a}}_i^{(l)}$ assuming that the bits that represent an element of the GF(q), where $q = 2^m$, are uniformly distributed and independent from each other.

- If the column $\tilde{\mathbf{a}}_i^{(l)}$ does not belong to the dual code $\mathcal{C}^{\perp}$, the variable $B_l(i)$, for all $i \in [\![1, l]\!]$, will follow a binomial distribution of parameters M and $1/q$ with a mean equal to M/q, denoted as $\mathcal{B}(M, 1/q)$.

- If the column $\tilde{\mathbf{a}}_i^{(l)}$ belongs to the dual code $\mathcal{C}^{\perp}$, the variable $B_l(i)$ will follow a binomial distribution with parameters M and P_i, denoted as $\mathcal{B}(M, P_i)$. The parameter P_i corresponds to the probability that a coefficient $\tilde{t}_i^{(l)}(j)$ of the column $\tilde{\mathbf{t}}_i^{(l)}$ is equal to 0 $\left(\text{i.e., } P_i = Pr\left[\tilde{t}_i^{(l)}(j) = 0 \mid \tilde{\mathbf{a}}_i^{(l)} \in \mathcal{C}^{\perp}\right]\right)$.

It is possible to limit the two behaviors of the variable $B_l(i)$ by computing an optimal threshold $\hat{\eta}_{\text{opt}}$ such that:

$$\begin{cases} \text{If } B_l(i) > \hat{\eta}_{\text{opt}} \text{ then } \tilde{\mathbf{a}}_i^{(l)} \in \mathcal{C}^{\perp} \\ \text{If } B_l(i) \le \hat{\eta}_{\text{opt}} \text{ then } \tilde{\mathbf{a}}_i^{(l)} \notin \mathcal{C}^{\perp} \end{cases} \tag{7}$$

where $\hat{\eta}_{\text{opt}} = \frac{M}{q} \cdot \eta_{\text{opt}}$ is a real in the interval $[0, M]$. The optimal threshold η_{opt} is able to minimize the probability of wrong detection of a column $\tilde{\mathbf{a}}_i^{(l)} \in \mathcal{C}^{\perp}$, denoted as P_{wd}, which corresponds to the sum of the false alarm probability, denoted as P_{fa}, and the probability of not detecting a theoretical dependent column, denoted as P_{nd}. The optimal threshold is determined by:

$$\begin{aligned} \eta_{\text{opt}} &= \arg\min_{\eta} (P_{\text{wd}}) = \arg\min_{\eta} (P_{\text{nd}} + P_{\text{fa}}) \\ &= \arg\min_{\eta} \left(1 + \sum_{j=\left\lfloor \frac{M}{q} \cdot \eta \right\rfloor + 1}^{M} \binom{M}{j} \cdot \right. \\ &\quad \left. \left[q^{-M} \cdot (q-1)^{M-j} - P_i^j \cdot (1-P_i)^{M-j}\right]\right) \end{aligned} \tag{8}$$

The normal distribution can be used to approximate the binomial probabilities of $B_l(i)$ when M is large:

- If $\tilde{\mathbf{a}}_i^{(l)} \in \mathcal{C}^{\perp}$:

$$B_l(i) \to \mathcal{N}\left(\mu_0, \sigma_0^2\right) \tag{9}$$

- If $\tilde{\mathbf{a}}_i^{(l)} \notin \mathcal{C}^{\perp}$:

$$B_l(i) \to \mathcal{N}\left(\mu_1, \sigma_1^2\right) \tag{10}$$

where $\mathcal{N}\left(\mu_0, \sigma_0^2\right)$ is the normal distribution of parameters $\mu_0 = M \cdot P_i$ and $\sigma_0^2 = M \cdot P_i \cdot (1 - P_i)$ and $\mathcal{N}\left(\mu_1, \sigma_1^2\right)$ corresponds to the normal distribution of parameters $\mu_1 = M/q$ and $\sigma_1^2 = M \cdot (q-1)/q^2$.

Henceforth, the optimal value of the threshold $\hat{\eta}$ minimizing the probability of wrong detection P_{wd} can be computed by:

$$\hat{\eta}_{\text{opt}} = \arg\min_{\hat{\eta}} \left(1 - \phi\left(\frac{\hat{\eta} - \mu_1}{\sigma_1}\right) + \phi\left(\frac{\hat{\eta} - \mu_0}{\sigma_0}\right)\right) \tag{11}$$

where $\phi(x)$ is the cumulative density function of the standard normal distribution:

$$\phi(x) = \frac{1}{\sqrt{2 \cdot \pi}} \cdot \int_{-\infty}^{x} e^{-\frac{t^2}{2}} \cdot dt \tag{12}$$

We can note that the optimal threshold $\hat{\eta}_{\text{opt}}$ depends on the parameters: M, q, and P_i. So, in order to delimit the two behaviors of the variable $B_l(i)$, it is necessary to compute the probability P_i.

Computation of the probability P_i

In the case of binary codes, the probability P_i has been calculated in [11]. But, it has never been studied in the general case of codes over GF(2^m). In fact, the computation of the parameter P_i is essential in order to detect the almost dependent columns in $\tilde{\mathbf{R}}_l$ by delimiting the two behaviors of the variable $B_l(i)$. Our aim is to investigate this probability in the case of non-binary codes. In the following, the theoretical study of P_i is presented.

For $l = n$ and i a position of a column $\tilde{\mathbf{a}}_i^{(l)}$ contained in $\mathcal{C}^{\perp}$, a coefficient $\tilde{t}_i^{(l)}(j)$ of the column $\tilde{\mathbf{t}}_i^{(l)}$ can be obtained, using (6), by:

$$\tilde{t}_i^{(l)}(j) = t_i^{(l)}(j) + \sum_{k=1}^{n} a_i^{(l)}(k) \cdot e_k^{(l)}(j) = \sum_{k=1}^{n} a_i^{(l)}(k) \cdot e_k^{(l)}(j) \tag{13}$$

where $t_i^{(l)}(j) = 0$ in the case of noiseless transmissions as explained previously. Indeed, the sum $\sum_{k=1}^{n} a_i^{(l)}(k) \cdot e_k^{(l)}(j)$ is null in this case because $e_k^{(l)}(j) = 0, \forall k \in \{1, \cdots, n\}$, and $\forall j \in \{1, \cdots, M\}$. However, in the case of noisy transmissions, the coefficients $e_i^{(l)}(j) \in \text{GF}(q)$ corresponds to the errors introduced by the noisy channel in the symbols $r_i^{(l)}(j) \in \text{GF}(q)$ in order to generate the noisy symbols $\tilde{r}_i^{(l)}(j) \in \text{GF}(q)$. Our aim is to determine P_i the probability of detecting a zero coefficient in the column $\tilde{\mathbf{t}}_i^{(l)}$ corresponding to having $\sum_{k=1}^{n} a_i^{(l)}(k) \cdot e_k^{(l)}(j) = 0$:

$$P_i = Pr\left[\sum_{k=1}^{n} a_i^{(l)}(k) \cdot e_k^{(l)}(j) = 0\right] \tag{14}$$

Let $N_i(l)$ be the minimum number of linear combinations of columns required to obtain $\tilde{\mathbf{t}}_i^{(l)}$. This number corresponds also to the Hamming weight of the column $\tilde{\mathbf{a}}_i^{(l)}$. Then, there could be positions among $N_i(l)$ where $e_i^{(l)}(j) = 0$. Thus, P_i can be defined as the probability of having $\sum_{k=1}^{s} a_i^{(l)}(k) \cdot e_k^{(l)}(j) = 0$ such that s is the number of positions among $N_i(l)$ where $e_i^{(l)}(j) \neq 0$:

$$P_i = Pr\left[X = 0\right] + \sum_{s=1}^{N_i(l)} Pr\left[X = s, \sum_{k=1}^{s} a_i^{(l)}(k) \cdot e_k^{(l)}(j) = 0\right] \tag{15}$$

where X is a random variable of the erroneous positions number among $N_i(l)$. Indeed, we show in Appendix that the probability P_i of having $\tilde{t}_i^{(l)}(j) = 0$ can be determined by:

$$P_i = \frac{1 + (q-1) \cdot \left(1 - \frac{p_e \cdot q}{q-1}\right)^{N_i(l)}}{q} \tag{16}$$

In the case of GF(2) (i.e., $q = 2$), this probability can be written as:

$$P_i = \frac{1 + (1 - 2 \cdot p_e)^{N_i(l)}}{2} \tag{17}$$

This expression corresponds to that used in [11].

In Figure 4, we represent the wrong detection probability P_{wd} as a function of $\hat{\eta}/M$ and p_e assuming $q = 2^3$, $w\left(\tilde{\mathbf{a}}_i^{(l)}\right) = 20$ and $M = 2,000$. For each value of p_e, the optimal threshold $\hat{\eta}_{\text{opt}}$ corresponding to a root of (11) is computed. From Figure 4, we can deduce that the threshold interval satisfying $P_{\text{wd}} \approx 0$ decreases when the value of p_e increases.

We can conclude that studying the behaviors of $B_l(i)$ in order to identify n is based on the calculation of the optimal threshold $\hat{\eta}_{\text{opt}}$. However, this threshold depends on the value of the error probability p_e which is unknown for the receiver. So, the need to estimate this parameter is a blocking step in the almost dependent columns method and also leads to a lack of robustness.

In order to address these problems, we propose a new iterative method based on the arithmetic mean of the variable $B_l(i)$ which do not depend on p_e and where the iterative process permits to improve the detection probability.

New iterative method based on the arithmetic mean of the variable $B_l(i)$

In this part, the proposed method based on the arithmetic mean of the number of zeros in the columns of the matrix $\tilde{\mathbf{T}}_l$ is described. We recall that the Gauss elimination described in Algorithm 1 should be applied in order to obtain $\tilde{\mathbf{T}}_l$. We show here that the identification of the parameter n by our proposed method does not depend on the error probability p_e. In this method, in order to improve the detection probability of n, an iteration process is introduced. We consider the idea of the iterative process proposed in [18,19]. The principle of this process is to perform random permutations on the rows of the matrix $\tilde{\mathbf{R}}_l$ in order to obtain a new virtual realization of the received data. These permutations permit to increase the probability to obtain non-erroneous pivots during the Gauss elimination.

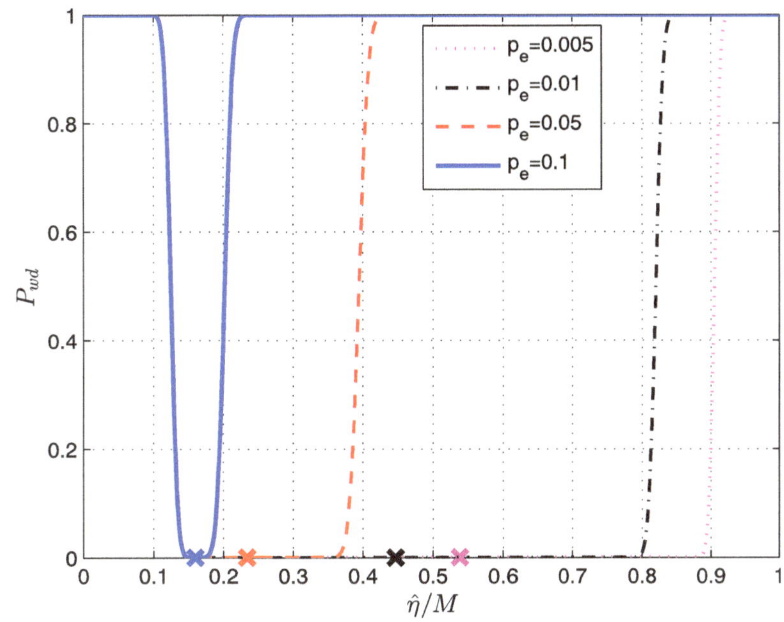

Figure 4 Probability P_{wd} according to $\hat{\eta}/M$ and p_e. The probability of wrong detection of almost dependent columns P_{wd} according to $\hat{\eta}/M$ and p_e is represented for $q = 2^3$, $M = 2{,}000$ and $w\left(\tilde{\mathbf{a}}_i^{(l)}\right) = 20$.

The arithmetic mean of the variables $B_l(i)$, $\forall i \in [\![1, l]\!]$, denoted E_l is defined by:

$$E_l = \frac{\sum_{i=1}^{l} B_l(i)}{l} \quad (18)$$

Property 1. *If $X_1, X_2, \cdots, X_m$ are independent random variables respectively following:*

$$\mathcal{N}\left(\mu_1, \sigma_1^2\right), \mathcal{N}\left(\mu_2, \sigma_2^2\right), \ldots, \mathcal{N}\left(\mu_m, \sigma_m^2\right)$$

the mean defined by $\frac{(X_1+X_2+\cdots+X_m)}{m}$ follows:

$$\mathcal{N}\left(\frac{\mu_1+\mu_2+\cdots+\mu_m}{m}, \frac{\sigma_1^2+\sigma_2^2+\cdots+\sigma_m^2}{m^2}\right) \quad (19)$$

We recall that the variable $B_l(i)$ which is the number of zeros in the i-th column of the matrix $\tilde{\mathbf{T}}_l$ has two possible behaviors depending on l:

- If $l \neq \alpha \cdot n$, for $\alpha \in \mathbb{N}$, the variable $B_l(i)$ follows a normal distribution $\mathcal{N}\left(\mu_1, \sigma_1^2\right)$ for all columns i of $\tilde{\mathbf{T}}_l$. In this case, using the property 1, the mean E_l will follow:

$$E_l \rightarrow \mathcal{N}\left(\mu_1, \frac{\sigma_1^2}{l}\right) \quad (20)$$

  We can note that the mean E_l will be close to M/q.

- If $l = \alpha \cdot n$, for $\alpha \in \mathbb{N}$:
  - If the i-th column is an almost dependent column, the variable $B_l(i)$ will follow the normal distribution of parameters $\mathcal{N}\left(\mu_0, \sigma_0^2\right)$.
  - If the i-th column is not an almost dependent column, the variable $B_l(i)$ will follow the normal distribution of parameters $\mathcal{N}\left(\mu_1, \sigma_1^2\right)$.

  Thereby, the mean E_l is given by:

$$E_l \rightarrow \mathcal{N}\left(\frac{Q(l)\cdot\mu_0 + k_l\cdot\mu_1}{l}, \frac{Q(l)\cdot\sigma_0^2 + k_l\cdot\sigma_1^2}{l^2}\right) \quad (21)$$

  where $Q(l)$ is the number of almost dependent columns in the matrix $\tilde{\mathbf{R}}_l$ such that:

$$Q(l) = \mathrm{Card}\left\{i \in [\![0, l]\!], B_l(i) > \hat{\eta}_{\mathrm{opt}}\right\} \quad (22)$$

  where Card(x) is the cardinal function which returns the set size. $k_l = l - Q(l)$ is the number of independent columns in the same matrix. In the noiseless environment, the mean E_l is stable at:

$$E_l = \frac{M\cdot(q\cdot(n-k)+k)}{q\cdot n} \quad (23)$$

We note two behaviors of E_l with respect to $l = \alpha \cdot n$ or $l \neq \alpha \cdot n$:

$$\begin{cases} \text{If } l \neq \alpha \cdot n \text{ then } E_l \leq \dfrac{M}{q} \\ \text{If } l = \alpha \cdot n \text{ then } E_l > \dfrac{M}{q} \end{cases} \quad (24)$$

The gap between these behaviors allows us to find the matrices which have the number of columns $l = \alpha \cdot n$.

Let $\mathcal{J}$ be a set of l-values where the gap $E_l - \frac{M}{q} > 0$:

$$\mathcal{J} = \left\{ l = 1, \cdots, l_{\max} | E_l - \frac{M}{q} > 0 \right\} \quad (25)$$

Thereby, the identified length of the code words will be such that:

$$\tilde{n} = \text{mode}(\text{diff}(\mathcal{J})) \quad (26)$$

where the functions diff($\mathbf{x}$) and mode($\mathbf{x}$) are defined by:

- Function diff($\mathbf{x}$): the output of this function is a vector of size $s - 1$ and it corresponds to the difference between two consecutive elements of the vector $\mathbf{x} = \left(x(1)\ x(2)\ \cdots\ x(s)\right)$:

$$\text{diff}(\mathbf{x}) = \left(x(2) - x(1)\ \cdots\ x(s) - x(s-1)\right) \quad (27)$$

- Function mode($\mathbf{x}$): this operation provides the value which has the highest occurrence in the vector $\mathbf{x}$.

The proposed iterative method of the code word length identification is summarized in the Algorithm 2.

Algorithm 2 The algorithm based on the arithmetic mean calculation

Require: $\tilde{\mathbf{r}}$, M, q and maximum number of iterations $\text{it}_{\max}$
Ensure: Identified code word length $\tilde{n}$
  Initialize the number of iterations $\text{it} = 1$
  Initialize the stop criterion $\text{end}_{\text{it}} = 0$
  while $\text{end}_{\text{it}} = 0$ **do**
    for $l = 1$ to $l_{\max}$ **do**
      Build matrix $\tilde{\mathbf{R}}_l$ of size $(M \times l)$
      $\tilde{\mathbf{R}}_l \rightarrow \tilde{\mathbf{T}}_l = \tilde{\mathbf{R}}_l \cdot \tilde{\mathbf{A}}_l$
      for $i = 1$ to l **do**
        Count $B_l(i)$
      end for
      Compute E_l
    end for
    Create the set $\mathcal{J}$ (25)
    if $\mathcal{J}$ is empty **then**
      if $\text{it} < \text{it}_{\max}$ **then**
        $\text{it} = \text{it} + 1$
        Permute randomly the rows of $\tilde{\mathbf{R}}_l$
      else
        $\text{end}_{\text{it}} = 1$
      end if
    else
      Determine $\tilde{n}$ (26)
      $\text{end}_{\text{it}} = 1$
    end if
  end while

Example 1. Let us consider the Reed-Solomon code, denoted $RS(15, 11)$, over GF(2^4) which is defined by: $n = 15$ and $k = 11$. The mean E_l normalized by M, which is set to 1,000, is represented in Figures 5 and 6. In Figure 5, a zero probability of error (i.e., $p_e = 0$) is considered. For

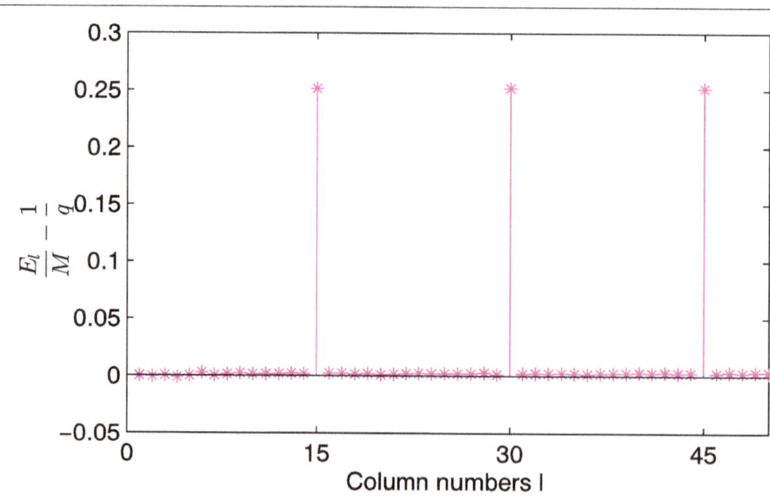

Figure 5 Gap between the mean E_l/M and $1/q$ of $RS(15, 11)$ over GF(2^4) for $p_e = 0$. The mean E_l normalized by $M = 1,000$ is represented in the case of $p_e = 0$.

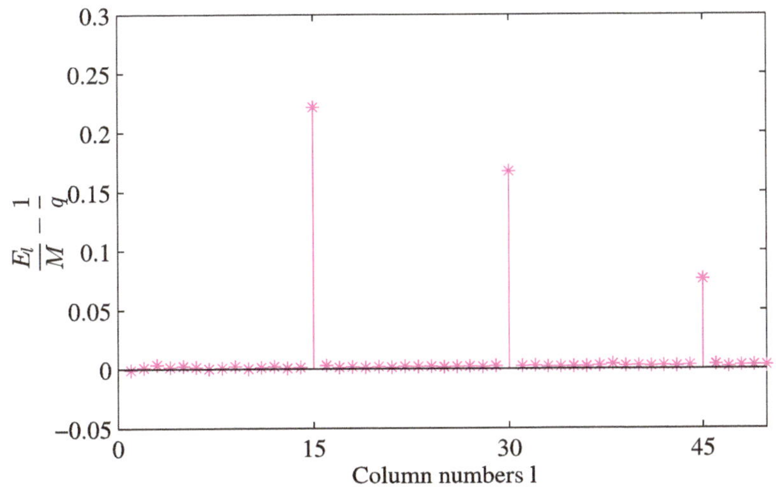

Figure 6 Gap between the mean E_l/M and $1/q$ of $RS(15, 11)$ over GF(2^4) for $p_e = 0.01$. The mean E_l normalized by $M = 1,000$ is represented in the case of $p_e = 0.01$.

$l \neq \alpha \cdot n$, we can verify that the mean E_l normalized by M is stable at $1/q = 0.0625$. For $l = \alpha \cdot n$, the mean E_l meets (23):

$$\frac{1}{q \cdot n} \cdot (q \cdot (n-k) + k) = 0.3125$$

So, the matrices of size $l = \alpha \cdot n$ have peaks for $\frac{E_l}{M} - \frac{1}{q} = 0.25 > 0$. In Figure 6, the gap $\frac{E_l}{M} - \frac{1}{q}$ is represented with respect to l when $p_e = 0.01$ for one iteration of our algorithm. According to (25), the set $\mathcal{J}$ is shown in Table 1. Henceforth, using (26), the identified length of the code words is $\tilde{n} = 15$.

Analysis and performances

The aim of our proposed algorithm is to blindly identify the length of non-binary code words in noisy environment. This purpose can be reached with an average complexity equal to $\mathcal{O}(M \cdot l_{\max}^3 \cdot \text{it}_{\max})$. Indeed, the proposed algorithm performs $((l_{\max} - 1) \cdot \text{it}_{\max})$ processes of Gaussian eliminations which have an average complexity equal to $\mathcal{O}(M \cdot l^2)$, where $l = 2 \cdots l_{\max}$. So, the average complexity is such that:

$$\mathcal{O}\left(M \cdot \sum_{\text{it}=1}^{\text{it}_{\max}} \sum_{l=2}^{l_{\max}} l^2\right) = \mathcal{O}(M \cdot l_{\max}^3 \cdot \text{it}_{\max}) \quad (28)$$

Table 1 Sizes of the matrices $\tilde{R}_l$ for $\frac{E_l}{M} - \frac{1}{q} > 0$

$l \in \mathcal{J}$	15	30	45
diff($\mathcal{J}$)		15	15

The sizes of matrices for which $\frac{E_l}{M} - \frac{1}{q} > 0$, $\mathcal{J}$, and the set diff($\mathcal{J}$) are given for $p_e = 0.01$ in the case of $RS(15, 11)$ over GF(2^4).

In order to analyze the performances of our blind identification method, the probability of correct detection of the code word length n is chosen as a performance criterion. In the simulations, our method is applied to the non-binary LDPC codes which became candidate for future communication systems. For each simulation, 2,000 Monte Carlo trials are run where the data symbols are randomly chosen at each trial. In this part, we focus on:

- the gain of the iteration process on the detection probability of n
- the performance comparison in the case of different channels
- the impact of increasing the Galois field dimension q on the detection probabilities of n
- the impact of increasing the code word length n on the detection probabilities for a given q

Gain of the iterative process

In our simulations, we consider a LDPC ($n = 6$, $k = 3$) over GF(4). Figure 7 shows the probability of detecting n according to p_e for one, three, five, and ten iterations. We can see that the gain between the first and the tenth iteration is significantly important. Indeed, for $p_e = 0.07$, with one iteration, the detection probability is equal to 0.76 and it becomes equal to 0.99 after 10 iterations. We can deduce that the iterative process improves significantly the detection performances of the blind identification method based on the mean calculation.

Performance comparison in the case of different channels

Let us illustrate the detection obtained by the proposed method for a LDPC ($n = 16, k = 8$) over GF(8) when an AWGN channel (the first channel) and a multipath Rayleigh channel associated to an AWGN channel (the

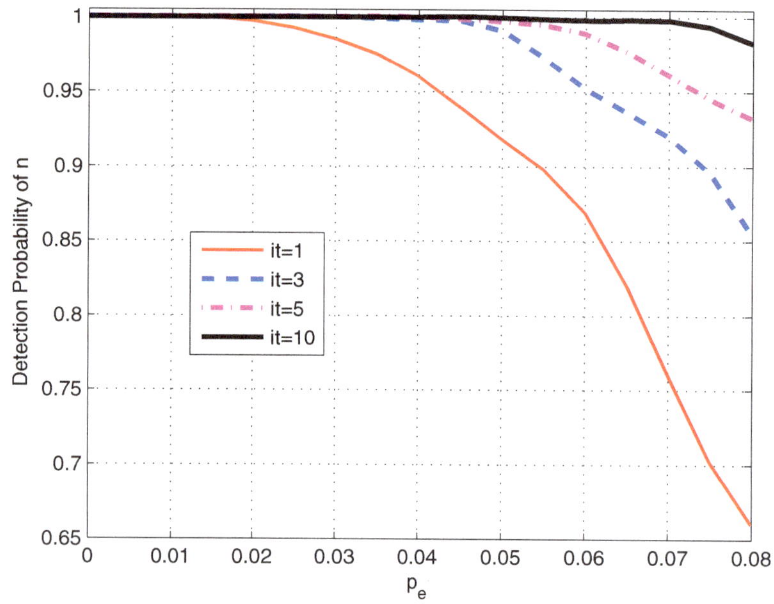

Figure 7 The detection probability of the method based on the mean calculation for LDPC ($n = 6, k = 3$) in GF(4). For LDPC ($n = 6, k = 3$) in GF(4), the probability of detecting n is depicted compared with the error probability p_e for one, three, five, and ten iterations.

second channel) are considered. In order to compensate and reduce the inter-symbol interference (ISI) caused by the multipath propagation, a linear mean square error (MSE) equalizer of length 20 was used.

We evaluate the performances of our method when the QAM or PAM modulation of order 8 (8-QAM and 8-PAM) is used to transmit the symbols coded by LDPC ($n = 16, k = 8$) over GF(8). In Figures 8 and 9, a comparison of performances of our blind identification method using 8-PAM or 8-QAM modulations in the case of an AWGN channel and a multipath channel with path number $L_{\text{path}} = 4$ and $\text{it}_{\text{max}} = 1$ is presented. In Figure 8, a comparison of the detection performances of our method in the case of AWGN channel is depicted. We can see that the proposed method for 8-QAM modulation gives better performances than for 8-PAM modulation when SNR<

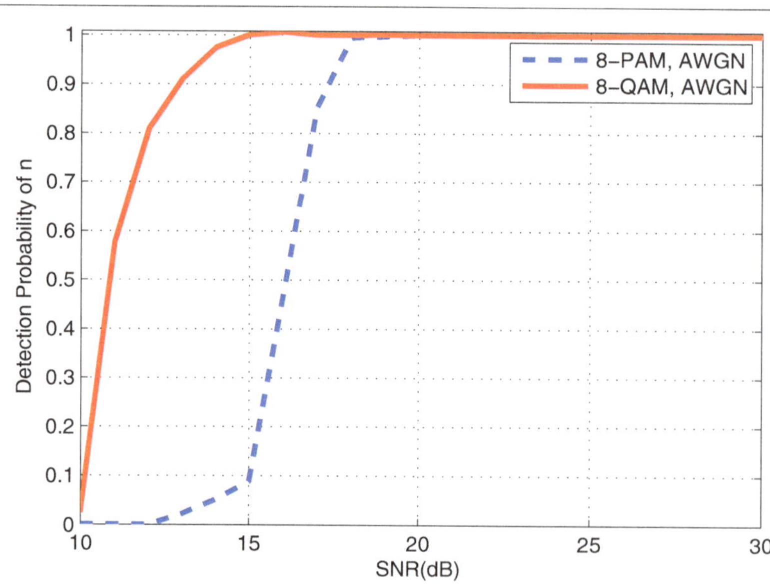

Figure 8 The detection probability of the method based on the mean calculation for LDPC ($n = 16, k = 8$) in the case of AWGN channel. For LDPC ($n = 16, k = 8$) over GF(8), the detection probabilities of the method based on the mean calculation in the case of AWGN channel are depicted when a 8-PAM and 8-QAM modulations are used.

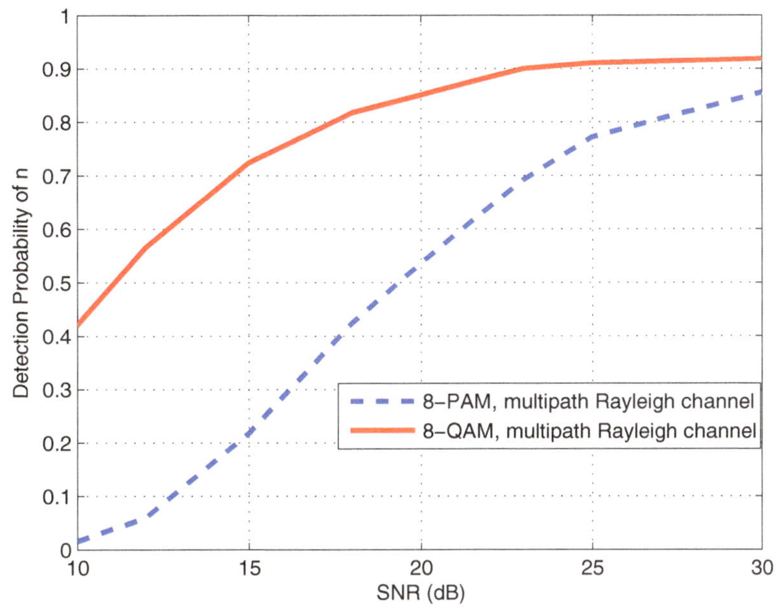

Figure 9 The detection probability of the method based on the mean calculation in the case of multipath Rayleigh channel. For LDPC ($n = 16, k = 8$) over GF(8), the detection probabilities of our method are depicted in the case of multipath Rayleigh channel with $L_{\text{path}} = 4$ when 8-PAM and 8-QAM modulations are used.

18 dB. The gain between both is equal to 5 dB. However, for SNR > 18 dB, the performances are similar and the detection probability is equal to 1. To obtain the detection probabilities presented in Figure 9, the modulated symbols by 8-PAM or 8-QAM modulations are transmitted in a quasi-static Rayleigh fading multipath channel with path number $L_{\text{path}} = 4$, then the received symbols are treated by the linear MSE equalizer of length 20. We can observe that, in the case of 8-QAM, our proposed method provides better performances than for 8-PAM. A gain equal to 5 dB is exhibited. We have chosen to evaluate our proposed methods in the worst case of 8-PAM modulation because our aim was to show that our method has the best performances even in the case of the PAM modulation.

In the following, the performance study of the impact of n and q on the proposed method is presented.

Impact of increasing q

Let us consider a LDPC ($n = 6$, $k = 3$), constructed in the Galois field GF(q), where $q = 4, 8, 16$. The matrices $\tilde{\mathbf{R}}_l$ are reshaped from $L = 30,000$ received symbols with $l = 2, \cdots, 30$ and $M = 1,000$. For each value of q, the method based on the mean calculation is applied to blindly identify the code word length of LDPC ($n = 6$, $k = 3$) over GF(q) when $\text{it}_{\max} = 1$. Figure 10 depicts the probability of detecting the correct n by our blind identification method according to the error probability p_e in the cases of GF(4), GF(8), and GF(16). This figure shows that the curve behavior is nearly similar for all $q = 4, 8, 16$. We can deduce that the method based on the mean calculation is slightly sensitive to the increase of the Galois field dimension q.

Impact of increasing n

To evaluate the detection performances of our blind identification method, the impact of increasing the code word length should be studied. In our simulations, we consider two LDPC codes over GF(8), a LDPC ($n = 6$, $k = 3$) and a LDPC ($n = 16$, $k = 8$). The matrices $\tilde{\mathbf{R}}_l$ are reshaped from $L = 64,000$ received symbols with $l = 2, \cdots, 64$ and $M = 1,000$. For each code, the method based on the mean calculation is applied to blindly identify the code word length n when $\text{it}_{\max} = 1$. Figure 11 shows the detection probabilities of n by the method based on the mean calculation. We can note that the increase of the code word length leads to lower detection performances with our proposed method. Indeed, for $p_e = 0.01$, the detection probability of the method of the mean calculation is constant and equal to 1 in the case of the two codes. For $p_e = 0.02$, the detection probability decreases from 0.99 to 0.94.

In order to show that our method works in the case of codes of a reasonable code word length, we computed the detection probability of the Reed-Solomon code RS ($n = 31, k = 25$) over GF(32) which corresponds to an equivalent code over GF(2) of length $m \cdot n = 5 \cdot 31 = 155$. For an error probability $p_e = 0.01$ and 1,000 trials

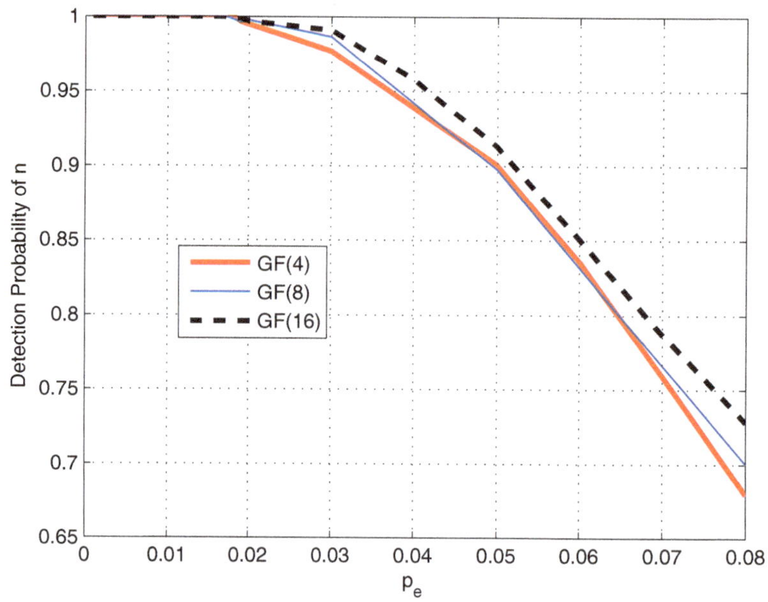

Figure 10 Impact of the Galois field dimension q on the detection probability of n by the proposed method considering $\text{it}_{max} = 1$. For LDPC ($n = 6, k = 3$), the probability of detecting the correct n by the method based on the arithmetic mean computation is depicted according to the error probability p_e in the cases of GF(4), GF(8), and GF(16).

of Monte Carlo, we obtained a detection probability of 0.87 for $\text{it}_{max} = 50$. This probability can be improved by increasing the number of iteration of our algorithm. For $\text{it}_{max} = 100$, we obtained a detection probability of 0.95.

Conclusions

In this paper, we have introduced an algorithm devoted to the blind identification of the code word length for a non-binary code in a noisy transmission environment. Using this algorithm, the code word length can be identified by

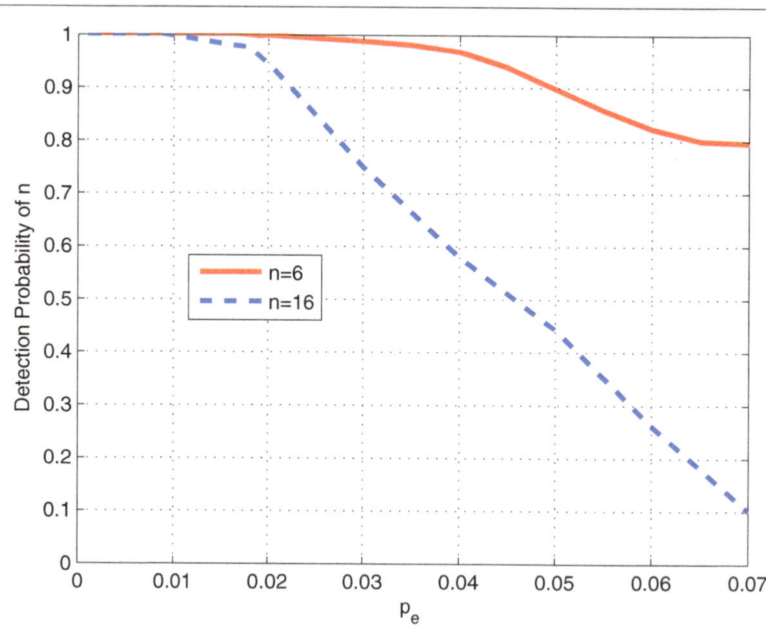

Figure 11 Impact of increasing n on the detection probability for LDPC codes. Of sizes $n = 6$ and $n = 16$ by using the proposed method considering $\text{it}_{max} = 1$. For LDPC ($n = 6, k = 3$) and LDPC ($n = 16, k = 8$) over GF(8), the probability of detecting the correct n by the method based on the arithmetic mean calculation is depicted according to the error probability p_e.

calculating the arithmetic mean of the number of zeros that occur in the columns of the matrix obtained by the Gauss elimination. We have shown that the proposed algorithm is robust because it does not require the estimation of error probability, is insensitive to the high order of Galois field, and has the best detection performances for the most of modulation types. Furthermore, this method provides better performances of detection when an iterative process is considered in order to increase the probability to obtain non-erroneous pivots during the Gauss elimination.

Our future work will focus on identifying the remainder of the non-binary code parameters as well as a parity check matrix, permitting to implement a generic decoder in a noisy environment. Furthermore, a method based on using soft information that allows us to improve the performances of the blind identification algorithms will be published soon [31].

Appendix

Proof of Equation 8

We define $\mathcal{H}_0$ and $\mathcal{H}_1$ by:

$$\mathcal{H}_0 \text{ if } \tilde{\mathbf{a}}_i^{(l)} \in \mathcal{C}^{\perp} \text{ and } \mathcal{H}_1 \text{ if } \tilde{\mathbf{a}}_i^{(l)} \notin \mathcal{C}^{\perp} \tag{29}$$

The two behaviors of $B_l(i)$ are limited in (7). The aim of this appendix is to demonstrate (8). In order to determine the probabilities of P_{fa} and P_{nd}, we should study the behaviors of the variable $B_l(i)$ according to the hypotheses $\mathcal{H}_0$ and $\mathcal{H}_1$:

Under the hypothesis $\mathcal{H}_0$: the variable $B_l(i)$ follows a binomial distribution $\mathcal{B}(M, P_i)$. So, the probability that $B_l(i)$ is greater than $\frac{M}{q} \cdot \eta$ is as follows:

$$Pr\left[B_l(i) > \frac{M}{q} \cdot \eta \mid \mathcal{H}_0\right] = \sum_{j=\lfloor \frac{M}{q} \cdot \eta \rfloor + 1}^{M} \binom{M}{j} \cdot P_i^j \cdot (1 - P_i)^{M-j} \tag{30}$$

Under the hypothesis $\mathcal{H}_1$: the variable $B_l(i)$ follows a binomial distribution $\mathcal{B}(M, 1/q)$. So, the probability that $B_l(i)$ is less than or equal to $\frac{M}{q} \cdot \eta$ is as follows:

$$Pr\left[B_l(i) \le \frac{M}{q} \cdot \eta \mid \mathcal{H}_1\right] = \sum_{j=0}^{\lfloor \frac{M}{q} \cdot \eta \rfloor} \binom{M}{j} \cdot \frac{(q-1)^{M-j}}{q^M} \tag{31}$$

Using these two probabilities, we will calculate the false alarm probability P_{fa}, the probability of not detecting a theoretical dependent column P_{nd} and the probability of detection P_{det}.

Calculation of the false alarm probability P_{fa}: this probability corresponds to decide that a column $\tilde{\mathbf{a}}_i^{(l)}$ belongs to a dual code $\mathcal{C}^{\perp}$ even thought in reality it does not belong. This probability can be determined by:

$$\begin{aligned} P_{\text{fa}} &= Pr\left[B_l(i) > \frac{M}{q} \cdot \eta \mid \mathcal{H}_1\right] \\ &= \sum_{j=\lfloor \frac{M}{q} \cdot \eta \rfloor + 1}^{M} \binom{M}{j} \cdot \frac{(q-1)^{M-j}}{q^M} \end{aligned} \tag{32}$$

Calculation of the probability of not detecting a theoretical dependent column P_{nd}: this probability corresponds to decide that a column $\tilde{\mathbf{a}}_i^{(l)}$ does not belong to $\mathcal{C}^{\perp}$ even thought in reality it belongs. This probability can be determined by:

$$\begin{aligned} P_{\text{nd}} &= Pr\left[B_l(i) \le \frac{M}{q} \cdot \eta \mid \mathcal{H}_0\right] \\ &= \sum_{j=0}^{\lfloor \frac{M}{q} \cdot \eta \rfloor} \binom{M}{j} \cdot P_i^j \cdot (1 - P_i)^{M-j} \end{aligned} \tag{33}$$

Calculation of the probability of detection P_{det}: this probability is defined by:

$$\begin{aligned} P_{\text{det}} = 1 - P_{\text{nd}} &= Pr\left[B_l(i) > \frac{M}{q} \cdot \eta \mid \mathcal{H}_0\right] \\ &= \sum_{j=\lfloor \frac{M}{q} \cdot \eta \rfloor + 1}^{M} \binom{M}{j} \cdot P_i^j \cdot (1 - P_i)^{M-j} \end{aligned} \tag{34}$$

Using (32) and (34), the optimal threshold can be determined by:

$$\begin{aligned} \eta_{\text{opt}} &= \arg\min_{\eta} (P_{\text{wd}}) = \arg\min_{\eta} (P_{\text{nd}} + P_{\text{fa}}) \\ &= \arg\min_{\eta} (1 + P_{\text{fa}} - P_{\text{det}}) \\ &= \arg\min_{\eta} \left(1 + \sum_{j=\left\lfloor \frac{M}{q} \cdot \eta \right\rfloor + 1}^{M} \binom{M}{j} \right. \\ &\qquad \left. \times \left[q^{-M} \cdot (q-1)^{M-j} - P_i^j \cdot (1 - P_i)^{M-j}\right]\right) \end{aligned} \tag{35}$$

Proof of the equation (16)

The probability P_i is initially expressed by (15). We denote $P_1(s) = Pr[X = s]$ and $P_2(s) = Pr$

$\left[\sum_{k=1}^{s} a_i^{(l)}(k) \cdot e_k^{(l)}(j) = 0\right]$ such that these two probabilities are independent. Henceforth, (15) becomes:

$$P_i = P_1(0) + \sum_{s=1}^{N_i(l)} P_1(s) \cdot P_2(s) \tag{36}$$

Assuming that the errors are independent from each other and uniformly distributed in $\mathrm{GF}(q)\backslash\{0\}$, the variable X follows a binomial distribution with parameters $N_i(l)$ and p_e. Thereby, the probability $P_1(s)$ is determined by:

$$P_1(s) = \binom{N_i(l)}{s} \cdot (p_e)^s \cdot (1 - p_e)^{N_i(l)-s} \tag{37}$$

The probability $P_2(s)$ is the probability of having $\sum_{k=1}^{s} a_i^{(l)}(k) \cdot e_k^{(l)}(j) = 0$ where $e_k^{(l)}(j) \in \mathrm{GF}(q)\backslash\{0\}$.

We demonstrate by the mathematical induction that the probability $P_2(s)$ can be expressed by:

$$P_2(s) = \frac{1 - P_2(s-1)}{q-1} \tag{38}$$

We have $P_2(0) = 1$ because there are no erroneous positions. In the case of a single erroneous position, we have $P_2(1) = 0$. However, considering the example of $\mathrm{GF}(2^2)$, the probability $P_2(s = 2)$ can be obtained by the matrix $\mathbf{M}$ whose the indexes of rows and columns correspond to non-zero elements of this field. The coefficients of this matrix correspond to the sum over $\mathrm{GF}(2^2)$ of the indexes of a row and a column.

$$\mathbf{M} = \begin{pmatrix} 0 & 3 & 2 \\ 3 & 0 & 1 \\ 2 & 1 & 0 \end{pmatrix} \tag{39}$$

If we have $\left(a_i^{(l)}(1), e_1^{(l)}(j)\right) \in \left(GF(2^2)^*\right)^2$, and $\left(a_i^{(l)}(2), e_2^{(l)}(j)\right) \in \left(GF(2^2)^*\right)^2$, the probability of having $a_i^{(l)}(1) \cdot e_1^{(l)}(j) + a_i^{(l)}(2) \cdot e_2^{(l)}(j) = 0$ will be $P_2(2) = 3/9 = 1/3$. The computed probability verifies (38).

We assume that (38) is verified for s, and we demonstrate it for $s+1$. If we have $\sum_{k=1}^{s+1} a_i^{(l)}(k) \cdot e_k^{(l)}(j) = 0$, we will have $e_{s+1}^{(l)} = -\frac{1}{a_i^{(l)}(s+1)} \cdot \sum_{k=1}^{s} a_i^{(l)}(k) \cdot e_k^{(l)}(j)$ that belongs to $\mathrm{GF}(q)^*$ with a probability equal to $1/(q-1)$. Therefore, the probability $P_2(s+1)$ is determined by:

$$\begin{aligned} P_2(s+1) &= Pr\left(e_{s+1}^{(l)} \in \mathrm{GF}(q)^*, \sum_{k=1}^{s} a_i^{(l)}(k) \cdot e_k^{(l)}(j) \neq 0\right) \\ &= Pr\left(e_{s+1}^{(l)} \in \mathrm{GF}(q)^*\right) \cdot Pr\left(\sum_{k=1}^{s} a_i^{(l)}(k) \cdot e_k^{(l)}(j) \neq 0\right) \\ &= \frac{\overline{P_2(s)}}{q-1} \\ &= \frac{1 - P_2(s)}{q-1} \end{aligned} \tag{40}$$

In order to simplify the expression of $P_2(s)$, a change of variable is done by considering $\varphi(s) = (q-1)^{s-1} \cdot P_2(s)$. When $P_2(s)$ is replaced by $\varphi(s)$, the expression (38) becomes:

$$\varphi(s) + \varphi(s-1) = (q-1)^{s-2} \tag{41}$$

Denoting $\rho(s) = (-1)^s \cdot \varphi(s)$, the expression (41) can be written as:

$$\rho(s) = \rho(1) + \sum_{i=0}^{s-1} (1-q)^i \tag{42}$$

but, the sum $\sum_{i=0}^{s-1} (1-q)^i$ is a geometric sequence of common ratio $1-q$. So, it can be written as:

$$\sum_{i=0}^{s-1} (1-q)^i = \frac{1 - (-1)^{s-1} \cdot (q-1)^{s-1}}{q} \tag{43}$$

The computation of $\rho(1)$ gives $\rho(1) = 0$. Therefore, using (43) and (41), the simplified expression of $P_2(s)$ is written as:

$$P_2(s) = \frac{(-1)^s + (q-1)^{s-1}}{q \cdot (q-1)^{s-1}} \tag{44}$$

Using (37) and (44), the overall probability P_i is given by:

$$\begin{aligned} P_i = \frac{1}{q} \cdot \Bigg(& \sum_{j=0}^{N_i(l)} \binom{N_i(l)}{j} \cdot p_e^j \cdot (1-p_e)^{N_i(l)-j} + (q-1) \\ & \times \sum_{j=0}^{N_i(l)} \binom{N_i(l)}{j} \cdot \left(\frac{-p_e}{q-1}\right)^j \cdot (1-p_e)^{N_i(l)-j} \Bigg) \end{aligned} \tag{45}$$

In order to simplify this equation, the Newton's binomial formula can be applied:

$$(Z+Y)^{N_i(l)} = \sum_{j=0}^{N_i(l)} \binom{N_i(l)}{j} \cdot Z^j \cdot Y^{N_i(l)-j}$$

Thus, the probability of having an element of the i-th column of $\tilde{\mathbf{T}}_l$ equal to 0 is determined by:

$$P_i = \frac{1 + (q-1) \cdot \left(1 - \frac{p_e \cdot q}{q-1}\right)^{N_i(l)}}{q} \tag{46}$$

Competing interests
The authors declare that they have no competing interests.

Author details
[1]Université Européenne de Bretagne, 5 Boulevard Laënnec, 35000 Rennes, France. [2]Université de Brest; CNRS, UMR 6285 Lab-STICC, 6 avenue Victor Le Gorgeu, 29238 Brest, France. [3]Université de Brest; CNRS UMR 6205, Laboratoire de Mathématiques Bretagne Atlantique, 6 avenue Victor Le Gorgeu, 29238 Brest, France.

References

1. MC Davey, D MacKay, Low-density parity-check codes over GF(q). IEEE Commun. Lett. **2**, 165–167 (1998)
2. JA Briffa, HG Schaathun, in *5th International Symposium on Turbo Codes and Related Topics*. Non-binary turbo codes and applications (IEEE Lausanne, 2008)
3. D Declercq, M Fossorier, Decoding algorithms for nonbinary LDPC codes over GF(q). IEEE Trans. Commun. **55**(4), 633–643 (2007)
4. L Barnault, D Declercq, in *Proceedings ITW*. Fast decoding algorithm for LDPC over GF(2^q) (IEEE, Paris, France, 2003), pp. 70–73
5. A Voicila, D Declercq, F Verdier, M Fossorier, P Urard, Low-complexity decoding for non-binary LDPC codes in high order fields. IEEE Trans. Commun. **58**(5), 1365–1375 (2010)
6. Yang Yu, W Chen, Design of low complexity non-binary LDPC codes with an approximated performance-complexity tradeoff. IEEE Commun. Lett. **16**(4), 514–517 (2012)
7. T Xia, HC Wu, Identification of nonbinary LDPC codes using average LLR of syndrome a posteriori probability. IEEE Commun. Lett. **17**(7), 1301–1304 (2013)
8. J Stern, A method for finding code words of small weight. Coding Theory Appl. **388**, 106–113 (1989)
9. A Canteaut, F Chabaud, A new algorithm for finding minimum-weight words in a linear code: application to McElieces cryptosystem and to narrow-sense BCH codes of length 511. IEEE Trans. Inf. Theory. **44**, 367–378 (1998)
10. A Valembois, Detection and recognition of a binary linear code. Discrete Appl. Math. **111**(1-2), 199–218 (2001)
11. M Cluzeau, in *2006 IEEE International Symposium on Information Theory*. Block code reconstruction using iterative decoding techniques (IEEE, Seattle, WA, 2006), pp. 2269–2273
12. M Cluzeau, M Finiasz, in *IEEE International Symposium on Information Theory 2009*. Recovering a code's length and synchronization from a noisy intercepted bitstream (IEEE, Seoul, 2009), pp. 2737–2741
13. M Côte, N Sendrier, in *IEEE International Symposium on Information Theory (ISIT)*. Reconstruction of convolutional codes from noisy observation (IEEE, Seoul, 2009), pp. 546–550
14. G Burel, R Gautier, in *IASTED International Conference on Communications, Internet and Information Technology*. Blind estimation of encoder and interleaver characteristics in a non cooperative context (ACTA Press, Scottsdale, AZ, USA, 2003)
15. Y Zrelli, R Gautier, M Marazin, E Rannou, E Radoi, Focus on theoretical properties of blind convolutional codes identification methods based on rank criterion. MTA Review. **XXII**(4), 213–234 (2012)
16. Y Zrelli, M Marazin, R Gautier, E Rannou, in *Proceedings of the International Conference on Computer Communication Networks*. Blind identification of convolutional encoder parameters over GF(2^m) in the noiseless case (IEEE, Maui, Hawaii, 2011)
17. G Sicot, S Houcke, in *IEEE International Conference on Acoustics, Speech and Signal Processing (ICASSP)*, vol. 3. Blind detection of interleaver parameters (IEEE, Philadelphia, Pennsylvania, 2005), pp. 829–832
18. G Sicot, S Houcke, J Barbier, Blind detection of interleaver parameters. Signal Process. **89**(4), 450–462 (2009)
19. M Marazin, R Gautier, G Burel, Blind recovery of k/n rate convolutional encoders in a noisy environment. EURASIP J. Wireless Commun. Netw. **2011**(168), 1–9 (2011)
20. Z Jing, H Zhiping, L Chunwu, S Shaojing, Z Yimeng, Information-dispersion-entropy-based blind recognition of binary BCH codes in soft decision situations. Entropy. **15**(5), 1705–1725 (2013)
21. A Hocquenghem, Codes correcteurs d'erreurs. Chiffres. **2**, 147–156 (1959)
22. RC Bose, DK Ray-Chaudhuri, On a class of error correcting binary group codes. Inf. Control. **3**(3), 68–79 (1960)
23. I Reed, G Solomon, Polynomial codes over certain finite fields. J. Soc. Ind. Appl. Math. **8**(2), 300–304 (1960)
24. M Baldi, M Bianchi, F Chiaraluce, R Garello, N Maturo, IA Sanchez, S Cioni, in *2013 IEEE Military Communications Conference*. Advanced coding schemes against jamming in telecommand links (IEEE, San Diego, CA, 2013), pp. 1220–1226
25. C Junbin, W Lin, L Yong, in *International Conference on Communications, Circuits and Systems*. Performance comparison between non-binary LDPC codes and Reed-Solomon codes over noise bursts channels, vol. 1 (IEEE, Hong Kong, China, 2005), pp. 1–4
26. B Zhou, L Zhang, J Kang, Q Huang, YY Tai, S Lin, M Xu, in *Information Theory and Applications Workshop*. Non-binary LDPC codes vs. Reed-Solomon codes (IEEE, San Diego, CA, 2008), pp. 175–184
27. version 8.8.0 Release 8 GT, LTE; Evolved Universal Terrestrial Radio Access (E-UTRA); Multiplexing and Channel Coding. The 3rd Generation Partnership Project 2, Technical Specification Group Radio Access Network (2010). http://www.3gpp.org, (2013)
28. M Marazin, R Gautier, G Burel, in *IEEE GLOBECOM Workshops*. Dual code method for blind identification of convolutional encoder for cognitive radio receiver design (IEEE, Honolulu, HI, 2009)
29. EM Moro, *Algebraic geometry modeling in information theory*. (World Scientific, Singapore, 2012)
30. E Anderson, Z Bai, C Bischof, S Blackford, J Demmel, J Dongarra, JD Croz, A Greenbaum, S Hammarling, A McKenney, D Sorensen, *LAPACK user's guide*. (SIAM, Philadelphia, 1999)
31. Y Zrelli, *Identification aveugle de codes correcteurs d'erreurs basés sur des grands corps de Galois et recherche d'algorithmes de type décision souple pour les codes convolutifs*. (PhD thesis, Université de Brest, France, 2013)

Continuous transmit in cognitive radio systems: outage performance of selection decode-and-forward relay networks over Nakagami-*m* fading channels

Mohammed Ridouani[1*], Aawatif Hayar[2] and Abdelkrim Haqiq[3]

Abstract

Cognitive radio is a promising technology that uses radio spectrum opportunistically and efficiently. In this paper, we present the performance of 'Always transmit' strategy, which consists of allowing the cognitive source to transmit over all time slots and in any time slot. Thus, the cognitive source will be able to operate continuously whenever either the primary user is active or absent. The proposed strategy is projected into dual-hop communication systems with decode-and-forward and relay selection over independent but not necessarily identically distributed Nakagami-*m* fading channels. We derive the closed-form expressions of the outage probability, and we analyze the performance. Simulation results show that the strategy 'Always transmit + Relay' outperforms other conventional cognitive radio strategies and the use of relays enhances the performance in deep fading channel.

Keywords: Continuous transmit in cognitive radio; Decode-and-forward relay; Spectrum sensing; Interference constraints; Interweave approach; Underlay approach; Nakagami-*m* fading; Outage probability

1 Introduction

Recently, significant attention has been given to cognitive radio as a new technology that uses radio spectrum opportunistically and efficiently and which allows new applications to share currently allocated spectrum while the quality of service (QoS) at primary side is guaranteed.

There are three approaches of cognitive radio used to access licensed or unlicensed band [1]: spectrum interweave, spectrum underlay, and spectrum overlay.

a- Spectrum interweave (interference avoiding): the secondary user (SU) uses the first part of time slot (iT) (first phase) for 'spectrum sensing' to detect spectrum hole (spectrum not currently used by primary user) and make a secondary user decision $H_s(i)$ at time slot (iT), and beginning transmission in the second phase, only if the result of sensing in the first phase is $H_s(i) = H_0$ (hole is detected, i.e., primary user is absent). In this approach, the SU does not interfere with primary user.

b- Spectrum underlay (interference controlling): the SU can transmit simultaneously with primary user over the same spectrum and through all time slots (iT) only if the interference generated by secondary transmitter (cognitive transmitter) at primary receiver is tolerable and controlled by an acceptable level, i.e., below a threshold or an outage probability limit that guarantees certain QoS for primary user's communication.

c- Spectrum overlay (interference mitigating): the SU (cognitive transmitter) can transmit simultaneously with primary user over the same spectrum. In this approach, cognitive transmitter can transmit with any power, part of its power and time slot (iT) used for relaying primary user's transmission, and the remainder of its power and time slot (iT) for own transmission.

Also, cooperative communications with amplify-and-forward (AF) or with decode-and-forward (DF) strategy have gained a great interest in the academic research circles. This is due to benefits over a fading channel, like combating fading and shadowing, and increasing the system capacity. In [2], the authors shown that the best relay selection scheme can achieve the same

*Correspondence: mohammed_ridouani@yahoo.fr
[1]Computer Sciences Department, GREENTIC/UH2C/EST, Casablanca, Morocco
Full list of author information is available at the end of the article

diversity-multiplexing tradeoff performance as the traditional protocols. As researches go deep and broad, some works introduce cooperative communication techniques into cognitive radio, to enhance the secondary user's data transmission and the reliability of primary user's detection.

One of the QoS parameters for secondary system is the continuity of service. Continuous transmission for secondary users is weakly addressed. Indeed, in [3,4], the SU transmits through underlay method either by controlling interference level [3,4] or outage probability limit. Nevertheless, underlay method underutilizes the spectrum, and SU does not transmit with high power even when primary user is absent. In [5,6], the SU transmits with the interweave method in the second phase only if the result of sensing in the first phase is $H_s(i) = H_0$. In [7], mixed strategies introduced for cognitive transmission in the second phase based on the sensing results in the first phase. Many studies in the literature on such mixed strategies have been conducted [8-13], in their scenarios, the secondary user transmits through interweave method in the second phase if the result of sensing in the first phase is $H_s(i) = H_0$ or transmits with underlay method in the second phase if the result of sensing in the first phase is $H_s(i) = H_1$ (hole is not detected, i.e., primary user is active). However, none of these papers have considered the transmission in two phases, SU does not transmit over all time slots, it transmits only in the second phase, which decreases the spectrum efficiency as some part of the time slot is used for sensing instead of data transmission.

In most of the above presented strategies, sensing is a key functionality that affects the spectral efficiency of SU due to the time dedicated to sensing. However, two different architectures could be adopted to operate spectrum sensing and communication by using single radio or dual radio [14,15]. Nevertheless, in the single radio architecture, no communication is allowed during sensing phase. Our work aims at exploiting sensing phase for transmission also optimizes the SU spectrum efficiency and thus adopts double radios architecture. To be able to do so without creating harmful interference to the primary user, we propose an underlay scheme with limited interference threshold.

In our previous works [16], where we have introduced our new strategy of transmission '*Always Transmit (AT): to enhance the cognitive sources of data transmission and to provide continuous transmit solution by using mixed strategies in cognitive radio context, which it performed via the combination of the underlay and interweave approaches depending on the activity of the primary system*' and have shown that it outperforms other conventional cognitive radio strategies for power transmission and interference management, in [17], where we have derived the closed form and analyzed the performance of outage probability over Rayleigh fading, and in [18], where we have studied the trade-off between spectrum sensing duration and the QoS at cognitive network while guaranteeing certain QoS at primary network controlled by the outage probability limit, over independent but not necessarily identically distributed Nakagami-m channels. In (M Ridouani, A Hayar, A Haqiq, Continuous-transmit in cognitive radio systems: perform sensing and transmission in parallel anytime. submitted to IEEE TWC), we have shown that the use of (average transmit P_{av}, instantaneous interference I) as power constraint is recommended for transmission, because it allows preserving the QoS of primary system at each time while guaranteing an outage performance limit for the secondary of about $\simeq$ 20% of maximum degradation, and due to fading and by using average transmit power constraint, the energy efficiency is improved.

Motivated by cooperative communications and our results in [17,18], in this paper, we extend our strategy 'AT' (the cognitive source (CS) will be able to transmit over all time slots and in any time slot, then, it will be able to operate continuously whenever either the primary user is active or absent) into cognitive relay selection. We derive the closed form of outage probability of relay selection in dual-hop transmission over independent but not necessarily identically distributed (i.n.i.d.) Nakagami-m channels, with integer values of fading severity parameter m.

Our main contributions are summarized as follows:

1) The continuity of service: we propose a scenario that enhances secondary user QoS in terms of outage probability by allowing it to transmit over *all time slots and in any time slots*. Therefore, the secondary can continue its transmission whether the primary user is active or absent.

2) We derive the closed-form expressions of the secondary user outage probability for transmission scenario over both phases and over independent but not necessarily identically distributed Nakagami-m fading channels. And we analyze for this considered scenario the impact of spectrum sensing duration and spectrum hole detection, the transmit and the interference power constraints, and the number of relays and fading on the secondary outage probability (QoS).

3) We show that the proposed strategy 'Always transmit + Relay' outperforms other conventional cognitive radio strategies.

This paper is organized as follows. In Section 2, we outline the system model (scenario, spectrum sensing, and the optimal power allocation) of the proposed scheme of cognitive transmission. The closed form of outage probability of cognitive relay selection over a Nakagami-m fading channel is given in Section 3. In Section 4, we present numerical results. Section 5 draws the conclusion.

2 System model

In this section, we describe the cognitive relay network of our scenario. As shown in Figure 1b, the primary destination (PD) coexists with the cognitive nodes (CS, cognitive destination (CD), and N potential cognitive relay (CR_k) $k = 1, .., N$). Each cognitive node has a dual-radio architecture, one radio chain is dedicated for spectrum sensing (see Section 2.1) while the other radio chain operates on a half-duplex mode, and it is dedicated for data transmission and reception.

Notice here that, if we look only at the second radio chain, we can see that it operates on a half-duplex mode. If we look at the cognitive node (which consists of a two radio chains), one radio is dedicated for spectrum sensing (reception), the second chain for communication (for example, transmitting data: transmission), we can see that it operates, globally, on a full-duplex mode.

In this model, we consider a half-duplex mode and a dual-hop communication system with regenerative mode for data transmitting. Also, it is assumed that there is no direct transmission between CS and CD. So the data transmission from CS to CD is subdivided into two steps. In step one, the CS transmits its data (one packet) to (CR_k) which is selected prior to a new transmission. In step two, the CR_k decodes the received data and then re-encodes and forwards it to a CD.

Also, we consider that the transmission of CS and CR_k process can move into two phases through a time slot (iT) as shown in Figure 1a. In the first phase (αT), the CS (resp. CR_k) transmits through underlay approach, in the meantime, each cognitive node senses the spectrum hole existence (activity of primary source (PS)). In the second phase ($(1 - \alpha)T$), the CS (resp. CR_k) transmits through interweave or underlay approach whether or not a hole is detected in the first phase, respectively.

The data (frame i) structure of the proposed system is shown in Figure 1a. The time duration of the transmission of each frame is T. Each frame consists of many packet transmission periods, and we assume that each packet transmission period is very small subject to the time slot T, which we allow to suppose that each packet transmitted from CS to the best cognitive relay CR_k and from CR_k to the CD is via the same transmission mode (underlay or interweave), and the best cognitive relay CR_k is selected to transmit all packets of a frame i during the time slot iT.

We assume that the channel changes slowly, so the channel remains constant during a one time slot. Also, we assume that the channels between any of the two nodes are independent but not necessarily identically distributed (i.n.i.d.) Nakagami-m channels, with integer values of fading severity parameter m, so that the probability distribution function (PDF) of the channel gains ($|g_{sp}|^2$, $|h_{SK}|^2$, $|g_{kp}|^2$, and $|h_{KD}|^2$) is statistically characterized by the Nakagami distribution [19], where g_{sp}, h_{SK}, g_{kp}, and h_{KD} are the fading coefficients of the channel from CS to PD, from CS to CR_k, from CR_k to PD, and from CR_k to CD, respectively, as shown in Figure 1. Let $X_k = |h_{SK}|^2$, $Y_k = |h_{KD}|^2$, $Z_0 = |g_{sp}|^2$, and $Z_k = |g_{kp}|^2$.

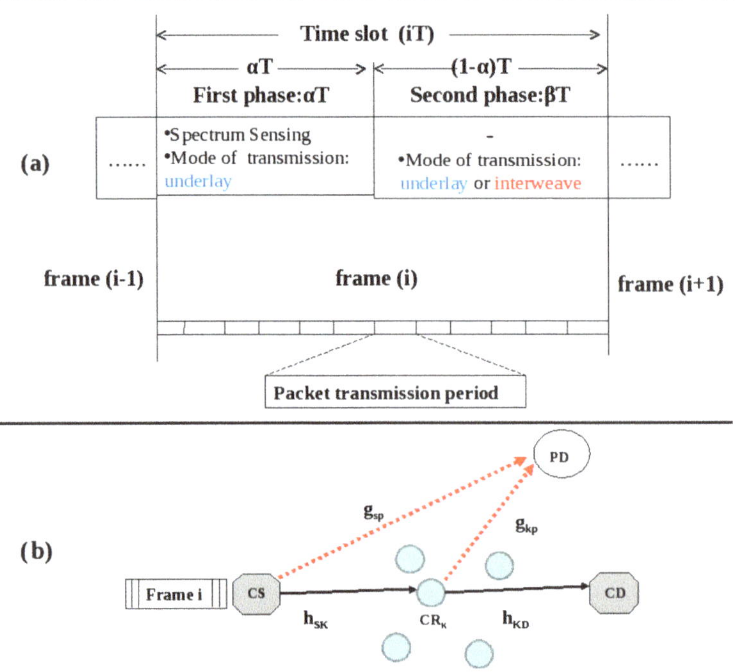

Figure 1 The mode of transmission and the allocation of time durations **(a)** and fading coefficient of the channels **(b)**.

Note that X_k, Y_k, and Z_k ($k = 0, 1, .., N$) follow the probability distribution $p_k(x)$, $p_k(y)$, and $p_k(z)$, respectively, and are independent of each other. Thus, the channel gain, γ ($\gamma = X_k, Y_k, Z_k$), is distributed according to a gamma distribution given by (1) with different fading parameters $\gamma_{xk}, \gamma_{yk}, \gamma_{zk}$ and fading severity parameters m_{xk}, m_{yk}, m_{zk}:

$$p_\gamma(\gamma) = \left(\frac{m}{\overline{\gamma}}\right)^m \frac{\gamma^{m-1}}{\Gamma(m)} \exp\left(-m\frac{\gamma}{\overline{\gamma}}\right), \gamma \geq 0, m \geq 1/2 \quad (1)$$

where $\overline{\gamma}$ is the average channel gain ($\overline{\gamma} = E[\gamma]$), m is the Nakagami fading severity parameter, and $\Gamma(.)$ is the gamma function [20].

The Nakagami fading model has proven useful for modeling multipath faded envelope in wireless channels and is widely accepted as best fit to model multipath propagation from severe to light fading. Indeed, it is known to accurately characterize a range of multipath and to model different fading conditions ranging from Rayleigh to strong Rician channels. As the fading parameter m tends to infinity, the Nakagami multipath fading channel converges to an AWGN channel.

For notation convenience, we use the notations as follows:

- k: the k-th cognitive node ($k = 0$ refer to CS and $k = 1, \ldots, N$ refer to CR_k).
- P_{SK} and P_{KD} are the transmit power of CS and CR_k, respectively.
- $H_p(i)$: denote whether or not the licensed band is occupied by PS in the first phase of time slot iT.
- H_0: the band is unoccupied by PS (and H_1: otherwise).
- CS and CR_k will make a decision $H_s(i)$ to transmit through interweave or underlay method in the second phase
- $H_s(i) = H_0$: considers the band is available, then, CS and CR_k transmit through interweave method
- $H_s(i) = H_1$: considers the band is unavailable, then, CS and CR_k transmit through underlay method.
- $P_a = Pr(H_p(i) = H_0)$: the probability that there is a hole (and $Pr(H_p(i) = H_1) = 1 - P_a$ otherwise).
- $P_f^k = Pr(H_s(i) = H_1/H_p(i) = H_0)$: the probability that the k-th cognitive node has a false alarm.
- $P_d^k = Pr(H_s(i) = H_1/H_p(i) = H_1)$: the probability that the k-th cognitive node detects the PS presence.
- ε: represent the fraction of a time slot utilized for data transmission in the first phase ($\varepsilon = \alpha$) and second phase ($\varepsilon = \beta = 1 - \alpha$)
- P_α: the probability to be in the first phase.
- $P_\beta = 1 - P_\alpha$: the probability to be in the second phase.

2.1 Spectrum sensing

Cooperative spectrum sensing can improve the performance of spectrum sensing by combating fading and shadowing and decreasing the detection time required. In this technique, each of $N + 1$ cognitive nodes (CS or CR_k) contributes to sense spectrum holes. To protect primary's communication when PS is active ($H_p(i) = H_1$), we use the OR rule [21] (spectrum is not available ($H_s(i) = H_1$) if any of $N + 1$ cognitive nodes sense the presence of PS) which minimizes the probability of interfering at PD.

Each cognitive node periodically senses the channel like in [5,6,22]. We assume that it first estimates the noise floor and distinguishes the rest of the information bearing samples for wireless signals [23]. After, it identifies the CS's signal (the signal transmitted through underlay approach in the first phase (αT)) [23]. And finally, if it detects a collision (i.e., the CS's signal is combined with other signal), then, it decides that the PS is active (a hole is not detected), else, it decides that PS is absent (a hole is detected).

We assume that the $N + 1$ cognitive nodes use the same spectrum sensing detector and sense the spectrum hole existence through the same duration (αT), then, the cooperative probabilities of detection P_d and false alarm P_f are as follows:

$$P_d = 1 - \prod_{k=0}^{N} \left(1 - P_d^k\right) \quad (2)$$

$$P_f = 1 - \prod_{k=0}^{N} \left(1 - P_f^k\right) \quad (3)$$

2.2 Optimal power allocation

In the underlay approach, the CS's (resp. CR_k) transmission power is constrained as $P_{SK}Z_0 \leq I$ (resp. $P_{KD}Z_k \leq I$), where I is the interference threshold that the interference generates by CS (resp. CR_k) on the PD remains below it, and is constrained as $P_{SK} \leq P$ (resp. $P_{KD} \leq P$), where P is the maximum transmission power.

In the interweave approach, there is just the maximum transmission power constraint, then, $P_{SK} \leq P$ and $P_{KD} \leq P$.

For each time slot iT, in the first phase (first fraction of time αT), the CS's (resp. CR_k) transmits through underlay approach, then, it transmits with a low power P_{US}^k (resp. P_{UR}^k). During the second phase ($(1 - \alpha)T$), the CS's (resp. CR_k) transmits through underlay approach with a low power P_{US}^k (resp. P_{UR}^k) or switch to interweave approach with a high power P_{IS}^k (resp. P_{IR}^k) based on the sensing result.

Subject to $P_{SK} \leq P$ (resp. $P_{KD} \leq P$) and $P_{SK}Z_0 \leq I$ (resp. $P_{KD}Z_k \leq I$), then, the optimal power P_{IS}^k (resp. P_{IR}^k)

and P_{US}^k (resp. P_{UR}^k) used by CS (resp. CR_k) in interweave and underlay respectively are as follow:

$$P_{IS}^k = P_{IR}^k = P \tag{4}$$

$$P_{US}^k = \min\left(P, \frac{I}{Z_0}\right) = \begin{cases} P, & Z_0 \leq \frac{I}{P} \\ \frac{I}{Z_0}, & Z_0 \geq \frac{I}{P} \end{cases} \tag{5}$$

$$P_{UR}^k = \min\left(P, \frac{I}{Z_k}\right) = \begin{cases} P, & Z_k \leq \frac{I}{P} \\ \frac{I}{Z_k}, & Z_k \geq \frac{I}{P} \end{cases} \tag{6}$$

According to the activity of PS and the sensing result, the synthesis of our scenario of transmission and the optimal power allocation in both phases are listed in Table 1.

3 Outage analysis of cognitive relay selection

3.1 Cognitive relay selection

If we select the k-th cognitive relay CR_k for data transmission in dual-hop, the instantaneous capacity in underlay approach C_k^U (resp. interweave approach C_k^I) between cognitive source and cognitive destination is given by [24]:

$$C_k^U = \min\left\{\varepsilon \log_2\left(1 + \frac{X_k P_{US}^k}{N_0}\right), \varepsilon \log_2\left(1 + \frac{Y_k P_{UR}^k}{N_0}\right)\right\} \tag{7}$$

$$C_k^I = \min\left\{\beta \log_2\left(1 + \frac{X_k P_{IS}^k}{N_0}\right), \beta \log_2\left(1 + \frac{Y_k P_{IR}^k}{N_0}\right)\right\} \tag{8}$$

In (7) and (8), we have assumed that any interference from the primary source is neglected (this can be possible if the primary transmitter is located far away from the secondary users, or the interference is represented by the noise term under an assumption that the primary transmitter's signal is generated by random Gaussian codebooks [4]). The coefficient ε (resp. β) is due to the fact that only ε (resp. β) fraction of a time slot is utilized for the CS's and CR_k's data transmission phase. And N_0 is the noise variance.

Substituting (5) and (6) in (7) and (4) in (8), we obtain:

$$C_k^U = \varepsilon \log_2\left(1 + \frac{P}{N_0} \min\left\{X_k \min\left(1, \frac{I}{PZ_0}\right), Y_k \min\left(1, \frac{I}{PZ_k}\right)\right\}\right) \tag{9}$$

$$C_k^I = \beta \log_2\left(1 + \frac{P}{N_0} \min\{X_k, Y_k\}\right) \tag{10}$$

We introduce two variables (W_k^I, W_k^U) for briefly. The short-hand notation is as follow:

$$W_k^U = \min\left\{X_k \min\left(1, \frac{I}{PZ_0}\right), Y_k \min\left(1, \frac{I}{PZ_k}\right)\right\} \tag{11}$$

$$W_k^I = \min\{X_k, Y_k\} \tag{12}$$

For the best relay selection scheme, and by using the max-min criterion [4,25] which maximizes the minimum of signal-to-noise ratios (SNRs) of the source-relay link and relay-destination link, we select the relay for which W_k^U (resp. W_k^I) is maximal in underlay (resp. interweave) approach.

Let:

$$W^U = \max_{1,..,N} W_k^U \tag{13}$$

$$W^I = \max_{1,..,N} W_k^I \tag{14}$$

then, consequently, we obtain for the instantaneous capacity in underlay approach C^U (resp. interweave approach C^I)

$$C^U = \max_{1,..,N} C_k^U = \varepsilon \log_2\left(1 + \frac{P}{N_0} W^U\right) \tag{15}$$

$$C^I = \max_{1,..,N} C_k^I = \beta \log_2\left(1 + \frac{P}{N_0} W^I\right) \tag{16}$$

3.2 Outage probability

Given a predetermined transmission rate of the source C_T, the outage probability mathematically defined as $P_{\text{out}} = Pr(C < C_T)$, where C is the capacity of channel.

The outage probability in underlay approach $P_{\text{out}}^{U,\varepsilon}$ (resp. interweave approach P_{out}^I) is given:

$$P_{\text{out}}^{U,\varepsilon} = Pr\left(C^U < C_T\right) \tag{17}$$

$$P_{\text{out}}^I = Pr\left(C^I < C_T\right) \tag{18}$$

Table 1 The synthesis of the transmission scenario and the optimal power allocation in both phases

Phase and fraction time	PS's activity	Sensing results	Mode of transmission	Related probability	CS's power	CR_k's power
First phase: αT	Active or absent	-	Underlay	P_α	$\min\left(P, \frac{I}{Z_0}\right)$	$\min\left(P, \frac{I}{Z_k}\right)$
Second phase: βT	Active: $H_p(i) = H_1$	$H_s(i) = H_1$	Underlay	$P_\beta(1 - P_a)P_d$	$\min\left(P, \frac{I}{Z_0}\right)$	$\min\left(P, \frac{I}{Z_k}\right)$
	Active: $H_p(i) = H_1$	$H_s(i) = H_0$	Interweave	$P_\beta(1 - P_a)(1 - P_d)$	P	P
	Absent: $H_p(i) = H_0$	$H_s(i) = H_1$	Underlay	$P_\beta P_a P_f$	$\min\left(P, \frac{I}{Z_0}\right)$	$\min\left(P, \frac{I}{Z_k}\right)$
	Absent: $H_p(i) = H_0$	$H_s(i) = H_0$	Interweave	$P_\beta P_a(1 - P_f)$	P	P

Substituting (15) in (17) and (16) in (18), we obtain:

$$P_{\text{out}}^{U,\varepsilon} = Pr\left(W^{U} < \frac{U_\varepsilon}{P}\right) \tag{19}$$

$$P_{\text{out}}^{I} = Pr\left(W^{I} < \frac{U_\beta}{P}\right) \tag{20}$$

where $U_\varepsilon = N_0(2^{\frac{C_T}{\varepsilon}} - 1)$
Hence, by using (11) and (12), we get:

$$P_{\text{out}}^{U,\varepsilon} = \prod_{k=1}^{N} Pr\left(W_k^{U} < \frac{U_\varepsilon}{P}\right) \tag{21}$$

$$P_{\text{out}}^{I} = \prod_{k=1}^{N} Pr\left(W_k^{I} < \frac{U_\beta}{P}\right) \tag{22}$$

Using Appendix (Equations 32 and 39) and replacing A by $\frac{U_\varepsilon}{P}$ and B by $\frac{I}{P}$, the closed form of (21) yields:

$$P_{\text{out}}^{U,\varepsilon} = \prod_{k=1}^{N}\left(1 - \Phi\left(m_{xk}, \gamma_{xk}, m_{z0}, \gamma_{z0}, \frac{I}{P}, \frac{U_\varepsilon}{P}\right)\right. \\ \left.\Phi\left(m_{yk}, \gamma_{yk}, m_{zk}, \gamma_{zk}, \frac{I}{P}, \frac{U_\varepsilon}{P}\right)\right) \tag{23}$$

Replace A by $\frac{U_\beta}{P}$ in Appendix (Equation 33), the closed form of (22) yields:

$$P_{\text{out}}^{I} = \prod_{k=1}^{N}\left(1 - \frac{\Gamma\left(m_{xk}, \frac{m_{xk}}{\gamma_{xk}}\frac{U_\beta}{P}\right)}{\Gamma\left(m_{xk}\right)} \frac{\Gamma\left(m_{yk}, \frac{m_{yk}}{\gamma_{yk}}\frac{U_\beta}{P}\right)}{\Gamma\left(m_{yk}\right)}\right) \tag{24}$$

Finally, the outage probability of our system is:

$$P_{\text{out}} = P_\alpha P_{\text{out}}^{\text{first-phase}} + (1 - P_\alpha) P_{\text{out}}^{\text{second-phase}} \tag{25}$$

where:

$$P_{\text{out}}^{\text{first-phase}} = P_{\text{out}}^{U,\alpha}$$

and:

$$\begin{aligned} P_{\text{out}}^{\text{second-phase}} &= \left[Pr(H_s(i) = H_0, H_p(i) = H_0) + Pr(H_s(i)\right. \\ &= \left.H_0, H_p(i) = H_1)\right] P_{\text{out}}^{I} \\ &\quad + \left[Pr(H_s(i) = H_1, H_p(i) = H_0) + Pr(H_s(i)\right. \\ &= \left.H_1, H_p(i) = H_1)\right] P_{\text{out}}^{U,\beta} \end{aligned} \tag{26}$$

which can also be written as:

$$\begin{aligned} P_{\text{out}}^{\text{second-phase}} &= \left[P_a(1 - P_f) + (1 - P_a)(1 - P_d)\right] P_{\text{out}}^{I} \\ &\quad + \left[P_a P_f + (1 - P_a) P_d\right] P_{\text{out}}^{U,\beta} \end{aligned} \tag{27}$$

Then, the closed form of outage probability of our strategy (Always transmit + Relay) is derived in (28):

$$\begin{aligned} P_{\text{out-AT-}R} &= P_\alpha \prod_{k=1}^{N}\left(1 - \Phi\left(m_{xk}, \gamma_{xk}, m_{z0}, \gamma_{z0}, \frac{I}{P}, \frac{U_\alpha}{P}\right)\right. \\ &\quad \left.\times \Phi\left(m_{yk}, \gamma_{yk}, m_{zk}, \gamma_{zk}, \frac{I}{P}, \frac{U_\alpha}{P}\right)\right) \\ &\quad + (1 - P_\alpha)\left(\left[P_a\left(1 - P_f\right) + \left(1 - P_a\right)\left(1 - P_d\right)\right]\right. \\ &\quad \times \prod_{k=1}^{N}\left(1 - \frac{\Gamma\left(m_{xk}, \frac{m_{xk}}{\gamma_{xk}}\frac{U_\beta}{P}\right)}{\Gamma(m_{xk})} \frac{\Gamma\left(m_{yk}, \frac{m_{yk}}{\gamma_{yk}}\frac{U_\beta}{P}\right)}{\Gamma(m_{yk})}\right) \\ &\quad + \left[P_a P_f + (1 - P_a) P_d\right] \\ &\quad \times \prod_{k=1}^{N}\left(1 - \Phi\left(m_{xk}, \gamma_{xk}, m_{z0}, \gamma_{z0}, \frac{I}{P}, \frac{U_\beta}{P}\right)\right. \\ &\quad \left.\left.\times \Phi\left(m_{yk}, \gamma_{yk}, m_{zk}, \gamma_{zk}, \frac{I}{P}, \frac{U_\beta}{P}\right)\right)\right) \end{aligned} \tag{28}$$

If CS and CR_k transmit just through interweave approach, (28) can reduced to (29):

$$\begin{aligned} P_{\text{out-}IN-R} &= P_\alpha + (1 - P_\alpha)\left(\left[P_a\left(1 - P_f\right) + \left(1 - P_a\right)\left(1 - P_d\right)\right]\right. \\ &\quad \times \prod_{k=1}^{N}\left(1 - \frac{\Gamma\left(m_{xk}, \frac{m_{xk}}{\gamma_{xk}}\frac{U_\beta}{P}\right)}{\Gamma(m_{xk})} \frac{\Gamma\left(m_{yk}, \frac{m_{yk}}{\gamma_{yk}}\frac{U_\beta}{P}\right)}{\Gamma(m_{yk})}\right) \\ &\quad \left.+ \left[P_a P_f + (1 - P_a) P_d\right]\right) \end{aligned} \tag{29}$$

If CS and CR_k transmit just through underlay approach, (28) can reduced to (30):

$$\begin{aligned} P_{\text{out-}UN-R} &= P_\alpha \prod_{k=1}^{N}\left(1 - \Phi\left(m_{xk}, \gamma_{xk}, m_{z0}, \gamma_{z0}, \frac{I}{P}, \frac{U_\alpha}{P}\right)\right. \\ &\quad \left.\times \Phi\left(m_{yk}, \gamma_{yk}, m_{zk}, \gamma_{zk}, \frac{I}{P}, \frac{U_\alpha}{P}\right)\right) \\ &\quad + (1 - P_\alpha)\left(\left[P_a P_f + (1 - P_a) P_d\right] \prod_{k=1}^{N}\right. \\ &\quad \times \left(1 - \Phi\left(m_{xk}, \gamma_{xk}, m_{z0}, \gamma_{z0}, \frac{I}{P}, \frac{U_\beta}{P}\right)\right. \\ &\quad \left.\left.\times \Phi\left(m_{yk}, \gamma_{yk}, m_{zk}, \gamma_{zk}, \frac{I}{P}, \frac{U_\beta}{P}\right)\right)\right) \end{aligned} \tag{30}$$

If there is no rely ($k = 0$), CS transmits data directly to PD without help of relays, (28) can reduced to (31):

$$\begin{aligned} P_{\text{out}-\text{AT}-NR} = P_\alpha & \left(1 - \Phi\left(m_{x1}, \gamma_{x1}, m_{z0}, \gamma_{z0}, \frac{I}{P}, \frac{U_\alpha}{P}\right)\right) \\ & + (1 - P_\alpha)\left(\left[P_a\left(1 - P_f^1\right) + (1 - P_a)\left(1 - P_d^1\right)\right]\right. \\ & \times \left(1 - \frac{\Gamma\left(m_{x1}, \frac{m_{x1}}{\gamma_{x1}} \frac{U_\beta}{P}\right)}{\Gamma(m_{x1})}\right) \\ & + \left[P_a P_f^1 + (1 - P_a) P_d^1\right] \\ & \left. \times \left(1 - \Phi\left(m_{x1}, \gamma_{x1}, m_{z0}, \gamma_{z0}, \frac{I}{P}, \frac{U_\beta}{P}\right)\right)\right) \end{aligned} \tag{31}$$

4 Numerical results

Computer simulations were carried out in order to validate the performance of the proposed scenario ('Always transmit + Relay') based on the outage probability. We analyze the impact of the fraction α of time slot used for spectrum sensing and other parameters (number of relay N, fading severity parameters (m_{xk}, m_{yk}, m_{zk}, m_{z0}), the transmit (resp. interference) power constraint P (resp. I)) over data transmission. Throughout simulations, we have assumed a perfect spectrum hole detection, thus, $P_f^k = 0.1(k = 1, \ldots, N)$ to use the cognitive channel with a higher chance with the interweave method, and $P_d^k = 0.9(k = 1, .., N)$ to protect the primary source transmission. We consider that P_α follows a uniform distribution, then, $P_\alpha = \alpha$ and $P_\beta = 1 - \alpha$. Also, in all simulations, we fix: $\gamma_{xk} = \gamma_{yk} = \gamma_{zk} = \gamma_{z0} = 1$ dB ($k = 1, .., N$), $N_0 = 1$, $P_a = 0.8$, $C_T = 0.5$, and we vary α, m, N, I, and P.

From the figures (Figures 2, 3, 4, 5, and 6), it has been observed that outage performances are degraded when α increases; in contrast, we see a higher reduction of outage probability, thus improving the system performance, when α is near 0. Indeed, when α is near 0, which is referred to as perfect sensing, the CS and CR_k transmit with underlay method just in short time αT in the first phase and can have more chance to transmit with the interweave method in a large amount of time $(1 - \alpha)T$ in the second phase. But when α increases, the CS and CR_k transmit in large time with the underlay method, and thus, the system is more probable to have an outage.

These figures also show that the outage probability decreases (which is the indication of performance improvement) as P, I, or N increase. Indeed, a) if the transmit power P at (CS and CR_k) increases, then, the SNRs increase which denote the increase of their capacity and finally the decrease of their outage probability, b) when the interference threshold I is increasing, the constraints are relaxed at the cognitive system side, so the system performance increases as I increases, c) more relays in fading channel network can combat fading and consequently, enhance the performance.

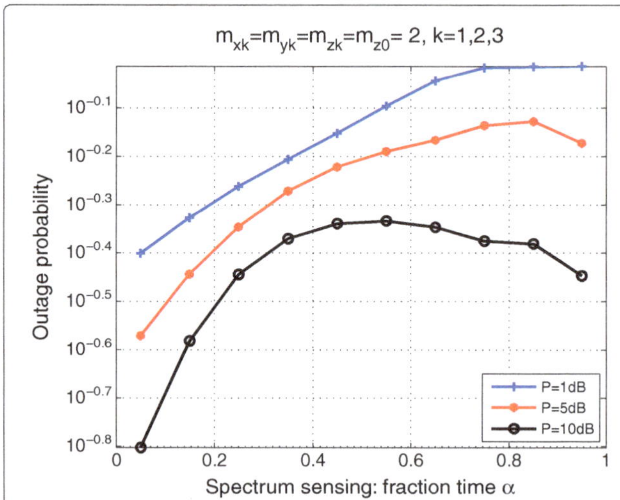

Figure 2 Outage probability versus spectrum sensing overhead $\boldsymbol{\alpha}$ for different values of the transmission power P at (CS and CR_k).

From Figure 5, we show that the outage probability decreases as probability of false alarm P_f decreases. Indeed, when P_f is small, which means that the system of detection is more reliable, which thus leads the CS to transmit with interweave method instead of underlay method when primary system is absent. Also, while looking at Figure 5, we can observe that the outage probability of secondary system converges to the same point when α go to 1, under a different value of P_f and P_d. In fact, when α go to 1, the CS expires the whole of the time slot (iT) for sensing and transmits through the whole of the time slot (iT) via underlay mode in the first phase and can

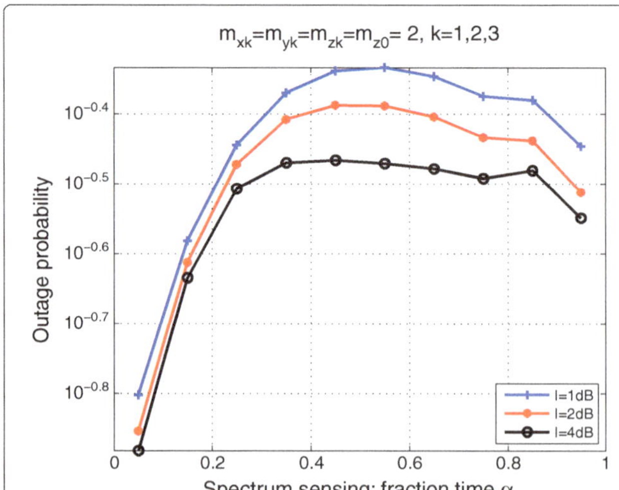

Figure 3 Outage probability versus spectrum sensing overhead $\boldsymbol{\alpha}$ for different values of the interference power constraint I.

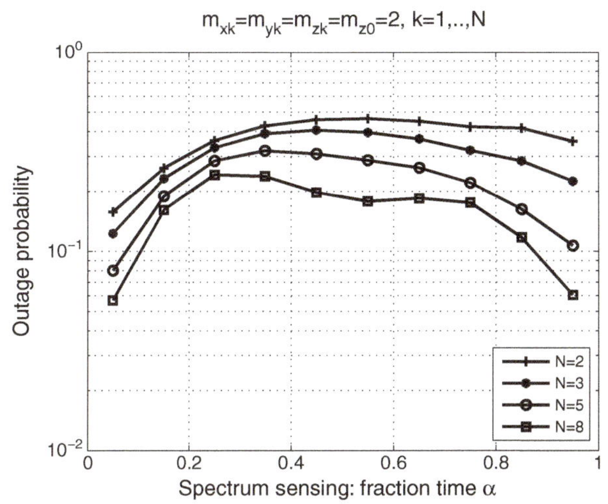

Figure 4 Outage probability versus spectrum sensing overhead **α** for different values of the number of relay *N*.

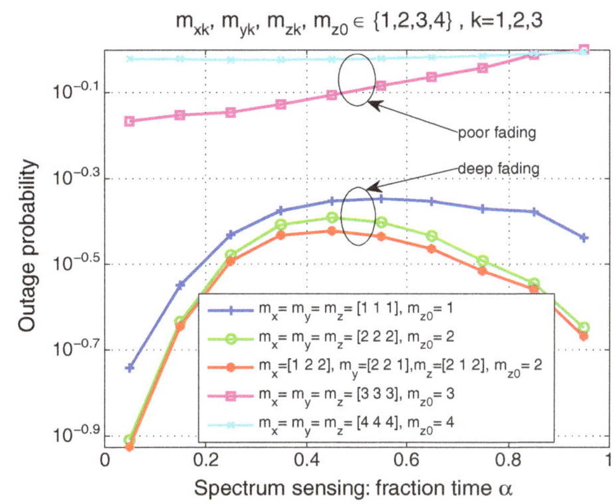

Figure 6 Outage probability versus spectrum sensing overhead **α** for different values of the fading severity parameters (m_{xk}, m_{yk}, m_{zk}, m_{z0}, k=1,2,3).

have no time in the second phase for transmitting through interweave mode whatever the probability value of detection P_d and/or of a false alarm P_f, and even if primary user is not detected or there is no false alarm.

From Figure 6, we can see the advantage of the use relays in deep fading channel. In contrast, in poor fading channel, the use relays decrease the performance.

An interesting result found from Figure 7 (resp. Figure 8) is that our strategy (Always transmit + Relay) outperforms other conventional cognitive radio strategies like interweave or underlay approach in deep (resp. poor) fading channel.

Also, Figure 7 (resp. Figure 8) presents the comparison of the outage probability between our strategies (Always transmit + Relay) and (Always transmit + No-Relay) in deep (resp. poor) fading channel. As we have shown in [16] that the strategy 'AT' enhances the performance, the performance of a cognitive system can be more improved by using relays in deep fading channel, and it recommended not to use the relays in poor fading channel.

5 Conclusion

We have presented 'Always transmit + Relay' strategy for dual-hop transmission with decode-and-forward and

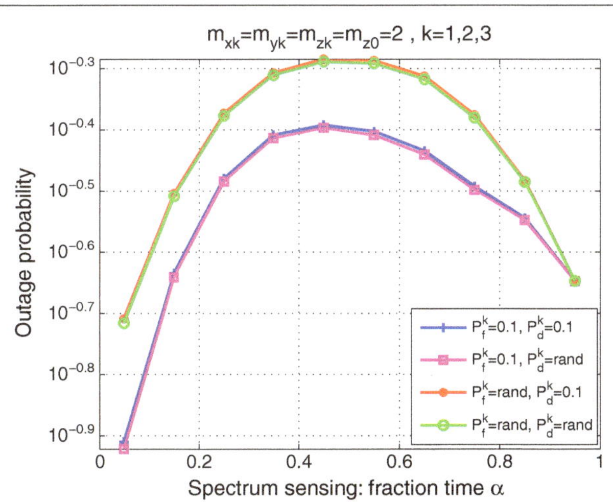

Figure 5 Outage probability versus spectrum sensing overhead **α** for a fixed/random values of the probabilities of false alarm and of detection.

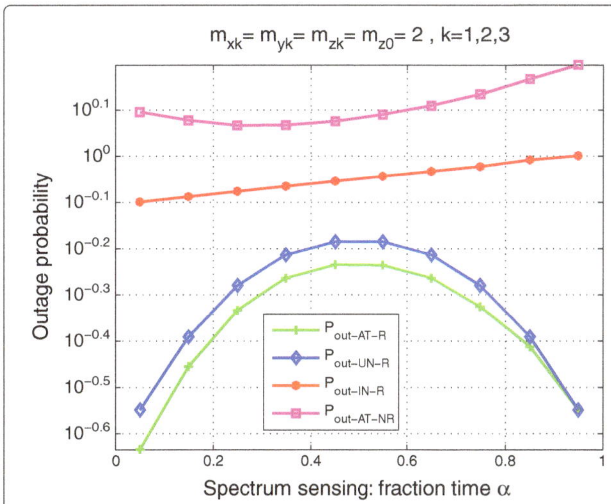

Figure 7 The outage probability of different strategies of transmission under a deep fading and a random average channel gain.

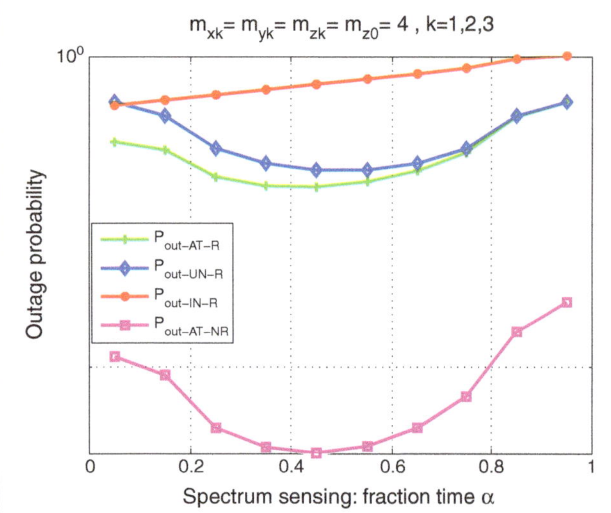

Figure 8 The outage probability of different strategies of transmission under a poor fading and a random average channel gain.

relay selection. The closed-form expressions of the outage probability have been derived over independent but not identically distributed Nakagami-m fading channels on all nodes. Simulation results have shown that the performance increases as either the transmit power P, the interference power constraint I, and the number of relays N increases and when spectrum sensing duration α is small (near to 0). Finally, we have shown that the strategy 'Always transmit + Relay' outperforms other conventional cognitive radio strategies and the use of relays enhances the performance in deep fading channel. A future work is dedicated to explore some new ideas on how to perform sensing and detecting the presence or the absence of the primary system while transmitting with underlay approach.

Appendix

Let X_k, Y_k, and $Z_k(k = 0, 1, .., N)$ follow the probability distribution $p_k(x)$, $p_k(y)$, and $p_k(z)$, respectively, and are independent random variables distributed according to gamma distribution given by the Equation (1) with different fading parameters γ_{xk}, γ_{yk}, γ_{zk} and fading severity parameters m_{xk}, m_{yk}, m_{zk}:

$$Pr\left(\min\left(X_k, Y_k\right) < A\right) = 1 - Pr\left(X_k > A\right) Pr\left(Y_k > A\right) \tag{32}$$

then:

$$Pr\left(\min\left(X_k, Y_k\right) < A\right) = 1 - \frac{\Gamma\left(m_{xk}, \frac{m_{xk}}{\gamma_{xk}} A\right)}{\Gamma\left(m_{xk}\right)} \frac{\Gamma\left(m_{yk}, \frac{m_{yk}}{\gamma_{yk}} A\right)}{\Gamma\left(m_{yk}\right)} \tag{33}$$

Let $t_k = X_k \min\left(1, \frac{B}{Z_k}\right)$, the cumulative distribution function of t_k

$$F_{t_k}(A) = Pr\left(t_k < A\right) = Pr\left(X_k \min\left(1, \frac{B}{Z_k}\right) < A\right) \tag{34}$$

$$F_{t_k}(A) = \int_{x=0}^{A} \int_{z=0}^{B} p_k(x) p_k(z) dx dz + \int_{x=0}^{\frac{AZ}{B}} \int_{z=B}^{\infty} p_k(x) p_k(z) dx dz \tag{35}$$

$$F_{t_k}(A) = \left(1 - \frac{\Gamma\left(m_{xk}, \frac{m_{xk}A}{\gamma_{xk}}\right)}{\Gamma\left(m_{xk}\right)}\right)\left(1 - \frac{\Gamma\left(m_{zk}, \frac{m_{zk}B}{\gamma_{zk}}\right)}{\Gamma\left(m_{zk}\right)}\right) + \frac{\Gamma\left(m_{zk}, \frac{m_{zk}B}{\gamma_{zk}}\right)}{\Gamma\left(m_{zk}\right)} - \int_{z=B}^{\infty} \frac{\Gamma\left(m_{xk}, \frac{m_{xk}AZ}{\gamma_{xk}B}\right)}{\Gamma\left(m_{xk}\right)} p_k(z) dz \tag{36}$$

Assuming m_x is integer and $a \geq 0$. And using the fact that $\Gamma\left(m_x, x\right) = \left(m_x - 1\right)! \, e^{-x} \sum_{i=0}^{m_x-1} \frac{x^i}{i!}$, we have:

$$\frac{\Gamma(m_x, az)}{\Gamma(m_x)} = \frac{\Gamma(m_x, az)}{(m_x - 1)!} = e^{-az} \sum_{i=0}^{m_x-1} \frac{a^i}{i!} z^i \tag{37}$$

With the help of the equation $\int_u^{\infty} x^n e^{-\mu x} dx = \mu^{-n-1} \Gamma(n+1, \mu u)$ ([19], equation (3.351.2)), we can get:

$$\int_{z=B}^{\infty} \frac{\Gamma(m_x, az)}{\Gamma(m_x)} p_k(z) dz = \left(\frac{\frac{m_z}{\gamma_z}}{\frac{m_z}{\gamma_z} + a}\right)^{m_z} \frac{1}{\Gamma(m_z)} \sum_{i=0}^{m_x-1} \left(\frac{a}{\frac{m_z}{\gamma_z} + a}\right)^{i} \times \frac{\Gamma\left(m_z + i, \left(\frac{m_z}{\gamma_z} + a\right) B\right)}{i!} \tag{38}$$

Let $a = \frac{m_{xk}A}{\gamma_{xk}B}$ and using (38), we can rewrite (36) as:

$$F_{t_k}(A) = 1 - \Phi\left(m_{xk}, \gamma_{xk}, m_{zk}, \gamma_{zk}, B, A\right) \tag{39}$$

where:

$$\Phi(m_{xk}, \gamma_{xk}, m_{zk}, \gamma_{zk}, B, A) = \frac{\Gamma\left(m_{xk}, \frac{m_{xk}A}{\gamma_{xk}}\right)}{\Gamma(m_{xk})} \left(1 - \frac{\Gamma\left(m_{zk}, \frac{m_{zk}B}{\gamma_{zk}}\right)}{\Gamma(m_{zk})}\right) - \left(\frac{\frac{m_{zk}}{\gamma_z}}{\frac{m_{zk}}{\gamma_{zk}} + \frac{m_{xk}A}{\gamma_{xk}B}}\right)^{m_{zk}} \frac{1}{\Gamma(m_{zk})} \sum_{i=0}^{m_{xk}-1} \left(\frac{\frac{m_{xk}A}{\gamma_{xk}B}}{\frac{m_{zk}}{\gamma_{zk}} + \frac{m_{xk}A}{\gamma_{xk}B}}\right)^{i} \times \frac{\Gamma\left(m_{zk} + i, \frac{m_{zk}}{\gamma_{zk}} B + \frac{m_{xk}}{\gamma_{xk}} A\right)}{i!} \tag{40}$$

Competing interests
The authors declare that they have no competing interests.

Author details
[1]Computer Sciences Department, GREENTIC/UH2C/EST, Casablanca, Morocco. [2]Computer Sciences Department, GREENTIC/UH2C/ENSEM, Casablanca, Morocco. [3]Department of Mathematics and Computer, GREENTIC/UH1/FST, Settat, Morocco.

References
1. A Goldsmith, SA Jafar, I Maric, S Srinivasa, Breaking spectrum gridlock with cognitive radios: an information theoretic perspective. Proc. IEEE. **97**(5), 894–914 (2009)
2. A Bletsas, H Shin, MZ Win, A Lippman, A simple cooperative diversity method based on network path selection. IEEE J. Sel. Areas Commun. **24**(3), 659–672 (2006)
3. L Le, E Hossain, Resource allocation for spectrum underlay in cognitive radio networks. IEEE Trans. Wireless Commun. **7**(12), 5306–5315 (2008)
4. J Lee, H Wang, JG Andrews, Outage probability of cognitive relay networks with interference constraints. IEEE Trans. Wireless Commun. **10**(2), 390–395 (2011)
5. Y-C Liang, Y Zeng, E Peh, AT Hoang, Sensing-throughput tradeoff for cognitive radio networks. IEEE Trans. Wireless Commun. **7**(4), 1326–1337 (2008)
6. Y Zou, J Zhu, B Zheng, Y-D Yao, An adaptive cooperation diversity scheme with best-relay selection in cognitive radio networks. IEEE Trans. Signal Process. **58**(10), 5438–5445 (2010)
7. X Kang, YC Liang, HK Garg, L Zhang, Sensing-based spectrum sharing in cognitive radio networks. IEEE Trans. Vehicular Tech. **58**(8), 4649–4654 (2009)
8. J Oh, W Choi, in *Vehicular Technology Conference Fall (VTC 2010-Fall) 2010 IEEE 72nd*. A hybrid cognitive radio system: a combination of underlay and overlay approaches, (2010), pp. 1–5
9. X Bao, P Martins, T Song, L Shen, Capacity of hybrid cognitive network with outage constraints. IET Commun. **5**(18), 2712–2720 (2011)
10. Y Zhang, PC Ching, Q Zhang, in *2013 IEEE International Conference on Acoustics, Speech and Signal Processing (ICASSP)*. Hybrid spectrum sharing with imperfect sensing in fading channels, (2013), pp. 5318–5322
11. X Tan, H Zhang, J Hu, Capacity maximisation of the secondary link in cognitive radio networks with hybrid spectrum access strategy. IET Commun. **8**(5), 1751–8628 (2014)
12. M Usman, Koo, Insoo "Access Strategy for Hybrid Underlay-Overlay Cognitive Radios With Energy Harvesting,". Sensors Journal, IEEE. **14**(9), 3164–3173 (Sept. 2014)
13. J Zou, H Xiong, D Wang, CW Chen, Optimal power allocation for hybrid overlay/underlay spectrum sharing in multiband cognitive radio networks. IEEE Trans. Vehicular Technol. **62**(4), 1827–1837 (May 2013)
14. G Vardoulias, J Faroughi-Esfahani, G Clemo, R Haines, in *Proc. Int. Conf. 3G Mobile Communication Technologies. Second International Conference on (Conf. Publ. No. 477)*. Blind radio access technology discovery and monitoring for software defined radio communication systems: problems and techniques (UK London, 2001), pp. 306–310
15. S Shankar, C Cordeiro, K Challapali, in *New Frontiers in Dynamic Spectrum Access Networks. DySPAN 2005. First IEEE International Symposium on*. Spectrum agile radios: utilization and sensing architectures (Baltimore, Maryland, USA, 2005), pp. 160–169
16. M Ridouani, A Hayar, in *20th International Conference on Telecommunications*. General scheme for always transmit solution in cognitive radio systems (Casablanca, Morocco, 2013), pp. 1–5
17. M Ridouani, A Hayar, in *Next Generation Networks and Services (NGNS)*. Outage probability based unified framework for always transmit solution in cognitive radio systems (Algarve, Portugal, 2012), pp. 59–65
18. M Ridouani, A Hayar, A Haqiq, in *1st International Workshop on Cognitive Cellular Systems (CCS)*. QoS of cognitive and primary networks and spectrum sensing tradeoff using always transmit strategy over Nakagami-m fading channels (Rhine River, Germany, 2014), pp. 1–5
19. M Nakagami, in *Statistical Methods in Radio Wave Propagation*. The m-distribution - a general formula of intensity distribution of rapid fading (England Pergamon Press, Oxford, 1960), pp. 3–36
20. IS Gradshteyn, IM Ryzhik, *Table of Integrals, Series, Products*, 4th ed. (Academic Press, San Diego, California, 2000)
21. E Peh, Y-C Liang, in *Wireless Communications and Networking Conference. WCNC 2007*. Optimization for cooperative sensing in cognitive radio networks (Hong Kong, 2007), pp. 27–32
22. A Mariani, S Kandeepan, A Giorgetti, A, Periodic Spectrum Sensing With Non-Continuous Primary User Transmissions. Wireless Commun. IEEE Trans. **14**(3), 1636–1649 (2015)
23. A Gorcin, H Arslan, Signal identification for adaptive spectrum hyperspace access in wireless communications systems. IEEE Communications Magazine. **52**(10), 134–145 (Oct. 2014)
24. JN Laneman, DNC Tse, GW Wornell, Cooperative diversity in wireless networks: efficient protocols and outage behavior. IEEE Trans. Inform. Theory. **50**(12), 3062–3080 (2004)
25. A Muller, J Speidel, in *IEEE International Conference on Communications ICC '08*. Relay selection in dual-hop transmission systems: selection strategies and performance results, (2008), pp. 4998–5003

9

A data dissemination mechanism for motorway environment in VANETs

Yang Yang[1*], Qian Liu[1], Zhipeng Gao[1], Xuesong Qiu[1], Lanlan Rui[1] and Xin Li[2]

Abstract

An efficient data dissemination mechanism is very critical for traffic safety and inquiring path information. All vehicles go straight and do not consider the effect of traffic lights during its running in the motorway. But in view of characteristic of motorway, the fast moving of vehicles will cause the frequent changes of network topology, which is a challenge for data dissemination for motorway environment in Vehicular *ad hoc* networks (VANETs). This paper proposes a novel data dissemination mechanism for VANETs which apply to broadcasting and on-demand unicast scenarios. We firstly propose a weighted *K*-nearest neighbors prediction algorithm with self-feedback to predict the current speed of a vehicle and construct a set of forwarders for emergency message broadcasting. To satisfy users' personal demands, the paper also designs a unicast mechanism based on information urgent level and the improved Greedy Perimeter Stateless Routing (GPSR) protocol. The simulation results show that our proposed method can achieve better performance with higher delivery ratio, lower number of forwarders, and less copies of messages.

Keywords: VANETs; Data dissemination; On demand; Delivery ratio

1 Introduction

Vehicular *ad hoc* networks (VANETs) are a special class of mobile *ad hoc* networks (MANETs) where nodes self-organize and self-manage in a distributed way. The networks consist of vehicles and/or roadside infrastructures that communicate with each other. Apart from conventional *ad hoc* networks, VANETs possess some special characteristics, such as highly dynamic topology and frequent disconnection [1,2].

In recent years, increasing research efforts have been put into VANETs, such as routing protocol, data dissemination, information security, etc. Among these research fields, the data dissemination mechanism has gained more attention as it determines whether users acquire on-demand services timely. Data dissemination in VANETs is frequently used as a way to disseminate messages (i.e., emergency accident, traffic information services, and advertisement) among a group of vehicles. Vehicles are equipped with the radio devices which allow the vehicles to communicate with others located within one-hop communication range. Then, the messages will be disseminated and shared among a group of moving vehicles [3].

In the motorway environment, there are two types of messages routed on the network layer in VANETs currently:

- *Cooperative Awareness Messages* (*CAM*). They contain basic status information like the current position, speed, acceleration, as well as the vehicle identifier. These parameter values are periodically exchanged by *hello* message between vehicles in one hop.
- *Decentralized Environmental Notification Messages* (*DENM*). These messages may be related to traffic condition information, query/response message, emergency warning information, etc. It always uses the multi-hop mechanisms to efficiently deliver these messages.

Generally, the works of data dissemination in VANETs are mainly focused on *DENM* messages. Consequently, for emergency warning information, they are disseminated using broadcasting scheme. The main advantage of broadcasting scheme is that a vehicle does not need to know a

* Correspondence: yyang@bupt.edu.cn
[1]State Key Laboratory of Networking and Switching Technology, Beijing University of Posts and Telecommunications, No. 10 Xi Tu Cheng Road, Beijing 100876, China
Full list of author information is available at the end of the article

route to a specific destination, and it spreads the emergency information rapidly. However, a large scale of broadcasting the same message will cause the problems of channel contention and packet collisions. Rebroadcasting due to transmission failure makes the problem even worse. This is called broadcast storm problem. To alleviate this problem, a common solution employed by most researchers is to reduce the number of redundant rebroadcast messages by only selecting part of vehicles for relaying messages. While as to simple user, he/she may have end-to-end communication demands. So, unicast is a more appropriate way for information inquiry demands. But some mechanisms have lower data delivery ratio when only considering the minimum path or deadline priority.

As vehicles move at high speed in the motorway, the topology formed by vehicles is always changing. The highly dynamic topology results in a frequent disconnection of link between two vehicles. Based on this, the *DENM* messages are classified into two types in the paper: the emergency message and the inquiry message. This paper proposes a data Dissemination Service for Broadcasting and on-Demand unicast in the VANETs (DSV-BD). We use different transmission methods to deliver different types of messages. For emergency messages, a proposed broadcasting algorithm can make the emergency messages cover a limited network domain rapidly. On the other hand, if some vehicles request real-time messages of personal interest, an on-demand unicast mechanism is proposed to reduce network load.

The rest of the paper is organized as follows. In Section 2, we investigate previous related works on data dissemination protocols in VANETs. The broadcasting algorithm for emergency message and the on-demand unicast mechanism for inquiry message are proposed in Section 3 and Section 4. In these two parts, we give the assumption of our system and introduce several terminologies and key mechanisms. Section 5 validates the solution by simulations. Finally, we give the conclusions of this paper in Section 6.

2 Related work

Flooding is a fundamental mechanism to implement the data dissemination in VANETs. Unfortunately, flooding introduces significant communication overhead due to redundant rebroadcasting. This will cause the well-known problem called broadcast storm problem. A common solution is to reduce the number of redundant rebroadcast packets. This is typically solved by selecting some special vehicles to relay the packet rather than rebroadcast it.

S.Y. Ni et al. [4] research five different classes of solutions: probabilistic schemes, counter-based schemes, distance-based schemes, location-based schemes, and cluster-based schemes for broadcast storm problem. Simulation results show that a counter-based scheme can reduce some redundant rebroadcasts when the host distribution is dense. While location-based scheme can achieve higher performance when GPS service is available. S. Panichpapiboon et al. [5] gave a review of variety data dissemination protocols for VANETs. The authors introduce a new metric called dissemination efficiency to discuss the performance evaluation of those protocols.

The urban multi-hop broadcast (UMB) protocol [6] is designed to solve the broadcast storm and the reliability in multi-hop broadcasting. The forward vehicle tries to select a far relay vehicle in the intended direction on a linear road through request to broadcast (RTB) packet and clear to broadcast (CTB) packet. In order to disseminate the message in all directions, a repeater vehicle is employed at the intersection to rebroadcast the message. UMB is inefficient in a sense that the next rebroadcast vehicle has to wait the longest before being able to transmit the CTB packet. This is because the longest black-burst duration is assigned to the next rebroadcast vehicle. G. Korkmaz et al. [7] proposed a fully *ad hoc* multi-hop broadcast protocol (AMB) for inter-vehicular networks. The AMB is an *ad hoc* extension of the UMB protocol. The protocol eliminates drawbacks of the UMB protocol by employing an efficient intersection broadcast mechanism. L. Wischof et al. [8] presented traffic information services. Road conditions and information on traffic jams are disseminated to vehicles and used for navigation and early warning.

Some researchers proposed a bio-inspired metadata harvesting algorithm [9]. The main goal of this algorithm is to effectively cover large urban areas and favor agent movements towards information patches where the concentration of metadata is higher. Zhao et al. [10] proposed a carry-and-forward mechanism for sparsely connected network. It can disseminate vehicular traffic messages when there is no other node between source and destination vehicles. S. Lee et al. [11] proposed a hybrid communication protocol due to the need of exploiting network infrastructure to avoid disconnections in sparse-traffic scenarios. Moreover, the hybrid approach is particularly suitable for comforting applications that use network infrastructures for advertisement dissemination. A prioritized handling of events can be achieved, for instance, by a priority queue as it has been applied in [12]. The authors set different priorities for different events, and then the emergency events have the highest priority than other events. The feasibility is limited due to the excessive control overhead to collect neighbor position including vehicles traveling in the opposite direction.

Greedy Perimeter Stateless Routing (GPSR), a novel routing protocol for vehicle *ad hoc* network that utilizes the positions of routers and a packet's destination to make packet-forwarding decisions [13]. GPSR algorithm uses a greedy algorithm to establish the route. Greedy algorithm is a method that the upstream node chooses a node that is nearest to the destination from its neighbors as its next hop. When the node finds that its own distance to the destination is the shortest compared with the other distances between its neighbors and the destination node, the node goes into a local minimum problem. Then, the GPSR algorithm goes into perimeter forwarding mode [14]. During this process, the *right-hand* rule combined with face-routing algorithm can solve the Local Minimum Problem. D.Q. Li et al. [15] proposed a new algorithm based on angle and distance (GPSR-AD) for solving the problem which GPSR may produce excessive unwanted route hops when spatial neighbor exists in the *ad hoc* network.

Although the abovementioned data dissemination mechanisms can reduce some network overheads, they do not have an appropriate solution for emergency and personal interested information simultaneously. Based on this, we propose a data dissemination mechanism based on emergency broadcasting and on-demand unicast for inquiry message.

3 Broadcasting-based data dissemination

The advantage of our proposed broadcasting scheme is that it abandons the arbitrary method used in the traditional VANETs that might cause a broadcast storm. The arbitrary method causes not only network congestion and subsequent loss of data but also a waste of network resource. Despite some improvements in epidemic protocol and UMB protocol that introduce a probabilistic resilient multicast scheme to avoid broadcast storm, broadcasting messages must be repeated periodically to ensure the message coverage rate in the network. Some improved algorithms such as the topology discovery-based data dissemination algorithms need to transfer the messages and find the next forwarder at the same time. But they need to set the waiting timer for timeout so that the requirements of rapid transferring cannot be satisfied.

In this paper, vehicles get information of their neighbors via hello messages. Before broadcasting, source vehicle adjusts its broadcast radius according to the neighbors' information, and then it chooses some appropriate neighbors to construct a forwarder set. The algorithm can effectively increase message coverage rate and avoid broadcast storm, as well as the network congestion.

3.1 Assumptions and definitions

In this paper, we have the assumptions as follows:

- Each vehicle knows its location through the global positioning system (GPS) service at any given time.
- Each vehicle is equipped with several sensors for monitoring any events and a radio that is capable of short-range wireless communication.
- All vehicles broadcast the hello messages periodically, so that they can know the information (e.g., the current position, speed) of their neighboring vehicles.
- No roadside infrastructure is available during the communication among the vehicles in motorway.
- All vehicles go straight and do not consider the effect of traffic lights during its running in motorway.

The neighbors of a vehicle are usually located in its communication range. We will define the senses by using Figures 1 and 2 as follows.

As is shown in Figure 1, the source vehicle *S* broadcasts a hello message at time t_1. The hello message will be received by vehicles *B*, *E*, *F*, *H*, *I*, and *J*. In Figure 2, the vehicles that have received the hello messages send a response message to the source vehicle at time t_2. But the source vehicle *S* will only receive the response messages from vehicles *B*, *E*, *F*, *H*, and *I*, because vehicle *J* moves out of the communication range of vehicle *S*. So, the neighbors of vehicle *S* are vehicles *B*, *E*, *F*, *H*, and *I*, not containing vehicle *J*.

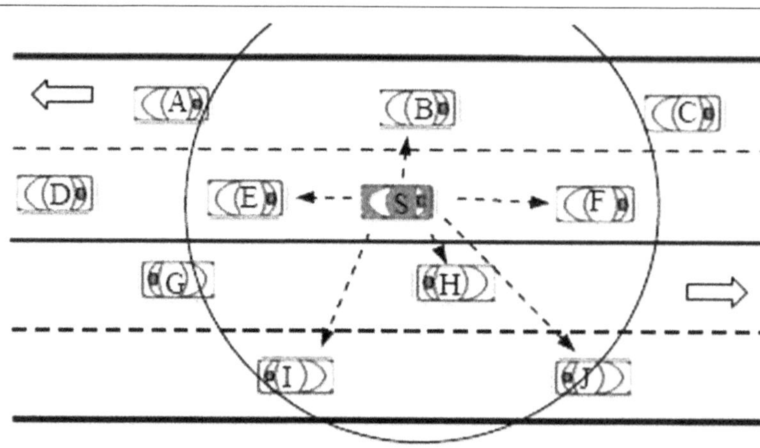

Figure 1 Source vehicle *S* broadcasts a hello message at time t_1 in the motorway.

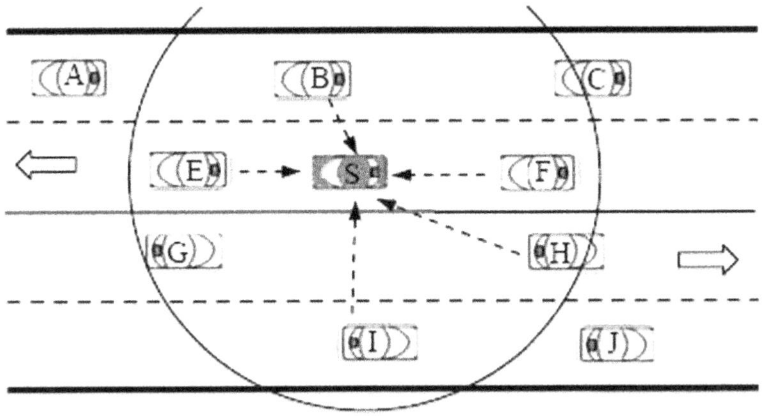

Figure 2 **Source vehicle S receives the response message from its neighbor vehicles at time t_2.**

The hello message includes vehicle's identity, position, velocity, broadcast radius, and time stamp. We define that the interval of broadcasting hello message behaves according to a uniform distribution U (0.5P, 1.5P). Here, the symbol P is an average of historical interval of broadcasting. Therefore, the time a vehicle node receives the hello message from other vehicle nodes is evenly distributed. By this way, each node can update their neighbors' information in time and avoid the broadcast storm as well.

We assume that an emergency event-driven message is generated by an observer vehicle when it passed by the spot where an emergency event occurs (e.g., car accident). The message includes the source vehicle ID, position, timestamp, event type, and forwarders' IDs. The field position stores the geographical coordinates of event, and the field forwarders' IDs indicates all the forwarders, and the emergency event message will be forwarded once it arrived at these nodes.

3.2 Prediction for vehicle state

In order to broadcast the emergency message, the source node needs to select part of its neighbors to forward the message. But the state information of neighbors always became outdated, and it may lead to the inaccuracy of the selection of the forwarders. In this section, we propose a weighted K-nearest neighbors prediction algorithm with self-feedback to predict the current speed of a vehicle. That is to predict the real-time status of a being predicted node by its K-nearest neighbors' historical data. The innovation of our algorithm is as follows.

1. The method for choosing the number of neighbors' historical data depends on the neighbors' corresponding historical coordinates. We assume that during a range of road, the road condition could be similar and the speed of vehicles could be stable.
2. The value of K is related to the number of neighbors that the being predicted node can detect. It is unreasonable to choose a fixed value of K because of the dynamic neighbors on different roads at different times. So, the principle taken in our algorithm is so long as the distance between neighbors' historical data and the being predicted node's historical data fluctuates within a certain scope, the corresponding neighbor can be regarded as a close neighbor.
3. Instead of using Euclidean distance or Mahalanobis distance, we adopt a new method that uses degree of dispersion to evaluate the distance between historical data, regardless of the dimension of historical velocity vectors.
4. A weighted prediction algorithm is adopted according to the distance between neighbors' historical data and the predicted node's historical data. A neighbor that is closer geographically to the predicted node would have a greater weight.
5. After prediction, the source node evaluates errors between the true value and the prediction value, so that it adaptively adjusts the judgment for being a close neighbor.

Assume that the information of neighbor node A is detected outdated by source node S at time t, then node S will go the steps of prediction as follows.

1. Extract the broadcast radius of node A from the latest hello message, called R_A. Then, traverse the neighbor set of S (include S). For each node X with timely information in the neighbor set, if the distance between A and X is less than R_A, and the velocity direction of X is the same as that of A, then regard X as a temporary neighbor of A and put it into a set called M.
2. For each node X in set M, node S extracts the historical velocity of X from its hello messages and maintains the historical records for a long term. It is considered in this paper that during a 100-m long section of road, the road condition could be

similar, and the speed of vehicles could be stable. So, we only focus on the historical velocity in a hello message that indicates the node's historical coordinate is within 100 m from the node's current location, and we call them valid historical velocity. Thus, the number of historical velocities extracted can be determined. Suppose there are i historical velocities S gets from node X's hello messages, then the historical velocity vector of X is $V_{Xh} = (v_{X1}, v_{X2},..., v_{Xi})$. Accordingly, the historical velocity vector of node A is $V_{Ah} = (v_{A1}, v_{A2},..., v_{Ai})$. If S cannot get enough historical hello messages of node A, then the dimension of the two velocity vectors is min (i_X, i_A), where i_X is the number of valid historical velocity of node X, and i_A is the number of historical velocity of node A that S can get. Notice that, for different nodes in set M, the dimension of historical velocity vector is different.

3. Because historical velocity vectors have different dimensions, Euclidean distance between vectors that have high dimensions would be greater than that having low dimensions. In order to evaluate the distance between neighbors' historical velocity and node A's historical velocity, we introduce degree of dispersion. A high discrete degree means a greater distance to node A. We can calculate the distance between node X and A by the following formula.

$$dis(X,A) = \sqrt{\frac{1}{i_X} * \sum_{j=1}^{i_X} (v_{Xj} - v_{Aj})^2} \tag{1}$$

where i_X indicates the dimension of historical velocity vectors constructed in step (2). The standard of being a close neighbor to node A is dis$(X, A) \le r * V_A$, where V_A is the velocity of node A in the latest hello message that node S received, and r is a threshold factor set to 0.1 in the first round prediction.

4. All the nodes in set M that satisfy dis$(X, A) \le r * V_A$ will be regarded as close neighbors to node A. Assume that there are K close neighbors, then the weight assigned to a close neighbor is as follows.

$$W_X = \frac{1}{u_X^2} \Big/ u2 \tag{2}$$

$$u_X^2 = dis^2(X,A) = \frac{1}{i_X} * \sum_{j=1}^{i_X} (v_{Xj} - v_{Aj})^2 \tag{3}$$

$$u^2 = \sum_{X=1}^{K} \frac{1}{u_X^2} \tag{4}$$

where u_X^2 reflects the distance between node X and node A. Evidently though, a higher weight will be assigned to a node that is closer to node A.

5. The final prediction for the real-time velocity of node A is:

$$V\tilde{A} = \sum_{X=1}^{K} \left(W_X * \frac{1}{i_X} * \sum_{j=1}^{i_X} v_{Xj} \right) \tag{5}$$

6. Define a variation factor $\delta = V_{\tilde{A}}/V_A$, where V_A represents the velocity of A in the latest hello message that S has received.

7. Predict the distance between A and S at time t.

$$dS\tilde{A} = \sqrt{(x\tilde{A} - x_S)^2 + (y\tilde{A} - y_s)^2} \tag{6}$$

where x_S and y_S represent the coordinates of node S at the calculation time. $x_{\tilde{A}}$ and $y_{\tilde{A}}$ are the prediction coordinates of node A and can be calculated as follows.

$$\frac{x\tilde{A} = x_A + V\tilde{A} - V_A}{2(t - t_A)} \tag{7}$$

$$y\tilde{A} = y_A \tag{8}$$

The parameter t_A represents the timestamp in the latest hello message of node A that S has received. x_A and y_A are the coordinates of node A at time t_A. We consider that node A and node S move in the same direction or in the opposite direction, so the y coordinate remains unchanged.

8. The prediction broadcast radius of node A at time t can be calculated as $R_{\tilde{A}} = R_A * \delta$, where R_A is the broadcast radius of A in the latest hello message that S has received. In the worst case, there is no eligible node in step (1), and then the original values of $dS\tilde{A}$ and $R_{\tilde{A}}$ will be retained.
9. Before the next prediction, node S can get node A's last hello message. Extract the velocity V'_A and regard it as a true value at the prediction time. Define the relative error $E = |V'_A - V\tilde{A}| / V'_A$ and adaptively adjusts the threshold factor r to reduce the relative error.

3.3 Construct the set of forwarders

In order to reduce the influence of broadcast storm, only a part of appropriate neighbors will be selected to forward the emergency message. The set of forwarders of S can be constructed as follows.

1. Define the set of forwarders of S called F_S, and then put the neighbor node of S which has the maximum distance to S into F_S.
2. Define the rest neighbors of S as a sequence $\{X_i \mid i = 1, 2,\ldots, n-1\}$, where n is the number of neighbors of S, and X_i is sorted by the distance between X_i and S from the greatest to smallest. The nodes in F_S is called Y_j ($j = 1, 2,\ldots, m$, where m is the number of nodes in F_S). For a given X_i, if the distance between X_i and Y_j is greater than the latest broadcast radius R_{Yj} for all Y_j ($j = 1, 2,\ldots, m$), then put X_i into F_S. Otherwise, abandon X_i.
3. Traverse all neighbors of S, and eventually, F_S contains all forwarders of S.

3.4 The algorithm for broadcasting emergency message

The source vehicle S receives hello messages from its neighbors at a random time and save the information into its neighbor list. If there is a new hello message from node A, then the information of A in the neighbor list will be updated. Since node S cannot obtain the future status of its neighbors, it will execute the prediction algorithm to make its neighbors' information more accurate.

Once an emergency event is detected by node S at time t, then the broadcasting for emergency messages starts as follows.

1. The source vehicle S traverses the neighbor list and extracts the information of its neighbors.
2. For each neighbor of S, if the difference between its timestamp (the timestamp field in its latest hello message) and current time is less than or equal to an overdue threshold dl, then the information of this neighbor node is considered as real time. Otherwise, S executes the prediction algorithm for this neighbor node. The threshold dl is recommended to be set to $P/2$, where P is an average of historical interval of broadcasting.
3. If all the neighbors' information is real time or has been updated by the predicted value, then S calculates its broadcast radius and constructs the set of forwarders F_S.
4. S constructs the emergency event message and fills the last field with the forwarder *IDs* in set F_S, and then it broadcasts this message.

The forwarders also need to broadcast the emergency event messages of source node S, so that each forwarder will execute the broadcasting algorithm mentioned above. Then, the emergency event message can spread rapidly in the network.

In the proposed broadcasting algorithm, the broadcasting radius can be adjusted according to its surroundings. By this way, on the one hand, the situation that some nodes cannot be covered because of a small broadcast radius can be avoided. On the other hand, the problem of signal attenuation caused by a too large broadcast radius can also be solved. Therefore, the proposed broadcasting algorithm improves the coverage rate of nodes in the network. We construct the forwarders set F_S, and the overall network can be covered using less nodes. The introduction of overdue threshold dl ensures that the information saved in neighbor lists can be closer to a true value. By analyzing the latest status of neighbors, we hope that the prediction can be more accurate.

4 The on-demand unicast algorithm for inquiry message

While as to simple user, he/she may have end-to-end communication demands. As all of vehicles need not respond the request messages from a specific individual, we design an on-demand unicast scheme for these messages. The fields of information request message must include the type of information (e.g., parking place, gas station location, traffic congestion information, etc.) and deadline of receiving reply. The message should be broadcasted in order to acquire timely reply while the reply is sent by unicast.

4.1 Reply sorting

When a source node has some reply messages to transmit, it should sort the messages to satisfy deadlines of different requestors. As a node only transmits a message at one time, the source selects a node according to the expected deadline. It is assumed that $E_{ij}(t)$ is the emergency level for reply message j to node i at time t.

$$E_{ij}(t) = \frac{T_{ij} - t - MT_{ij}}{\frac{1}{n} * \sum_{k=1}^{n} (T_{kj} - t)} \tag{9}$$

$\{P_1, P_2,\ldots, P_n\}$ is the set of destination nodes (requestors). For each node P_i, T_{ij} is the deadline for transmitting message j to node P_i ($i = 1,2,\ldots,n$). MT_{ij} is the transmission time for message j to node P_i. The source evaluates the transmission time for each requestor. The more the value $E_{ij}(t)$ is, the sooner the message j is sent. The overhead of this sorting process is related to the number of messages the source node needs to reply, and the time computational complexity of formula (9) is $O(n)$, where n is the number of destination nodes. Through ranking the sequence for messages, the source vehicle can decide which message should be sent at first.

4.2 Data dissemination route protocol

In the paper, the algorithm is concerned to data distribution in the motorway. In the original GPSR algorithm, the data distribution process can be described as follows:

1. Greedy mode. An upstream node selects some of its neighbor nodes which is nearest to the destination and regards it as the next hop. When the node finds that its own distance to the destination is the shortest path compared with the distance between each of neighbors and the destination node, then the algorithm turns into perimeter forwarding mode.
2. Perimeter forwarding mode. During this process, the *right-hand* rule combined with face-routing algorithm is applied to find the next hop. Though the *right-hand* rule can ensure the upstream node to find its next hop, the choice of the next hop is accompanied by arbitrariness, which may lead to more route hops.

In order to reduce computation complexity and avoid excessive unwanted route hops caused by *right-hand* rule, we maintain the greedy mode and modify the perimeter forwarding mode to select an appropriate next hop node whose distance and angle are near the destination. When it turns into perimeter forwarding mode, then the node will calculate the next hop neighbor. We will find the neighbor of N_i with minimum D_{i+1}.

$$D_{i+1} = \frac{dis(N_{i+1,}G)}{dis(N_{i,}G)} + \frac{\angle N_{i+1}N_iG}{\angle XN_iG} \tag{10}$$

In Figure 3a, node G is the destination, and N_i is the current forwarding node. The expressions dis(N_{i+1},G) and dis(N_i,G) respectively depict the distance between N_{i+1} and N_i to G. The node X is an intersection of two circles whose centers are respectively N_i and G. According to the formula (10), we find the neighbor of N_i with minimum D_{i+1}. Then, the corresponding node N_{i+1} is the next hop node. By this algorithm, when greedy mode fails, the current forwarding node chooses a next hop node which is closer to the destination in the respects of distance and angle [14]. So, the multi-hop and fallback problems in GPSR can be solved.

In Figure 3b, we can observe the significant advantage of GPSR-AD clearly. The arrows mean the information stream direction. Node D is the destination, and when the message reaches node C, there is no other node closer to node D than node C, so the algorithm turns into perimeter forwarding mode. The forward routing path is *C-K-E-F-G-H-I-J-L-D* according to *right-hand* rule in the original GPSR. But by formula (10), the forward routing path can be simplified to *C-H-I-J-L-D*. It is obvious that GPSR-AD algorithm reduces route hops and improves the efficiency.

4.3 Demand-driven data dissemination mechanism

The data distribution is designed as follows:

1. If a source node S receives more than one inquiry message, then S should calculate emergency level for each message and determine the order of message delivery. According to the locations of all the requestors, S classifies them into *forward* and *backward* groups. The former consists of the requestors locate in front of the source, and the latter includes nodes at the back (Figure 4 is an example of the location group). The angle θ is the angle between the positive Y-axis and the line from S to any requestors. If $\sin\theta \geq 0$, then the requestor is a *forward* node, or else is a *backward* node. In Figure 4, node E, F, and G are *forward* nodes, and node A, B, C, and D are *backward* nodes.
2. Node S forms two destination address lists according to *forward* and *backward* groups, and then constructs two reply messages filled with the destination address lists and message types (specifically, parking place, gas station location, traffic congestion information, etc.). When S starts to transmit the reply messages, it selects the nearest requestor (e.g., A_1) from the address list as the destination and transmits the packet using improved GPSR algorithm.
3. When the reply message arrives to A_1, node A_1 acquires the message content, deletes itself from the list, and selects a nearest requestor as next destination to forward the message.
4. Any forwarding node should calculate the urgent level to determine the order of message delivery if it needs to forward more than one inquiry message. If any request exists in the destination list, the algorithm returns to step (3), or else this type of message is sent completed.
5. If a requestor receives the reply before the deadline, then the procedure of *request-response* is successful, or else it repeats the request in view of its demands.

5 Performance evaluation

The number of vehicles in actual world is various, and the road structure is complex, so we choose a type of motorway model as shown in Figure 4. All vehicles go straight and cannot change driveway during its running. The vehicle velocity is randomly set but still limited by traffic density. In addition, the traffic lights are not considered in motorway. In our simulation, the number of vehicles ranges from 100 to 700 and the velocity from 60 to 100 km/h. The MAC protocol is 2 Mbps 802.11*p*. The proposed algorithms are coded in MATLAB.

5.1 Simulation of broadcasting of emergency message

Our algorithm (DSV-BD) is compared with the UMB protocol and the epidemic scheme. In these two algorithms, broadcasting radius is set to 50 *m*. The simulation is repeated for 30 times, and the results are the average of those experiments. There are three indicators in our simulations:

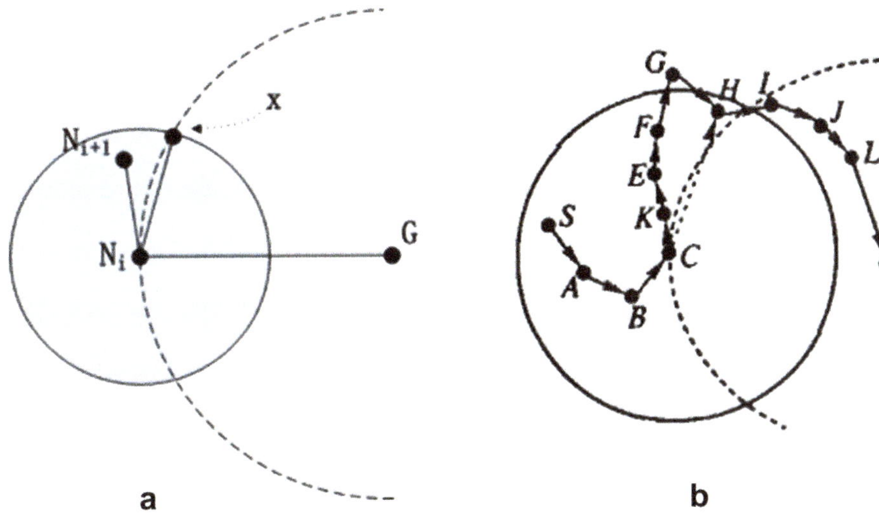

Figure 3 Illustration of a formula and comparison between GPSR and GPSR-AD. **(a)** The illustration for formula (10); **(b)** Comparison between GPSR and GPSR-AD.

Delivery ratio: a probability defined as the number of message receivers divided by the total number of vehicles in the networks.

Number of forwarders: the number of vehicles which forward the message in a round of broadcasting. Generally, a smaller number of forwarders indicate less delay of message and lower network loads.

Number of message copies: the number of message copies in a round of broadcasting. Once a forwarder receives a message, it generates a copy. This indicator reflects the network congestion level and influences the delivery ratio and transmission delays.

It is known that the traffic density is varying by different roadsides and different times. So, we study the performance of the algorithm for various traffic densities by varying the total number of vehicles in the network. The size of this simulation scenario is 1,000 m × 50 m. We assume that the number of vehicles increases from 100 to 700.

Also, message coverage area has an influence on three abovementioned indicators. It reflects the distribution and spreading ability of emergency messages. We adjust the length of road from 500 to 3,500 m to simulate the change of message coverage area. In this simulation scenario, the number of vehicles is set to 500 and remains unchanged.

5.1.1 Delivery ratio

Figure 5a shows the comparison of data delivery ratio in the three algorithms in the case of different number of vehicles. When the number of vehicles is lower, delivery ratio is about 70.8% in these algorithms. But as the number of vehicles increases, delivery radio increases linearly over 90.3%. This is because as the number of vehicles increases, the number of forwarders increases accordingly, so that the emergency message can cover more vehicle nodes. However, the delivery ratio of DSV-BD is higher

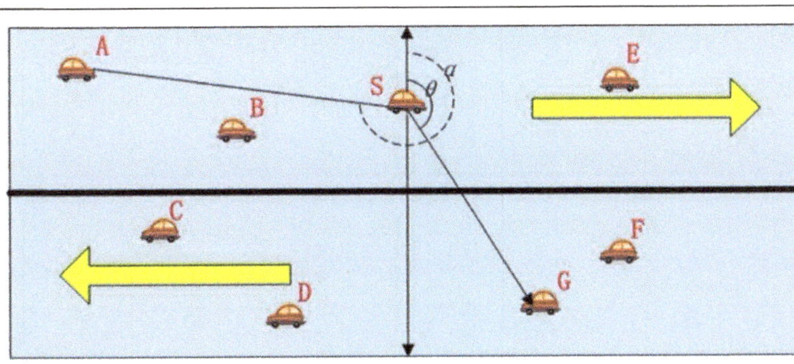

Figure 4 An example of location group.

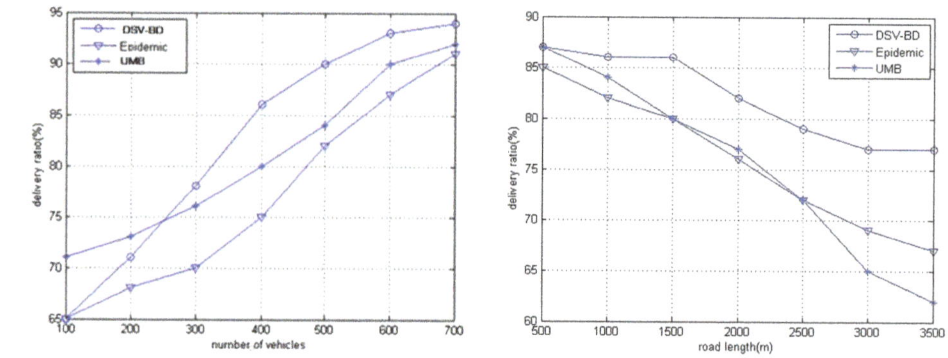

a b

Figure 5 Comparison of data delivery ratio with different number of vehicles and different message coverage areas. (a) Data delivery ratio with different number of vehicles. **(b)** Data delivery ratio with different message coverages.

than other protocols especially when the number of vehicles is higher. The average delivery radio of DSV-BD is 6.92% higher than epidemic and 1.95% higher than UMB when the number of vehicles is 700.

Figure 5b shows the comparison of data delivery ratio in the three algorithms in the case of different message coverage area. From the chart, we can see that data delivery ratio has a sharp decrease due to data collisions. This is because as the length of road increases, the message coverage area increases accordingly. So, that delay of messages increases and leads to message dropping. By comparison, the delivery ratio of DSV-BD has the slowest descending speed and is 7.90% higher than that of epidemic and 8.91% higher than that of UMB on average.

5.1.2 Number of forwarders

Figure 6a shows that in case of the different number of vehicles, the number of forwarders in broadcasting varies. It increases linearly as the number of vehicles increases. Also, it is clearly observed that DSV-BD only needs fewer forwarders compared with other two algorithms. The fitting calculations for the three slopes of the line are DSV-BD 0.0317, epidemic 2.156, and UMB 2.86. The number of forwarders in our algorithm rises slowly and has a distinct advantage.

In Figure 6b, with increasing length of road, the number of forwarders increases linearly. The fitting calculations for the three slopes of the line are DSV-BD 0.143, epidemic 0.488, and UMB 0.854. It is obvious that our algorithm has fewer forwarders than the other two algorithms, so the transmission time and network loads can be reduced.

5.1.3 Number of message copies

Figure 7a shows the comparison of number of message copies in the three algorithms in the case of the different number of vehicles. The number of message copies reflects the congestion status of networks, and the network congestion influences the transmission delay directly. In DSV-BD algorithm, less forwarders lead to less message copies. As seen from the chart, with increasing of the

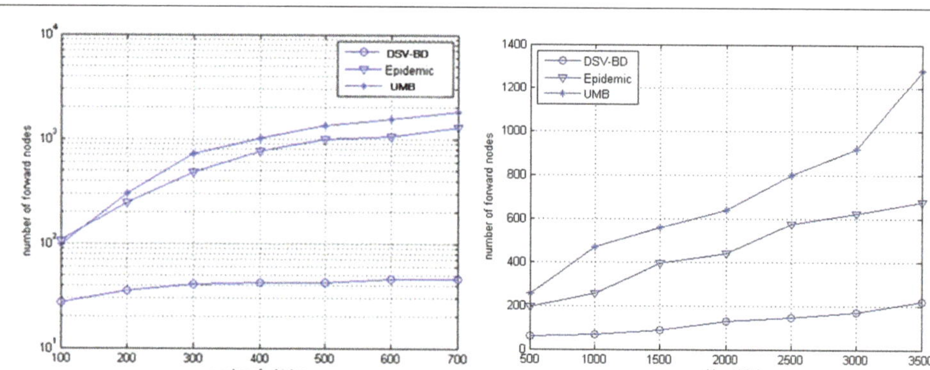

a b

Figure 6 Comparison of different number of forwarders and different message coverage areas. (a) Number of forwarders with different number of vehicles. **(b)** Number of forwarders in different message coverages.

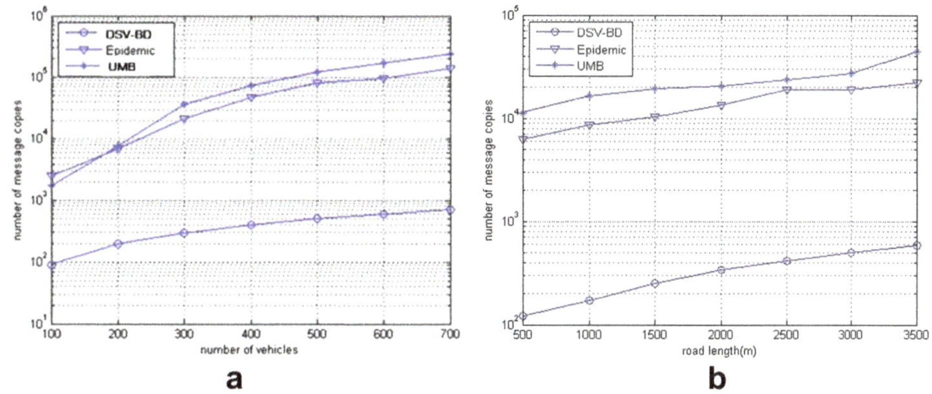

Figure 7 Comparison of different number of message copies and different message coverage areas. (a) Number of message copies with different number of vehicles. **(b)** Number of message copies with different message coverages.

number of vehicles, number of message copies of epidemic and UMB methods increase obviously.

In Figure 7b, the number of message copies increases as the length of road increases. By comparison, the growth rates are rapidly increasing in epidemic and UMB, but slowly in our algorithm. The number changes from 2.2% to 5.1%. So our method can reduce the network load effectively.

5.2 Simulation of on-demand unicast algorithm

The two compared algorithms are involved:

- Greedy: the upstream forwarders always choose the requestor which has the shortest distance to them, and the reply messages will be sent according to this order.
- Deadline-pri: the upstream forwarders always choose the requestor which has a smaller deadline, and the reply messages will be sent according to this order.

The indicator involved is delivery ratio, which means the probability defined as the number of message receivers divided by the total number of vehicles in the networks. We aim at the impact of the number of requestors. The deadline of each requestor obeys uniform distribution from 0 to 6 s. The simulation is repeated for 30 times, and the results are the average of those experiments. Figure 8 shows the comparison of data delivery ratio in these three algorithms in the case of different number of requestors.

As the number of requestors increases, the delivery ratios of three algorithms decrease generally. In terms of the two compared algorithms, the more the requestors, the higher the queuing time is. This leads to higher probability of timeout. As to DSV-BD, the increase of requestors means the increase of hops in the delivery path and more timeout happens due to increasing delivery time. As is shown from the figure above, the delivery ratio of DSV-BD is 17.3% higher than that of greedy and 15.9% higher than that of deadline-pri, which shows the higher delivery reliability with various numbers of requestors.

6 Conclusions

The high-speed moving of the vehicles in motorway environment results in the highly dynamic topology of networks, which brings out challenges for data dissemination of VANETs. In this paper, we propose a new broadcasting algorithm and an on-demand unicast algorithm for VANETs. In the proposed broadcasting algorithm, there are two innovation points: 1) predict the status of nodes by an improved *K*-nearest neighbors' prediction and 2) construct the set of forwarders to alleviate the influence of broadcast storm. In the on-demand unicast algorithm, the source vehicle node only sends at most two messages, and all the requestors can receive the messages they need. The results show that our method can achieve better performance with higher delivery ratio and less number of message copies.

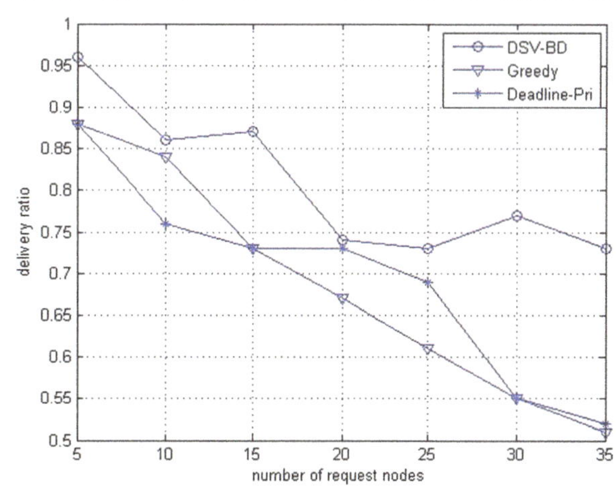

Figure 8 Data delivery ratio in different number of requestors.

Competing interests
The authors declare that they have no competing interests.

Acknowledgements
This work was partly supported by the NSFC (61372108), Ph.D. Programs Foundation of Ministry of Education of China (No. 20110005110011), Fundamental Research Funds for the Central Universities (No. 2014RC1102), and Beijing Higher Education Young Elite Teacher Project(YETP0474).

Author details
[1]State Key Laboratory of Networking and Switching Technology, Beijing University of Posts and Telecommunications, No. 10 Xi Tu Cheng Road, Beijing 100876, China. [2]China Electronic Systems Engineering Corporation, Xiaotun Road, Beijing 100039, China.

References
1. K Rakesh, D Mayank, A Framework for Handling Local Broadcast Storm Using Probabilistic Data Aggregation in VANET. Wireless Personal Communication **72**, 315–341 (2013)
2. W Shangguang, F Cunqun, H Ching-Hsien, S Qibo, Y Fangchun, A Vertical Handoff Method via Self-selection Decision Tree for Internet of Vehicles, in *IEEE System Journal*, 2014. doi: 10.1109/JSYST.2014.2306210
3. S Michael, M Imad, Spatial distribution and channel quality adaptive protocol for multihop wireless broadcast routing in VANET. IEEE Trans Mobile Comput **12**(4), 722–734 (2013)
4. SY Ni, YC Tseng, YS Chen, JP Sheu, The Broadcast Storm Problem in a Mobile Ad Hoc Network. Wireless Networks **8**, 153–167 (2002)
5. S Panichpapiboon, W Pattara-atikom, A Review of Information Dissemination Protocols for Vehicular Ad Hoc Networks. Communications Surveys & Tutorials IEEE **99**, 1–15 (2011)
6. Korkmaz, G., Ekici, E., Ozguner, F., and Ozguner, U.: Urban Multi-hop Broadcast Protocol for Inter-vehicle Communication Systems. In: Proc. 1st ACM International Workshop on Vehicular Ad Hoc Networks. (VANET'04), Philadelphia, USA, pp. 76–85 (2004)
7. Korkmaz, G., Ekici, E., and Ozguner, F.: An Efficient Fully Ad-hoc Multi-hop Broadcast Protocol for Inter-vehicular Communication Systems. In: Proc. IEEE International Conference on Communications. (ICC'06), Istanbul, Turkey, pp. 423-428 (2006)
8. L Wischof, A Ebner, H Rohling, Information Dissemination in Self Organizing Intervehicle Networks. Intelligent Transportation Systems, IEEE Transactions on **6**, 90–101 (2005)
9. L Uichin, M Eugenio, G Mario, B Paolo, L Pietro, L Kang-Won, Bio-inspired multi-agent data harvesting in a proactive urban monitoring environment. Ad Hoc Networks **7**(4), 725–741 (2009)
10. J Zhao, G Cao, VADD: Vehicle-Assisted Data Delivery in Vehicular Ad Hoc Networks. Proceedings of IEEE International Conference on Computer Communications (INFOCOM'06), 2006, pp. 1–12
11. Lee, S., Pan, G., Park, J., Gerla, M., and Lu, S.: Secure Incentives for Commercial Ad Dissemination in Vehicular Networks. In: ACM international symposium on Mobile ad hoc networking and computing (Mobihoc), New York, NY, USA, pp. 150–159 (2007)
12. M Röckl, P Robertson, Data Dissemination in Cooperative ITS from an Information-Centric Perspective. International Conference on Communications (ICC), Cape Town, South Africa, pp. 1–6 (2010)
13. Karp, B., and Kung, H.T.: GPSR: Greedy Perimeter Stateless Routing for Wireless Networks. In: Proceedings of the Annual International Conference on Mobile Computing and Networking (MobiCom'00), Boston, USA, pp. 243–254 (2000)
14. Hill, J., Szewczyk, R., Woo, A., Hollar, S., Culler, D., and Pister, K.: System Architecture Directions for Networked Sensors. In: International Conference on Architectural Support for Programming Languages and Operating Systems (ASPLOS), New York, NY, USA, pp. 93–104 (2000)
15. L Daoquan, L Haiyan, C Qiguang, W Huaicai, New routing algorithm based on geographical location: GPSR-AD. J Comput Appl **29**(12), 3215–3217 (2009)

10

Proactive channel access in cognitive radio networks using statistical radio environment maps

Chamara N Devanarayana[1*] and Attahiru S Alfa[2]

Abstract

Inclusion of statistical knowledge of the primary user (PU) channel usage had shown to be beneficial in dynamic spectrum access. Motivated by this fact, this paper investigated the importance of collecting and using statistics on neighboring secondary users (SUs) in selecting channels in addition to the knowledge of PU channel usage. The paper assumed that PU traffic characteristics of the channels are included in the radio environment map in the form of probabilistic suffix trees, which is a sequence predictor based on Markov property. In the proposed method, an intelligent sequence hopping-based common control channel and a carrier sense multiple access (CSMA)/collision avoidance (CA)-based medium access control (MAC) protocol were introduced. As shown in the paper, selecting channels using statistics of both the neighboring SUs and PUs reduced the number of packet collisions compared to a scheme which only uses PU statistics. Furthermore, the simulation results showed that the scheme proposed had better throughput performance with respect to both the random channel selection scheme and the scheme which only uses PU statistics while having less training complexity.

Keywords: Cognitive radio; Prediction methods; Predictive control; Artificial intelligence

1 Introduction

The vast growth in wireless applications in the past decade has pushed the FCC to search for efficient spectrum allocation mechanisms to avoid spectrum scarcity [1]. The measurement operation carried out by the FCC's Enforcement Bureau in 2002 gave some insight into the real cause of the widely accepted notion of spectrum scarcity [2]. The recommendations made by the spectrum efficiency working group, based on these findings, consisted of promoting flexible use of spectrum development and deployment of advanced technologies and promoting secondary markets for spectrum [2]. A technology known as cognitive radio (CR) was found to have the capability to implement those recommendations [1].

However, the secondary usage of spectrum requires the CRs to opportunistically utilize the spatiotemporally available primary user (PU) spectrum. This utilization brings about several problems as follows:

1. The problem of imperfect sensing (false alarms and missed detection) in the physical layer.
2. Unavailability of a common control channel which is always available.
3. Deciding on which channels to sense to find an opportunity efficiently.
4. Efficient selection of a different channel to continue with the communication when the PU appears in the currently used one (spectrum mobility).
5. Efficient sharing of the spectrum with other secondary users (SUs).

Proactive spectrum access (PSA) can be used as a solution for the third and fourth problems mentioned above. Furthermore, using the statistical information on the PU channel usage, highly available sequence hopping (SH)-based common control channels (CCCs) can be designed. One of the solutions to the problem of imperfect sensing is cooperative sensing. In [3], the authors discuss a semi-distributed cooperative spectrum sensing scheme which

*Correspondence: umdevana@myumanitoba.ca
[1]Department of Electrical and Computer Engineering, University of Manitoba, E1-551 EITC, 75A Chancellors Cir., R3T 5V6 Winnipeg, MB, Canada
Full list of author information is available at the end of the article

reduces this problem. In Sections 1.3 and 1.2, we discuss some available solutions for the problems 2 and 5, respectively. This paper is an extension to the work in [4]. In that paper, we only discussed the usage of probabilistic suffix tree (PST) as a prediction mechanism, where there was only a single SU trying to access the channel.

In the scheme presented, the radio environment map (REM) is enhanced with statistical knowledge of the PU channel usage as in [5]. Furthermore, due to PU channels being highly dynamic, it may not be possible to query the REM for information on the current state of the channels. Even if the channels are not highly dynamic, it might not be possible to query the REM before every time slot due to the control channel saturation problem. Considering these facts, in the scheme presented, each transmitting SU queries the REM once for the probabilistic information before the transmission. Then, the SU will cache the probabilistic information to be used until the end of the communication between itself and the receiver while carrying out periodic sensing. In the work, we presented in this paper, the REM learned the PU channel usage distributions from the past channel usage data using a scheme called the PST algorithm [6], which belongs to the class of variable-order Markov models (VMMs). PSTs only increased the length of the conditioning strings if there was evidence that higher order gives a better prediction result with high probability, although we set a maximum value for the memory in the algorithm [7]. Therefore, the model takes less memory and does not need much bandwidth to transmit. In [8], it was observed that in some channels, the predictability cannot be improved by increasing the Markov order and the largest improvement for other channels happens when going from zeroth order to first order. But the authors admitted that the predictability did not monotonically decrease with the Markov order. In their results, there were evidence for the existence of channels which are more predictable using higher Markov order, for example, 10% of the channels they tested had an optimal Markov order of 8. Therefore, the PST scheme is a good candidate when a CR operates in diverse wireless channels, because of its self-regulation of the Markov order and the convenience of training. Furthermore, the PST is capable of predicting the channel availability conditioned on a string of past channel observations which are of variable length. The second property mentioned in the above line was very useful in our case since the length of consecutive observations the SUs possessed was varying according to the sensing schedule. In a scenario where a carrier sense multiple access (CSMA)/collision avoidance (CA)-based medium access control (MAC) scheme is used, collisions are inevitable. Therefore, time slots become wasted. To alleviate the network from this issue, we used the statistics collected by each SU on the number of successes and failures on each used channel in a given window of time slots when selecting a set of channels to transmit data. We believe that this is the first time SU channel access data are used in PSA. In this scheme, the channel conditions were not considered. It was ignored to keep the model simple and to clearly show the advantage of using available statistics on PUs and SUs. The channel conditions can be obtained by channel probing after the handshake or from the REM itself, and it will enhance the capacity obtained without doubt.

1.1 Existing schemes on PSA

In the literature, there are mainly three categories of PSA schemes as follows: analytic formulations ([9-11]), heuristic schemes based on assumed channel usage ([12-15]), and heuristic schemes based on channel usage learning ([4,16-18]). Most of the learning methods like neural networks (e.g.,[16]) and binary time series models (e.g.,[18]) in the literature fixed the value of the memory to be used in prediction. The hidden Markov models (HMMs) (e.g.,[17]) fixed the number of states in the Markov chain. Although being very good predictors, the training complexities of neural networks, HMMs, and binary time series models are a lot higher than those of the PST, and thus, they cannot be trained online. The analytic formulations (e.g.,[9-11]) used variants of Markov decision process to calculate the sensing and access schedule to maximize the SU reward while keeping the PU interference probability under a given threshold. Although they can calculate the optimum policy, they become impractical even for a small number of channels because of the complexity of policy calculation. The heuristic schemes proposed in the literature although being simple can lead to more SU packet collisions, because all the SUs use the same schedule only varied because of imperfect sensing. Thus, they lead to inefficient channel usage in a CSMA/CA-based protocol.

1.2 Existing medium access control schemes

In [19], the MAC schemes in the literature were categorized according to the spectrum access mode as contention-based, time-slotted, and hybrid. Our scheme falls in the category of hybrid spectrum access protocols. Furthermore, in [20], they were categorized according to the level of coordination performed as direct access-based (DAB) protocols and dynamic spectrum allocation (DSA)-based MAC protocols. In DAB protocols, only the user pair involved in the communication take part in coordinating the access. While in DSA protocols, users coordinate with the neighbors to achieve a network-wide optimum. Although being optimal, DSA protocols have scalability issues. The MAC scheme that we proposed is a DAB with proactive access. The difference in our scheme from these MAC schemes is that being proactive and switching channels inside a time slot when the channel is not available. Interested readers are referred to the survey papers [19] and [20] and the references therein for more information.

1.3 Existing common control channel design schemes

The CCCs proposed for overlay networks can be divided into three main categories [21] as follows:

1. Sequence hopping-based CCC
2. Group-based CCC [22]
3. Dedicated CCC

We used a sequence-based CCC for the initial handshake between two SUs and an SU and REM. The advantage of SH-based CCCs over the group-based ones is having less coordination burden in a highly dynamic PU spectrum access scenario. Then, compared to the dedicated CCCs, it has less probability of being saturated. Our scheme is different from the ones in the literature (e.g.,[23-25]) because of using PU stationary channel usage probabilities in deciding on the channel to be used. The scheme proposed in [25] also uses PU stationary channel usage probabilities to rank the channels. In that paper, each individual SU made a ranking table of PU channels according to their activity. Then, using a linear and a parabolic function, the number of occurrences of each channel in a sequence of fixed length L was decided such that channels with lower activity had more appearances. Because of the deterministic nature of the number of occurrences of the channels, the arrangement of channels in the sequence by each SU plays a huge role on the channel load, which was not discussed in detail in that paper.

One of the drawbacks of the SH-based schemes is the burden of synchronization. Many papers considering decentralized scenario in the literature took the assumption of the network being synchronized (e.g., [23-26]). In [23], although they claimed to have alleviated the network from the need for synchronicity, they assumed the switching boundaries to be synchronous. In other words, the delay between the two sequences is an integer multiple of the time slot length. Our scheme could also handle that form of asynchronicity. For more information on how this synchronicity is maintained, the readers are referred to Section 4.4.

1.4 Main contribution

The main objective of this paper was to find solutions to the MAC layer problems faced by SUs utilizing PU channel usage models acquired from the REM and the SU channel usage individually observed by each SU. These MAC problems are the problem of designing a highly available control channel, the problem of efficiently searching for spectrum holes, the problem of spectrum mobility, and the problem of spectrum sharing among SUs. The contributions of this paper are listed below:

- A synchronized channel hopping-based common control channel was proposed for the initial handshake.
- The theoretical properties of this common control channel were derived.
- A stochastic channel set selection scheme for sensing and access was proposed taking both the PU channel usage and SU channel usage into consideration.
- A probabilistic suffix tree-based prediction scheme was proposed to arrange the channels from best availability to least availability in each time slot.

The rest of this paper is organized as follows. We present the system model used in Section 2. In Section 3, we present a brief explanation of the prediction scheme used in this paper. Then, we discuss the MAC used in this paper in Section 4 followed by the simulation results and conclusion in Sections 5 and 6, respectively.

2 System overview

In the system investigated, we have a centralized time slotted *primary network* (PN) and a decentralized *secondary network* (SN) which opportunistically utilizes the PN resources. In this secondary network, a REM exists which has the ability to collect the PN channel occupancy details. REM is assumed to possess records of past channel usage information of N channels, which are collected from the SUs and a dedicated set of sensors as in [5]. In the event that channel occupancy of PUs change and the statistics are recalculated, they are distributed by the REM to the registered users using the CCC.

The SUs register themselves with the REM and obtain the PU channel usage information. They were assumed to be synchronized among themselves and with the REM. This synchronization is vital in having a universally quiet period (QP) among the SUs. The universally QP enables the use of a less complex sensing method like energy detection as opposed to matched filter detection or cyclostationary detection. The SUs were assumed to be equipped with a widely tunable antenna, a sensing unit and a radio unit which is capable of sensing all the N primary channels and transmitting on the same. The SUs and the dedicated sensors were assumed to be capable of differentiating an SU signal from a PU signal since they are synchronized and have a QP at the beginning of each time slot. Further details on this topic are explained in Section 4.

We assumed unlicensed operation of SUs without any QoS guarantees, which was one of the two DTV band secondary spectrum usage schemes proposed in [27]. In this paper, we adopted the convention of denoting the *busy* period with 1 and the *idle* period with 0. The channel state distribution learning scheme used by the REM is explained in the next section.

3 Channel modeling and prediction

The most commonly used Markov chain in the literature is the one-step Markov chain. Suppose we let the states of the Markov chain be labeled by the window of last m observations. Then, even though the Markov chain does only a one-step prediction, that prediction depends on the last m observations. For example, let '1' and '0' represent the busy and idle states of a particular channel, a fixed-order Markov model of order m has strings of channel observations of the past m time slots as its state labels. Thus, there are 2^m number of states in this particular case. For example, if we assume $m = 2$, then we have $2^2 = 4$ states, and the state labels are $\{00, 01, 10, 11\}$. For $m = 3$, we have $2^3 = 8$ states having state labels $\{000, 001, 010, 011, 100, 101, 110, 111\}$. Since the number of states grows exponentially with m, only Markov chains with lower order can be practically used [6]. In variable-order Markov models as opposed to fixed-order Markov models, we only consider the states which are highly abundant on the training sequence and have a significant impact on the next observation. Therefore, the number of states in a variable-order Markov model is usually less than those in a fixed-order Markov models of the same order [28]. The PST introduced in [6] belongs to the category of variable-order Markov models. A summary of the algorithm is given below for the convenience of the reader.

If we take a variable-order Markov model of order $D = 2$, its states are labeled with binary strings of length $l \leq 2$ (strings are binary since the observation space is binary). The set of states in variable-order Markov model in this case is a subset S of $\Lambda = \{0, 1, 00, 10, 01, 11\}$, chosen by an algorithm run on a past observation sequence. But a sequence in this subset S cannot be a suffix of another sequence in the same set, for example, both 1 and 01 cannot be in set S. A suffix s' of a string s is defined as $s' \in \{s_i s_{i+1} ... s_l | 1 \leq i \leq l\} \cup \{e\}$ where $s = \{s_1 s_2 ... s_l\}$ and e is the null string. In our case $s_i \in \{1, 0\}\ \forall i \in \{1, 2, ..., l\}$. The other requirement about the state space S is that for every string $s \in S$, if the probability of having $\sigma \in \{1, 0\}$ after s ($Probability(\sigma|s)$) is greater than zero, there should be a unique string $\hat{s}$ which is a suffix of $s\sigma$ in S. Where $s\sigma$ is the the string having σ attached to the end of the string s. As mentioned earlier, to find out the states that matter in the variable-order Markov model, we used an algorithm called the PST algorithm, which was introduced in [6]. To use this algorithm, we needed a training sequence of sufficient length which represents channel behavior under normal circumstances, in our case, it was the channel state data gathered by the REM which are binary sequences. An algorithm to discover the *suffix set* S and build the PST using a given training sequence of length m is explained in the papers [6,28]. An algorithmic explanation of the PST algorithm is given in Appendix A for the convenience of the reader. In this paper, the symbol on the right hand side end of any time series string of channel state observations was assumed to be the most recent one.

3.1 Prediction using the probabilistic suffix tree

For a given PST $\bar{T}$ with maximum order D and the channel observations for the previous k ($0 \leq k \leq D$) consecutive slots, we traverse the tree starting from the node e which denotes the empty string. An example tree calculated according to the parameters given in Table 1 is shown in Figure 1. In this parameter list, the first 5 entries are the parameters of the PST algorithm, while σ_1^{10} is a training sequence of length 10. This traversal is done according to the past k channel observations starting with the most recent channel observation and going backwards in time in the past k channel states. Traversing the tree, we will reach a leaf node either using all the k channel state observations or partially using them. When this leaf node is reached, we use the probability distribution of the occurrence of an *idle* slot or a *busy* slot for that particular node as the probability of occurrence of the next channel state. For example in Figure 1, the values given in parentheses, $\{p_{\text{idle}}, p_{\text{busy}}\}$, are the probabilities of the occurrence of an idle slot and a busy slot, where node label gives the longest suffix of the observation string present in the model. In the next section, we will discuss the MAC protocol and the CCC design used.

4 MAC protocol

As discussed in the related work section, most MAC protocols in the literature did not take into account the channel usage characteristics of the PUs, with [9,25] being exceptions. But in those schemes too, the effect of SU access was not taken into account. Therefore, if all the SUs tried to access the optimum channel, the packet collisions bring the throughput down. In this section, we

Table 1 Threshold values used in algorithm and an example training sequence

Description	Value
P_{min}	0.006
α	0
r	1.05
D	3
γ	0.0006
σ_1^{10}	1110010000

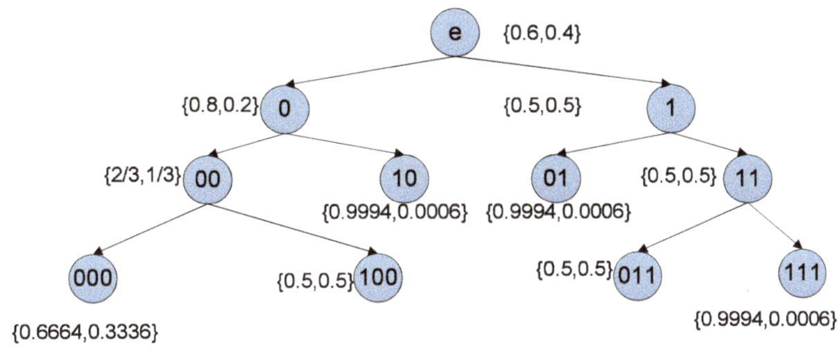

Figure 1 Example probabilistic suffix tree.

discuss how the PU channel usage statistics and the SU channel usage statistics could be used towards increasing the throughput of SUs. First, we explain how the PU channel statistics can be used in the CCC sequence selection process. Then, we explain how both the PU and SU statistics can be used in the data transmission process.

4.1 Proactive common control channel

Most SH CCC schemes in the literature searched for idle channels in a predetermined or random order. So each SU selected channels based on that channel scan order and listened to them to identify the state of the channel. If the channel was idle, the SU used it and if not, the search continued till it finds one and the receiver channel selection coincided with it. In the scheme presented, to reduce the interruptions to the primary and secondary users, we ordered the channels in the ascending order of activity. Thus, the SUs selected the channel with the highest expected availability more often, saving them from interfering with the PU and high variability in the idle channel search delay.

Let p_i denote the stationary probability of PU channel i being idle, where $i \in \{1, 2, \ldots, N\}$. This probability is given by the distribution of events at node e. This node e represents empty string in the PST. Then, Algorithm 1 was used to generate the channel sequence ϕ_j^t, where ϕ_j^t is the channel selected by SU j at time slot t.

Algorithm 1 Control channel selection algorithm

```
Initialize φ_j^t = ∅
u ⇐ rand()
for k = {1, 2, 3, ..., N} do
    if u ≤ (Σ_{l=1}^{k} p_i) / (Σ_{i=1}^{N} p_i) then
        φ_j^t = k
        break;
    end if
end for
```

In the above scheme, it can be seen that the SUs select the channels with less PU activity more often. Furthermore, due to the randomness in the selection process, the packet collisions will be low. The downside to this channel selection method is not having a guaranteed rendezvous time as the schemes [24] and [23].

4.1.1 *Analytic formulation of the average time to rendezvous*

The time to rendezvous is an important performance metric because it decides the number of attempts an SU has to make to successfully complete a handshake with the receiver which in turn dictates the packet delays. Therefore, time to rendezvous should be as small as possible. In order to derive a formula for the average time to rendezvous for a transmitting SU, first, the probability of successfully contacting the receiver should be calculated. The method of calculation is explained below.

First, the following assumptions were made: the contention window size is CW, all backoff time values are equiprobable with probability $\frac{1}{\mathrm{CW}+1}$ and the total number of SUs is M. The vulnerable time is given by t_v. This is the minimum time gap between the smallest backoff time window and the second smallest backoff time window of contending neighbors on the same channel to avoid a packet collision. In other words, this is the average time duration it takes for nodes in the CSMA/CA network to identify the channel is busy after the first node transmits. This time is equal to the sum of the transmission time duration of the request to send (RTS) packet and the propagation time, in the unit of sub-time slots (smallest synchronous time unit) [29]. The false alarm probability is given by symbol p_{fa} and the probability of selecting channel $k (\leq N)$ is given by p_k^s. According to the CCC selection algorithm above, $p_k^s = \frac{p_k}{\sum_{i=1}^{N} p_i}$. The probability of channel access by an SU is given by a, and the cumulative distribution of the random backoff time slots generated is given by $F(\cdot)$. Using these parameters, the formula for the probability of successful handshake, θ_s, was formulated as given in Equation 1.

$$\theta_s = \sum_{i=1}^{N} p_i p_i^s \left[\sum_{r=0}^{M-1} (1-p_{\text{fa}}) \binom{M-1}{r} (ap_i^s)^r (1-ap_i^s)^{M-r-1} \times \sum_{\tau=0}^{\text{CW}} (1-F(\tau+t_v))^r \frac{1}{\text{CW}+1} \right] p_i^s(1-a) . \tag{1}$$

The steps of the derivation of this equation are given below.

Proof. First let us find the conditions that should be met in order for an SU to succeed in the handshake. A given SU j can succeed in the handshake if all the conditions below are satisfied:

- SU j has a packet to transmit which is given by probability a.
- The SU j selects channel i which is given by the probability p_i^s.
- Channel i is idle, given by probability p_i.
- Channel i is sensed as idle by j, given by $(1 - p_{\text{fa}})$.
- SU j wins the contention.
- The receiver has selected channel i and is not transmitting, given by $p_i^s \times (1 - a)$.

For the SU j to win the contention, the backoff duration of it should be less than the minimum of all the other SUs trying to transmit in the same channel by atleast the vulnerable time t_v. In this paper, we assumed all the SUs can hear each other. If the cumulative distribution function of the backoff duration is given by $F(\cdot)$ and SU j's backoff duration is τ, the probability, $R(\tau + t_v)_r$, that r number of transmitting SUs backoff times are more than $\tau + t_v$ is given by:

$$R(\tau + t_v)_r = (1 - F(\tau + t_v))^r . \tag{2}$$

Then, the distribution of r SUs trying to access channel i can be derived using the binomial distribution with probability of success $a \times p_i^s$ and total number of other SUs $M - 1$ as given in Equation 3.

$$B\left(M-1, ap_i^s\right) = \binom{M-1}{r} (ap_i^s)^r \left(1 - ap_i^s\right)^{M-r-1} . \tag{3}$$

Equation 2 gives the probability of winning the contention conditioned on, the backoff time τ of SU j, the number of other SUs trying to access channel i which is r and the given SU j is trying to transmit on channel i. Since the backoff time is uniformly distributed in $[0, \text{CW}]$, we took the expectation of $R(\tau+t_v)_r$ with respect to the backoff duration and the number of other users trying to access channel i. The result is the probability of an SU winning the contention which is given in Equation 4.

$$\zeta_s(i) = \sum_{r=0}^{M-1} \binom{M-1}{r} (ap_i^s)^r \left(1 - ap_i^s\right)^{M-r-1} \times \sum_{\tau=0}^{\text{CW}} (1 - F(\tau + t_v))^r \frac{1}{\text{CW}+1} . \tag{4}$$

Then, multiplying $\zeta_s(i)$ by the probability of the receiver being tuned into channel i, ap_i^s, we get the probability of successful handshake in channel i. Then, finally, we took the expectation of it w.r.t. all the channels where the probability of selecting a given channel was $p_i\,(1 - p_{\text{fa}})\,p_i^s$ which gave us the result given in Equation 1.

□

Then, the probability distribution for the number of time slots to achieve rendezvous $TR(q = l)$ was calculated as given in Equation 5.

$$TR(q = l) = (1 - \theta_s)^{l-1}\,\theta_s . \tag{5}$$

4.1.2 *Channel load in the SH control channel*

Since this scheme uses a stochastically selected control channel, only the average channel load can be calculated. The probability of r users out of M SUs accessing channel i can be given by $B(M, ap_i^s)$ which was defined in Equation 3. Then, using $B(M, ap_i^s)$, an equation for the maximum average channel load was formulated as shown in Equation 6.

$$L\left(M, \underline{\mathbf{p}}^s, a\right) = \max_{i=\{1,2,\cdots,N\}} \sum_{r=1}^{M} r \cdot B(M, ap_i^s) . \tag{6}$$

4.1.3 *Degree of rendezvous*

This metric gives the number of minimum overlaps between any two channel sequences in a SH-based CCC. In a stochastic channel selection scheme, although a number for the degree of rendezvous cannot be given, the probability for two sequences of length Q to have a minimum of k overlaps can be calculated. Equation 7 gives an expression for this probabilistic degree of rendezvous.

$$\text{DR}(Q, k, \underline{\mathbf{p}}^s) = \sum_{i=1}^{N} p_i^s \sum_{r=k}^{Q} \binom{Q}{r} (p_i^s)^{2r} \left(1 - (p_i^s)^2\right)^{Q-r} . \tag{7}$$

4.2 Channel set selection for the data transmission

In predicting the status of a channel using the PST, we need to sense the channels. To make the CR less expensive, we used a single half-duplex transceiver and did energy detection-based narrow band sensing. Therefore, in our scheme, only sequential sensing was possible and there is an upper bound m, to the number of channels an

SU can sense in a given time slot. This upper bound is the time needed to fail in transmitting $m-1$ times and trying transmitting for the mth time, so that there is time for contention window, RTS and clear to send (CTS) messages, data, and the ACK. Therefore, the selection of a set of channels which are both utilized less by the PU and the other SUs is necessary to achieve a better throughput. We used a channel set selection scheme which chose the channels which are highly available for a given SU j with high probability. To find this channel set, we have to calculate the probability of winning the contention on a given idle channel i.

4.2.1 *Calculating the probability of winning the contention*

This is a subjective usage indicator on an SU j of interest. To calculate this parameter, each SU kept track of the instances it tried to access channel i, $\epsilon_{ij}^{\text{tot}}$ and the instances it acquired channel i, $\epsilon_{ij}^{\text{acq}}$. Then, for the channels that the SU has been accessing, we calculated c_{ij} using the ratio between the two, $\frac{\epsilon_{ij}^{\text{acq}}}{\epsilon_{ij}^{\text{tot}}}$. For the channels that the SU has not tried to access, these parameter values were calculated based on the probability a channel is successfully accessed by an SU, given that each SU chooses each channel with equal probability $\frac{1}{N}$. Thus, for the unused channel i and SU j, the parameter c_{ij} was calculated as given in Equation 8.

$$c_{ij} = \sum_{r=0}^{M-1} (1-p_{\text{fa}}) \binom{M-1}{r} \left(\frac{a}{N}\right)^r \left(1-\frac{a}{N}\right)^{M-r-1} \times \sum_{\tau=0}^{\text{CW}} (1-F(\tau+t_v))^r \frac{1}{\text{CW}+1}. \quad (8)$$

4.2.2 *Channel set selection for the data transmission*

Using the above calculated parameter c_{ij} and the stationary probability of channel i being idle, p_i, we selected the channel set using Algorithm 2.

Algorithm 2 Channel set selection algorithm

1. Initialize $\mathcal{S}_j = \emptyset$
2. **while** $|\mathcal{S}_j| < m-1$ **do**
3. $\quad \bar{\mathcal{S}}_j = \{1, 2, 3, \ldots, N\}/\mathcal{S}_j$
4. $\quad k = \arg\max_{i \in \bar{\mathcal{S}}_j} \{p_i c_{ij}\}$
5. $\quad \mathcal{S}_j = \mathcal{S}_j \cup \{k\}$
6. **end while**

In the following section, we discuss how the communication between two SUs happen.

4.3 Communication between two SUs

In the scheme presented, the communication between two nodes happens in two stages. First, an initial handshake between SUs is completed, and then, the SUs communicate data between each other. In the first stage, the transmitting SU generates the channels to be used at each time slot t, ϕ_j^t, using Algorithm 1. Then, it generates the channel set $\mathcal{S}_j$ to be used for data transmission, using Algorithm 2. After that, the transmitting SU u starts the handshake phase with receiver v. This is explained in the next sub-section.

4.3.1 *Handshake between the sender and the receiver*

Prior to the handshake with the receiver, in the QP, the transmitter u senses the channel list $\mathcal{S}_j$ first and then senses the control channel ϕ_j^t. Then, in the backoff period, it calculates the statistical availability of the channel set $\mathcal{S}_j \cup \phi_j^t$ at time slot $t+1$ using the sensing results and the PST models acquired. Then, the channels which have the probability of being idle greater than 0.5 are arranged in the descending order. This ordered list H_u^{t+1} is piggybacked on the RTS packet. If the intended receiver happened to listen to the same channel (i.e., $\phi_u^t = \phi_v^t$) and was able to decode the packet, it transmits a CTS packet. Then, the data communication is done using the same channel ϕ_j^t till the end of the time slot and the receiver sends an ACK. At this stage, the handshake is considered to be successful. If the CTS get timed out, the procedure is continued in the next time slot. A diagram showing an example of transmitter slot structure at handshake phase is given in Figure 2.

4.3.2 *Data transmission after the successful handshake*

If the handshake was successful at time slot t, the receiver v and transmitter u, knowing the channels to be used in access, sense the channel set H_u^{t+1} in the quiet period of time slot $t+1$. This sensing is done starting from the worst to the best (i.e., in the reverse order). Therefore, after sensing the best channel, the transmitter u generates the backoff time and if sensed idle, it will send the RTS with the sensed channel status data on set H_u^{t+1} piggybacked on it. Then, if this RTS was received at the receiver v, it replies with a CTS with sensed channel status data on set H_u^{t+1} piggybacked on it. Then, after the data transmission starts at the end of the time slot, the receiver v sends an ACK if the transmission was successful. If the CTS was not received at the transmitter, the transmitter will try the next channel in sequence H_u^{t+1} after the CTS times out. This goes on until the list H_u^{t+1} gets exhausted within time slot $t+1$. We set this time out period to be equal to the sum of CTS transmission time and propagation delay and the time to transmit the header of the data part and propagation delay. If the transmitter u was unable to contact the receiver v for an entire time slot, then the initial

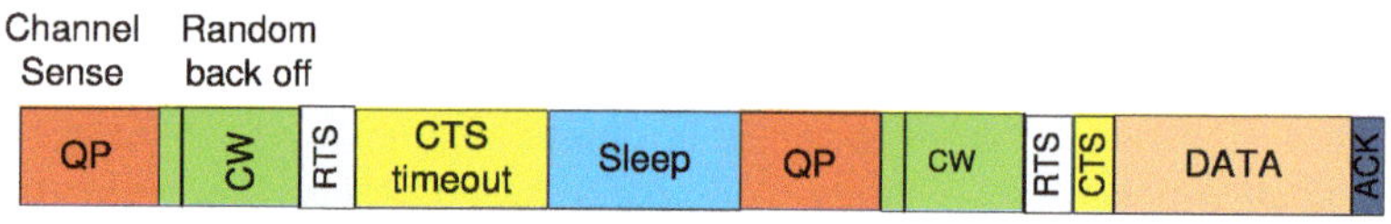

Figure 2 An example of transmitter slot structure at handshake phase.

handshake has to be redone, because the channel status at both ends of the current slot are not known to both u and v.

If the RTS and CTS exchange was successful, then the sensing outcomes of the channels at both ends are available to both u and v. Then, both u and v fuse the sensing results using OR decision fusion and save this result for each channel i separately in a buffer s^i_{uv}, where $i \in \mathcal{S}_j \cup \phi^t_j$. Each user u clears the buffer s^i_{uv} when the channel i is not sensed or the communication between each other is over in that time slot. Then, this buffer content at time slot $t+1$ is used to predict the channel status at time $t+2$. Then, both u and v make their ordered lists H^{t+2}_u and H^{t+2}_v. Both of these lists are the same. Therefore, in time slot $t+2$, the communication is possible. This process goes on till the communication between u and v is complete. A diagram showing an example of transmitter slot structure at data transmission phase is given in Figure 3.

4.4 Registration of new users, synchronization, and model updates

We assume that the REM transmits beacons every T time slots on all the available PU channels. This beacon transmission starts at the QP of the SUs; therefore, it is easy for the SUs to detect this beacon. This beacon contains a synchronizing bit sequence and encoded channel models. The SUs who has registered with the REM can decode these models. In these beacons, the periodic channel hopping pattern of the REM is mentioned so that the SUs newly arriving to the network can follow the REM hopping pattern and communicate with it to get registered. A newly arriving SU should scan the spectrum till it receives the beacon. The period T was determined as the minimum of the time for 5% of SUs to have a clock drift of $\frac{\text{QP}}{2}$ and the inter-arrival time between two new user arrivals to have 95% probability of arrival. Since all the registered SUs get synchronized with the REM from time to time, the assumption of synchronicity holds true.

5 Simulation

In this section, we carry out the numerical evaluation of the proposed scheme. In order to create the channel model according to the probabilistic suffix tree scheme, we needed a training sequence. Real data was not available to us, so we created a set of training data as strings of '1's and '0's. For this, we used a renewal process with *busy* time and an *idle* time generated according to a pseudo self-similar Markovian arrival process (MAP). The used MAP is capable of capturing the correlation between the different channel states. For the simulation purpose, we developed the MAP as given in [30,31]. As mentioned above, this MAP exhibited what is known as 'pseudo long-range-dependent self-similar characteristics'. Self-similarity means that the burstiness of traffic is the same over different time scales. In other words, if we measure the traffic in a particular link and plot, the traffic as a graph of the average amount of packets per 1, 10, 100, 1,000s, etc., the graph will only be scaled in amplitude and the shape will remain the same. The structure of the Markovian arrival process given in [30] has burstiness similar to Ethernet traffic. The MAP distribution is defined by a stochastic matrix D as defined in Equation 9, where D_0 and D_1 are sub-stochastic matrices containing the state transition probabilities to phases within the idle state and busy state, respectively, and d_{01} (d_{10}) contains the probabilities of transition to a phase in busy (idle) state from a phase in idle (busy) state. We fixed the size of the D matrix to be 6×6. The parameters of the MAP were calculated such that the channel utilization are 0.2,0.4, and 0.8, and the mean burst length is 10 slots. The theoretical properties of this MAP distribution can be found in [30]:

$$D = \begin{bmatrix} D_0 & d_{01} \\ d_{10} & D_1 \end{bmatrix} \tag{9}$$

Before training the models, we changed the *busy* slots to *idle* with probability P_m and *idle* slots to *busy* with

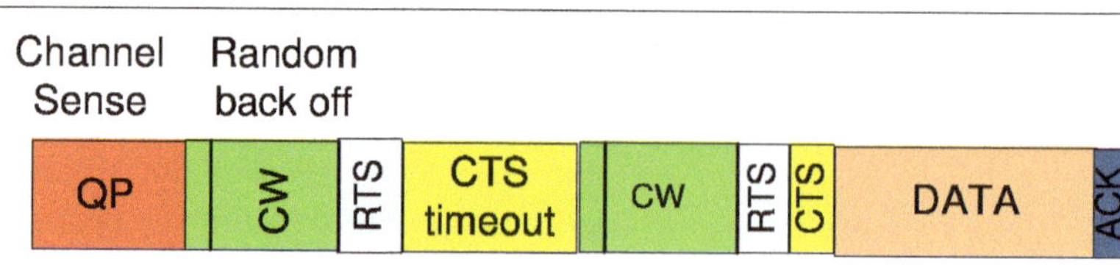

Figure 3 An example of transmitter slot structure at data phase.

probability P_f in the channel state sequence generated. The false alarm probability P_f was calculated for a given missed detection probability, $P_m = 0.1$, assuming an AWGN channel with mean SNR 5 dB at the SU. We then simulated the channel access of 5 and 10 SUs using 10 channels, while varying the channel access probability values a from 0.1 to 0.9.

We used the parameters given in Table 2 when training the PST and in the simulation setup. For comparison, we used a scheme from the literature called the HMM technique given in [17] for predicting the channel state in the random channel selection scheme and the channel selection scheme which uses the channels with lowest PU channel access probability. The results for the random channel selection scheme is named in the graphs as *Random* while the results for the scheme which only uses PU statistics is named as *PU only*. The scheme introduced in this paper is named as *Both stats* in the graphs. Since the HMM model in [17] did not provide any information on the number of states in the HMM or the memory length of sequences they used to determine the parameters, we chose ten states since training can be done in a reasonable time. The length of the string used for training was 20,000 for both PST and HMM schemes.

5.1 Initial handshake

In both the random channel set selection scheme and the channel set selection scheme based on PU activity, we used the control channel selection scheme introduced in [23] (*RCCH-Sync*). In Figures 4 and 5, we plot the average number of time slots taken by each scheme to achieve rendezvous for five and ten SUs, respectively. One can see that our scheme performs as equally well as the *RCCH-Sync* scheme. For this comparison, we used the scheme developed for synchronized networks in [23]. In that scheme, the synchronization offset between the transmitter and the receiver could not even have values which are integer multiples of the time slot length. Our scheme can handle that form of synchronization issue. In [23], authors also developed schemes which can withstand those kind of time offsets, but they had worse time to rendezvous than the synchronous one. In the same figures, we plot the theoretical results. One can see that the theoretical result is quite higher than the simulated value. The reason behind it is, when deriving this formula, we assumed the channels are accessed by the SUs according to the control channel selection scheme in all the time slots, while in the simulation, once the handshake is done, the channels are accessed according to a different distribution which allows SUs to balance its load among all the channels due to the use of SU statistics in data phase channel selection. Furthermore, in the data phase, the receiver

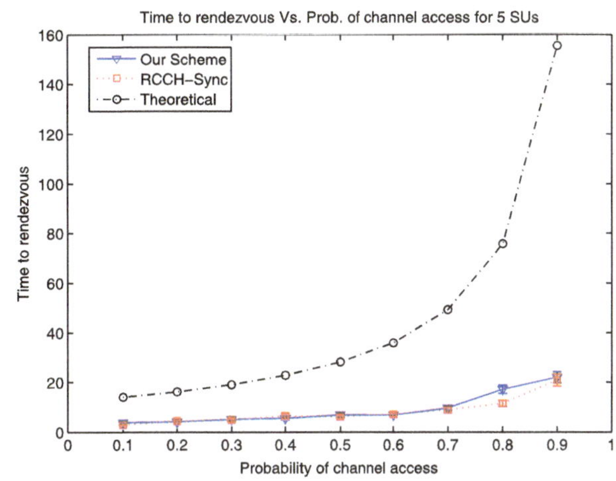

Figure 4 Average number of control slots vs. SU channel access probability for five SUs.

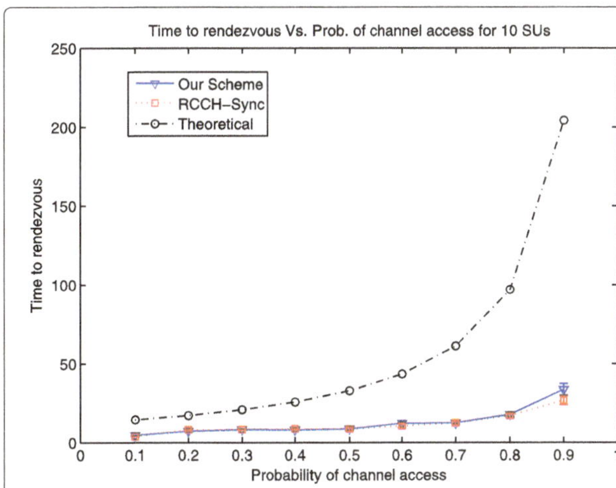

Figure 5 Average number of control slots vs. SU channel access probability for ten SUs.

Table 2 Simulation parameters

Parameter	Value
D	10
P_{min}	0.006
α	0
r	1.05
γ	0.0006
N	10
Number of SUs	20
k	3
Channels 1 to 4 PU utilization	0.2
Channels 5 to 7 PU utilization	0.4
Channels 8 to 10 PU utilization	0.8
Probability of missed detection	0.1

is tuned to the channel the transmitter is using and it is not transmitting. Therefore, these happen with probability 1. But in the derivation of the theoretical result, we associated them with the respective probabilities associated with them in the control phase. Ideally, only the number of users in the handshake phase and the channels free from both PU activity and the SU data transmission activity should be used in this calculation. But due to the complex interactions of SUs, this is not feasible.

We used an exponential backoff scheme for the CSMA/CA access of the SUs. We limited the backoff exponent to 4 for the handshake phase and 3 for the data phase. The time duration details for the control phase is given in Table 3.

The duration of the RTS, CTS, ACK, and DATA parts of the frame for a given data rate R, D bytes of payload, S bytes of channel sensing information is given in Equations 10, 11, 12, and 13. The expressions are taken from [29].

$$t_{\mathrm{RTS}} = 20\ \mu s + \frac{(22 + (20 + S) \cdot 8)}{R}, \quad (10)$$

$$t_{\mathrm{CTS}} = 20\ \mu s + \frac{(22 + (14 + S) \cdot 8)}{R}, \quad (11)$$

$$t_{\mathrm{ACK}} = 20\ \mu s + \frac{(22 + 14 \cdot 8)}{R}, \quad (12)$$

$$t_{\mathrm{data}} = 20\ \mu s + \frac{(22 + (28 + D) \cdot 8)}{R}. \quad (13)$$

5.2 Data phase

In the simulations, once the initial handshake is done, each time slot is divided into three sub-slots. This number is equal to the number of channel switches allowed at each time slot. Then, an SU who acquired the channel can transmit a packet in each sub-slot only if the PU is not present. If an SU transmitted when a PU is present, it is called an interfered sub-slot. This performance measure is plotted against the probability of SU channel access in Figures 6 and 7 for five and ten SUs, respectively. For the random and PU activity-based schemes, we used an HMM for the prediction and in our scheme, we used PST scheme for prediction. In this figure, we can see that our scheme has higher interference than the PU statistics-based one and random one for all SU user channel access probabilities. But the probability of interference to PU is less than 0.005 which is insignificant. A possible reason behind this is, due to the usage of the SU statistics based on past contention failures, a given SU overestimates the level of competition for the highly available channels and try to aggressively use the channels with high PU activity. Fine tuning of the data collection window of contention failures and successes should be able to reduce this problem.

Table 3 Time duration for time slot

Parameter	Value
Transmission rate	1 Mbps
Sensing time per channel	0.05 ms
Channel switch time	1 μs
Slot duration	4.615 ms
CW_{min}	15
Slot time	9 μs
Propagation time	5 μs

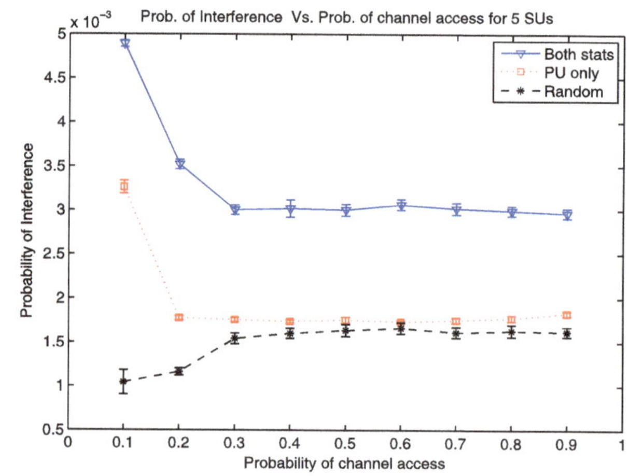

Figure 6 Probability of interference to PU vs. SU channel access probability for five SUs.

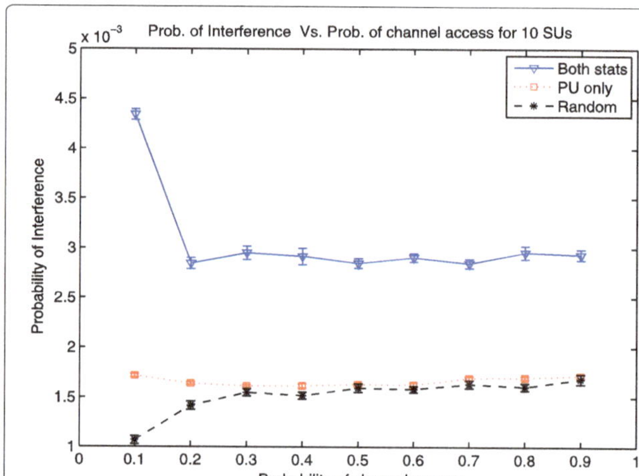

Figure 7 Probability of interference to PU vs. SU channel access probability for ten SUs.

In this simulation, the probability with which an SU begins a transmission was distributed according to the Bernoulli distribution with $a = \{0.1, 0.2, \ldots 0.9\}$ being the probability of success. Then, as long as the number of packets to be transmitted is above zero, packets were added to the transmission queue at the beginning of each main slot according to the discrete uniform distribution having the set of outcomes $\{1, 2, 3\}$. The transmission was assumed to have completed when the queue gets empty. Then, the transmission probability is calculated for each SU by taking the ratio between the number of packets successfully transmitted and the number of attempts made. The graph depicting the fluctuation of the probability of successful transmission is given in Figures 8 and 9 which are plotted for five and ten SUs, respectively. In these figures, we can see that our scheme performs better in ten SU scenarios but becomes inferior to the performance of the random channel selection method when SU transmission probability is 0.4 in the scenario having five SUs. The possible reason behind this is, when the number of transmission attempts are low, the probability of contention failures is not large enough to penalize a channel with low PU activity from getting selected; therefore, most SU select those ending up with higher collisions. Therefore, the probability of success is low. Evidence for this reasoning can be seen in Figures 10 and 11. In these graphs, we plot the probability of the transmissions of two or more SUs colliding with each other for five and ten SUs, respectively. The low probabilities of successful transmission for our scheme in Figures 8 and 9 corresponds to the high collision probabilities in Figures 10 and 11, respectively. Furthermore, although the collision probability of our scheme is more than that of the random scheme, our scheme has better probability of successful transmission. This is due to the fact that our scheme uses channels with both less PU channel activity and less SU channel activity. Since we calculate the SU activity using previous access results, it is not perfect. This imperfectness causes higher collisions than the random scheme which uses all the channels equally, but since it stays biased towards the channels with low PU activity, it results in a better probability of successful transmission.

Finally, in Figures 12 and 13, we plot the average number of SUs, using the same channel at the same time, which is a measure of channel load. The higher the channel load, the higher will be the number of collisions and thus lower throughput.

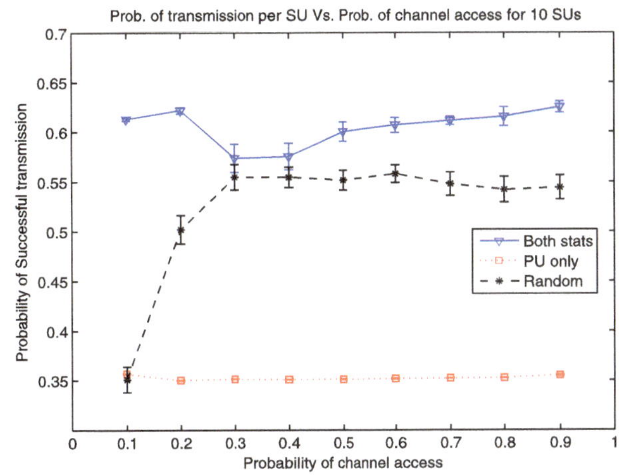

Figure 9 Probability of successful transmission per SU vs. SU channel access probability for ten SUs.

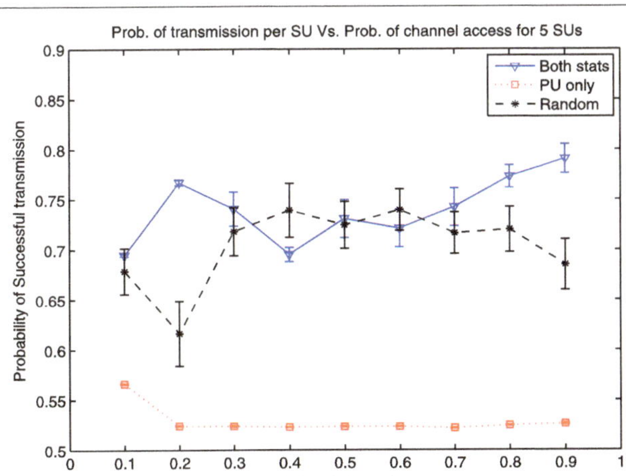

Figure 8 Probability of successful transmission per SU vs. SU channel access probability for five SUs.

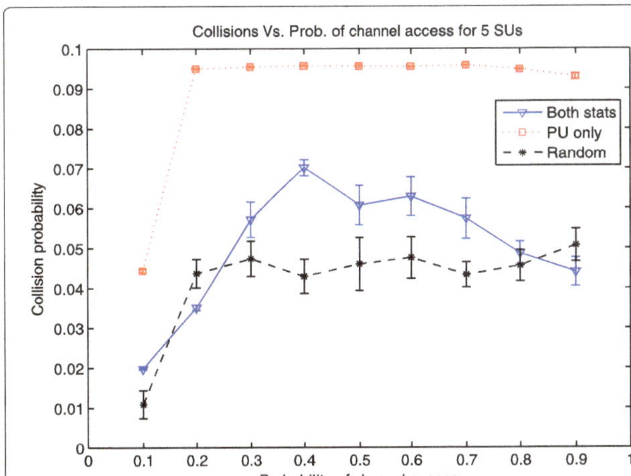

Figure 10 Probability of collision among SUs vs. SU channel access probability for five SUs.

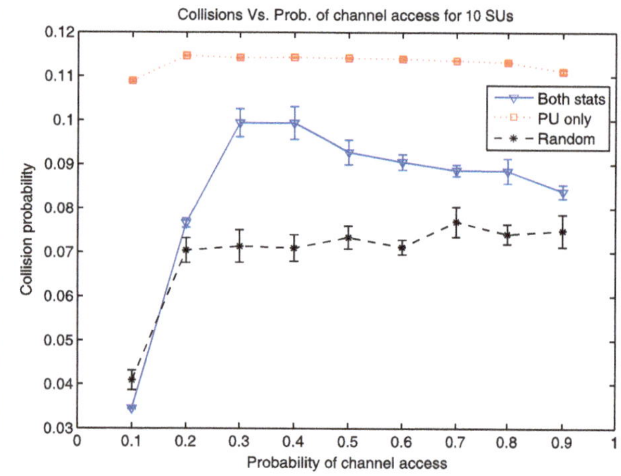

Figure 11 Probability of collision among SUs vs. SU channel access probability for ten SUs.

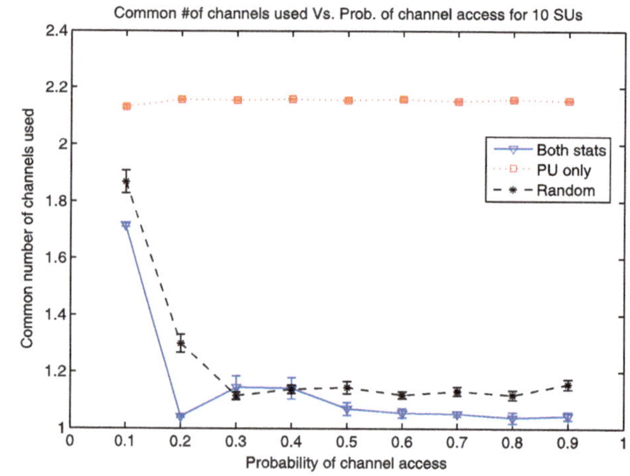

Figure 13 Average number of channels simultaneously used by SUs vs. SU channel access probability for ten SUs.

6 Conclusion

In this paper, we presented a proactive channel access MAC for an *ad hoc* CR network having a REM for initial setup and showed how this MAC can be used to keep the interference experienced by the PUs at a low level while providing better throughput for the SUs. The novel feature of this research was the incorporation of the SU statistics when selecting a channel set to be used. Furthermore, we introduced a SH CCC and derived the theoretical properties of it. This CCC used PU channel access probabilities when selecting the hopping pattern in the control phase. Then, in the data transmission phase, we used a prediction scheme called the PST which has a very low training and memory complexity. We tested our scheme in a scenario with a MAP-based channel state distribution with different channel utilization levels with imperfect sensing and showed that the impact of the channel set selection method has on the throughput and interference to the PUs. In the simulations, our channel set selection scheme had better probability of successful transmission than the random channel set selection scheme and the scheme based only on PU statistics at high SU traffic scenarios while having a PU interference probability less than 0.005. So when considering all of these factors, we can conclude that our MAC scheme is candidate worthy of consideration in proactive channel access in an *ad hoc* setting similar to the one the schemes we tested on.

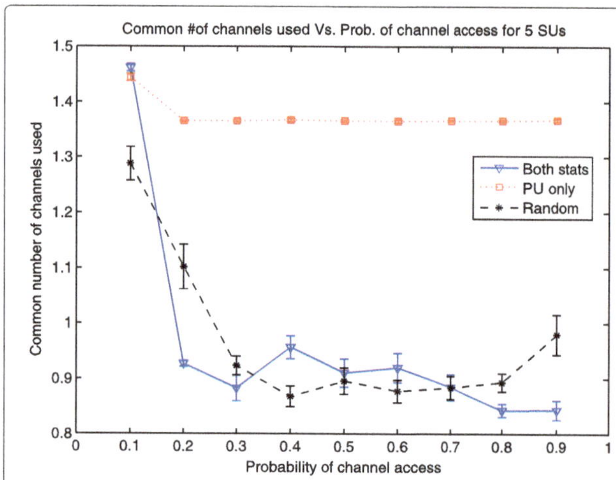

Figure 12 Average number of channels simultaneously used by SUs vs. SU channel access probability for five SUs.

Appendix A
PST learning algorithm

This algorithm consists of two phases. The main purpose of the algorithm was to find out a hypothesis which resembles the distribution which generated the sequence, where the Kullback-Leiber distance [32] between the actual and the empirical distribution is less than ϵ per state with probability $1 - \delta$ where $0 < \epsilon < 1$ and $0 < \delta < 1$ [6]. The Algorithms 3 and 4 mentioned below are based on [6,28]. A training sequence of sufficient length which represented the general channel usage was used by this algorithm to generate an accurate hypothesis on the distribution of the channel states. In the algorithm, empirical probabilities of the occurrence of any string s, $\tilde{P}(s)$, and the occurrence of a particular observation σ given the observation history sequence, $\tilde{P}(\sigma|s)$, were used to create the hypothesis. The formulae for calculating these which were taken from [6] are given in Equations 14 and 15, respectively. We assumed that a training sequence $\sigma_1^m = \sigma_1\sigma_2\ldots\sigma_m$ of length m having consecutive observations of the channel states is provided.

$$\tilde{P}(s) = \frac{1}{m - D + 1} \sum_{j=D}^{m-1} \chi_j(s), \tag{14}$$

$$\tilde{P}(\sigma|s) = \frac{\sum_{j=D}^{m-1} \chi_{j+1}(s\sigma)}{\sum_{j=D}^{m-1} \chi_j(s)}, \tag{15}$$

where $\chi_j(s)$ is an indicator function which takes value 1 when $s = \sigma_{j-|s|+1} \ldots \sigma_j$ and 0 otherwise, $|s|$ is the length of string s and D is the order of the model. As an example, let $\sigma_1^{10} = 1110010000$ and $s = 00$. When $j = 2$, $\chi_2(00) = 0$ since $\sigma_1\sigma_2 = 11$ is not equal to $s = 00$. When $j = 5$, $\chi_5(00) = 1$, since $\sigma_4\sigma_5 = 00 = s$. All Markov predictors depend on the property that the probability distribution of the next symbol can be approximated by conditioning it on the previous D symbols, which is called by the name 'short memory principle' [7]. If we chose the memory D to be smaller than required, the distribution is incapable of capturing all the dependencies between the symbols, which degrades prediction efficiency [7]. On the other hand, if we chose it to be too high, it over-fits the training sequence and gives a higher prediction error.

The probabilistic suffix tree algorithm has a favorable property of avoiding the over-fitting of the model to the training sequence if the maximum memory kept is too high. Therefore, we can afford to keep D higher than the required since the algorithm takes care of it [7]. But this advantage does not come free. For the algorithm to take care of this issue, more tunable parameters were introduced. This algorithm is governed by five parameters namely $D, P_{\min}, \alpha, r$, and γ. These parameters control the upper bound of the number of states in the variable-order Markov model. The first parameter is the maximum memory of the variable-order Markov model, denoted by D. Value of D is the maximum number of ones and zeros in the strings in set S used to name the states of the Markov chain or the maximum level to which the tree is grown. The second parameter is the minimum probability of abundance $P_{\min}$ in the training sequence, of any binary string s' of length $|s'|(\leq D)$, for s' to be a member of the tree. This is a necessary condition, but it is not sufficient. The third parameter α denotes the minimum value the conditional probability $\tilde{P}(\sigma|s')$ (see Equation 15) can take, for a binary string s' of length $|s'|(\leq D)$ and a symbol $\sigma \in \{1, 0\}$. This is also a necessary condition a string s' should meet in order for it to be a node on the tree. The fourth parameter r is a threshold which determines whether the string s contributes additional information in predicting the next symbol σ than its longest suffix $\hat{s}$ which is a node on the tree (if $s = \sigma_1\sigma_2 \ldots \sigma_{k-1}\sigma_k$, then $\hat{s} = \sigma_2 \ldots \sigma_{k-1}\sigma_k$). The final parameter γ is the probability assigned to $\tilde{P}(\sigma|s)$, if it is zero for any symbol σ and state s. Phase 1 of the algorithm is given in Algorithm 3.

Algorithm 3 PST learning algorithm
Phase 1

```
Initialize T̄ and S̄ : T̄ is a binary tree having a root node
e denoting empty string & S̄ is an empty set
S̄ ⇐ {σ|σ ∈ {1,0} and P̃(σ) ≥ P_min}
while S̄ ≠ ∅, pick any s ∈ S̄ and do
  remove s from S̄
  if ∃σ ∈ {1,0} s.t P̃(σ|s) ≥ α and P̃(σ|s)/P̃(σ|suffix(s)) > r then
    add the node corresponding to s with all the
    nodes on the path starting from the deepest node
    in T̄ which is a suffix of s to s, to T̄
  end if
  if |s| < D then
    Σ' = {1,0}
    while Σ' ≠ ∅ do
      remove σ' from Σ'
      if P̃(σ's) ≥ P_min then
        add σ's to S̄
      end if
    end while
  end if
  Continue to next s' ∈ S̄
end while
```

As one can see in the above algorithm, we started with a single node tree representing the null string e and as the algorithm progressed, we added nodes which were mandatory for the distribution to be correct. In the algorithm, we add a node v labeled by a string s to the tree $\bar{T}$ if the following criteria were satisfied by the string s. First, the empirical probability of the occurrence of string s in the training sequence $\tilde{P}(s)$ should be larger than a threshold $P_{\min}$. Because of this requirement, we avoided the exponential growth in the number of strings to be tested in the algorithm. Then, we checked whether the probability of occurrence of a symbol $\sigma \in \{1, 0\}$ after the string s, $\tilde{P}(\sigma|s)$ is greater than a threshold α. Finally, we checked to see whether $\tilde{P}(\sigma|s)$ is greater than the probability of getting σ after the longest suffix suffix(s) of s. We checked this by taking the ratio $\frac{\tilde{P}(\sigma|s)}{\tilde{P}(\sigma|\text{suffix}(s))}$ and finding out whether it is greater than r where $r > 1$. We added the extended version of s which is σs where $\sigma = \{1, 0\}$ to the set of strings $\bar{S}$ if the probability of occurrence of σs in the training sequence is greater than $P_{\min}$, immaterial of whether s is included in the tree or not. This was done because it is possible to have a string s' which has a conditional distribution substantially different from its parent.

Parent of a node in our tree was labeled by the longest suffix of that particular node.

After performing Algorithm 3 on the training sequence, there is a possibility that, for some of the nodes, the probability of occurrence of either 1 or 0 after the string s labeling that node ($\gamma_s(1)$ or $\gamma_s(0)$) is 0. Furthermore, there is a possibility that the internal nodes of the tree $\bar{T}$ may only have one of the children with respect to the occurrence of either 1 or 0. We corrected those problems in the phase 2 of the algorithm given in Algorithm 4 [6].

Algorithm 4 PST learning algorithm
Phase 2

1: Initialize $\hat{T}$ to be $\bar{T}$
2: Extend $\hat{T}$ by adding all missing sons of internal nodes.
3: For each node in $\hat{T}$ labeled by string s calculate the conditional probability $\gamma_s(\sigma)$ for each $\sigma = \{1, 0\}$ using Equation:
4: $\gamma_{s'}(\sigma) = \tilde{P}(\sigma|s')(1 - 2\gamma) + \gamma$,
5: where $s' = s$ if s is in $\bar{T}$ else s' is the longest suffix of s in $\bar{T}$.

Although we talked about a Markov chain, the algorithm grows a tree. In this tree, the leaf nodes gave the respective state labels of the equivalent Markov chain. But in the case that the destination state after a transition from a given state is ambiguous, we should add some additional states. Since the tree is sufficient for the predictions to be done, we do not explain it here, the readers are referred to [6] for more information about constructing the equivalent Markov chain.

Competing interests
The authors declare that they have no competing interests.

Acknowledgements
The work reported in this paper is financially supported by the Natural Sciences and Engineering Research Council of Canada.

Author details
[1] Department of Electrical and Computer Engineering, University of Manitoba, E1-551 EITC, 75A Chancellors Cir., R3T 5V6 Winnipeg, MB, Canada.
[2] Department of Electrical and Computer Engineering, University of Manitoba, E3-504A EITC, 75A Chancellors Cir., R3T 5V6 Winnipeg, MB, Canada.

References
1. FCC, ET Docket No 03-222 Notice of proposed rule making and order, December 2003 (2003)
2. Federal Communications Commission Spectrum Policy Task Force Report of the Spectrum Efficiency Working Group (2002)
3. L Tan, L Le, Joint cooperative spectrum sensing and MAC protocol design for multi-channel cognitive radio networks. EURASIP J. Wireless Commun. Netw. **2014**(1), 101 (2014). doi:10.1186/1687-1499-2014-101
4. C Devanarayana, AS Alfa, in *Global Telecommunications Conference (GLOBECOM 2011), 2011 IEEE*. Predictive channel access in cognitive radio networks based on variable order Markov Models, (2011)
5. F Bouali, O Sallent, J Perez-Romero, R Agusti, in *Cognitive Radio Oriented Wireless Networks and Communications (CROWNCOM), 2011 Sixth International ICST Conference On*. Strengthening radio environment maps with primary-user statistical patterns for enhancing cognitive radio operation, (2011), pp. 256–260
6. D Ron, Y Singer, N Tishby, The power of amnesia: learning probabilistic automata with variable memory length. Mach. Learn. **25**(2-3), 117–149 (1997). doi:10.1007/BF00114008
7. D Katsaros, Y Manolopoulos, Prediction in wireless networks by Markov chains. IEEE Wireless Commun. **16**, 56–64 (2009). doi:10.1109/MWC.2009.4907561
8. C Song, D Chen, Q Zhang, in *Communications (ICC) 2010 IEEE International Conference On*. Understand the predictability of wireless spectrum: a large-scale empirical study, (2010), pp. 1–5. doi:10.1109/ICC.2010.5502054. http://ieeexplore.ieee.org/stamp/stamp.jsp?tp=&arnumber=5502054&isnumber=5501741
9. Q Zhao, L Tong, A Swami, Y Chen, Decentralized cognitive MAC for opportunistic spectrum access in ad hoc networks: A POMDP framework. IEEE J. Selected Areas in Commun. **25**(3), 589–600 (2007). doi:10.1109/JSAC.2007.070409
10. Y Chen, Q Zhao, A Swami, Joint design and separation principle for opportunistic spectrum access in the presence of sensing errors. Inf. Theory IEEE Trans. **54**(5), 2053–2071 (2008). doi:10.1109/TIT.2008.920248
11. Q Zhao, S Geirhofer, L Tong, BM Sadler, Opportunistic spectrum access via periodic channel sensing. Signal Process. IEEE Trans. **56**(2), 785–796 (2008). doi:10.1109/TSP.2007.907867
12. J Gu, WS Jeon, JM Kim, Proactive frequency-hopping dynamic spectrum access against asynchronous interchannel spectrum sensing. Vehicular Technol. IEEE Trans. **62**(8), 3614–3626 (2013). doi:10.1109/TVT.2013.2246595
13. Y Song, J Xie, Prospect: A proactive spectrum handoff framework for cognitive radio ad hoc networks without common control channel. IEEE Trans. Mobile Comput. **11**(7), 1127–1139 (2012). doi:10.1109/TMC.2011.140
14. H Kim, KG Shin, Adaptive mac-layer sensing of spectrum availability in cognitive radio networks. Technical report, University of Michigan, (2006). http://www.eecs.umich.edu/techreports/cse/2006/CSE-TR-518-06.pdf
15. L Yang, L Cao, H Zheng, Proactive channel access in dynamic spectrum networks. Phys. Commun. Elsevier. **1**(2), 103–111 (2008). doi:10.1016/j.phycom.2008.05.001
16. VK Tumuluru, P Wang, D Niyato, in *IEEE International Conference on Communications*. A neural network based spectrum prediction scheme for cognitive radio, vol. 294, (2010), pp. 1–5. doi:10.1109/ICC.2010.5502348. http://ieeexplore.ieee.org/stamp/stamp.jsp?tp=&arnumber=5502348&isnumber=5501741
17. I Akbar, W Tranter, in *Proceedings 2007 IEEE SoutheastCon*. Dynamic spectrum allocation in cognitive radio using hidden Markov models: Poisson distributed case, (2007), pp. 196–201. doi:10.1109/SECON.2007.342884
18. S Yarkan, H Arslan, in *2007 IEEE 66th Vehicular Technology Conference*. Binary time series approach to spectrum prediction for cognitive radio, (2007), pp. 1563–1567. doi:10.1109/VETECF.2007.332
19. P Ren, Y Wang, Q Du, J Xu, A survey on dynamic spectrum access protocols for distributed cognitive wireless networks. EURASIP J. Wirel. Commun. Netw. **2012**(1), 60 (2012). doi:10.1186/1687-1499-2012-60
20. A De Domenico, EC Strinati, M Di Benedetto, A survey on mac strategies for cognitive radio networks. Commun. Surv. Tutorials IEEE. **14**(1), 21–44 (2012). doi:10.1109/SURV.2011.111510.00108
21. BF Lo, A survey of common control channel design in cognitive radio networks. Phys. Commun. Elsevier. **4**(1), 26–39 (2011). doi:10.1016/j.phycom.2010.12.004
22. BF Lo, IF Akyildiz, AM Al-Dhelaan, Efficient recovery control channel design in cognitive radio ad hoc networks. Vehicular Technol. IEEE Trans. **59**(9), 4513–4526 (2010). doi:10.1109/TVT.2010.2073725
23. G-Y Chang, W-H Teng, H-Y Chen, J-P Sheu, Novel channel-hopping schemes for cognitive radio networks. Mobile Comput. IEEE Trans. **13**(2), 407–421 (2014). doi:10.1109/TMC.2012.260

24. K Bian, J-M Park, R Chen, Control channel establishment in cognitive radio networks using channel hopping. Selected Areas Commun. IEEE J. **29**(4), 689–703 (2011). doi:10.1109/JSAC.2011.110403
25. C Cormio, KR Chowdhury, Common control channel design for cognitive radio wireless ad hoc networks using adaptive frequency hopping. Ad Hoc Netw. **8**(4), 430–438 (2010). doi:10.1016/j.adhoc.2009.10.004
26. S-S Tan, J Zeidler, B Rao, Opportunistic spectrum access for cognitive radio networks with multiple secondary users. Wireless Commun. IEEE Trans. **12**(12), 6214–6227 (2013). doi:10.1109/TWC.2013.102313.130125
27. H Bogucka, M Parzy, P Marques, JW Mwangoka, T Forde, Secondary spectrum trading in tv white spaces. Commun. Mag. IEEE. **50**(11), 121–129 (2012). doi:10.1109/MCOM.2012.6353691
28. R Begleiter, R El-yaniv, G Yona, On prediction using variable order Markov models. J. Artif. Intell. Res. **22**, 385–421 (2004)
29. B Jang, ML Sichitiu, IEEE 802.11 saturation throughput analysis in the presence of hidden terminals. Netw. IEEE/ACM Trans. **20**(2), 557–570 (2012). doi:10.1109/TNET.2011.2165322
30. S Robert, J-YL Boudec, New models for pseudo self-similar traffic. Perform. Eval. **30**(1-2), 57–68 (1997). doi:10.1016/S0166-5316(96)00055-7
31. A Alfa, V Pla, J Martinez-Bauset, V Casares-Giner, in *NETWORKING 2011 Workshops. Lecture Notes in Computer Science*, ed. by V Casares-Giner, P Manzoni, and A Pont. Discrete time analysis of cognitive radio networks with saturated source of secondary users, vol. 6827 (Springer Berlin, Heidelberg, 2011), pp. 3–12
32. TM Cover, JA Thomas, *Elements of information theory*, 2nd edn. (Wiley-Interscience, New York, New York, USA, 2006)

11

Systematic construction of common channel hopping rendezvous strategies in cognitive radio networks

Erik Ortiz Guerra[1], Vitalio Alfonso Reguera[1], Richard D Souza[2*], Evelio G Fernández[3] and Marcelo E Pellenz[4]

Abstract

Cognitive radio networks (CRNs) have emerged as a promising paradigm that can solve the shortage of spectrum resources, providing the ability to adapt and opportunistically exploit the spectrum holes. The rendezvous process allows cognitive users to find a common channel and establish a communication link. In this paper, we focus on the design of fast blind rendezvous algorithms to guarantee that every node should be able to rendezvous in all common available channels. We follow a systematic approach by first proposing a role-based algorithm that ensures maximum rendezvous diversity and then extending it to a common strategy through the use of multiples radios. Both proposals guarantee rendezvous under symmetric and asymmetric models. Simulation results show that our proposals outperform other recently developed rendezvous protocols with similar approaches in terms of expected time to rendezvous and maximum time to rendezvous.

Keywords: Cognitive radio; Blind rendezvous; Rendezvous diversity; Channel hopping sequences

1 Introduction

Recent spectrum measurements indicate that much of licensed spectrum is underutilized most of the time and unlicensed spectrum has been overcrowded due to the growth of wireless applications. Cognitive radio networks (CRNs) have been proposed to solve the shortage of usable spectrum and improve the efficiency in the usage of licensed spectrum [1,2].

In CRNs, each cognitive device, or secondary user (SU), has one or more radios capable of sensing the spectrum and detecting portions that are not being used by licensed users, or primary users (PU). To establish a communication link, two SUs should meet on a common channel and exchange handshake information, this process is referred to as *rendezvous* and is essential in the CRNs operation [3]. Because of the different locations of SUs, the set of spectrum holes, also called available channel set, may not be the same for each SU. If the set of available channels is the same for all SUs, then we are dealing with a symmetric model; otherwise, the model is said to be asymmetric (i.e., different SUs perceive different available channels).

A simple way to solve the rendezvous problem is by using a common control channel (CCC) [4]. In this approach, one of the available channels is assigned as the CCC and serves as the rendezvous channel. Some works have proposed static and dynamic CCC selection [5-10]. Although they have good theoretical performance, their main drawback is that a CCC constitutes a single point of failure, experiencing congestion when the number of PUs increases and facing the possibility that the CCC is occupied by a PU [11,12]. Another approach is to use channel hopping (CH) sequences, in which each SU visits the network channels following a predefined strategy in order to achieve rendezvous in any of the available channels. This process is referred to as *blind rendezvous* and overcomes the drawbacks of CCC.

Additionally, according to the SUs' behavior, there are different types of rendezvous strategies [13]. If SUs have preassigned roles (e.g., transmitter or receiver), they can follow different algorithms to generate their CH sequences and perform a role-based (RB) strategy. These strategies have been proven to be optimal in minimizing the time to rendezvous, and the sequences generated by

*Correspondence: richard@utfpr.edu.br
[2]Federal University of Technology, Paraná (UTFPR), Avenida Sete de Setembro, Curitiba, Brazil
Full list of author information is available at the end of the article

each SU are different. On the other hand, the preassigned role is not possible in some scenarios, and in these cases, the SUs must perform a common strategy (CS) to generate their CH sequences. Unlike the RB strategy, in CS the sequences generated by each SU are identical. Mixed or hybrid strategies are also possible in which the SUs use the same algorithm to generate the CH sequences but the generated sequences are different (e.g., by using a random seed). Furthermore, CS-CH sequences can be deterministic, in which case they are periodically repeated, or can be generated following a random pattern. The latter approach has been used in [12,14]. The authors in [12] show that in scenarios with intermittent PU activity the use of deterministic sequences offers no advantages over random CH algorithms. However, the main drawback of the random solutions is that they cannot guarantee the rendezvous in finite time under any circumstance. The analysis of random sequences is out of the scope of this paper, whose main focus is on the design of deterministic sequences.

Most works regarding CH rendezvous [14-29] consider that the time is divided into slots and the time to rendezvous (TTR) is defined as the number of time slots needed to successfully rendezvous after all SUs have begun the CH sequences. Moreover, the expected TTR (ETTR) and the maximum TTR (MTTR) are two important metrics commonly used to evaluate these strategies and are often computed when all channels are available. The authors in [15,16] adopt the term maximum conditional TTR (MCTTR) to refer to the MTTR when not all channels are available to all SUs. Some authors [15,17,27,30] also consider in the design of CH sequences the number of channels in which the rendezvous may occur (*rendezvous diversity*). However, most of them only make a qualitative analysis of this parameter. A metric for evaluating the rendezvous diversity can be found in [15,19,21]. As expected, in the asymmetric model, the algorithms with greater rendezvous diversity perform better than the mechanisms where the rendezvous is only achieved in a few channels.

In this paper, we focus on the design of deterministic CH sequences based algorithms to solve the blind rendezvous problem. Our main purpose is to develop a fast rendezvous algorithm which guarantees that SUs may meet in all available channels. We follow a systematic approach by first proposing a *full diversity channel hopping* (FDCH) role-based algorithm that ensures maximum rendezvous diversity, and then extending it to a common strategy through the use of multiples radios. We include a theoretical analysis as well as extensive simulations for two-user and multi-user CRNs under symmetric and asymmetric models, in order to demonstrate that the proposed strategies outperform other recently proposed ones.

The remainder of this paper is organized as follows. Related work is reviewed in Section 2. The system model and problem formulation is presented in Section 3. In Section 5, we propose the FDCH role-based (FDCH-RB) and FDCH common strategy (FDCH-CS) algorithms and present their theoretical performance analysis. Section 6 extends our proposal to the multi-user scenario. Simulation results are presented in Section 7. Finally, Section 8 concludes the paper.

2 Related works

Several algorithms have been proposed to solve the blind rendezvous issue. In [17], the asynchronous channel hopping (ACH) algorithm is proposed where the authors focus on designing CH sequences that ensure rendezvous in all available channels and use the term *rendezvous diversity* to denote the number of channels in which the rendezvous may occur. First, they present a transmitter and a receiver sequence (i.e., a RB strategy) and then propose a CS strategy that allows each SU to alternate between these sequences, as to guarantee rendezvous and ensure maximum rendezvous diversity (i.e., rendezvous in all available channels). Recently, in [18], we proposed a simple role-based (SRB) strategy for the symmetric model that outperforms the ACH(RB) algorithm. However, the SRB is unable to guarantee rendezvous in the asymmetric model.

The authors in [15] present four quorum-based channel hopping (QCH) strategies. These algorithms are role based. The M-QCH and L-QCH are described under the assumption of global time synchronization between SUs (i.e., synchronous environment), while A-QCH and A-MOCH do not need time synchronization between SUs. Although A-QCH and A-MOCH work in asynchronous environments, the A-MOCH ensures maximum rendezvous diversity while A-QCH does not. Like ACH(RB) and A-MOCH, our approach considers a role-based strategy which ensures maximum rendezvous diversity but with a different sequence construction scheme that guarantees rendezvous in both, synchronous and asynchronous, environments. Also we extend our proposal to a common strategy by the use of two radios.

Two RB strategies called rendezvous couple channel hopping (RCCH) and asynchronous rendezvous channel hopping (ARCH) are proposed in [19]. The first one considers that the SUs have global clock synchronization while the other assumes an asynchronous scenario and both ensure maximum rendezvous diversity when the number of network channels is even. Also this paper presents the symmetric asynchronous rendezvous channel hopping (SARCH) which is a CS strategy but not ensures maximum diversity. In our proposal, no global clock synchronization is assumed, neither the number of network channels is constrained.

As in ACH, A-MOCH, RCCH, and ARCH algorithms, the authors in [20,21] consider the rendezvous diversity in the design of an efficient channel hopping (ETCH) algorithm. This algorithm has two variants; first, the authors propose SYNC-ETCH which considers synchronous environments. Next, the ASYN-ETCH algorithm is proposed which does not assume the existence of global clock synchronization. Both ensure maximum rendezvous diversity. The main drawback of the ASYN-ETCH is that it exhibits a greater MTTR than that of other similar proposals [19].

In [22] three algorithms called generated orthogonal sequence (GOS) based, modular clock (MC), and modified modular clock (MMC) are presented. The GOS and MC algorithms only work under the symmetric model while MMC provides asymmetric model support. In terms of ETTR, the MC and MMC performance is better than that of GOS, but they do not guarantee rendezvous.

Other CS-CH sequence algorithms are introduced in [23] and [24]: the deterministic rendezvous (DRSEQ) and the channel rendezvous (CRSEQ). Although both schemes guarantee rendezvous, none of them guarantees maximum rendezvous diversity. The DRSEQ algorithm only works under the symmetric model and outperforms GOS, MC, and CRSEQ in this scenario.

The aforementioned algorithms consider that the SUs follow a CH sequence for visiting each of the channels in the network and once in the selected channel perform spectrum sensing to identify whether the channel is free of PU activity. This strategy can be effective when the channels used by the PUs change dynamically. However, in scenarios where these channel changes do not occur or occur very slowly, other mechanisms are more suitable. In this paper, we consider that during the rendezvous attempt the PUs operate in a fixed set of channels and that the SUs have prior knowledge of the state of each of the network channels[a]. With this knowledge, some authors [16,25-29] provide each SU with the ability to attempt to rendezvous only in a channel previously identified as available (free of PU interference).

In [25] the ring walk (RW) rendezvous algorithm and two variants, RW1 and RW2, are proposed. Both work under symmetric and asymmetric models, guarantee multi-user rendezvous, and outperform GOS, MC, and MMC in terms of ETTR. The performance of RW2 is better than RW1, but none of them ensures maximum rendezvous diversity.

The authors in [26] propose the jump stay (JS) algorithm which guarantees rendezvous under symmetric and asymmetric models, supports multi-user rendezvous, and ensures maximum rendezvous diversity. In [27] the same authors report the enhanced jump stay (EJS) algorithm that also guarantees rendezvous in all available channels and improves over JS in terms of ETTR and MTTR for the asymmetric model but slightly increases the MTTR in the symmetric model. The fast rendezvous channel hopping (FRCH) algorithm is introduced in [16] and is shown to outperform JS in terms of ETTR. In [28] we propose the short sequence-based (SSB) rendezvous algorithm which outperforms the FRCH and EJS algorithm in terms of ETTR and MTTR but, like FRCH, does not guarantee rendezvous in all available channels.

Except in the ACH(RB), SRB, RCCH, and ARCH in all other protocols described above, all SUs use the same algorithm to generate the CH sequences. However, in the SYNC-ETCH, ASYN-ETCH, MC, MMC, JS, EJS, and RW algorithms, each SU uses different parameters to generate its CH sequence. Thus, even when using a common algorithm, the jump sequences generated by each SU are different. These algorithms are an hybrid variant between CS and RB strategies. Note that they do not need preassigned roles, but the hopping sequences are different for each SU. One way to generate different CH sequences is that each SU uses random selected parameters, as this is the approach used in the MC, MMC, JS, and EJS algorithms. On the other hand, in the RW algorithm, each SU uses its ID as a seed to generate the CH sequences.

Most of the reported algorithms [16-28] implicitly consider that each SU is equipped with one radio. However, the use of multiple radios can help to reduce the TTR. This view is presented in [29] where a role-based parallel sequence (RPS) is proposed to exploit multiple radios and reduce the ETTR in a two-user scenario. The SUs use an hybrid strategy to generate the CH sequences which ensure rendezvous and maximum rendezvous diversity. The use of multiple radios allows each radio to use a different CH sequence, so that each SU assumes more than one role simultaneously. Our approach is similar to that described in [29], but with different CH sequences, providing better performance in most practical scenarios as can be seen in Section 7.

Table 1 shows a brief comparison of some of the above mentioned algorithms considering the asymmetric model support, rendezvous diversity, and rendezvous strategy. Also we include the theoretical results in terms of ETTR, MTTR, and MCTTR as provided by the respective authors.

3 System model and problem formulation

We consider a CRN consisting of $M \geq 2$ SUs which coexist with some PUs in the same geographical area. The PUs are the licensed users of the spectrum, which can be divided into $N \geq 2$ non-overlapping channels, labeled as $0, 1, \ldots N-1$, and these labels are the same for all SUs. We assume that the time is divided into slots of equal duration. For any time slot $t \in \mathbb{N}$ [b] let $S_u(t)$, $S_u : \mathbb{N}^* \rightarrow \{0, \ldots, N-1\}$, be the CH sequence for SU

Table 1 Summary of representative rendezvous protocols

Protocol	ETTR	MTTR	MCTTR	Asymmetric model support	Maximum rendezvous diversity	Strategy
ACH(RB) [17]	Unknown	Unknown	Unknown	✓	✓	RB
A-MOCH [15]	Unknown	$N^2 - N + 1$	N^2	✓	✓	RB
SRB [18]	$\frac{N-1}{2}$	$N - 1$	Unknown			RB
ARCH [19]	Unknown	$2N - 1$*	N^2*	✓	✓*	RB
ASYN-ETCH [21]	$\frac{2N^2+N}{N-1}$	Unknown	Unknown	✓	✓	Hybrid
GOS [22]	$\frac{N^4+2N^2+6N-3}{3N^2+3N}$	$N(N+1)$	Unknown			CS
MC [22]	$\frac{2P^2}{P-1}$	Unknown	Unknown			Hybrid
MMC [22]	Unknown	Unknown	Unknown	✓		Hybrid
DRSEQ [23]	Unknown	$2N + 1$	Unknown			CS
CRSEQ [24]	Unknown	Unknown	$P(3P - 1)$	✓		CS
RW2 [25]	$(N-1)(\ln^n + \frac{1}{2})$	$n(N - 1)$	$2n(N + 1 - G)(N - 1)$	✓		Hybrid
JS [26]	$\frac{5P}{3} + 3$	$3P$	$3NP(P - G) + 3P$	✓	✓	Hybrid
EJS [27]	$\frac{3P}{2} + 3$	$4P$	$4P(P + 1 - G)$	✓	✓	Hybrid
FRCH [16]	Unknown	$2N + 1$	$N(2N + 1)$**	✓		CS
SSB [28]	$\frac{2(N-1)^2}{2N-1}$	$2N - 2$	$(N - 1)(2N - 1)$	✓		CS
RPS*** [29]	(1)	(2)	Unknown	✓	✓	Hybrid

N, number of available channels. P, smallest prime greater or equal than N for MC and CRSEQ, but $N + 1$ for JS and EJS. n, number of nodes in the network. G, number of common available channels for two-user scenario. *These results are valid when N is even, please see Table one in [19] for further details. **This expression is only valid for some values of N, please see [16] for further details. ***In the RPS algorithm, each SU is equipped with more than one radio. (1): ETTR $= \left\lceil \frac{P}{\max(u,v)-1} \right\rceil + \frac{\left(\left\lceil \frac{P}{\max(u,v)-1} \right\rceil - 1\right)^2}{2\left\lceil \frac{P}{\min(u,v)-1} \right\rceil}$. (2): MTTR $= \left\lceil 2\left\lceil \frac{P}{\max(u,v)} \right\rceil - 1 \right\rceil$. u, v, number of radios of SUs attempting to rendezvous. ARCH(RB), asynchronous channel hopping (role based). A-MOCH, asynchronous maximum overlapping channel hopping. SRB, simple role-based. ARCH, asynchronous rendezvous channel hopping. ASYN-ETCH, asynchronous efficient channel hopping. GOS, generated orthogonal sequences. MC, modular clock. MMC, modified modular clock. DRSEQ, deterministic rendezvous sequence. CRSEQ, channel rendezvous sequence. RW2, ring walk 2. JS, jump stay. EJS, enhanced jump stay. FRCH, fast rendezvous channel hopping. SSB, short sequence-based. RPS, role-based parallel sequence.

u, $u \in \{1, 2, \ldots M\}$. Let $A_u \subseteq \{0, 1, 2, \ldots, N-1\}$ be the set of available channels for SU u, where a channel is said to be available for SU communication if PU activity is not detected in this channel. Like [16,25-27] we consider that each SU know its set of available channels (i.e., A_u) before the rendezvous process begins. Let G be the number of commonly available channels for all SUs (i.e., $G = |\bigcap_{\forall u} A_u|$, where $\bigcap_{\forall u} A_u$ is the intersection between the set of available channels for all SUs and $|\cdot|$ is the cardinality operator). Note that $G \neq 0$ is a necessary condition for a feasible rendezvous solution.

According to the channels' availability, we consider both symmetric and asymmetric models. In the symmetric model, the SUs perceive the same available channels, so that $A_i = A_j, \forall i,j \leq M$. For the asymmetric model, different SUs might perceive different available channels.

We also assume that the SUs have not global time synchronization and each one has a local independent clock. We emulate slot boundary synchronization following a similar approach as in [23,26]. Let τ be the minimum duration for exchanging the necessary handshake information. In [23] the authors proved that since the SUs may not be time synchronized, the duration of a time slot of at least 2τ ensures that the overlap of any two slots is sufficient to complete the rendezvous process. As proved in [26], this is equivalent to consider that the SUs are synchronized with respect to slot boundaries. Moreover, we consider that the SUs can start its CH sequences at different time slots.

We consider important to highlight that even if two SUs are in the same channel, the communication between them can fail due to fading, contention with other SUs or some other situations that affect the quality of the channels. Then, if the CH sequence only guarantees that SUs meet in few channels, the rendezvous will be conditioned by the state of those channels and eventually will degrade the TTR performance. Thus, in addition to ensuring rendezvous, it is also important to diversify the channels where rendezvous may occur (i.e., to maximize rendezvous diversity). To compute the rendezvous diversity, we introduce a new metric, the *rendezvous diversity index*, which is a variation of the *degree of rendezvous* proposed in [19].

Definition 1. In a CRN with G common available channels for SUs, if C: $|C| \leq G$ is a set of channels in which the rendezvous occurs, the rendezvous diversity index can be computed as: $D = \frac{|C|}{G}$.

The rendezvous diversity is maximized when $|C| = G$ which minimizes the risk of rendezvous failures. For some rendezvous protocols [16,22-25,28], the cardinality of C is a function of the channels in which each SU starts the rendezvous process. Accordingly, we consider $\overline{D}$ as the mean value of the rendezvous diversity index and when $\overline{D} = 1$ we say that the rendezvous scheme ensures maximum rendezvous diversity.

4 Basic idea for CH sequence construction

Similar to [18,24], we allocate T discrete points in a ring, as shown in Figure 1, such that each point represents one of the N network channels. In order to maximize rendezvous diversity, all channels appears at least once in the ring, so that $T \geq N$. Note that this is an abstraction in order to clarify the analysis in the rest of this paper. In the CRN context, two SUs are initially placed in any of the T positions in the ring. In the literature, there are some classical rendezvous algorithms that work over continuous rings, as the OP-DIR algorithm [13], in which those intending to rendezvous jump in opposite directions over the ring. However, since the SUs jump at discrete positions in the CRN context, the use of such classical rendezvous algorithm is not effective in all cases. Next, some considerations regarding this issue are discussed in detail.

Consider that two SUs, denoted by A and B, sequentially run through the T positions in Figure 1 in opposite directions (i.e., SU A runs clockwise and SU B runs anticlockwise, for instance), one position at a time with the same speed. Then, the CH sequences can be described as:

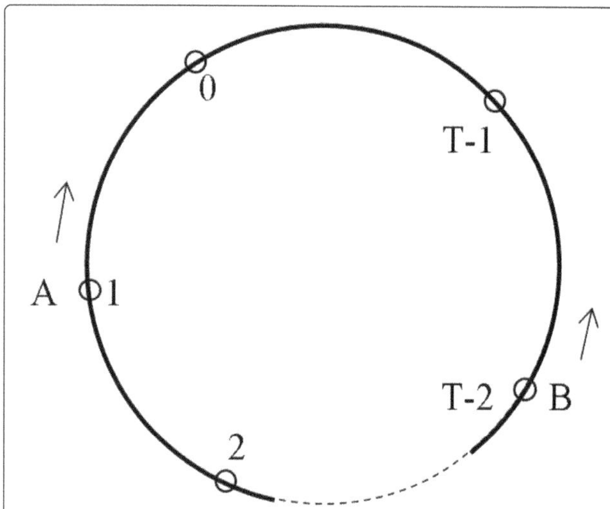

Figure 1 Channel distribution.

$$S_A(t) = (S_A(0) - t) \bmod T, \quad (1)$$

$$S_B(t) = (S_B(0) + t) \bmod T, \quad (2)$$

where $S_A(0)$ and $S_B(0)$ are the initial positions in the ring for SUs A and B, respectively, and $a \bmod b$ is the non-negative remainder of the division of a by b.

Let $d(t) : 0 \leq d(t) < T$ be the distance between the SUs which can be computed as a number of nodes between SUs A and B at time slot t, such that:

$$d(t) = \begin{cases} S_A(t) - S_B(t), & S_A(t) \geq S_B(t) \\ T + S_A(t) - S_B(t), & S_A(t) < S_B(t). \end{cases} \quad (3)$$

Theorem 1. *If all channels are available for the SUs and $d(0)$ is even, the rendezvous occurs in $t_r = \frac{d(0)}{2}$ time slots.*

Proof. For the rendezvous to occur at time slot $t_r = \frac{d(0)}{2}$, we must have $S_A(t_r) = S_B(t_r)$. If $S_A(0) \geq S_B(0)$, from (3) we have that $S_B(0) = S_A(0) - d(0)$. Then, with the help of (2)

$$S_B(t_r) = \left(S_A(0) - d(0) + \frac{d(0)}{2}\right) \bmod T \quad (4)$$
$$= \left(S_A(0) - \frac{d(0)}{2}\right) \bmod T = S_A(t_r).$$

If $S_A(0) < S_B(0)$, then from (3) $S_B(0) = T + S_A(0) - d(0)$, and consequently

$$S_B(t_r) = \left(T + S_A(0) - d(0) + \frac{d(0)}{2}\right) \bmod T \quad (5)$$
$$= \left(T + S_A(0) - \frac{d(0)}{2}\right) \bmod T = S_A(t_r).$$

Alternatively, $S_A(t_r) = S_B(t_r)$ implies $d(t_r) = 0$. By noting that in each time slot both SUs sequentially jump in opposite directions (see Figure 1), if $d(t) \neq 1$ then in the next time slot the distance between them decreases by 2, so that $d(t+1) = d(t) - 2$ (i.e., the distance between SUs decreases twice the number of elapsed time slots). Therefore, the time to rendezvous can be calculated as $t_r = \frac{d(0)}{2}$ for $d(0)$ even. □

Following the above reasoning, it is easy to see that, if $d(0)$ is odd, after $\frac{d(0)+1}{2}$ time slots the SUs transpose their positions without meeting each other. Once the SUs transpose their positions, the new resulting distance is $T - 1$. Now, if T is even, the new distance, just after the SUs transpose their positions, is odd and the SUs indefinitely run through the ring without ever meeting. To overcome this problem, in what follows we restrict the analysis by making T odd.

Theorem 2. *If all channels are available for the SUs and $d(0)$ and T are odd, rendezvous occurs in $t_r = \frac{d(0)+T}{2}$ time slots.*

Proof. Given that $d(0)$ is odd, from the above discussion, we have $d(\frac{d(0)+1}{2}) = T - 1$. Since T is odd, $T - 1$ is even and from Theorem 1 we can write $d(\frac{d(0)+1}{2} + \frac{T-1}{2}) = 0$ which implies $t_r = \frac{d(0)+T}{2}$. □

4.1 Rendezvous diversity analysis

To compute the rendezvous diversity of the CH sequences presented in Equations 1 and 2, let us recall the example in Figure 1, where it can be seen that the rendezvous occurs in a single channel on every lap around the ring. Also given that Equations 1 and 2 exhibit a periodic behavior with period T and that $d(t)$ is a linear combination of $S_A(t)$ and $S_B(t)$ we have $d(0) = d(kT)$, $k \in \mathbb{N}^*$, where the value of k identifies the number of turns of each SU around the ring (thus $k = 0$ identifies the first round, $k = 1$ the second one and so on). This means that the rendezvous channel is the same in all sequences periods and therefore this strategy does not guarantee maximum rendezvous diversity. Therefore, next we discuss the proposed algorithm which aims at maximizing the rendezvous diversity while minimizing the TTR.

5 Proposed algorithm

As proved in Theorems 1 and 2 and according to the *rendezvous diversity index* definition, in order to guarantee finite MTTR and maximize the rendezvous diversity, two conditions must be met: (i) T **must be odd** and (ii) $|C| = G$. With this in mind we make

$$T = \begin{cases} N, & \text{for } N \text{ odd} \\ N+1, & \text{for } N \text{ even.} \end{cases} \quad (6)$$

Note that, in order to make T odd, in the case of N even an additional node is placed in the ring of Figure 1 so that one of the channels is listed twice. We consider important to highlight that even when the authors in [19] also use a ring channel distribution, they restrict their analysis for the case of N even (i.e., the number of networks channels must be even). Also, the construction of the sequences presented in this section differ greatly from that of the sequences described in [19].

In what follows we propose the full diversity channel hopping algorithm with two variants, one that is role based (FDCH-RB) and another that is a common strategy (FDCH-CS). We first present the role based rendezvous algorithm and, as in [15,17-19], we assume that the SUs attempting to rendezvous have different preassigned roles which are allocated before the rendezvous process starts. In the sequel, we use the terms transmitter and receiver in order to differentiate the role of each SU. In our proposal, the transmitter runs clockwise and jumps in each time slot. The receiver runs anticlockwise and jumps for $T - 1$ time slots but remains at the last node for one additional time slot before starting to jump for the next $T - 1$ time slots. This behavior of the receiver ensures maximum rendezvous diversity as we will discuss in Section 5.2.

Moreover, we consider that each sequence replaces the unavailable channels by an available one. Then, let $S_T(t)$ and $S_R(t)$ be the CH sequences of the transmitter and the receiver, respectively. The procedure for sequences construction are shown in Algorithms 1 and 2 where $S_T(0)$, $S_R(0)$ denote the initial channels and A_T, A_R are the set of available channels of the transmitter and the receiver, respectively.

Algorithm 1: Transmitter sequence

Data: $t, N, A_T, S_T(0)$
if *N is odd* **then**
  $T = N$
else
  $T = N + 1$
$S_T(t) = (S_T(0) - t) \bmod T$
if $S_T(t) = N$ **then**
  $S_T(t) = 0$
if $S_T(t) \notin A_T$ **then**
  call replacement operation
return $S_T(t)$

Algorithm 2: Receiver sequence

Data: $t, N, A_R, S_R(0)$
if *N is odd* **then**
  $T = N$
else
  $T = N + 1$
$S_R(t) = \left(t + S_R(0) - \left\lfloor \frac{t}{T} \right\rfloor\right) \bmod T$
if $S_R(t) = N$ **then**
  $S_R(t) = 0$
if $S_R(t) \notin A_R$ **then**
  call replacement operation
return $S_R(t)$

An example of sequences for even and odd values of N is shown in Figure 2. Due to the lack of synchronization between SUs, each one can begin its CH sequence at different time slots. The time to rendezvous is measured taking as starting point the time slot where both sequences have already begun. In Figure 2a, $N = 4$ and the additional channel is marked as 4(0). The first number is used to measure the distance between SUs and the number in parentheses is the actual channel selected in the CH

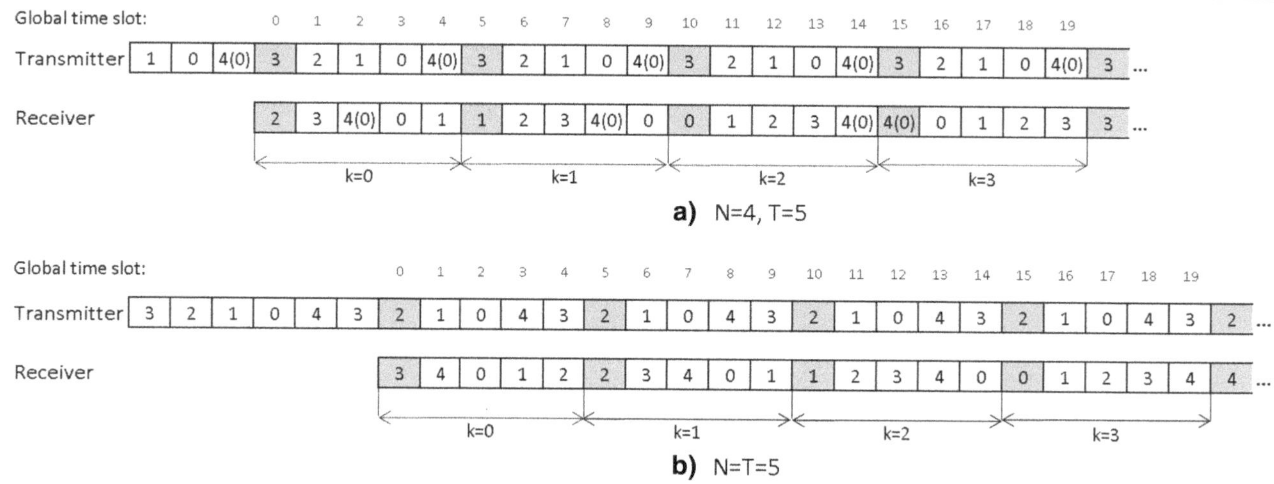

Figure 2 Transmitter and receiver sequences. **(a)** N is even; **(b)** N is odd.

sequence. Since both SUs use the same additional channel in every turn around the ring, rendezvous takes place.

5.1 ETTR and MTTR performance

By using the mathematical framework presented in [28], the ETTR can be computed as:

$$\mathrm{ETTR} = \frac{1}{T} \sum_{d(0)=0}^{T-1} \min\left\{t_r : S_T(t_r) = S_R(t_r)\right\}. \tag{7}$$

By Theorems 1 and 2, the sequences intersect for first time after

$$t_r = \begin{cases} \frac{d(0)}{2}, & \text{for } d(0) \text{ even} \\ \frac{d(0)+T}{2}, & \text{for } d(0) \text{ odd} \end{cases} \tag{8}$$

time slots. Therefore, (7) can be rewritten as:

$$\begin{aligned} \mathrm{ETTR} &= \frac{1}{T}\left\{\sum_{d(0)=1}^{\frac{T-1}{2}} \frac{2d(0)}{2} + \sum_{d(0)=1}^{\frac{T-1}{2}} \left[\frac{2d(0)-1+T}{2}\right]\right\} \\ &= \frac{1}{T}\left\{2\sum_{d(0)=1}^{\frac{T-1}{2}} d(0) + \sum_{d(0)=1}^{\frac{T-1}{2}} \frac{T-1}{2}\right\} \\ &= \frac{T-1}{2}. \end{aligned} \tag{9}$$

Additionally, from (8) the maximum time to rendezvous occurs when $d(0)$ is odd and equal to $T-2$, then by making $d(0) = T-2$ in (8) we have MTTR $= T-1$. The greatest value of MTTR for the proposed FDCH-RB algorithm happens when $T = N+1$, in such case MTTR $= N$. Note that this value is comparable with the MTTR for the SRB algorithm and smaller than that of the MTTR for the A-MOCH and ARCH algorithms (see Table 1). We consider important to highlight that the ETTR and MTTR as computed by the above equations are valid when all N channels are available for the SUs. In the next section, we analyze the MCTTR as the maximum TTR in scenarios where the number of common available channels is less than N.

5.2 Rendezvous diversity and MCTTR

As described at the beginning of Section 5 and shown in Figure 2b, at every time slot, the transmitter jumps to the next position in the sequence. However, the receiver follows the jump pattern for $T-1$ time slots and then stays in the same channel for one time slot before the new jump period. As a consequence, the distance, as defined in (3), between SUs at the beginning of each lap around the ring increases by one with respect to the previous lap, except when the distance has its maximum value in which case is reduced to zero, that is

$$d((k+1)T) = (d(kT)+1) \bmod T, \tag{10}$$

and by solving the recurrence relation (10) we obtain

$$d(kT) = (d(0)+k) \bmod T. \tag{11}$$

Note that $d(kT)$ takes values in the range from 0 to $T-1$ being an infinite sequence of truncated discrete ramp functions concatenated together. Then, after T laps of the SUs around the ring (i.e., from $k=0$ to $k=T-1$), $d(kT)$ has swept all possible integer values in its output range. Also, it is important to highlight that the proposed algorithm guarantees rendezvous for each lap around the ring. Let t_r^k be the TTR in the kth lap around the ring, measured from the beginning of that lap. Then, using (8) and (11), t_r^k can be written as

$$t_r^k = \begin{cases} \frac{d(kT)}{2}, & \text{for } d(kT) \text{ even} \\ \frac{d(kT)+T}{2}, & \text{for } d(kT) \text{ odd.} \end{cases} \quad (12)$$

By evaluating the transmitter sequence in t_r^k we have

$$S_T(t_r^k) = (S_T(0) - t_r^k) \bmod T. \quad (13)$$

Note that each different value of $d(kT)$ implies a different value of t_r^k in the range from 0 to $T-1$. Hence, after T turns around the ring, the rendezvous will have occurred in all possible channels according to (13). In other words, in every lap around the ring, the initial distance between SUs changes, modifying the next time to rendezvous and the resulting rendezvous channel. This means that FDCH-RB guarantees maximum rendezvous diversity.

Theorem 3. *The MCTTR for the FDCH-RB algorithm is* $MCTTR = T^2 - 1$.

Proof. If not all channels are available for the SUs, due to the maximum rendezvous diversity property, in the worst case the rendezvous occurs in lap $k = T - 1$. By noting that at the start of lap k, the elapsed time since the beginning of the rendezvous process is kT, the maximum conditional TTR (i.e., for the worst case) can be computed as $MCTTR = (T-1)T + MTTR$. Since, as derived in Section 5.1, in a lap where rendezvous is guaranteed the MTTR is $T-1$, we have $MCTTR = T^2 - 1$. □

The greatest value of MCTTR happens when N is even, in such case $T = N + 1$ and MCTTR $= N^2 + 2N$. This value is greater than that of the MCTTR for A-MOCH and ARCH algorithms. However, for N odd the MCTTR of the proposed FDCH-RB algorithm is MCTTR $= N^2 - 1$ which is slightly smaller than the MCTTR for the A-MOCH algorithm. Note that the ARCH algorithm only ensures rendezvous when N is even (see Table 1).

Additionally, the jump pattern of the transmitter causes that the channel $S_T(t + kT)$ is the same for any value of k. However, as a consequence of the jump-stay pattern of the receiver, $S_R(t + kT)$ sweeps all possible values in its output range for $0 \leq k \leq T - 1$. Then, in each lap around the ring to the same channel $S_T(t + kT)$, corresponds a different channel $S_R(t + kT)$ (e.g., time slots 1, 6, 11, and 16 in Figure 2a). Exploiting this feature, if the SUs are under the symmetric model, the unavailable channels are replaced by one channel in A_u (i.e., the set of available channels seen by SU u) in the following T laps to the ring as described in line 2 of Algorithm 3, where the notation $Au(i)$ is used to indicate the ith element of Au. Note that for the symmetric model $G = |A_u|$ and SUs always replace unavailable channels by a commonly available one.

Algorithm 3: Replacement operation

Data: T, A_u
1 $k = \lfloor \frac{t}{T} \rfloor$
2 $i = \lfloor \frac{k}{T} \rfloor \bmod |A_u|$
3 $S_u(t) = A_u(i)$
4 **return** $S_u(t)$

However, under the asymmetric model, $G \neq |A_u|$ and therefore, the replacement operation described above offers no advantage. In this case, we propose a random *replacement operation*. Note that since the proposed strategy ensures maximum rendezvous diversity, the random nature of the replacement operation do not compromise the MCTTR. The random replacement operation is described in Algorithm 4.

Algorithm 4: Replacement operation

Data: A_u
$i = \text{rand}(1, |A_u|)$
$S_u(t) = A_u(i)$
return $S_u(t)$

5.3 Removing the preassigned roles

The CH sequences construction described in the previous section gume slots, the ETTR can be computed as:arantees rendezvous and ensures maximum rendezvous diversity but requires that the roles be preassigned, which may not be possible in all scenarios. To overcome this drawback, the authors in [17] propose that each SU uses its ID to alternate between the role of transmitter and receiver. Although this proposal guarantees rendezvous, the performance in terms of ETTR of the CS strategy is worse than the RB strategy.

To solve the problem of allocation of roles, we equipped the SUs with two radios that operate simultaneously. One radio follows the transmitter sequence and the other the receiver sequence. Since the SUs implement the same algorithms, they perform a CS strategy. The appropriate use of more than one radio improves the ETTR and MTTR as was concluded in [29] and as we are going to show in the sequel. Figure 3 shows the basic idea behind our maximum diversity channel hopping common strategy (FDCH-CS) algorithm. Let $S_{Tu}(t)$ and $S_{Ru}(t)$ denote the transmitting and receiving CH sequences followed by the SU u, and $S_u(0)$ the initial position of SU u on the ring.

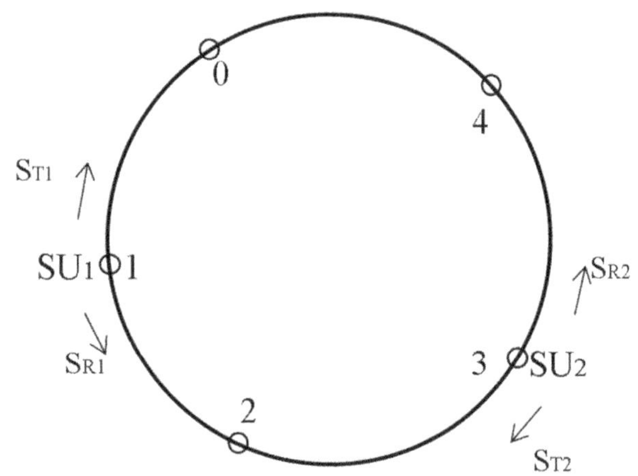

Figure 3 Common strategy. $N = 5, d_{12}(0) = 3, d_{21}(0) = 2$.

Since each SU runs both sequences simultaneously, the rendezvous is expected to be achieved between the transmitter sequence of one SU and the receiver sequence of the other (i.e., $S_{T1}(t) = S_{R2}(t)$ or $S_{T2}(t) = S_{R1}(t)$). Then, there exist two different values of $d(0)$: $d_{12}(0)$ for $S_{T1}(0)$ with respect to $S_{R2}(0)$ and $d_{21}(0)$ for $S_{T2}(0)$ with respect to $S_{R1}(0)$. Algorithm 5 shows the proposed common strategy (FDCH-CS).

Algorithm 5: Common strategy

Data: $t, N, A_u, S_u(0)$
1 S_{Tu} = Transmitter sequence $(t, N, A_u, S_u(0))$
2 S_{Ru} = Receiver sequence $(t, N, A_u, S_u(0))$
3 **return** S_{Tu}, S_{Ru}

Theorem 4. *If two SUs use the FDCH-CS algorithm to generate their CH sequences, then: MTTR* $= \frac{T-1}{2}$ *and ETTR* $= \frac{(T-1)(T+1)}{4T}$.

Proof. Considering that the SUs begin the rendezvous process at the same time and from Figure 3, it can be noted that $d_{12}(0) + d_{21}(0) = T$. Given that T is odd, always one of the distance values between SUs ($d_{12}(0)$ or $d_{21}(0)$) is even. We refer to this value as $d_p(0)$. Note that even when the SUs do not begin the rendezvous process at the same time, $0 \leq d_p(0) \leq T - 1$, so that the worse case occurs when the SUs begin the rendezvous process at the same time and the maximum value of $d_p(0)$ is $T - 1$ and by Theorem 1 MTTR $= \frac{T-1}{2}$.

In Figure 3 $d_{21}(0) = d_p(0) = 2$, however, if SU_2 was at node 4 in the ring, then $d_{12}(0) = 2$, $d_{21}(0) = 3$ and also $d_p(0) = 2$. This analysis leads to a value of $d_p(0)$ that is the same for two different positions of SU_2 in the ring. Then, considering that $d_p(0) = 0, 2, 4, \ldots T - 1$, the probability of $d_p(0)$ assuming a given value is

$$\Pr\left[d_p(0)\right] = \begin{cases} \frac{1}{T}, & \text{for } d_p(0) = 0 \\ \frac{2}{T}, & \text{for } d_p(0) > 0 \text{ and } d_p \text{ even.} \end{cases} \quad (14)$$

As the SUs intersect for the first time after $t_r = \frac{d_p(0)}{2}$ time slots, the ETTR can be computed as:

$$\begin{aligned} \text{ETTR} &= \frac{2}{T} \sum_{d_p(0)=1}^{\frac{T-1}{2}} \frac{2d_p(0)}{2} \quad (15) \\ &= \frac{1}{T}\left[\left(\frac{T-1}{2}\right)\left(\frac{T-1}{2}+1\right)\right] \\ &= \frac{(T-1)(T+1)}{4T}. \end{aligned}$$

□

Theorem 5. *If two SUs use the FDCH-CS algorithm to generate their CH sequences, then MCTTR* $= T^2 - 1$.

Proof. Because the SUs follow $S_T(t)$ and $S_R(t)$ sequences simultaneously, they meet twice in every lap around the ring. As we also considered in the proof of Theorem 4, the rendezvous happens for the first time at $t_r = \frac{d_p(0)}{2}$. Furthermore, there is a distance value between SUs that is odd ($T - d_p(0)$). Considering the second term of (8), the SUs rendezvous for the second time at $t_s = \frac{2T - d_p(0)}{2}$.

The rendezvous channels can be identified as: $c_1 = S_T(t_r)$ and $c_2 = S_R(t_s)$ for the first and second rendezvous, respectively. With the help of Algorithms 1 and 2, c_1 and c_2 can be rewritten as:

$$\begin{aligned} c_1 &= \left(S_T(0) - \frac{d_p(0)}{2}\right) \bmod T, \quad (16) \\ c_2 &= \left(S_R(0) + \frac{2T - d_p(0)}{2}\right) \bmod T \\ &= \left(T + S_R(0) - \frac{d_p(0)}{2}\right) \bmod T. \end{aligned}$$

Considering that the remainder operation is non-negative and that $S_{Tu}(0) = S_{Ru}(0) = S_u(0)$ it can be seen that $c_2 = c_1$. Then, although SUs meet twice at every lap around the ring, the rendezvous happens in the same channel and according to the proof of Theorem 3 MCTTR $= T^2 - 1$. □

Let us remark that in our CS strategy each user follows the S_T and S_R sequences simultaneously; therefore, as in the RB strategy, maximum rendezvous diversity is also guaranteed. Also we consider important to highlight that the FDCH-CS algorithm, in terms of ETTR and MTTR, performs better than the FDCH-RB algorithm. This is an expected result due to the fact that FDCH-CS uses two radios.

6 Multi-user rendezvous algorithm

The proposed FDCH-CS algorithm can be extended to a multi-user scenario. The purpose is to achieve rendezvous for M SUs by multiple pair-wise rendezvous. Once two SUs rendezvous in one of the channels, they interchange information to synchronize their parameters and, after that, the two SUs generate an identical CH sequence.

Each SU u has two parameters to determine the jump sequence: the initial channel ($S_u(0)$) and the current time slot since the beginning of its sequence (t_u). After SU_1 and SU_2 successfully rendezvous, each one updates its parameters according to the following strategy: $S_1(0) = S_2(0) = \min\{S_1(0), S_2(0)\}$ and $t_1 = t_2 = \min\{t_1, t_2\}$. Afterward, the two SUs generate an identical CH sequence.

As an illustrative example, consider the scenario shown in Figure 4 where P_u is the parameter set of SU_u and the lines between users denote that they are within the communication range of each other (*i.e.*, they are neighbors). Since two users may have different available channels, the replacement operation described in Section 5.2 may cause that each user replaces unavailable channels by a different one. This replacement operation may broken the previous parameter synchronization.

To illustrate this phenomenon, let us consider that SU_1 and SU_2 achieve rendezvous and synchronize their parameters, so they visit the commonly available channels in the same time slots. Afterward, if SU_2 performs the replacement operation and eventually achieves rendezvous with SU_3, the parameters synchronization between SU_2 and SU_3 may break the previous synchronization between SU_2 and SU_1. Although the pair-wise rendezvous has been proposed in [25,26], these works do not consider the problem of breaking the previous parameter synchronization. To avoid this problem, let us define ChS_u as the set of potentially channels for parameter synchronization and at the beginning of rendezvous process $ChS_u = A_u$. When a rendezvous between two SUs occurs, they only synchronize their parameters if the rendezvous channel belongs to $ChS_1 \cap ChS_2$. Let us remark that the only purpose of ChS_u is to ensure the parameters' synchronization between all SUs so, like in the two-user scenario, each SU uses its perceive available channels (*i.e.*, A_u) to perform the replacement operation. The proposed multi-user strategy is show in Algorithm 6.

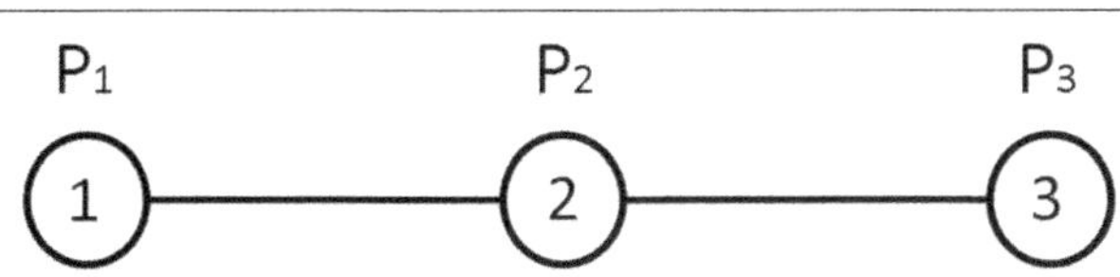

Figure 4 Multi-user scenario.

Algorithm 6: Multi-user strategy

Data: $t, T, N, A_u, S_u(0)$
$ChS_u = A_u$
S_{Tu} = Transmitter sequence $(t, N, A_u, S_u(0))$
S_{Ru} = Receiver sequence $(t, N, A_u, S_u(0))$
Attempt rendezvous on channels S_{Tu} and S_{Ru}
if *rendezvous success* **then**
  receive ChS_1 from the neighbor
  if *rendezvous channel belongs to* $ChS_u \cap ChS_1$ **then**
    receive $S_1(0)$ and t_1 from the neighbor
    $S_u(0) = \min\{S_u(0), S_1(0)\}$
    $t = \min\{t, t_1\}$
    $ChS_u = ChS_u \cap ChS_1$
return S_{Tu}, S_{Ru}

In order to compute the ETTR, MTTR, and MCTTR in the multi-user scenario, let H: $H \in \mathbb{N}$ be the network diameter denoting the maximum number of hops between two SUs (e.g., in Figure 4 the more remote SUs are SU_1 and SU_3, so $H = 2$).

Theorem 6. *In a CRN with network diameter H and $M \geq 2$ SUs, the MTTR and MCTTR for all SUs to achieve rendezvous are* $ETTR_M \leq H\left(\frac{(T-1)(T+1)}{4T}\right)$, $MTTR_M = H\left(\frac{T-1}{2}\right)$ *and* $MCTTR_M = H\left(T^2 - 1\right)$, *respectively.*

Proof. Without loss of generality, consider the scenario of Figure 4 with $S_1(0) < S_2(0) < S_3(0)$ and $t_3 < t_2 < t_1$. In this situation, the three SUs rendezvous when all of them update their parameters to $S_1(0)$ and t_3. For this to happen $H = 2$ pair-wise rendezvous are required (i.e., SU_1 with SU_2 and later, SU_2 with SU_3). Note that independently of the order in which the rendezvous between SUs occurs, always are needed H pair-wise rendezvous. By Theorem 4, in the FDCH-CS strategy, two neighboring SUs rendezvous in at least $\frac{T-1}{2}$ time slots. Then, for any H, all SUs in the network rendezvous in at least $MTTR_M = H\left(\frac{T-1}{2}\right)$.

By Theorem 4 the ETTR for two SUs is $\frac{(T-1)(T+1)}{4T}$. Considering the above analysis and the properties of expected value, then $ETTR_M \leq H\left(\frac{(T-1)(T+1)}{4T}\right)$. When not all channels are available, two SUs achieve rendezvous in at least $MCTTR = T^2 - 1$ (see Theorem 5). Then, like in the previous analysis, all SUs in the network rendezvous in at least $MCTTR_M = H\left(T^2 - 1\right)$. □

7 Simulation

Simulations are used to evaluate the performance of the proposed strategies. The simulations cover two-user and multi-user scenarios under symmetric and asymmetric models. We consider a CRN with at most 45 potentially available channels for SUs ($N \leq 45$), each SU performs spectrum sensing to determine its set of available channels (A_u). To evaluate the ETTR and the MTTR, we compute the average TTR and the maximum TTR of 10^5 independent runs, respectively. In each run, the SUs are initially placed in a randomly selected channel.

7.1 Impact of rendezvous diversity

To evaluate the impact of rendezvous diversity, we assume that all channels are available for the SUs ($N = |A_u| = G = 45$). We also consider that in αN, $0 \leq \alpha \leq 1$ channels, the communication between SUs may fail with a certain probability ρ, so that even if a pair of SUs jump to one of these channels at the same time slot, the rendezvous between them may fail[c] with probability ρ. Table 2 shows the mean of rendezvous diversity index for some of the protocols in Table 1[d] Note that while RW2, DRSEQ, SSB, and FRCH only guarantee rendezvous in 5% of the channels, EJS, ACH(RB), and the proposed schemes ensure maximum rendezvous diversity.

To show the impact of maximum rendezvous diversity, let ETTR_ρ and ETTR_0 be the ETTR for $\rho > 0$ and for $\rho = 0$, respectively. Thus, we introduce the *increased ETTR* (I-ETTR), as a new metric that can be calculated as

$$\text{I-ETTR} = \text{ETTR}_\rho - \text{ETTR}_0. \tag{17}$$

Figure 5 shows the I-ETTR as a function of ρ when the communications between SUs may fail in 10% of the channels ($\alpha = 0.1$). It is worth noting that the I-ETTR increases faster with ρ for protocols with $\overline{D} < 1$. This result can be extended for $\alpha > 0.1$, and although the I-ETTR for all protocols increase, the maximum rendezvous diversity protocols still have better performance, with a smaller increase rate in the I-ETTR. These results show that maximum rendezvous diversity protocols have a certain advantage in realistic scenarios.

Table 2 Mean of rendezvous diversity index ($\overline{D}$)

$\overline{D}$	Protocols
0.05	RW2 [25], DRSEQ [23], SSB [28], and FRCH [16]
1.00	EJS [27], ACH(RB) [17], RPS [29], FDCH-RB, and FDCH-CS

RW2, ring walk 2. DRSEQ, deterministic rendezvous sequence. SSB, short sequence-based. FRCH, fast rendezvous channel hopping. EJS, enhanced jump stay. ARCH(RB), asynchronous channel hopping (role based). RPS, role-based parallel sequence. FDCH-RB, full diversity channel hopping- role based. FDCH-CS, full diversity channel hopping- common strategy.

7.2 Two-user symmetric scenario

In this section, we first consider a scenario like the one used in [25,26,29], in which all channels are available for the SUs. Therefore, $N = |A_u| = G = 45$, as in the previous section, but only the protocols with the best results in this scenario are included.

Table 3 shows the analytical and simulation results for the selected protocols. Note that we divide the table in four parts. The first one collects the results for DRSEQ, SSB, and FRCH algorithms which use a single radio and the SUs perform a CS strategy to generate identical sequences but, due to the lack of synchronization between SUs, each one starts its CH sequence at different time slots. In the second part, we include the RW2 and EJS algorithms which use a single radio and the SUs perform a hybrid strategy. In the RW2 algorithm, each SU uses its identification number (i.e., node ID) as a seed to generate the CH sequence. However, in the EJS algorithm, the seed is a random selected value between one and the smallest prime number greater than or equal to the number of network channels. Next, in the third part, we present the algorithms that also use a single radio but the SUs use a RB strategy (i.e., the proposed FDCH-RB and ACH(RB)). Finally, in the last part, RPS and the proposed FDCH-CS are included. Both of them consider that the SUs have two radios. It is important to note (see Table 1) that in the RPS algorithm, each SU uses a hybrid strategy to generate the CH sequence. Also, the authors of RPS exploit the use of more than two radios; however, this fact increases the complexity of secondary devices. Hence, we only consider the simpler case in which each SU has two radios.

Note that, although RB strategies in general should outperform the CS and hybrid approaches, the ETTR and MTTR performance of ACH(RB) is worse than that of EJS, which is an hybrid strategy. Moreover, it is also worth noting that in terms of ETTR and MTTR, the analytical and simulation results of the proposed FDCH-RB algorithm match very well and that FDCH-RB outperforms the ACH(RB) strategy and also the other strategies. On the other hand, the analytical results for EJS and RW2 provided by their authors are only upper bounds.

In the case of protocols which consider that SUs have two radios, the performance of RPS in terms of ETTR is better than the proposed FDCH-CS. However, in terms of MTTR, the performance of RPS is almost twice worse than the proposed FDCH-CS. Also note that the analytical and simulation results of the proposed FDCH-CS algorithm match very well, while for the RPS algorithm the analytical result for ETTR and MTTR are the same (please see Theorem 4 in [29] for more information about this result).

The scenario described above assumes that the network channels are free of PU activity, but in the context of cognitive radio, this consideration may be unrealistic. In the

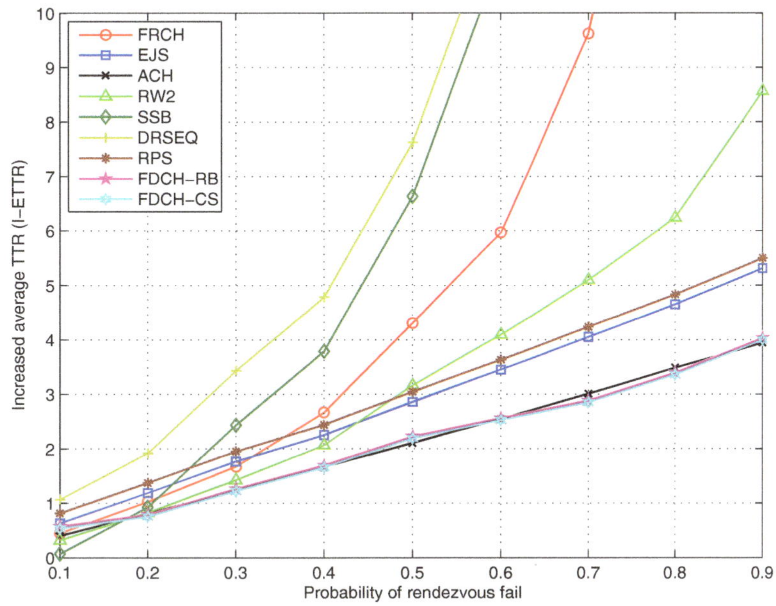

Figure 5 Impact of rendezvous diversity. $N = 45$ and $\alpha = 0.1$.

sequel we consider that the SUs share the same available channels, but not all of them are free of PU activity (i.e., $|A_u| = G$ and $|A_u| \leq N$). Note that this is a typical scenario for a CRN in which SUs are in a small area. Figures 6 and 7 show the ETTR and MTTR results in this scenario for single and two-radio protocols, respectively.

Table 3 Analytical and simulation results for $N = G = 45$

Protocol	ETTR		MTTR	
	Analytical	Simulation	Analytical	Simulation
Common strategies				
DRSEQ [23]	–	44.3	91	91
SSB [28]	43.5	43.2	88	88
FRCH [16]	–	45.1	91	91
Hybrid strategies				
RW2 [25]	52.5	44.1	88	88
EJS [27]	73.5	26.0	188	142
Role based strategies				
ACH (RB) [17]	–	42.7	–	185
Proposed FDCH-RB	22	22	44	44
Multiple radios				
RPS [29]	47	9.9	47	42
Proposed FDCH-CS	11.2	11.1	22	22

ETTR, expected time to rendezvous. MTTR, maximum time to rendezvous. DRSEQ, deterministic rendezvous sequence. SSB, short sequence-based. FRCH, fast rendezvous channel hopping. RW2, ring walk 2. EJS, enhanced jump stay. ARCH(RB), asynchronous channel hopping (role based). FDCH-RB, full diversity channel hopping- role based. RPS, role-based parallel sequence. FDCH-CS, full diversity channel hopping- common strategy.

In Figure 6, we only consider the protocols that perform the replacement operation and attempt to rendezvous at previously identified available channels. The RW2 algorithm presents better results than the EJS algorithm when $G \leq 14$; however, like FRCH and SSB, its ETTR rapidly grows when the number of common available channels increases. For $G > 14$ the EJS performs better than RW2, SSB, and FRCH. The proposed FDCH-RB outperforms the RW2, EJS, FRCH, and SSB protocols. Let us remark that in the proposed FDCH-RB the SUs that attempt to rendezvous need preassigned roles.

In the case of protocols which consider that the SUs have two radios, for $G \leq 42$ the proposed FDCH-CS clearly outperforms the RPS algorithm. For $G > 35$ the performance of RPS slightly improves and when almost all of the channels are commonly available to SUs ($42 \leq G \leq 45$), the performance of the RPS algorithm is better than the proposed FDCH-CS.

Figure 7 shows that the MTTR of the FRCH, the RW2, and the SSB algorithms increases faster than for the others. For $G \geq 12$ the performance of all of them is worse than the EJS algorithm and the proposed RB algorithm (FDCH-RB) performs better than FRCH, RW2, SSB, and EJS. When each SU is equipped with two radios, the MTTR performance of the proposed FDCH-CS is twice better than the RPS algorithm.

7.3 Two-user asymmetric scenario

Under the asymmetric model, we consider that each SU has 22 available channels ($|A_u| = 22$) and $4 \leq G \leq |A_u|$. In this scenario, each SU jumps to all available channels

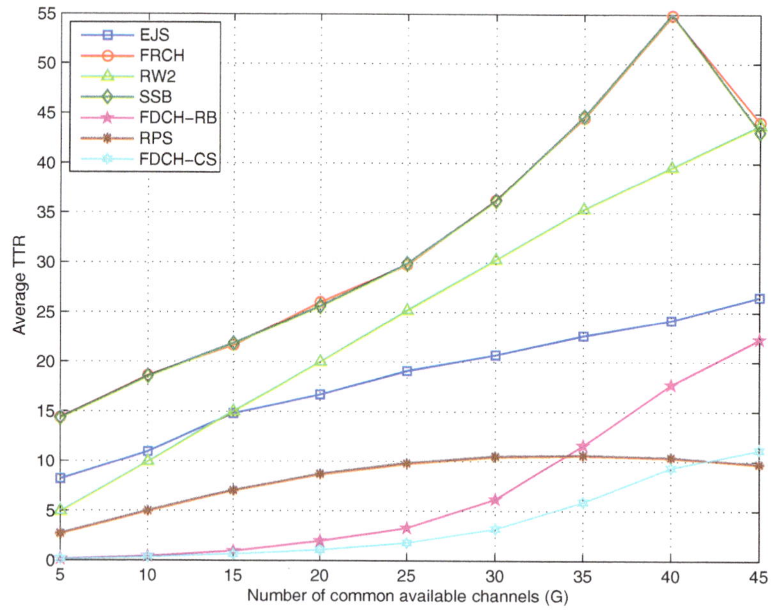

Figure 6 Average TTR for two-user symmetric scenario.

(i.e., all channels in A_u) but can only achieve rendezvous in G channels. The results in Figures 8 and 9 show that for all the compared algorithms the ETTR and MTTR performance improves when the number of common available channels increases.

Figure 8 shows that unlike the symmetric model, the performance of SSB is better than FRCH and EJS, specially for small values of G. The performance of RW2 in terms of ETTR is better than FRCH, SSB, and EJS. The proposed FDCH-RB outperforms the other compared protocols. When each SU has two radios, it is worth noting that for $G < 16$ the proposed FDCH-CS algorithm outperforms the RPS algorithm. If the number of commonly available channels is close to the amount of available channels seen by each SU (i.e., $G \approx |A_u|$), the performance of RPS and FDCH-CS is similar.

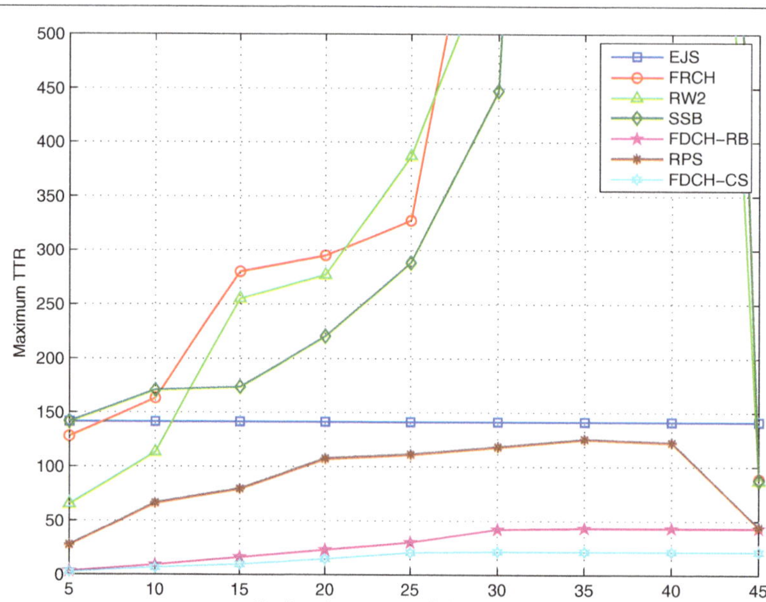

Figure 7 Maximum TTR for two-user symmetric scenario.

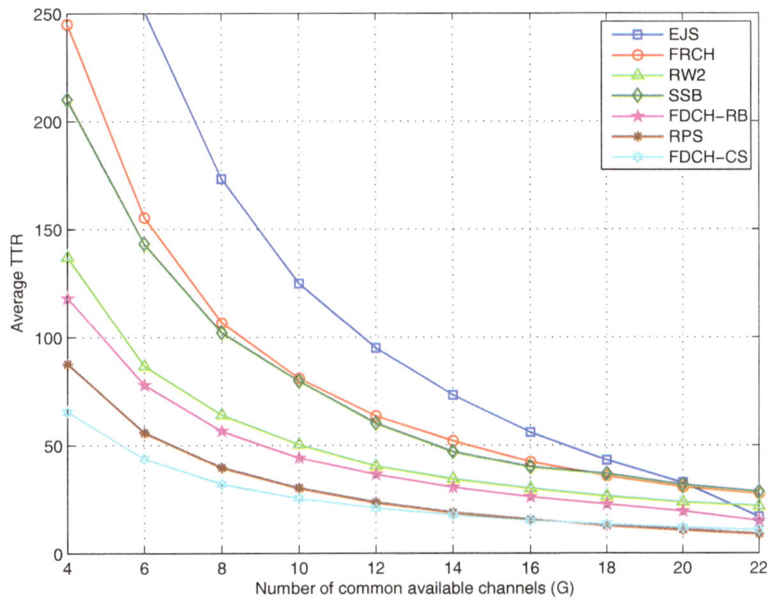

Figure 8 Average TTR for two-user asymmetric scenario.

In terms of MTTR, Figure 9 shows that the EJS algorithm has the worse results among the compared algorithms. The FRCH and SSB algorithms have similar performance and are outperformed by the RW2 algorithm. Again, as was expected, the proposed RB algorithm (FDCH-RB) outperforms the EJS, FRCH, SBB, and RW2 algorithms. When the SUs have two radios, the proposed FDCH-CS performs better than RPS for $G \leq 14$ and for $G > 14$ their results are similar.

7.4 Multi-user scenario

We compare the proposed FDCH-CS algorithm with RW2 [25] and JS [26] (which were extended to this scenario by their authors). We assume a network with $M = 10$ SUs, each one having θN available channels. As in [25,26] we set $\theta = 0.8$ and consider the symmetric and asymmetric models. For the asymmetric model, we repeatedly generate the available channels for each user until at least one common available channel for all SUs is ensured.

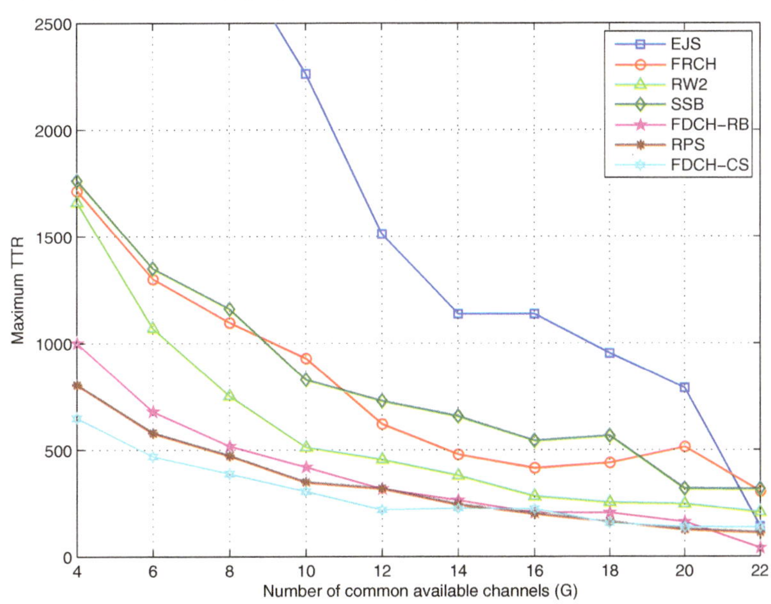

Figure 9 Maximum TTR for two-user asymmetric scenario.

According to Figure 10, in both models when the number of channels in the network increases, the ETTR for the three algorithms increases. However, in the asymmetric scenario (Figure 10b) the ETTR of the RW2 algorithm grows faster than JS and the proposed FDCH-CS. Note that in the proposed FDCH-CS algorithm each SU has two radios, while in RW2 and JS algorithms each SU has a single radio. Thus, it is reasonable that the proposed FDCH-CS outperforms RW2 and JS in the multi-user scenario.

8 Conclusions

In this paper, we focus on the design of CH sequences that ensure maximum rendezvous diversity and low ETTR for SUs in CRNs. We consider two-user and multi-user rendezvous under symmetric and asymmetric scenarios. Through simulation, we analyze the impact of the rendezvous diversity index in the performance of rendezvous algorithms in CRNs. The results show that the protocols with maximum rendezvous diversity exhibit a better behavior in realistic scenarios.

Two new rendezvous algorithms are proposed, the FDCH-RB and the FDCH-CS. In the first one, the SUs require preassigned roles, while the FDCH-CS algorithm removes this constraint by considering that each SU has two radios, which improves the ETTR and MTTR performance but requires more hardware resources. The theoretical analysis as well as the simulation results show that the proposed algorithms ensure maximum rendezvous diversity and in most scenarios of practical interest outperform other recently proposed schemes.

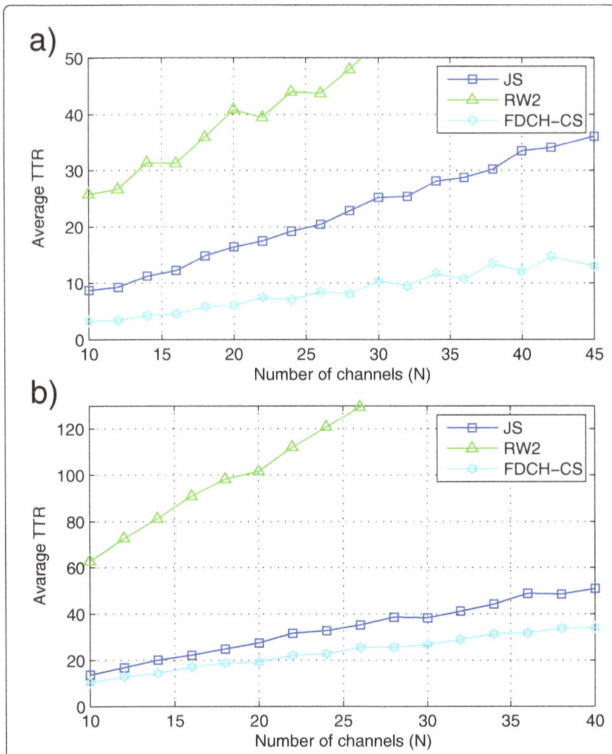

Figure 10 Average TTR for multi-user scenario. $|A_u| = 0.8N$ and $M = 10$. **(a)** Symmetric scenario. **(b)** Asymmetric scenario.

Endnotes

[a]How to perform spectrum sensing is out of the scope of this paper. Interesting surveys of spectrum sensing techniques are presented in [31,32]

[b]In this paper, we use the notation $\mathbb{N}^*$ to refer to as natural numbers including the zero and $\mathbb{N}$ to refer to as natural numbers without including the zero.

[c]Even when PUs signals can be perfectly detected, the communication between two SUs may fail because of the hidden transmitter, the exposed transmitter, and the hidden receiver. We refer the reader to [33] for a detailed discussion on these topics.

[d]The protocols that are not shown in Table 2 have worse results.

Competing interests

The authors declare that they have no competing interests.

Acknowledgements

This work was supported by CAPES, CNPq and Araucaria Foundation (Brazil) and MES (Cuba).

Author details

[1]Central University of Las Villas (UCLV), Camajuaní Highway Km 5 1/2, Santa Clara, Cuba. [2]Federal University of Technology, Paraná (UTFPR), Avenida Sete de Setembro, Curitiba, Brazil. [3]Federal University of Paraná (UFPR), Rua XV de Novembro, 1299 - Centro, Curitiba, Brazil. [4]Pontifical Catholic University of Paraná (PUC), Rua Imaculada Conceicao, 1155 Curitiba, Brazil.

References

1. IF Akyildiz, W-Y Lee, MC Vuran, S Mohanty, NeXt generation/dynamic spectrum access/cognitive radio wireless networks: a survey. Comput. Netw. **50**(13), 2127–2159 (2006). doi:10.1016/j.comnet.2006.05.001
2. IF Akyildiz, W-Y Lee, KR Chowdhury, CRAHNs: Cognitive radio ad hoc networks. Ad Hoc Netw. **7**(5), 810–836 (2009). doi:10.1016/j.adhoc.2009.01.001
3. LA DaSilva, I Guerreiro, in *2008 3rd IEEE symposium on new frontiers in dynamic spectrum access networks*. Sequence-based rendezvous for dynamic spectrum access (Chicago, 2008), pp. 1–7. doi:10.1109/DYSPAN.2008.52
4. Z Htike, CS Hong, S Lee, The life cycle of the rendezvous problem of cognitive radio ad hoc networks: A Survey. J. Comput. Sci. Eng. **7**(2), 81–88 (2013). doi:10.5626/JCSE.2013.7.2.81
5. A Alshamrani, X Shen, L-L Xie, A Cooperative MAC with efficient spectrum sensing algorithm for distributed opportunistic spectrum networks. J. Commun. **4**(10), 728–740 (2009)
6. Q Chen, Y-C Liang, M Motani, W-C Wong, A two-level MAC protocol strategy for opportunistic spectrum access in cognitive radio networks. IEEE Trans. Vehicular Technol. **60**(5), 2164–2180 (2011). doi:10.1109/TVT.2011.2141694
7. C Cordeiro, K Challapali, D Birru, in *First IEEE International symposium on new frontiers in Dynamic Spectrum Access Networks, 2005. DySPAN 2005.* IEEE 802.22: the first worldwide wireless standard based on cognitive radios (Baltimore, 2005), pp. 328–337. doi:10.1109/DYSPAN.2005.1542649

8. L Ma, X Han, C-C Shen, in *First IEEE international symposium on new frontiers in Dynamic Spectrum Access Networks, 2005. DySPAN 2005*. Dynamic open spectrum sharing MAC protocol for wireless ad hoc networks (Baltimore, 2005), pp. 203–213. doi:10.1109/DYSPAN.2005.1542636
9. J Jia, Q Zhang, X Shen, HC-MAC: A hardware-constrained cognitive MAC for efficient spectrum management. IEEE J. Selected Areas Commun. **26**(1), 106–117 (2008). doi:10.1109/JSAC.2008.080110
10. C Cormio, KR Chowdhury, A survey on MAC protocols for cognitive radio networks. Ad Hoc Netw. **7**(7), 1315–1329 (2009)
11. BF Lo, A survey of common control channel design in cognitive radio networks. Phys. Commun. **4**(1), 26–39 (2011). doi:10.1016/j.phycom.2010.12.004
12. J Misic, VB Misic, Probabilistic vs. sequence-based rendezvous in channel-hopping cognitive networks. IEEE Trans. Parallel Distributed Syst. **PP**(99), 1–1 (2013). doi:10.1109/TPDS.2013.192
13. S Alpern, S Gal, *The Theory of Search Games and Rendezvous*, 1st edn. (Kluwer Academic, Boston, 2003), p. 319
14. C Cormio, KR Chowdhury, Common control channel design for cognitive radio wireless ad hoc networks using adaptive frequency hopping. Ad Hoc Netw. **8**(4), 430–438 (2010). doi:10.1016/j.adhoc.2009.10.004
15. B Kaigui, P Jung-Min, C Ruiliang, Control channel establishment in cognitive radio networks using channel hopping. IEEE J. Selected Areas Commun. **29**(4), 689–703 (2011). doi:10.1109/JSAC.2011.110403
16. G-Y Chang, J-F Huang, A fast rendezvous channel-hopping algorithm for cognitive radio networks. IEEE Commun. Lett. **17**(7), 1475–1478 (2013). doi:10.1109/LCOMM.2013.060513.130471
17. K Bian, Park J-mJ, Maximizing rendezvous diversity in rendezvous protocols for decentralized cognitive radio networks. IEEE Trans. Mobile Comput. **12**(7), 1294–1307 (2013)
18. EO Guerra, VA Reguera, RD Souza, G Brante, EMG Fernandez, Simple role-based rendezvous algorithm for cognitive ad hoc radio networks. Electron. Lett. **50**(3), 182–184 (2014). doi:10.1049/el.2013.2994
19. G-Y Chang, W-H Teng, H-Y Chen, J-P Sheu, Novel channel-hopping schemes for cognitive radio networks. IEEE Trans. Mobile Comput., 1–1 (2012). doi:10.1109/TMC.2012.260
20. Y Zhang, Q Li, G Yu, B Wang, in *2011 proceedings IEEE INFOCOM*. ETCH: Efficient Channel hopping for communication rendezvous in dynamic spectrum access networks (Shanghai, 2011), pp. 2471–2479. doi:10.1109/INFCOM.2011.5935070
21. Y Zhang, G Yu, Q Li, H Wang, X Zhu, B Wang, Channel-hopping-based communication rendezvous in cognitive radio networks. IEEE/ACM Trans. Netw. **PP**(99), 1–1 (2013). doi:10.1109/TNET.2013.2270443
22. NC Theis, RW Thomas, LA Dasilva, Rendezvous for cognitive radios. IEEE Trans. Mobile Comput. **10**(2), 216–227 (2011). doi:10.1109/TMC.2010.60
23. D Yang, J Shin, C Kim, Deterministic rendezvous scheme in multichannel access networks. Electron. Lett. **46**(20), 1402 (2010). doi:10.1049/el.2010.1990
24. J Shin, D Yang, C Kim, A channel rendezvous scheme for cognitive radio networks. IEEE Commun. Lett. **14**(10), 954–956 (2010). doi:10.1109/LCOMM.2010.091010.100904
25. H Liu, Z Lin, X Chu, Y-W Leung, in *2010 IEEE/ACM Int'l conference on green computing and communications*. Ring-walk based channel-hopping algorithms with guaranteed rendezvous for cognitive radio networks, (2010), pp. 755–760. doi:10.1109/GreenCom-CPSCom.2010.30
26. H Liu, Z Lin, X Chu, Y Leung, Jump-stay rendezvous algorithm for cognitive radio networks. IEEE Trans. Parallel Distributed Syst. **23**(10), 1867–1881 (2012)
27. Z Lin, H Liu, X Chu, Y-W Leung, Enhanced jump-stay rendezvous algorithm for cognitive radio networks. IEEE Commun. Lett. **17**(9), 1742–1745 (2013). doi:10.1109/LCOMM.2013.071013.131029
28. V Reguera, E Guerra, R Souza, E Fernandez, G Brante, Short channel hopping sequence approach to rendezvous for cognitive networks. Commun. Lett. **18**(2), 1–4 (2014). doi:10.1109/LCOMM.2013.112713.132338
29. L Yu, H Liu, Y-W Leung, X Chu, Z Lin, in *2013 IEEE International Conference on Communications (ICC)*. Multiple radios for effective rendezvous in cognitive radio networks (Budapest, 2013), pp. 2857–2862. doi:10.1109/ICC.2013.6654974
30. K Bian, J-M Park, R Chen, in *Proceedings of the 15th annual international conference on Mobile computing and networking - MobiCom '09*. A quorum-based framework for establishing control channels in dynamic spectrum access networks, vol. 25 (New York, 2009). doi:10.1145/1614320.1614324
31. T Yucek, H Arslan, A survey of spectrum sensing algorithms for cognitive radio applications. IEEE Commun. Surv. Tutorials. **11**(1), 116–130 (2009). doi:10.1109/SURV.2009.090109
32. F Khan, K Nakagawa, in *2013 World Congress on Computer and Information Technology (WCCIT)*. Comparative study of spectrum sensing techniques in cognitive radio networks (Sousse, 2013), pp. 1–8. doi:10.1109/WCCIT.2013.6618728
33. A De Domenico, E Calvanese Strinati, M-G Di Benedetto, A survey on MAC strategies for cognitive radio networks. IEEE Commun. Surv. Tutorials. **14**(1), 21–44 (2012). doi:10.1109/SURV.2011.111510.00108

12

Base-band involved integrative modeling for studying the transmission characteristics of wireless link in railway environment

Dawei Li[1,2*], Junhong Wang[1,2], Meie Chen[1,2], Zhan Zhang[1,2] and Zheng Li[1,2]

Abstract

Base-band involved integrative modeling method (BIMM) is proposed for studying the transmission characteristics and bit error rate of the wireless communication, in which the transmitting and receiving antennas, wave propagation environment, and the modulation and demodulation modules are modeled and simulated integratively. Comparing with the conventional wave propagation method used in prediction of field coverage, BIMM is capable of taking the interaction between antenna and environment into consideration, and can give the time-domain waveforms of signals in different places of the wireless link, including those in the base-band modules. Therefore, the distortion reason of signal and in what place it happens can be found, and the bit error rate of the wireless communication can be analyzed. The BIMM in this paper is implemented by finite-difference time-domain (FDTD) method and is applied to the wireless link of railway communication. The effect of the electric spark generated by the power supplying network of the express train on the transmitting property and bit error rate is analyzed.

Keywords: Integrative modeling method; Wireless communication link; FDTD; Base-band signal; Railway environment

1 Introduction

With the rapid development and wide application of wireless communication, the prediction of the wave propagation and field coverage of the wireless link becomes important in characterizing the radio channel of the wireless communication, especially for the case of complicated environment, such as the railway environment. Usually, measurement is taken to get the field coverage of wireless communication systems, but measurement is expensive and time consuming and cannot give realistic results in some cases. For example, the results measured by an antenna in free space (normally used in measuring field strength) cannot reflect the realistic case of an express train system, in which the antenna is installed on top of the locomotive and is influenced by the locomotive body. Besides the measurement, numerical methods are also used in prediction of field coverage. There are basically two kinds of numerical methods used, one is the high-frequency method and the other is the full-wave method. The typical high-frequency method used is the ray-tracing method [1-4], which can give results with acceptable accuracy for simple and regular environments [3,4]. However, ray-tracing method is not suitable for analyzing the field coverage of the wireless link in complicated environments, so full-wave method is used in such environments. The typical full-wave method used is the finite-difference time-domain (FDTD) method [5,6], which has been used in studying the wave propagation in the micro-cell of mobile communication [7], between different floors of buildings [8], and in indoor environment [9]. It has also been combined with ray-tracing method to study the effect of wall on the indoor field coverage [10-12]. But these works did not consider the interaction between the transmitting/receiving antennas and environments and can only give the field distribution (field parameters). In our previous work, integrative modeling method (IMM) is proposed for characterizing the wireless link in complicated environments [13,14]. In IMM, the wireless link is decomposed into three parts, namely the transmitting antenna with its neighboring environment,

* Correspondence: 11111001@bjtu.edu.cn
[1]The Key Laboratory of All Optical Network and Advanced Telecommunication Network of Ministry of Education, No.3 Shangyuancun, Haidian District, Beijing, China, 100044
[2]The Institute of Lightwave Technology, Beijing Jiaotong University, No.3 Shangyuancun, Haidian District, Beijing, China, 100044

the wave propagation environment between transmitting and receiving antennas, and the receiving antenna with its neighboring environment. These three parts can be integratively modeled and simulated by full-wave method only (small problem) or by hybrid methods involving full-wave methods and high-frequency methods (large problem). The difference between the IMM and other available methods is that IMM can not only give the field distribution in environment but also give the voltages/currents of input and output signals of the antennas (circuit parameters). With IMM, the effect of the environment on the output signal and the reason of signal distortion can be analyzed. In [13,14], FDTD is used to model and simulate the transmitting/receiving antennas together with their neighboring environments, and ray-tracing method is used to calculate the wave propagation in the rest environment.

However, IMM can only deal with the RF link of the wireless system and can give the reason of distortions of RF signals. But how these distortions affect the base-band signal? And how the environment influences the bit error rate of the wireless system? These are the motivations of this paper. In this paper, the modeling and analyzing of RF link (IMM) and base-band modules are combined together, and we call it the base-band involved integrative modeling method (BIMM). By this method, the time-domain waveform of signal at different places of the wireless link can be obtained, including those intermediate signal waveforms in base-band modules. So the reason of signal distortion and in what place it happens can be found, and the bit rate error of the complete wireless link can be studied. To our knowledge, there are no similar available works that can be used to analyze the relationship between distortion of base-band signal and environment.

In this paper, the wireless link of the railway communication is studied. Global system for mobile communications railway (GSM-R)-modulating signal, which is extensively used in express train, is adopted as the excitation signal. The detail description of the problem and the exciting signal involving the base-band information are given in Section 2. In Section 3, several examples are given to show the efficiency of the BIMM. Conclusions are drawn in Section 4.

2 Problem description and method

2.1 Problem description

The complete wireless communication link defined in this paper consists of modulation and demodulation modules, transmitting antenna, receiving antenna, and radio propagation environment, as shown in Figure 1.

Digital base-band signal modulates the carrier, and the modulated RF signal transmits to the space through the transmission antenna. When the wireless link environment is small enough, the transmitting and receiving antennas and the environment can be directly involved into the FDTD meshes by setting corresponding parameters at different mesh grids. The FDTD iteration will produce full-wave effects including the reflection, diffraction, and refraction between the antennas and environment. These effects will impact on the time-domain waveforms of the output signal of the receiving antenna. When the wireless link environment is too large to deal with by available computer resource, FDTD method could combine with ray-trace method to solve long-range propagation problem as mentioned in [13]. In this paper, we only consider the small-scale wireless link environment, so only FDTD method is used. The RF signal output from the receiving antenna is restituted through the demodulation module. By comparing the difference between input and output base-band signals, the transmission property of the whole wireless link and the error of bit rate can be studied.

2.2 RF signal generation and base-band signal recover

Actually, modulation mode of the base-band signal can be different for different systems. Gaussian-filtered minimum shift keying (GMSK) [15] is now taken as the main modulation mode in the GSM-R system for wireless communication in railway. The GMSK modulation can

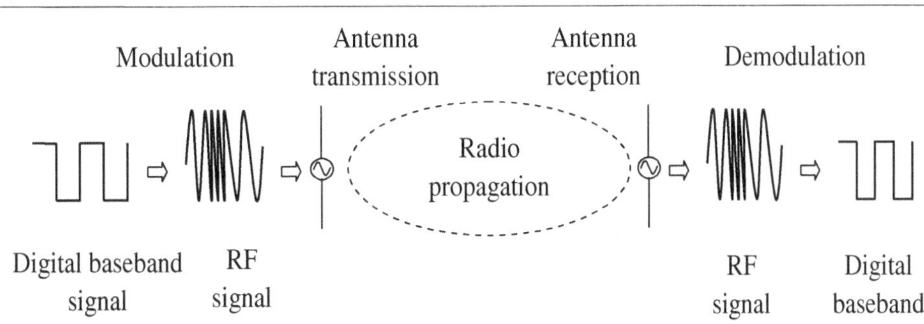

Figure 1 Schematic diagram of the wireless communication link.

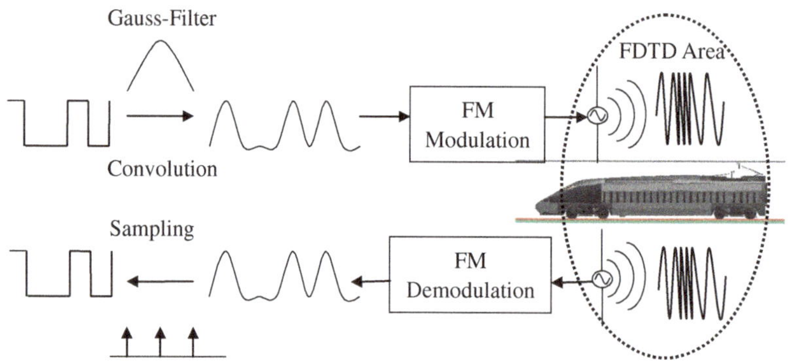

Figure 2 RF signal generation structure.

be realized by a pre-modulation low-pass filter (a general Gaussian low pass filter (LPF)) and a frequency modulation (FM) modulator. The Gaussian low pass filter can be represented by:

$$h(t) = \exp\left(\frac{-t^2}{2\delta^2 T^2}\right) / \left(\sqrt{(2\pi)}\cdot\delta T\right) \tag{1}$$

$$\delta = \sqrt{\ln(2)}/(2\pi BT) \tag{2}$$

where B is the bandwidth of the LPF and T is the symbol period. BT is a typical parameter of GMSK, and different BTs present different correlations between adjacent symbols. Due to the limitation of computer resource, the rate of symbols in this paper is set to 5.4166×10^7 bps.

Further, the pulse response of the base-band signal is expressed by the following equation:

$$g(t) = a_i \mathrm{rect}\left(\frac{t}{T}\right) * h(t) \quad (a_i = -1,\ 1) \tag{3}$$

The output signal from Gaussian LPF is modulated by FM algorithm by:

$$x(t) = \sqrt{\frac{2E_c}{T}}\cdot \cos\left(2\pi f_0 t + \int_{-\infty}^{t} g(\tau)\,d\tau + \phi_0\right) \tag{4}$$

where E_c is the energy of single bit, f_0 is the center frequency of carrier, and ϕ_0 is the random phase of carrier. In this paper, the orthogonal frequency modulation is used and the center frequency is set to 900 MHz.

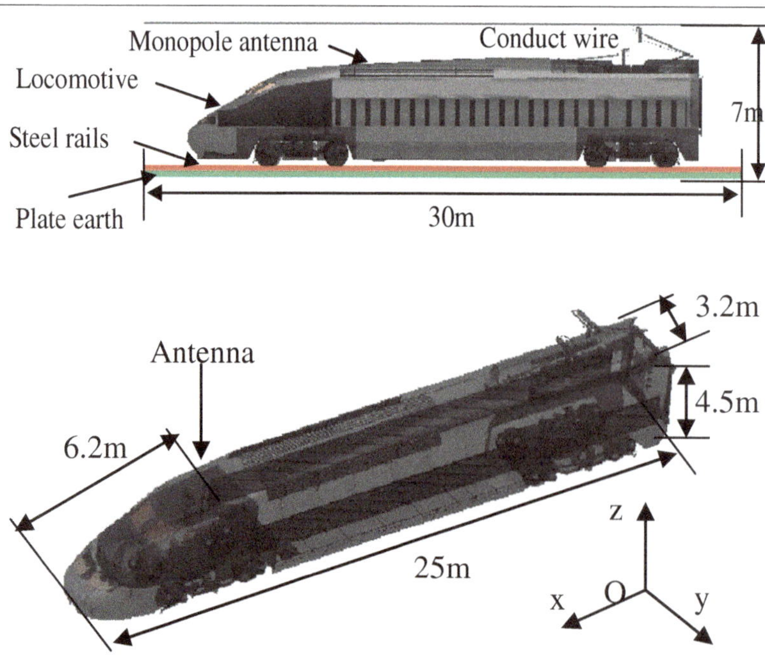

Figure 3 Modeling of locomotive and its environment.

Table 1 Parameters of materials used in the model

Model name	ε_r	μ_r	σ (S/m)
Locomotive body	1.0	1.0	2.494×10^7
Vehicle windows	5.5	1.0	0.0
Power supply wire	1.0	1.0	2.494×10^7
Steel rails	1.0	1.0	1.1×10^6
Earth plane	14.0	1.0	0.01

The received RF signal is demodulated by the orthogonal coherent demodulation method. It is multiplied by the orthogonal carriers respectively first:

$$x_{Rx}(t)\cdot\cos(2\pi f_0 t) = \frac{1}{2}(\cos(2\pi f_0 t)+1)\cos\left(\int_0^t g(\tau)\,d\tau\right) - \frac{1}{2}\sin(2\pi f_0 t)\sin\left(\int_0^t g(\tau)d\tau\right) \quad (5)$$

$$x_{Rx}(t)\cdot\sin(2\pi f_0 t) = \frac{1}{2}\sin(2\pi f_0 t)\cos\left(\int_0^t g(\tau)\,d\tau\right) - \frac{1}{2}(1-\cos(2\pi f_0 t))\sin\left(\int_0^t g(\tau)\,d\tau\right) \quad (6)$$

and then passing through the LPF to extract the phase information in the modulated signal. After then, the signal is divided into two parts, that is, the quadrature signal and the in-phase signal:

$$I(t) = -\frac{1}{2}\sin\left(\int_0^t g(\tau)\,d\tau\right) \quad (7)$$

$$Q(t) = \frac{1}{2}\cos\left(\int_0^t g(\tau)\,d\tau\right) \quad (8)$$

Finally, the base-band signal can be obtained by the two signals and their differentials using following equation:

$$g(t) = -4(I'(t))Q(t) - I(t)Q'(t) \quad (9)$$

Digital signal can be recovered by sampling this signal. Figure 2 shows the modulation and demodulation process from base-band to base-band.

2.3 Simulation scheme

In this paper, a small scale wireless link environment is modeled using FDTD algorithm, which includes the transmitting and receiving antennas and the propagation environment. To further enhance the simulation speed, we adopted the parallel algorithm. Combining message passing interface (MPI) with FDTD to realize parallel computation has been proven to be an efficient way in improving the computing speed. The parallel algorithm of FDTD utilizes a one-cell overlap region to exchange the information between adjacent sub-domains, and only

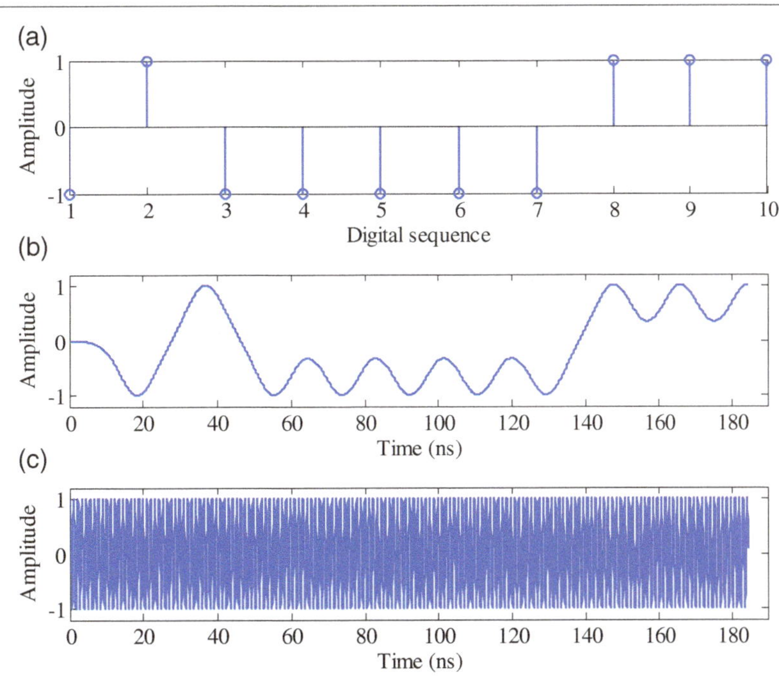

Figure 4 Generation of the base-band signal and RF signal. **(a)** shows the base-band digital data sequence. **(b)** shows the base-band signal after filtering. **(c)** shows the RF signal after modulation.

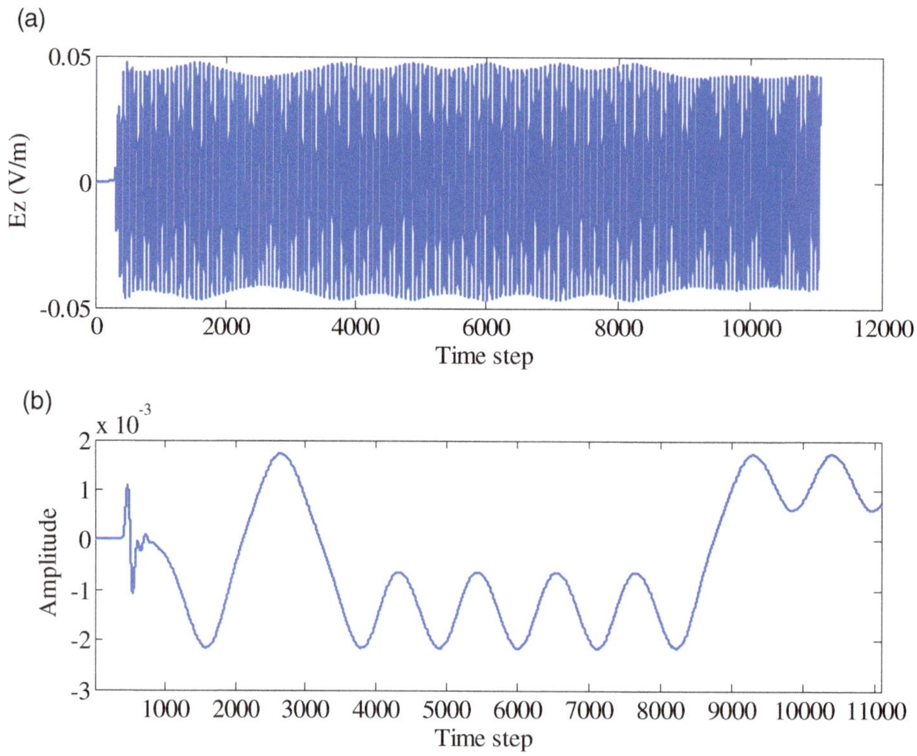

Figure 5 Received signal in free space. **(a)** and **(b)** show the RF signal received by the monopole antenna in free space (without locomotive) and the demodulated base-band signal, respectively.

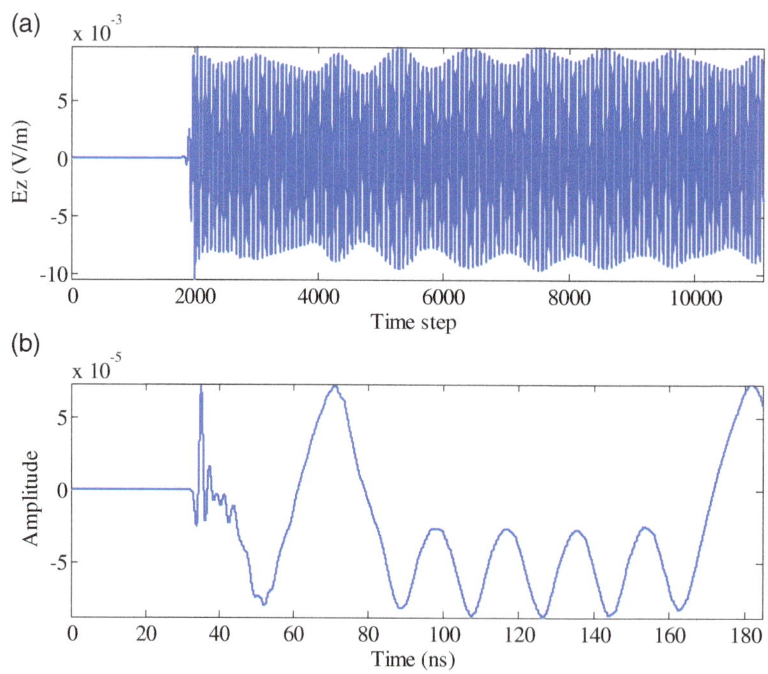

Figure 6 Received signal by locomotive antenna. **(a)** and **(b)** show the RF signal received by locomotive antenna and the demodulated base-band signal.

the tangential magnetic fields are exchanged at each time step [16].

After all the parameters are set, the base-band signal is modulated into RF signal which excites the transmission antenna and starts the FDTD computation. The receiving antenna output the voltage amplitude in each time step of FDTD iteration for post-processing. This output RF signal involves the environment effect, such as the multi-path effect and the interference between antennas and nearby materials, and the demodulated base-band signal also involves the environment effect. Using this base-band-involved integrative modeling method, the entire wireless communication link can be simulated accurately and the reason of bit error can be found.

3 Implementation and results

3.1 Modeling of antennas and environment

In order to show the capability of BIMM in analysis of base-band signal transmission and bit rate error of wireless system, an example is given, which involves a locomotive with antenna mounted on top, two steel rails, earth plane, and a conducting wire over the locomotive for power supply. Figure 3 shows the detail description of these models drawn in CAD. The electromagnetic parameters for these models are listed in Table 1. In this example, a dipole antenna is used as the transmitting antenna, and a monopole antenna mounted on the top of locomotive is used as the receiving antenna. These two antennas are located not far away, so they are involved into the FDTD computational domain together with the environment.

Table 2 Parameters of the electromagnetic spark used in this paper

Parameters	Values
U_0^p	2.962×10^4 V
t_0^p	4.5 ns
f_0^p	500 MHz
τ_0^p	4.0 ns

The center frequency of the antennas is set to 900 MHz, and the frequency band covers the signal bandwidth. The length and cross section of the transmitting half-wavelength dipole are 150 and 25 × 25 mm, respectively, and the feeding gap between two arms of the dipole is 10 mm. The length and cross section of the monopole mounted on the locomotive top is 70 and 25 × 25 mm, respectively, and the feeding gap between locomotive top plane and monopole is 10 mm. The dipole antenna is located 1 m above the locomotive and with a distance of 14.3 m (in the rear of the locomotive) to the monopole antenna. Non-uniform grid technique and parallel FDTD algorithm are utilized. The general grid size of FDTD is set to 30, 30, and 30 mm in the x, y, z directions, respectively, while the non-uniform grid size in the antenna and nearby region is set to 12.5, 12.5, and 10 mm, respectively. The locomotive is located along the x-direction. The surrounding environment is

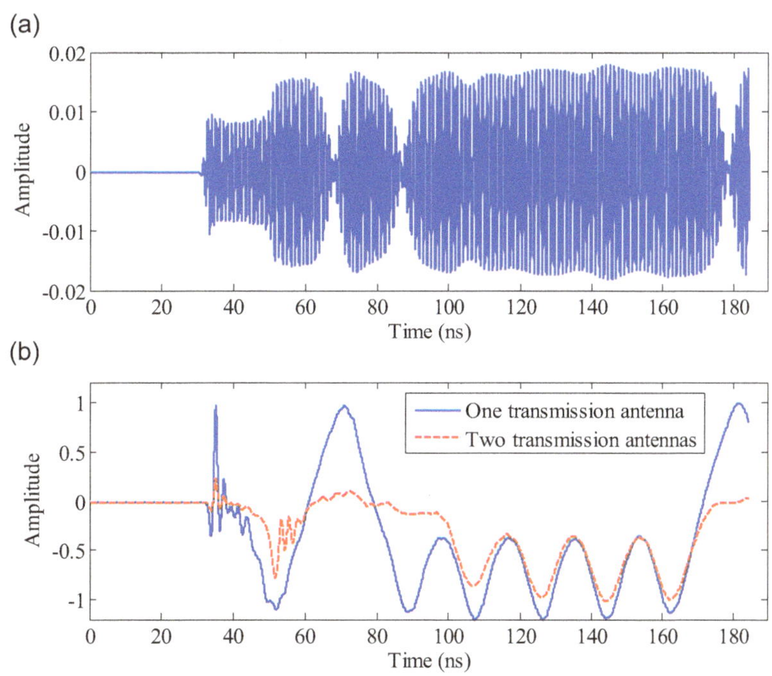

Figure 7 Received signal. (a) and **(b)** show the received RF signal and demodulated base-band signal for the case of two dipoles radiating simultaneously at different places.

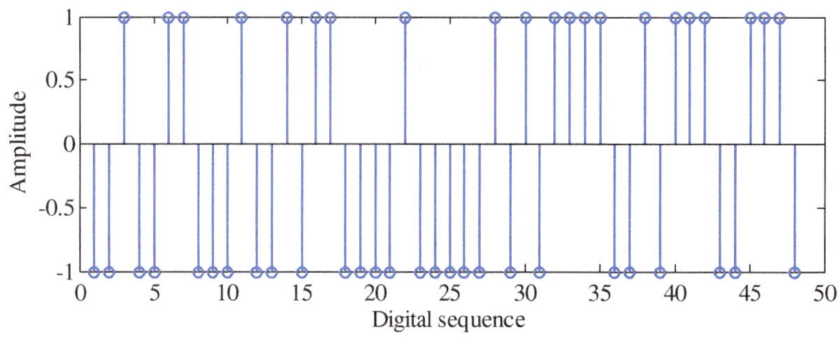

Figure 8 Input digital signal of the simulation.

divided into 1,109 × 218 × 312 ΔS grids. The feeding gap of the locomotive antenna is located at the coordinates of (755, 105, 190 ΔS), and the feed gap of the transmitting dipole is located at (105, 172, 254 ΔS). Δt, the time step of FDTD, equals 16.667 ps, and the total computation time steps are set to 11,070.

3.2 Impact of environment on transmission signal

In this paper, a sequence of data is modulated by the GMSK scheme. The data rate of GMSK used in this paper is 5.4166×10^7 bps, and parameter *BT* in Equation 2 is set to 0.5. The sequence of the digital data and the waveform of the base-band signal after filtering are shown in Figure 4.

In order to see the impact of the antenna itself on the transmission of RF signal, the receiving monopole antenna is placed individually in free space (without locomotive and environment) first, and the received signal is shown in Figure 5. From this figure, we can see that the envelope of the received RF signal is slightly undulated as shown in Figure 5a, but this does not have much influence on the demodulated base-band signal as shown in Figure 5b. Therefore, the effect of antenna can be ignored.

The monopole is then put on the locomotive, and the received signal with the effect of the locomotive and environment is shown in Figure 6. This time, a larger fluctuation of amplitude of the RF signal can be observed. But the demodulated base-band signal is still very fine, as shown in Figure 6b.

In realistic application, co-channel interference is usually inevitable. In order to simulate the multi-path effect

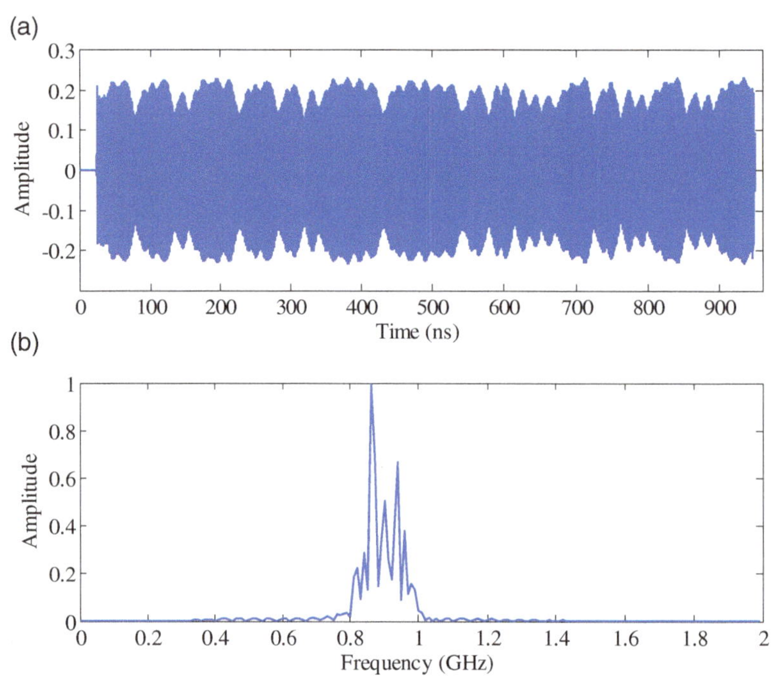

Figure 9 Received signal without spark effect. (a) and **(b)** show the RF signal received by the dipole without effect of electric spark.

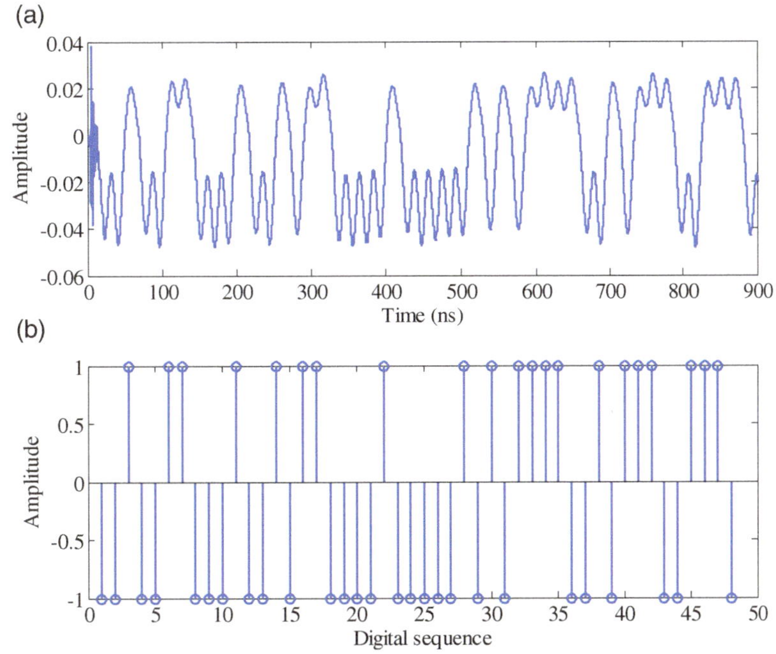

Figure 10 Base-band signal. (a) and **(b)** show the demodulated base-band and digital signals of system without effect of electric spark.

on the receiving signal, another dipole antenna is introduced in the front of the locomotive with a distance of 9.3 m to the locomotive antenna. The two dipole antennas are excited by the modulated signals with the same amplitude. The signal received by the locomotive antenna is shown in Figure 7a. Comparing with that in Figure 6a, the waveform of received RF signal is distorted significantly. The demodulated base-band signal is also distorted, as shown in Figure 7b. This will influence the judgment of the digital signal.

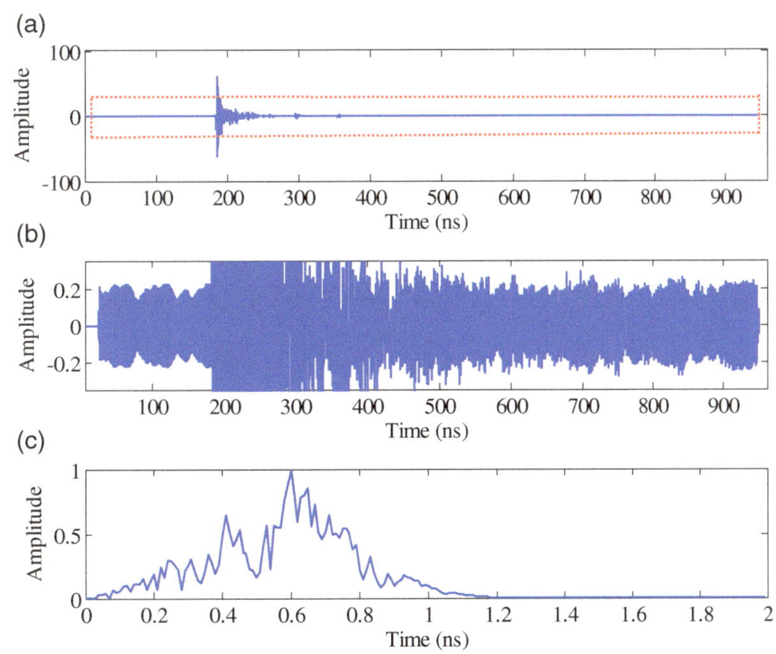

Figure 11 Received signal affected by spark. (a), **(b)**, and **(c)** show the RF signal and its frequency spectrum which received by the dipole with effect of electric spark. **(b)** is the enlargement of **(a)**.

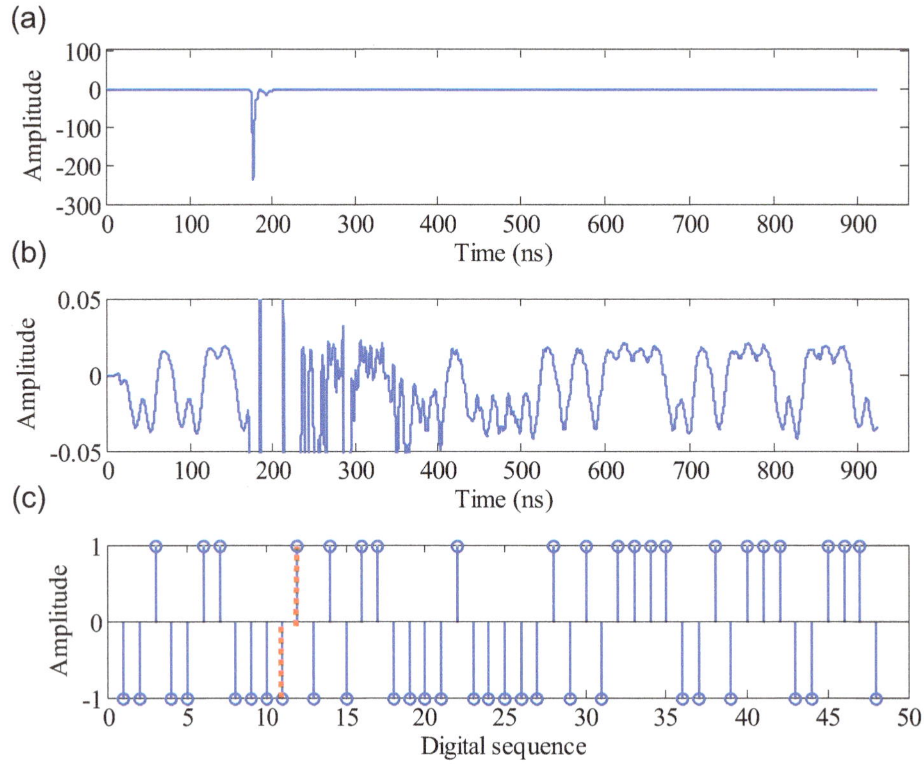

Figure 12 Base-band signal affected by spark. (a), **(b)**, and **(c)** show the demodulated base-band and digital signals of system with effect of electric spark. **(b)** is the enlargement of **(a)**.

3.3 Impact of EMI on base-band digital signal

It is the normal case for the running train system that the power supply network will generate electric spark. This spark actually is an electromagnetic pulse, which interferes with the transmitting signal. In order to find the impact of this interference on the transmitted digital signal, the generation of the electric spark and its propagation are introduced into the FDTD simulation together with the RF signal. In this example, we let the locomotive antenna transmits and the dipole antenna receives. Assuming that the electric spark has the form of modulated Gaussian pulse:

$$U_p^{n'} = E_z^n \Delta z = -U_0^p \cos(2\pi f_0^p n' \Delta t) \exp\left(-4\pi (n'\Delta t - t_0^p)^2 / (\tau_0^p)^2\right) \quad (10)$$

where n is the iterating time step of FDTD, and n' represents the time step at which the spark begin to generate. U_0^p and f_0^p are the amplitude and fundamental frequency of the spark, respectively. t_0^p represents the time moment at which pulse peak occurs. τ_0^p is related to the bandwidth, and bandwidth is equal to $2/\tau_0^p$. Table 2 gives the values of the parameters.

The digital signal input to the simulation system is shown in Figure 8. The amplitude of the modulated signal is set to 1 V. The FDTD mesh size is 30 mm which balances the computing time and accuracy. The influence of the electromagnetic spark on the transmitted signal can be analyzed by comparing the receiving signals without and with electrical spark. Figure 9 shows the received RF signal and the corresponding frequency response of the system without spark, and the demodulated base-band signal and recovered digital signal are shown in Figure 10. Comparing Figure 10b with Figure 8, we can see that although the RF waveform in Figure 9 seems not perfect, the recovered digital sequence is correct. Figure 11 shows the received RF signal and corresponding frequency response of the system with the effect of the electromagnetic spark, and the demodulated base-band signal and recovered digital signal are shown in Figure 12. From Figure 11, we can see that the transmitted RF signal is submerged in the pulse within the interference period of the spark, and the frequency spectrum of the received RF signal is spread. From Figure 12, we can see that the base-band signal waveform in which period the spark appears is distorted seriously, which further influences the sampling and judgment, and results in bit errors, as shown by the dashed line in Figure 12c.

It is also found from Figure 12b that the influence of the spark on sampling and judgment lasts more than 200 ns although the spark only lasts 9 ns. This is due to the ringing of the spark pulse on the antenna. In addition, comparing Figure 12c with Figure 8, we can see that the impact of high-intensity electromagnetic pulse on bit error seems not so great (because only two errors occur). However, if comparing base-band waveforms in Figure 12b and Figure 10a carefully, it is found that the judgment of the symbols is at high risk in the period from 200 to 400 ns, because the waveform is significantly distorted. In other words, the accuracy of the symbol judgment cannot be guaranteed between the ninth symbol and the 20th symbol.

4 Conclusions

Estimation of the transmission characteristics of wireless link is an important work especially for the wireless communication in complicated environment, such as the high-speed railway environment. But conventional method can only give the field coverage and cannot give the effect of environment on the transmission property of the wireless link. This paper proposed a base-band involved integrative modeling method which can not only take the interaction between antennas and environment into consideration but also give the effect of environment on the base-band signal and bit error rate of the wireless communication. Railway environments are taken as examples to show the implementation and efficiency of the method, and the results show that the proposed method is effective in finding the reason of waveform distortion of base-band signal and can give the explanation of the occurring of bit errors.

Competing interests

The authors declare that they have no competing interests.

Acknowledgments

This work was supported in part by NSFC under grant no. 61331002 and in part by the National Key Basic Research Project under grant no. 2013CB328903.

References

1. MF Iskander, Z Yun, Propagation prediction models for wireless communication systems. IEEE Trans. Microw. Theory Tech. **50**(3), 662–673 (2006). doi:10.1109/22.989951
2. CF Yang, BC Wu, CJ Ko, A ray-tracing method for modeling indoor wave propagation and penetration. IEEE Trans. Antenna Propag. **46**(6), 907–919 (1998). doi:10.1109/8.686780
3. Z Ji, BH Li, HX Wang, HY Chen, Efficient ray-tracing methods for propagation prediction for indoor wireless communications. IEEE Antenna Propag. Magazine **43**(2), 41–49 (2001). doi:10.1109/74.924603
4. MS Sarker, AW Reza, K Dimyati, A novel ray-tracing technique for indoor radio signal prediction. J. Electromagnet. Waves Appl. **25**(8–9), 1179–1190 (2011). doi:10.1163/156939311795762222
5. K Yee, Numerical solution of initial boundary value problems involving Maxwell's equations in isotropic media. IEEE Trans. Antenna and Propag. **14**(3), 302–307 (1966). doi:10.1109/TAP.1966.1138693
6. SD Gedney, An anisotropic perfectly matched layer absorbing medium for the truncation of FDTD lattices. IEEE Trans. Antennas Propag. **44**(12), 1630–1639 (1996). doi:10.1109/8.546249
7. G Rodriguez, Y Miyazaki, N Goto, Matrix-based FDTD parallel algorithm for big areas and its applications to high-speed wireless communications. IEEE Trans. Antennas Propag. **54**(3), 785–796 (2006). doi:10.1109/TAP.2006.869895
8. AC Austin, MMJ Neve, GB Rowe, RJ Pirkl, Modeling the effects of nearby buildings on inter-floor radio-wave Propagation. IEEE Trans. Antennas Propag. **57**(7), 2155–2161 (2009). doi:10.1109/TAP.2009.2021965
9. A Alighanbari, CD Sarris, Rigorous and efficient time-domain modeling of electromagnetic wave propagation and fading statistics in indoor wireless channels. IEEE Trans. Antennas Propag. **55**(8), 2373–2381 (2007). doi:10.1109/TAP.2007.901992
10. M Thiel, K Sarabandi, 3D-wave propagation analysis of indoor wireless channels utilizing hybrid methods. IEEE Trans. Antennas Propag. **57**(5), 1539–1546 (2009). doi:10.1109/TAP.2009.2016710
11. M Thiel, K Sarabandi, A hybrid method for indoor wave propagation modeling. IEEE Trans. Antennas Propag. **56**(8), 2703–2709 (2008). doi:10.1109/TAP.2008.927548
12. Y Wang, SN Safieddin, SK Chaudhuri, A hybrid technique based on combining ray tracing and FDTD methods for site-specific modeling of indoor radio wave propagation. IEEE Trans. Antennas Propag. **48**(5), 743–754 (2000). doi:10.1109/8.855493
13. S Pu, JH Wang, Z Zhang, Estimation for small-scale fading characteristics of RF wireless link under railway communication environment using integrative modeling technique. Prog. Electromagn. Res. **106**, 395–417 (2010). doi:10.2528/PIER10042806
14. S Pu, J-H Wang, *Research on the Receiving and Radiating Characteristics of Antennas on High-Speed Train using Integrative Modeling Technique* (Proceedings of 11th Asia Pacific Microwave Conference,(APMC), Singapore, 2009), pp. 1072–1075
15. K Murota, K Hirade, GMSK modulation for digital mobile radio telephony. IEEE Trans. Com. **29**(7), 1044–1050 (1981). doi:10.1109/9780470546543.ch20
16. WH Yu, YJ Liu, T Su, N-T Hunag, R Mittra, A robust parallel conformal finite-difference time-domain processing package using the MPI library, antennas and propagation magazine. IEEE. **47**(3), 39–59 (2005). doi:10.1109/MAP.2005.1532540

LEAUCH: low-energy adaptive uneven clustering hierarchy for cognitive radio sensor network

Errong Pei[1*], Haozhe Han[1], Zehua Sun[1], Bin Shen[1] and Tianqi Zhang[2]

Abstract

The integration of wireless sensor network (WSN) and cognitive radio (CR) technology enables a new paradigm of communication: cognitive radio sensor networks (CRSN). The existing WSN clustering algorithm cannot consider the advantage of channel resource brought by CR function in CRSN, and the CR network (CRN) clustering algorithm is designed based on the infinite energy nodes; thus both algorithms cannot operate with energy efficiency in CRSN. The paper proposes a low-energy adaptive uneven clustering hierarchy for CRSN, which can not only consider the advantage of the channel resource in reducing the energy consumption but also employ uneven clustering method for balancing the energy consumption among the cluster heads under multiple hops transmission means. Simulation results show that compared with the existing several typical clustering algorithms including WSN and CRSN clustering algorithms, low-energy adaptive clustering hierarchy (LEACH), HEED, energy-efficient unequal clustering (EEUC), cognitive LEACH (CogLEACH), and distributed spectrum-aware clustering (DSAC), the proposed algorithm can not only efficiently balance the energy consumption among cluster heads and network load in CRSN but also remarkably prolong the network lifetime.

Keywords: Cognitive radio sensor network; Channel resource; Uneven clustering; Energy consumption

1 Introduction

The existing wireless sensor network (WSN) operates in the public unlicensed spectrum band, which has become increasingly crowd because of the emergence of the vast wireless communication technologies. Meanwhile, the increasing serious mutual interference caused by the coexistence of heterogeneous wireless systems in the public unlicensed spectrum band has become a bottleneck problem which greatly limits the further development of WSN.

In order to solve the above both problems in WSN, the cognitive radio sensor networks (CRSN) is proposed, where implement dynamic spectrum access can be used [1-3]. Introducing cognitive radio (CR) technology into WSN makes the sensor nodes possess the ability of spectrum sensing. WSN can thus operate over the idle licensed spectrum band, which can not only reduce the collision probability of channels but also increase spectrum utilization and enlarge the operate spectrum band. The network throughput can further be increased, and the communication delay can also be shortened [4].

Since a large number of sensor nodes with limited energy may be randomly deployed in harsh environments and operate by forming a network in an *ad hoc* manner. Such a scenario requires an energy-efficient routing protocol that accounts for scalability. Clustering is shown to achieve such constraints and generally enlarges the network lifetime [5].

However, CRSN brings a new technical challenge in terms of its routing algorithm. On the one hand, the existing routing strategies in CR network (CRN) mainly focus on overcoming the spectrum scarcity and increasing the spectrum utilization by providing the combination scheme of the spectrum and route. However, they fail to consider the energy scarcity problem and hardware limitation problem inherited from the traditional WSN [6-9]. On the other hand, the existing clustering routing algorithms in non-CR WSN mainly focus on minimizing

*Correspondence: cqpeier@qq.com
[1] Engineering Research Center of Communication Network Testing Technologies, Chongqing University of Posts and Telecommunications, No. 2, Chongwen Rd, Nan'an Dist, Chongqing 400065, China
Full list of author information is available at the end of the article

energy consumption in nodes. However, they do not consider the spectrum sensing and spectrum management problem [10-13]; thus, they cannot exploit the advantage of spectrum resource brought by CR function in reducing energy consumption, and further, they cannot efficiently operate in CRSN.

Recently, CRSN has attracted much attention of researches from the world, and some literatures including routing algorithms are also published [6,14-24]. Taking account of the shortcomings of the existing clustering routing algorithm in WSN and CRN, the paper proposes a low-energy adaptive uneven clustering hierarchy (LEAUCH) for CRSN, which can not only consider the advantage of channel resources brought by cognitive function in CRSN but also exploit the uneven clustering method based on the channel resources. More specially, in the proposed algorithm, the number of idle channels of each node is taken as its weight and the nodes with more idle channels are elected as candidate cluster head (CH) nodes. Based on the idea of the uneven clustering method, there are fewer members in the clusters near the sink. In this way, the energy of CHs near sink can be saved, and further more energy can be used for forwarding data, which can balance energy consumption among CHs under multiple hops transmission means in CRSN. In addition, when selecting the next hop node, the CHs consider not only its relative distance to the sink but also residual energy in the candidate nodes. Experimental results show that the proposed algorithm can not only be suitable for the CRSN, effectively balance the energy consumption in CHs under multiple hop transmission means, but also optimize the energy consumption of each node in the network, and further remarkably prolong the network lifetime.

The paper is organized as follows. Section 2 presents related work. Section 3 presents network model and LEAUCH algorithm. Section 4 is a performance analysis of LEAUCH. Section 5 is a simulation analysis of the proposed algorithm. The last section is a conclusion of the paper.

2 Related work

Low-energy adaptive clustering hierarchy (LEACH) [25] is regarded as the most representative traditional algorithms. However, it can be only suitable for WSN operating in the unlicensed spectrum band, and thus, it cannot operate with energy efficiency in CRSN.

The literature [14] proposed a distributed spectrum-aware clustering (DSAC) algorithm based on the traditional K-means clustering algorithm with group-wise constraint. The algorithm initially takes each node as CHs and then merges CHs in each iteration until the number of CHs reaches a theoretically optimal number. However, a large number of information is intensively exchanged between nodes and CHs, which lead to the great waste of the energy in the nodes.

The literature [15] proposed a cognitive LEACH (CogLEACH) for CRSN that uses the number of vacant channels as a weight in the probability of each node to become a CH and that can prolong the network lifetime compared with LEACH algorithm. However, the algorithm does not consider the balance of energy consumption among CHs under multiple hops transmission means, which may lead to the premature death of the nodes near the sink because of their excessive energy consumption (CHs near the sink need to frequently forward data to the sink).

The literature [22] proposed a spectrum-aware cluster-based energy-efficient multimedia (SCEEM) routing protocol for CRSNs, which can support the quality of service (QoS) and energy-efficient routing by limiting the participating nodes in route establishment. The proposed protocol in the literature is thus a cross layer routing protocol and only suitable for wireless multimedia sensor networks (WMSNs) application scenario, which comprised of sensor devices equipped with audio and visual information collection modules, can have the ability to retrieve multimedia data, store or process data in real-time, correlate and fuse multimedia data originated from heterogeneous sources, and wirelessly transmit collected data to desired destinations. Moreover, WMSNs are designed for those real-time applications which demand strict deadline, low delay, high throughput, and reliability as well as those non-real time applications which require high or medium bandwidth, loss intolerance, etc. However, most deployed WSNs measure physical phenomena like temperature, pressure, humidity, or location of objects. In general, most of those applications have low bandwidth demands and are usually delay tolerant. Therefore, the proposed protocol in the literature cannot be suitable for the CRSNs.

The literature [16] proposed an event-driven clustering algorithm. The qualified nodes are determined based on the distance from sensor nodes to the event occurrence point and the sink. CHs are selected among the qualified nodes according to node degree, available channels, and the distance to the sink in their neighborhood. The clusters in the scenario are immediately dismissed after finishing data transmission, and all nodes enter the sleeping state again in order to save the energy. Therefore, the proposed algorithm in the literature is only confined to event-driven CRSN, which cannot be suitable for other scenarios such as the time-triggered CRSN scenario.

In order to prolong the network lifetime, prior research works mainly focus on balancing energy consumption among nodes as cluster members (CMs). However, they neglect the problem of balancing energy consumption among CHs under multiple hops transmission means.

In energy-efficient unequal clustering (EEUC) algorithm [17], the authors proposed an uneven-sized clustering method to balance the energy consumption among CHs. However, the EEUC algorithm does not take into account the advantage brought by CR technology in CRSN, so its CHs selection method can only apply to WSN. Apparently, the algorithm cannot efficiently operate in CRSN.

3 Network model and algorithm description

3.1 Network model

In the paper, we consider the scenario where there are N randomly deployed CRSN nodes collecting data periodically and P primary users (PU). Let s_i denote the Ith node, and the corresponding set of nodes is denoted by $S = \{s_1, s_2, \cdots, s_N\}$. Let $C = \{1, 2, \cdots, m\}$ denote the available channels in the network and C_i the idle channels available to node s_i. We assume:

1. The sink is located outside the square observation area. Once deployed, CRSN node and the sink will no longer be moved;
2. All nodes are homogeneous and possess the function of information fusion. Each node has a unique identification (ID);
3. Each node has the ability of spectrum sensing and can correctly detect the available channels in the surroundings;
4. All nodes can adjust its transmit power to save energy based on its distance from the receiver;
5. The entire network has a network-wide common control channel (CCC) [26].

The CR technology is introduced into sensor nodes in CRSN, so the traditional LEACH algorithm must be improved to adapt the new scenario. For purpose of making the number of CMs near the sink relatively small, the CRSN is divided into many uneven clusters based on the determined CHs, as shown in Figure 1. In Figure 1, the circles of different sizes represent the different competitive ranges of the CHs and the lines with arrow represent multiple hop data transmission among CHs.

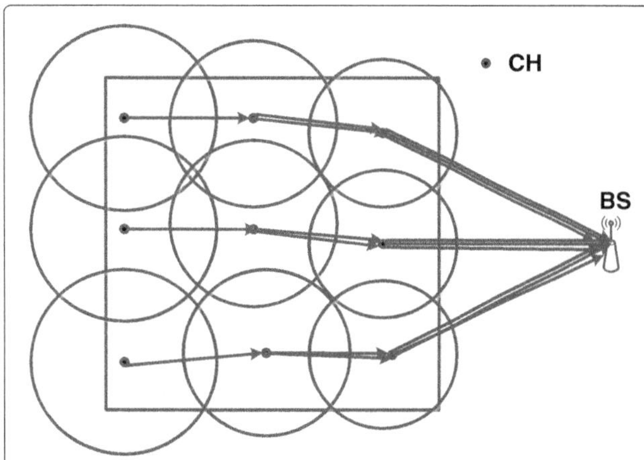

Figure 1 Uneven clustering routing algorithm.

3.2 Determination of the candidate CHs

In the regular LEACH, $P_{i(t)}$ is chosen such that $E\{\#CHs\} = k$, where [14]

$$P_{i(t)} = \begin{cases} \frac{k}{N-k*(r \bmod \frac{N}{K})} & : C_i(t) = 1 \\ 0 & : C_i(t) = 0 \end{cases}$$

where N denotes the total number of nodes, k denotes the desired number of CHs (on average) per round, r denotes the current round number, and $C_i(t)$ is an indicator function determining whether or not node i has been a CH for the most recent $r \bmod (N/K)$ rounds. This formula of $P_{i(t)}$ maintains that each node is selected as a CH once per cycle of operation and thus balancing the load between nodes and extending time of the first node death.

The number of idle channels detected in each node is considered as an important factor in choosing CHs in the paper. If node i detects more idle channels than node j, it has more opportunity to find a common channel with nearby nodes, which may result in more opportunity to set up a cluster with a common channel.

A form of $P_{i(t)}$ should be constructed such that:

- $E\{\#CHs\} = k$.
- A node with more idle channels is more likely to become a CH.

According to the literature [10], the probability of a node to become CH is:

$$P_{i(t)} = \min\left(k\frac{c_i}{\sum_{j=1}^{N} c_j}, 1\right)$$

where c_i is the number of available channels. In the paper, the nodes with probability of $P_{i(t)} > 0.4$ is selected as candidate CHs.

3.3 Determination of CHs

The nodes with $P_{i(t)} > 0.4$ are selected as the candidate CHs and start to compete for CHs; other nodes enter sleeping mode until the end of the competition. Let s_i be one candidate CH, and its competition radius R_c can be calculated on the basis of the distance from the sink. During the competition, if s_i wins, then all other candidate nodes within its competition radius R_c quit the competition for CHs.

Figure 2 is the topology structure of candidate CHs. In Figure 2, the circles with different sizes represent competition radius of candidate CHs. In accordance with the rule of determining CHs, s_1 and s_2 can become final CHs at the same time; however, s_3 and s_4 cannot become final

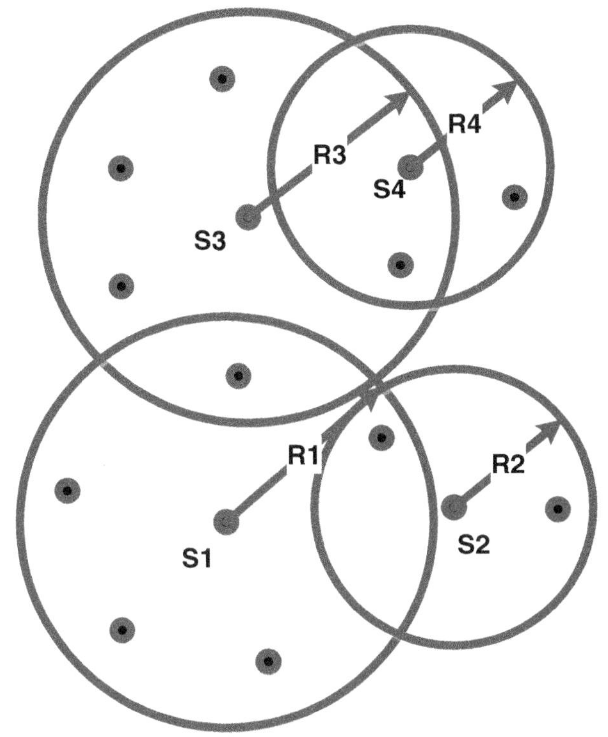

Figure 2 Competition radiuses of the candidate CHs.

CHs at the same time because s_4 falls in the range of the competition radius of s_3.

In the proposed algorithm, the clusters near the sink have short competition radius and fewer CMs, so that their CHs can use more energy to communicate among clusters and further balance the energy consumption among CHs.

Let R_c^0 be the maximum value of competition radius of candidate CHs, then the competition radius of the candidate CHs can be determined by $(1-c)\,R_c^0$, where c is the parameter used for controlling its value and $c \in [0, 1]$. Therefore, the competition radius R_c of a candidate node S_i can be determined by [17]:

$$R_c = \left(1 - c\frac{d_{\max} - d\,(s_i, BS)}{d_{\max} - d_{\min}}\right) R_c^0$$

where $d_{\max}$ and $d_{\min}$ denote the maximum and minimum of the distance from the nodes to the sink respectively, and $d\,(s_i, BS)$ denotes the distance between the node s_i and sink. The competition radius decreases linearly with the distance from the nodes to the sink. For example, $c = 1/4$, the range of the competition radius is $3/4R_c^0 \sim R_c^0$.

Definition 1. In CHs competition algorithm of LEAUCH, the set of adjacent CHs of the candidate CH s_i is denoted by S_{s_i} and

$$S_{s_i} = \left\{s_j \,|s_j \text{ is candidate CH}, d\left(s_i, s_j\right) < \max\left(R_{s_i}, R_{s_j}\right)\right\}$$

In the algorithm, each candidate node broadcasts their own competition message in the same power including node ID, competition radius R_c, and current residual energy, and the broadcast radius can be set to R_c^0 in order to save energy.

After receiving the broadcast message, each candidate CH forms its adjacent candidate CH set and then decides whether it becomes the CH or not. CHs can be decided according to the following step:

1. The node s_i with the highest residual energy in S_{s_i} become CH, then broadcast it;
2. If node s_i receives the winning message from s_j ($s_j \in S_{s_i}$), s_i quits the competition and broadcasts the quit message;
3. If node s_i receives the quit message from s_j ($s_j \in S_{s_i}$), then s_i will remove s_j from the set of its adjacent CHs.

3.4 Clusters establishment

Nodes that have not participated in the election will wake up from sleep after the CHs are determined and join their own clusters according to the following steps.

1. Node s_i is one of the CHs in a round, and C_i is the idle channel detected by node s_i. Then, CH s_i broadcast their own message after winning the election, including ID and available channel lists;
2. Node C_j within the range of s_i has one or more same element with C_i . s_j sends messages including ID and available channel to request to join the cluster once it receives the CH information of s_i;
3. s_i records IDs and available channels of requesting nodes and selects the C_{iq}, a common channel that most requesting nodes and s_i share, then broadcast it;
4. Ordinary nodes decide to join the cluster in which the node spends the minimum energy in communicating with CH, then sends joining messages to inform the CH.

3.5 Stabilization phase

CMs in the cluster receive slot time which is distributed from CHs by TDMA and then transmits collected data to the CHs in turn. The CHs fuse the collected data from CMs and forward to the sink.

4 Experimental results and analysis

In the section, the simulation experiments on the proposed algorithm are conducted using MATLAB. The energy consumption model is derived from the literature [7]. This experiment makes a comparison between the proposed LEAUCH and the other four algorithms, LEACH [25], DSAC [14], cognitive LEACH (CogLEACH)

[15], and EEUC [17], in terms of CH energy consumption, network load balance, and network lifetime.

In the paper, we consider the scenario where 200 cognitive sensor nodes are randomly distributed in an 200 m × 200 m area, which share five available channels with five primary users. Some other simulation parameters are assumed as shown in Table 1.

E_{elec} denotes the energy consumption of the circuit board when the nodes receive or transmit wireless data, ε_{fs} denotes the energy amplification coefficient where $d \leq d_0$ in the free space, and ε_{mp} denotes the energy amplification coefficient where $d > d_0$ in multipath attenuation model. In addition, $P_{i(t)} \geq 0.4$, $R_c^0 = 90$, c = 0.5.

4.1 The comparison of total energy consumption of CHs

Generally speaking, CHs undertake more tasks than other nodes in communication, which lead to more energy consumption of CHs than other nodes. The performance of a clustering routing algorithm can thus be evaluated by CHs energy consumption, as shown in Figure 3. Note that the 15 rounds are taken randomly from simulations.

It can be observed from Figure 3 that the CH energy consumption in LEAUCH is lower than that of the other protocols. CHs in LEACH are selected randomly and inefficiently, which makes CHs energy consumption fluctuate dramatically. In addition, the communications in and between the clusters are processed through single hop and the data do not fused necessarily, which make CHs energy consumption in LEACH much higher than others.

Both DSAC and CogLEACH consider the residual energy of nodes, and the distance between nodes and their information is transmitted over idle channels. It can be seen from the Figure 3 that CHs energy consumptions in DSAC and CogLEACH are relatively lower. However, both algorithms are based on the uniform clustering method which leads to more energy consumption of CHs near the sink. At the same time, both algorithms easily produce the isolated nodes, which make them be inferior to LEAUCH.

Table 1 Simulation parameter setting

Parameter	Value
Parameter scene range	200 m × 200 m
Position of sink	(250, 100)
CRSN node number	200
Primary user	5
Available channel	5
Initial energy of	
Cognitive sensor node	0.5 J
E_{elec}	50 nJ/bit
ε_{fs}	10 pJ/bit/m^2
ε_{mp}	0.0013 pJ/bit/m^2
d_0	87 m
Data package size	2,000 bits

EEUC protocol considers the distribution density of nodes and their residual energy as well as other factors when selecting CHs, which makes the selected CHs be relatively ideal and distribute in a relatively even manner. The CH energy consumption is thus relatively low.

The LEAUCH algorithm proposed in the paper can dynamically choose the best idle channels for communications, which can decrease the possibilities of the waiting and conflict caused by competing for channels among nodes, and the CHs energy consumption can thus be decreased. Furthermore, The LEAUCH also employs uneven clustering method, which can also contribute to balancing the energy consumption among CHs. Therefore, LEAUCH can obtain the most performance in the CHs energy consumption.

4.2 The comparison of network load balance

For the purpose of improving the energy efficiency and prolonging the network lifetime, the energy consumption among CHs should be kept relatively balanced. Figure 4 shows the comparison of network load balance among several algorithms.

It can be seen from Figure 4 that LEAUCH and EEUC algorithms are better in load balance, LEACH protocol is the worst, and the DSAC and CogLEACH algorithms are medium.

The selection of CHs in LEACH is random, which makes its CH energy consumption in each round extremely unsteady. Also, CHs communicate with the sink directly in LEACH, which make the network load more unbalance.

In DSAC protocol, all nodes are initialized to CHs. During the merging of CHs, the energy consumption of information interaction between nodes and CHs is relatively high. Besides, node isolation problem can easily be caused. However, by introducing the cognitive function, information interaction in DSAC can operate in other available channels, which makes its network load be in medium.

The size of the clusters in CogLEACH is uniform, which may lead to the premature death of CH near the sink because CHs near the sink need to not only process data in the cluster but also forward data from other clusters. At the same time, the problem of the isolated nodes is not considered in the algorithm. Therefore, the CogLEACH obtains relatively poor performance in network load balance.

Both LEAUCH and EEUC consider the density of nodes and the distance between CHs and sink, which make the both algorithms have better performance in network load balance. Furthermore, considering the advantage brought

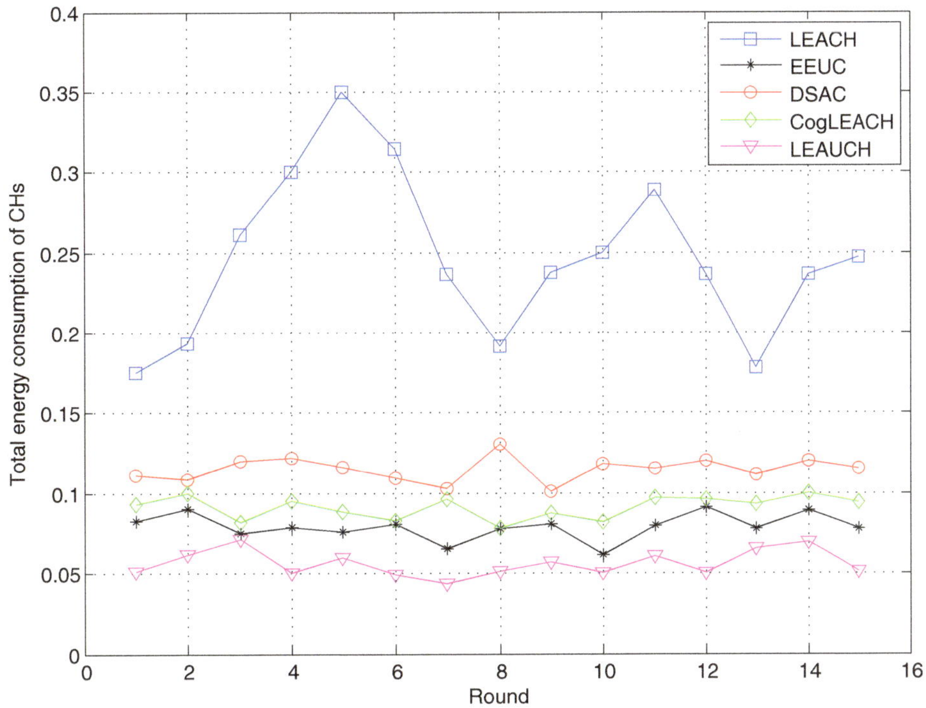

Figure 3 The comparison of total energy consumption of CHs.

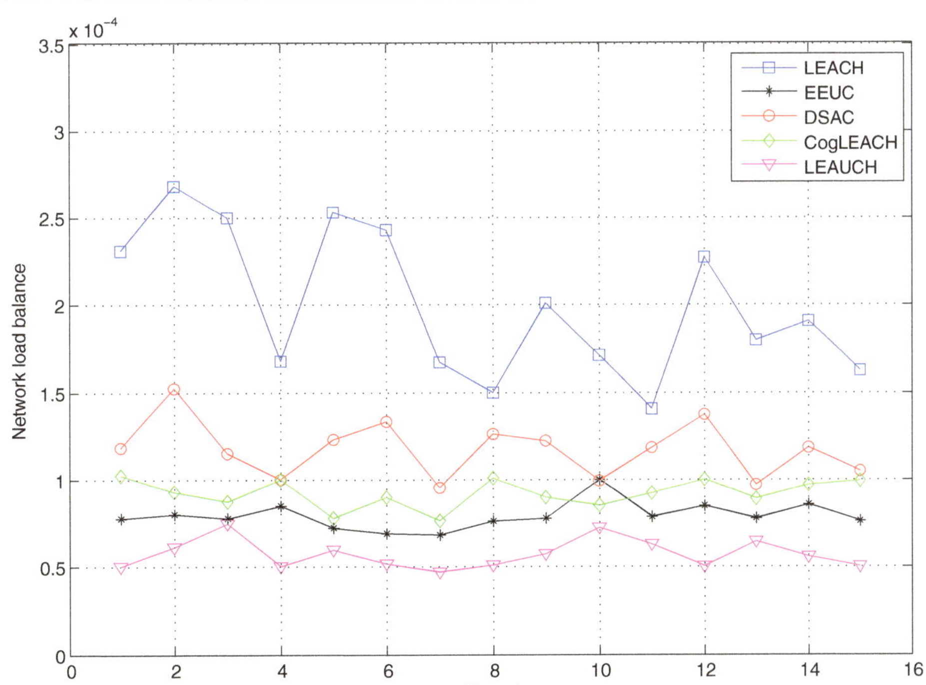

Figure 4 The comparison of network load balance.

by CR technology in CRSN makes LEAUCH superior to EEUC.

4.3 The comparison of network lifetime

Figure 5 is a comparison of the number of residual living nodes among the five algorithms. It can be seen from Figure 5 that the nodes in LEAUCH remain to have longer living time than the other four algorithms. The time span between the time of the first dead node and that of the last dead node is a reflection of the energy balance among nodes. The shorter the time span is, the better the performance of energy balance among nodes is and thus the more efficient the use of the energy is. Therefore, it can be seen from Figure 5 that LEAUCH can not only prolong the network lifetime but also remain the most performance of energy consumption among all nodes. It can also be seen from the figure that the network lifetime in DSAC and CogLEACH is longer than other protocols due to its cognitive function.

4.4 Performance analysis of throughput and delay

Most deployed WSNs measure physical phenomena like temperature, pressure, humidity, or location of objects. In general, most of those applications have low bandwidth demands and are usually delay tolerant. Therefore, the proposed routing protocol in the paper mainly focuses on the improvement of energy efficiency (energy consumption of nodes and the network lifecycle), while the performance of throughput and delay can be theoretically analyzed as follows:

Throughput: Apparently, LEAUCH has a significant improvement in the obtained throughput due to the spectrum-awareness property compared with other non-CR clustering routing protocols as CogLEACH in literature [9]. Nodes in non-CR WSNs may suffer from severe packet drop due to the competition for channels and interference with other systems operating on the same spectrum band, which results in a severe throughput degradation. Both of the proposed LEAUCH in the paper and CogLEACH are cluster-based routing protocols for CRSN; however, they have the similar data transmission process and almost same throughput.

Delay: Non-CR clustering routing algorithms operate in traditional WSN, and SNs transmit data using public spectrum band. There may exist huge amount of concurrent data transmission in public spectrum band and results in collisions, which may cause great transmission delay. The CR-based clustering routing algorithm (for example, CogLEACH and the proposed LEAUCH) in CRSN introduces cognitive function, and thus, the SNs in the CRSN can transmit data in the distributed idle channel to avoid data collision. Therefore, generally speaking, the spectrum-awareness clustering routing protocols including CogLEACH and the proposed LEAUCH in the paper have less delay relative to traditional clustering routing protocols, and obviously, CogLEACH and the proposed LEAUCH have almost same time delay.

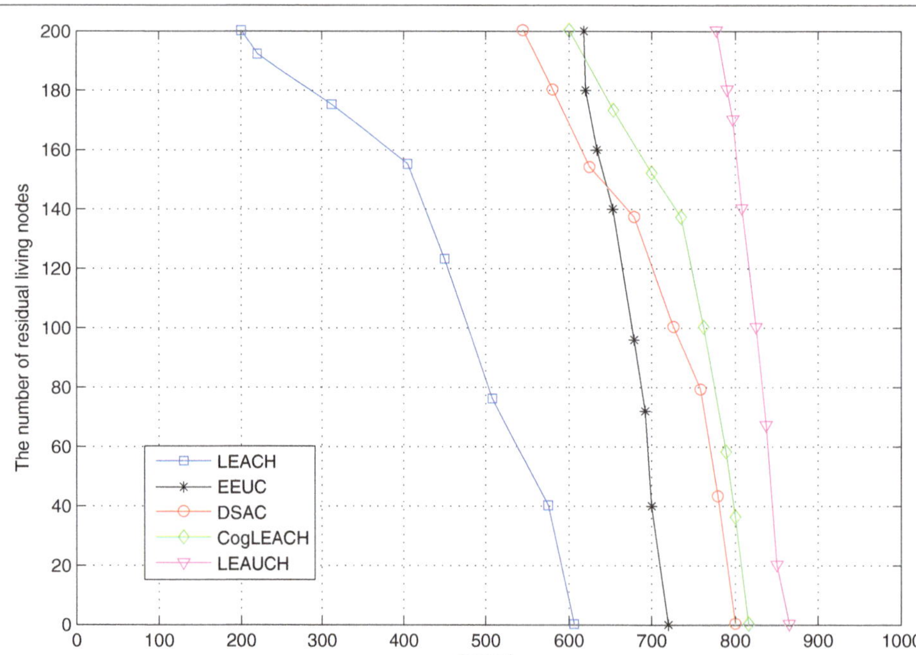

Figure 5 The comparison of the number of residual living nodes in the network.

5 Future works

The proposed routing protocol in this paper mainly focuses on the improvement of energy efficiency of sensor nodes and further increase the lifetime of WSN where the sensors are triggered periodically to measure physical phenomena like temperature, pressure, humidity, or location of objects. In general, most of those applications have low bandwidth demands and are usually delay tolerant. In future works, some application scenarios which require less delay, packets loss rate, and more throughput will also be considered, and thus, more complicated network topology is further considered to evaluate their performance.

6 Conclusions

The paper proposed the LEAUCH algorithm CRSN, which takes the number of idle channels of each node as its weight and chooses the nodes with more idle channels as candidates CHs. In addition, the proposed algorithm employs the uneven clustering method, which can balance energy consumption among CHs under multiple hops means. Experimental results show that the proposed algorithm can obtain the best performance in terms of network lifetime, the energy consumption of CHs, and network load balance.

Competing interests

The authors declare that they have no competing interests.

Acknowledgements

This work is partially supported by NSFC (61379159, 61201205, 61371164), Chongqing Education Commission Science and Technology Research Project (KJ130513), Chongqing Basic and Cutting-edge Project (CSTC2013jcyjA40020, CSTC2011jjjq40002), Chongqing Graduate Research and Innovation Project (CYS14149).

Author details

[1] Engineering Research Center of Communication Network Testing Technologies, Chongqing University of Posts and Telecommunications, No. 2, Chongwen Rd, Nan'an Dist, Chongqing 400065, China. [2] Chongqing Key Laboratory of Signal and Information Processing, Chongqing University of Posts and Telecommunications (CQUPT), No. 2, Chongwen Rd, Nan'an Dist, Chongqing 400065, China.

References

1. D Cavalcanti, S Das, JF Wang, K Challapali, in *Proceedings of 17th International Conference on Computer Communications and Networks*. Cognitive radio based wireless sensor networks, (2008), pp. 1–6
2. B Akan, OB Karli, O Ergul, Cognitive radio sensor networks. IEEE Netw. **23**(4), 34–40 (2009)
3. KLA Yau, P Komisarczuk, PD Teal, in *IEEE 34th Conference on Local Computer Networks*. Cognitive radio-based wireless sensor networks: conceptual design and open issues, (2009), pp. 955–962
4. GP Joshi, SY Nam, SW Kim. Cognitive radio wireless sensor networks: applications, challenges and research trends.SENSORS. **13**(9), 11196–11228 (2013)
5. B Mamalis, D Gavalas, C Konstantopoulos, G Pantziou, Clustering in wireless sensor networks.RFID and Sensor Networks: Architectures, Protocols. Security and Integrations, 324–353 (2009)
6. ZY H Z Zhang, Zhang, Yuen Chau, Energy-efficient spectrum-aware clustering for cognitive radio sensor networks. Chinese Sci. Bull. **57**(28-29), 3731–3739 (2012)
7. M Youssef, M Ibrahim, M Abdelatif, L Chen, AV Vasilakos.Routing metrics of cognitive radio networks: a survey. IEEE Commun. Surv. Tutorials. **16**(1), 92–109 (2014)
8. HG T G Jiang, AV Wang, Vasilakos, QoE-driven channel allocation schemes for multimedia transmission of priority-based secondary users over cognitive radio networks. IEEE J. Selected Areas Commun. **30**(7), 1215–1224 (2012)
9. SS Byun, I Balashingham, AV Vasilakos, HN Lee, Computation of an equilibrium in spectrum markets for cognitive radio networks. IEEE Trans. Comput. **63**(2), 304–316 (2014)
10. F Akyildiz, WY Lee, MC Vuran, S Mohanty. Next generation dynamic spectrum access/cognitive radio wireless networks: a survey.Comput Netw. **13**(50), 2127–2159 (2006)
11. N Chilamkurti, S Zeadally, A Vasilakos, V Sharma, Cross-layer support for energy efficient routing in wireless sensor networks. J. Sensors. **2009** (2009)
12. K Lin, JJPC Rodrigues, HW Ge, NX Xiong, XD Liang, Energy efficiency QoS assurance routing in wireless multimedia sensor networks. IEEE Syst. J. **5**(4), 495–505 (2011)
13. HJ Cheng, ZHSu R Guo, NX Xiong, WZ Guo, Service-oriented node scheduling schemes with energy efficiency in wireless sensor networks. Int. J. Distributed Sensor Netw. **2014** (2014)
14. H Zhang, Z Zhang, H Dai, R Yin, XM Chen, in *IEEE Global Telecommunications Conference*. Distributed spectrum-aware clustering in cognitive radio sensor networks, (2011), pp. 1–6
15. RM Eletreby, HM Elsayed, MM Khairy, in *International Conference on Cognitive Radio Oriented Wireless Networks and Communications*. CogLEACH: a spectrum aware clustering protocol for cognitive radio sensor networks, (2014), pp. 179–184
16. M Ozger, OB Akan, in *Proceedings IEEE INFOCOM*. Event-driven spectrum-aware clustering in cognitive radio sensor networks, (2013), pp. 1483–1491
17. CF Li, M Ye, GH Chen, J Wu, in *IEEE International Conference on Mobile Ad hoc and Sensor Systems*. An energy-efficient unequal clustering mechanism for wireless sensor networks, (2005), pp. 604–612
18. AO Bicen, OB Akan, Reliability and congestion control in cognitive radio sensor networks. Ad Hoc Netw. J. (Elsevier). **9**(7), 1154–1164 (2011)
19. B Gulbahar, OB Akan.Information theoretical optimization gains in energy adaptive data gathering and relaying in cognitive radio sensor networks. IEEE Trans. Wireless Commun. **11**(5), 1788–1796 (2012)
20. GA Shah, VC Gungor, OB Akan, A cross-layer QoS-aware communication framework in cognitive radio sensor networks for smart grid applications. IEEE Trans. Ind. Inf. **9**(3), 1477–1485 (2013)
21. GA Shah, OB Akan, Performance analysis of CSMA-based opportunistic medium access protocol in cognitive radio sensor networks. Ad Hoc Netw. J. (Elsevier). **15**, 4–13 (2014)
22. GA Shah, F Alagoz, E Fadel, OB Akan, A spectrum-aware clustering for efficient multimedia routing in cognitive radio sensor networks. IEEE Trans. Vehicular Technol. **63**(7), 3369–3380 (2014)
23. GA Shah, OB Akan, Cognitive adaptive medium access control in cognitive radio sensor networks. IEEE Trans. vehicular Technol. **64**(2), 757–767 (2015)
24. V Esmaeelzadeha, R Berangia, ES Hosseinia, OB Akanb, *Stochastic backlog and delay bounds of generic rate-based AIMD congestion control scheme in cognitive radio sensor networks*, (2015). Pervasive and Mobile Computing
25. W Heinzelman, A Chandrakasan, H Balakrishnan, An applicationspecific protocol architecture for wireless microsensor networks. IEEE Trans. Wireless Commun. **1**(4), 660–670 (2002)
26. GA Shah, OB Akan, in *IEEE International Conference on Communications*. Spectrum-aware cluster-based routing for cognitive radio sensor networks, (2013), pp. 2885–2889

Practical designs for out-of-band emission suppression and adjacent channel interference rejection for orthogonal frequency division multiplexing-based cognitive radios

Juan Fang* and I-Tai Lu

Abstract

Opposite to the conventional wisdom, practical considerations for out-of-band emission (OOBE) suppression and adjacent channel interference (ACI) rejection in orthogonal frequency division multiplexing (OFDM) -based cognitive radio systems are not completely alike. For example, ACI will increase after discarding the cyclic prefix (CP) because the interference signal is also truncated during the CP removal process. Furthermore, effective OOBE is much larger than the estimated OOBE if there is a timing offset (TO) between the primary and secondary users. To the best of our knowledge, these and some other practical issues have not been well addressed. Various important practical issues (including multipath delay spread, high peak-to-average power ratio (PAPR), spectral efficiency loss, high complexity, sensitivities to frequency and timing offsets, non-contiguous and dynamically changing spectrum, CP removal, spectral containment, etc.) concerning OOBE suppression and ACI rejection are studied in this paper. Numerical results show that none of the existing approaches can deal with all practical issues satisfactorily. Hybrid combinations of frequency-domain approaches (spectral precoding and PAPR precoding) and time-domain approaches (windowing and filtering) are developed to maintain the merits and mitigate the drawbacks of each constituent approach in order to deal with these practical challenges. It is shown that a combination of filtering and precoding provides the best OOBE suppression and a combination of windowing and precoding provides the best ACI rejection characteristics in our numerical examples.

Keywords: OOBE; ACI; Cognitive radio; OFDM

1 Introduction

Cognitive radio (CR), as a promising solution to the spectrum congestion problem brought by the rapid advance of wireless communication techniques and the drastic increase of wireless devices, has drawn a lot of attentions recently (e.g., see [1–3]). In CR systems, the secondary users (SUs) are allowed to dynamically access to the spectrum resources, which are not occupied by the primary users (PUs) and could be non-contiguous, without causing interference to the PUs or other SUs transmitted over adjacent channels. However, it is hard to achieve synchronization or any type of coordination between PUs and SUs or among SUs since they transmit independently. Therefore, on one hand, waveform shaping for spectral containment by a SU transmitter is needed to minimize its out-of-band emission (OOBE) in order to reduce interference to its neighboring PUs and/or SUs. On the other hand, an efficient adjacent channel interference (ACI) rejection approach that can reduce the ACI from neighboring PUs and/or SUs is needed at a SU receiver. Although ACI rejection is equally important as OOBE suppression, there are much more work on OOBE suppression than ACI rejection. This may be due to the fact that, for the conventional approaches (windowing and filtering), the principle of ACI rejection is very similar to that of OOBE rejection. However, this is not always true. Take orthogonal frequency division multiplexing (OFDM) as an example. If the ACI rejection

* Correspondence: jfang1985@gmail.com
NYU WIRELESS, Department of Electrical and Computer Engineering, Polytechnic School of Engineering of New York University, 6 MetroTech Center, Brooklyn 11201, USA

is done before the removal of cyclic prefix (CP) at the OFDM receiver without special considerations, ACI will regenerate because the processed interference signal is also truncated during the CP removal process. In addition, since OFDM signal is demodulated symbol by symbol, a timing offset (TO) between the desired and interfering transmitters can enhance the *equivalent* OOBE of the interfering transmitter greatly. These issues have not been addressed so far. Thus, considerations of OOBE suppressions and ACI rejection are not all alike, and it is important to find out which approach is better for OOBE suppression and which approach is better for ACI rejection.

OFDM technology, which divides the total bandwidth into several orthogonal subbands, provides the flexibility for deactivating specific subcarriers simply by using zeros as the corresponding input values. In addition, OFDM has several favorable properties like high spectrum efficiency, robustness to channel fading, multipath delay spread tolerance, efficient digital signal processing implementation, and granular resource allocation, etc. (for example, in [4, 5]). Because of these merits, OFDM has been considered as a candidate for CR in the first CR-based standard IEEE 802.22 [6]. However, OFDM has shortcomings such as high peak-to-average power ratio (PAPR), high sensitivity to frequency offset (FO) and TO, large inter-carrier interference (ICI) at high Doppler's, and large OOBE and/or ACI originating from large side lobes of the rectangular pulse shape. Among these shortcomings, the high OOBE and/or ACI is the main technical challenge for CR applications (although the other shortcomings may induce OOBE and/or ACI as well).

Various approaches have been proposed to suppress OOBE of OFDM signals. These approaches can be classified into two categories: the time-domain and frequency-domain approaches. The time-domain approaches include windowing [7] and filtering [8] where the OOBE suppression processes are carried out in the time domain (after the inverse fast Fourier transform (IFFT) operation). The frequency-domain approaches include subcarrier weighting [9], carrier cancellation [10, 11], and spectral precoding [12, 13], where the OOBE suppression processes are carried out in the frequency domain (before the IFFT operation). As mentioned before, the time-domain OOBE suppression approaches (windowing and filtering) at the transmitter can also be used at the receiver to reject ACI [14, 15]. However, it is not obvious that the spectral decoder at the receiver (which corresponds to the frequency-domain spectral precoding at the transmitter for OOBE suppression) can also reject ACI as well.

Each of the abovementioned approaches has merits and drawbacks. For the time-domain examples, the windowing approach has low complexity and no spectral efficiency loss. It can be used to suppress OOBE or reject ACI as long as the roll-off portion of the windowing function is long enough. However, a long roll-off portion will induce inter-symbol interferences (ISI) and subsequently seriously degrade the bit error rate (BER) performance if the CP length is not long enough. Therefore, the performance of windowing approach is limited by the CP length. Like windowing, the filtering approach has no spectral efficiency loss, and its performance is limited by the CP length. However, unlike windowing, it is less sensitive to the TO between PU and SUs when it is used to suppress OOBE of OFDM signal and has high complexity when the spectrum is non-contiguous or dynamically changing. For the frequency-domain example, spectral precoding approach is very efficient for reducing OOBE and can be easily extended to reduce PAPR simultaneously by combining with PAPR precoding [16]. However, it suffers spectral efficiency loss and has high computational complexity when the number of subcarriers is large or when the spectrum is non-contiguous or dynamically changing. In addition, its BER performance will degrade when there is TO between PU and SU. In summary, none of the existing approaches can deal with all practical issues and a study of these approaches under various practical conditions would be very valuable.

However, in most of published works for OOBE/ACI suppression, these existing approaches are evaluated in their favorable conditions (which are not practical) and therefore the suitability of these approaches in practical situations has not been properly addressed. For examples, firstly, in most of publications about the time-domain approaches, the CP length was oftentimes extended so long that the effects of delay spread is minimized. But the CP length is fixed in practice (e.g., in Long-Term Evolution (LTE) [17]), and a large delay spread will cause both ISI and ICI if a windowing or filtering approach is employed. Secondly, for the spectral precoding approach, the complexity and implementation issues for broadband channels have not been addressed. But, in reality, the CR system is usually broadband. Thirdly, the effects of other impairments (such as PAPR, TO, FO, etc.) on OOBE or ACI are usually not assessed. However, these impairments may cause significant increases of OOBE and ACI. Fourthly, the implications of non-contiguousness and time variations of the available spectrum on the complexity issues are mostly unaddressed. However, the available spectrum in CR is usually non-contiguous and dynamically changing.

As the frequency-domain approach (precoding) and the two time-domain approaches (windowing and filtering) have complementary characteristics, a hybrid combination of these approaches may be advantageous if it can maintain the merits and mitigate the drawbacks of these approaches. In this paper, hybrid approaches which combine the frequency-domain spectral coding and either one of the two time-domain approaches (windowing and filtering) are proposed to deal with various abovementioned practical issues. Detailed analyses are made on the complexity, spectral efficiency, spectral containments, PAPR and BER of the windowing, filtering, and spectral precoding approaches, and their possible combination schemes under practical conditions (broadband channels with radio impairments such as FO and TO). Simulation results show that the proposed hybrid approaches indeed maintain the merits and mitigate the drawbacks of their constituent approaches. Particularly, since windowing process can be carried out conveniently to deal with the CP removal issue, a hybrid combination of precoding and widowing is more appropriate for ACI rejection. On the other hand, since filtering is less sensitive to TO, a hybrid combination of precoding and filtering is more appropriate for OOBE suppression for OFDM systems.

This paper is organized as follows. In Section 2, we present the frequency and time-domain OOBE suppression approaches and the hybrid approaches. In Sections 3 and 4, we deal with various practical issues and present numerical examples. Conclusion remarks are made in Section 5. The notation of this paper is as follows. All boldface letters indicate vectors (lowercase) or matrices (uppercase). $\boldsymbol{A}^H$ and $\boldsymbol{A}^T$ stand for the conjugate transpose and transpose of $\boldsymbol{A}$, respectively. $[\boldsymbol{A}]_{i,j}$ denotes the element of $\boldsymbol{A}$ at the ith row and jth column. a^* and $E(a)$ denotes the conjugate and expectation of a, respectively. diag [...] denotes the diagonal matrix with elements [...] on the main diagonal. $\boldsymbol{I}_r$ represents an r by r identity matrix. $a \otimes b$ denotes the convolution of a and b.

2 Formulation

2.1 Conventional CP-OFDM transmit signal

The lth time-domain conventional CP-OFDM transmit symbol $\phi_l(t)$ of N contiguous subcarriers can be expressed as:

$$\phi_l(t) = \sum_{n=\mathcal{N}_0}^{\mathcal{N}_0+N-1} d_{l,n} p_n(t-lT), lT \le t < (l+1)T \quad (1)$$

where $d_{l,n}$ is the lth data symbol transmitted on the nth subcarrier, and

$$p_n(t) = e^{j2\pi\frac{n}{T_s}t} g(t) \quad (2)$$

The pulse shape function

$$g(t) = \begin{cases} 1, -T_{CP} \le t < T_s \\ 0, \quad \text{otherwise} \end{cases}, \quad (3)$$

where T_s and T_{CP} is the symbol and the CP durations, respectively. In (1), $T = T_{CP} + T_s$ is the effective symbol duration.

2.2 General framework for OOBE and ACI suppression approaches

Both time-domain and frequency-domain approaches can be used to suppress the OOBE at the transmitter and reject the ACI at the receiver. The block diagram in Fig. 1 is used to describe not only the time-domain and frequency-domain approaches but also the hybrid approaches which combine the processes in both frequency and time domains.

2.2.1 OOBE suppression at the transmitter

For the frequency-domain OOBE suppression approach at the transmitter, the precoded data vector $\boldsymbol{s}_l = \left[s_{l,\mathcal{N}_0}, \ldots, s_{l,\mathcal{N}_0+N-1}\right]^T$ with N correlated elements is obtained by precoding the M ($M \le N$) uncorrelated elements of the original data vector $\boldsymbol{d}_l = \left[d_{l,\mathcal{N}_0}, \ldots, d_{l,\mathcal{N}_0+M-1}\right]^T$ by a semi-unitary precoding matrix $\boldsymbol{G}$ with dimension of $(N \times M)$,

$$\boldsymbol{s}_l = \boldsymbol{G}\boldsymbol{d}_l, \quad \text{with } \boldsymbol{G}^H\boldsymbol{G} = \boldsymbol{I}_M \quad (4)$$

Instead of $\boldsymbol{d}_l$, $\boldsymbol{s}_l$ is fed into IFFT to generate the lth OFDM symbol. Note that, in essence, only M data elements are transmitted per OFDM symbol in the available frequency band, reserving $N - M$ degrees of freedom for shaping the spectrum of the transmitted signal.

For the time-domain OOBE suppression approach at the transmitter, either a bandpass filter $f_{tx}(t)$ or a windowing function $w_{tx}(t)$ is employed and the lth transmit signal is

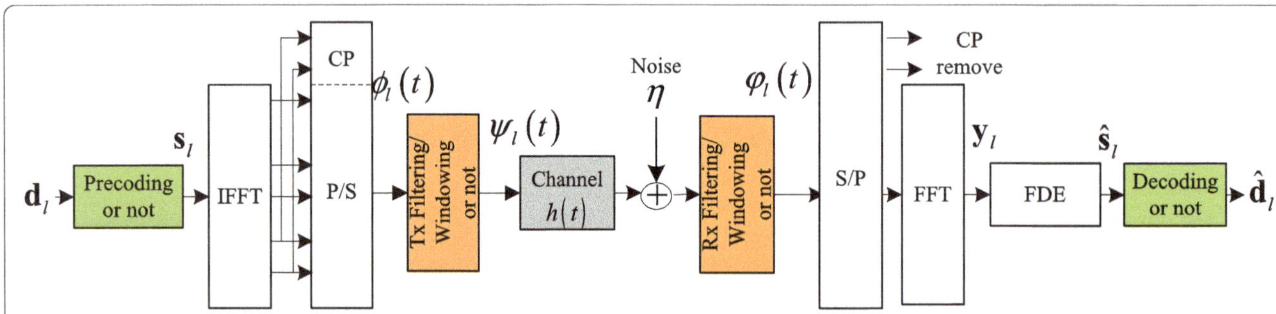

Fig. 1 A block diagram of the multicarrier modulation transceiver

$$\psi_l(t) = \begin{cases} \phi_l(t) \otimes f_{\text{tx}}(t) & \text{filtering} \\ \phi_l(t) w_{\text{tx}}(t - lT) & \text{windowing} \end{cases} \quad (5)$$

2.2.2 Channel and equalizer

The lth signal received at the receiver is

$$z_l(t) = \psi_l(t) \otimes h(t) + \eta(t), \quad (6)$$

where $h(t)$ is the channel impulse response and $\eta(t)$ is the additive white Gaussian noise with power spectral density (PSD) N_0. Considering a filtering or windowing process at the front end of the receiver as shown in Fig. 1, the resultant signal is

$$\varphi_l(t) = \phi_l(t) \otimes h_{\text{eff}}(t) + \eta_{\text{eff}}(t) \quad (7)$$

where h_{eff} and η_{eff} are effective channel (see (12)) and effective noise (see (13)), respectively. If the CP duration T_{CP} is larger than the delay spread of the effective channel, the symbol can be estimated at the receiver without ISI. Using (4), the data $\boldsymbol{y}_l$ can be written as

$$\boldsymbol{y}_l = \boldsymbol{H}\boldsymbol{G}\boldsymbol{d}_l + \boldsymbol{\Theta} = \boldsymbol{H}\boldsymbol{s}_l + \boldsymbol{\Theta} \quad (8)$$

where

$$\begin{aligned} \boldsymbol{y}_l &= \left[y_{l,0}, y_{l,1}, \ldots, y_{l,N-1}\right]^T \\ \boldsymbol{H} &= \text{diag}[H_0, H_1, \ldots, H_{N-1}], H_n = \int_0^{T_s} h_{\text{eff}}(t) e^{-j2\pi \frac{n}{T_s} t} dt \\ \boldsymbol{\Theta} &= [\Theta_0, \Theta_1, \ldots, \Theta_{N-1}]^T, \Theta_n = \int_0^{T_s} \eta_{\text{eff}}(t) e^{-j2\pi \frac{n}{T_s} t} dt \end{aligned} \quad (9)$$

If a minimum mean square error (MMSE) frequency-domain equalizer (FDE) is used, the estimated $\hat{s}_l$ can be obtained by

$$\hat{s}_l = \boldsymbol{D}_{\text{MMSE}} \boldsymbol{y}_l, \quad (10)$$

where

$$\boldsymbol{D}_{\text{MMSE}} = E\{\boldsymbol{s}_l \boldsymbol{s}_l{}^H\}(\hat{\boldsymbol{H}})^H \left(\hat{\boldsymbol{H}} E\{\boldsymbol{s}_l \boldsymbol{s}_l{}^H\}(\hat{\boldsymbol{H}})^H + E\{\boldsymbol{\Theta}\boldsymbol{\Theta}^H\}\right)^{-1} \quad (11)$$

2.2.3 ACI rejection at the receiver

For the time domain ACI rejection approach at the receiver, either a bandpass filter $f_{\text{rx}}(t)$ or a windowing function $w_{\text{rx}}(t)$ is employed, the effective channel in (7) is

$$h_{\text{eff}}(t) = \begin{cases} h(t) \otimes f_{\text{tx}}(t) \otimes f_{\text{rx}}(t) & \text{filtering} \\ h(t) w_{\text{tx}}(t - lT) w_{\text{rx}}(t - lT) & \text{windowing} \end{cases} \quad (12)$$

and the effective noise in (7) is

$$\eta_{\text{eff}}(t) = \begin{cases} \eta(t) \otimes f_{\text{rx}}(t) & \text{filtering} \\ \eta(t) w_{\text{rx}}(t - lT) & \text{windowing} \end{cases} \quad (13)$$

When filters and windowing functions are used at the receiver as the time-domain process to reject ACI, the effective noise PSD in (11) are

$$\begin{aligned} &E(\Theta_n \Theta_n^*) = |F_{rx}(n)|^2 N_0, F_{rx}(n) = \int_0^{T_s} f_{rx}(t) e^{-j2\pi \frac{n}{T_s} t} dt \text{ and} \\ &E(\Theta_n \Theta_n^*) = N_0 \int_0^{T_s} |w_{rx}(t)|^2 dt, \end{aligned} \quad (14)$$

respectively.

The frequency-domain ACI rejection at the receiver, corresponding to the OOBE suppression at the transmitter, is implemented using $\boldsymbol{G}^H$ and the estimated $\hat{\boldsymbol{d}}_l$ can be obtained by the hard decision made after decoding

$$\hat{\boldsymbol{d}}_l = \text{decision}\left\{\boldsymbol{G}^H \hat{\boldsymbol{s}}_l\right\} \quad (15)$$

Let $E(\boldsymbol{d}_l \boldsymbol{d}_l^H) = E_0 \boldsymbol{I}_M$ where $E_0 = E\left(d_{l,m} d_{l,m}^*\right)$, for every l and m, is the data energy. With (4), $\boldsymbol{D}_{\text{MMSE}}$ in (11) can be written as:

$$\boldsymbol{D}_{\text{MMSE}} = E_0(\boldsymbol{G}\boldsymbol{G}^H)(\hat{\boldsymbol{H}})^H \left(\hat{\boldsymbol{H}}(\boldsymbol{G}\boldsymbol{G}^H)(\hat{\boldsymbol{H}})^H E_0 + E\{\boldsymbol{\Theta}\boldsymbol{\Theta}^H\}\right)^{-1} \quad (16)$$

2.2.4 Various approaches

The block diagram in Fig. 1 is a general framework for not only the hybrid approaches but also the individual time or frequency-domain approaches. When filters are employed at both the transmitter and receiver as the time-domain process, the hybrid approach is named filtered-precoded-OFDM (F-P-OFDM). Similarly, when windowing functions are employed at both the transmitter and receiver as the time-domain process, the hybrid approach is named windowed-precoded-OFDM (W-P-OFDM). If there is no time-domain process,

$$h_{\text{eff}}(t) = h(t) \text{ and } \eta_{\text{eff}}(t) = \eta(t) \quad (17)$$

the hybrid approaches (F-P-OFDM and W-P-OFDM) are reduced to precoded-OFDM (P-OFDM). If there is no frequency-domain process,

$$\boldsymbol{G} = \boldsymbol{I}_N \quad (18)$$

the hybrid approach F-P-OFDM is reduced to filtered-OFDM (F-OFDM) and the hybrid approach W-P-OFDM is reduced to windowed-OFDM (W-OFDM). If both (17) and (18) are satisfied and neither time or frequency-domain process is employed, the approaches are reduced to the conventional CP-OFDM. Due to the constraint on paper length, many other possible combinations are not included in this paper.

2.3 Frequency-domain approaches: P-OFDM

Two different precoding techniques, spectral precoding and PAPR precoding, are employed to suppress OOBE and PAPR, respectively. In order to distinguish these two precoded-OFDM (P-OFDM), they are named as spectral precoded (SP)-OFDM and PAPR precoded (PP)-OFDM in this paper. Our research in [16] shows that these two precoding techniques can be combined to achieve satisfactory spectral containments and PAPR performance simultaneously. (The spectrum containment of a waveform is satisfactory when the stopband attenuation is larger than 55 dB, as in the spectral mask requirement of Federal Communications Commission (FCC) TV white space when the effective isotropic radiated power is +36 dBm [18].) The combination of SP-OFDM and PP-OFDM will be denoted as PMix-OFDM. Also, for convenience, let P-OFDM represent general precoded-OFDM which includes PP-OFDM, SP-OFDM, and PMix-OFDM.

The main idea of spectral precoding **Gs** is to project the M-dimensional data symbol vector, $\boldsymbol{d}_l$, into an N-dimensional linear subspace [19] that ensures the desired spectral containment. Denote the frequency-domain representation of $\phi_l(t)$ in (1) at frequency f as

$$X(f) = \sum_{n=0}^{N-1} s_{l,n} P_n(f) \tag{19}$$

with

$$P_n(f) = \frac{1}{T}\int_{-T_{CP}}^{T_s} p_n(t)e^{-j2\pi f t}dt = \frac{e^{j\pi\left(\frac{n}{T_s}-f\right)(T_s-T_{CP})}}{\pi\left(\frac{n}{T_s}-f\right)T}\sin\left(\pi\left(\frac{n}{T_s}-f\right)T\right) \tag{20}$$

Using matrix expressions, the frequency responses $\boldsymbol{x}$ at K selected out-of-band frequencies (called notched frequencies), $\{f_k, k = 0, 1, \ldots, K-1\}$, can be written as

$$\boldsymbol{x} = \boldsymbol{P}\boldsymbol{s}_l = \boldsymbol{P}\boldsymbol{G}_s\boldsymbol{d}_l,\ \boldsymbol{x} = \begin{bmatrix} X(f_0) \\ X(f_1) \\ \vdots \\ X(f_{K-1}) \end{bmatrix},$$

$$\boldsymbol{P} = \begin{bmatrix} P_0(f_0) & \cdots & P_{N-1}(f_0) \\ \vdots & \ddots & \vdots \\ P_0(f_{K-1}) & \cdots & P_{N-1}(f_{K-1}) \end{bmatrix} \tag{21}$$

OOBE suppression can then be done by projecting $\boldsymbol{d}_l$ into a linear subspace with the form:

$$\mathcal{L}(\boldsymbol{P}) = \{\boldsymbol{s}_l\,|\boldsymbol{P}\boldsymbol{s}_l = \boldsymbol{0}\} \tag{22}$$

The solution to this problem is to use a semi-unitary precoding matrix **Gs** whose columns constitute an orthogonal basis of $\mathcal{L}(\boldsymbol{P})$. Performing the singular value decomposition (SVD) on $\boldsymbol{P}$:

$$\boldsymbol{P} = \boldsymbol{U}\boldsymbol{\Sigma}\boldsymbol{V}^H \tag{23}$$

where $\boldsymbol{U}$ is a $K\times K$ unitary matrix, $\boldsymbol{\Sigma}$ is a $K\times N$ diagonal matrix containing the singular values of $\boldsymbol{P}$ in non-increasing order on its diagonal and $\boldsymbol{V}$ is an $N\times N$ unitary matrix whose columns are $\boldsymbol{v}_0, \boldsymbol{v}_1, \cdots, \boldsymbol{v}_{N-1}$. The last M columns of $\boldsymbol{V}$ are collected to construct the $N\times M$ precoding matrix

$$\boldsymbol{G}_S = [\mathbf{v}_{N-M+1} \cdots \quad \mathbf{v}_{N-2} \quad \mathbf{v}_{N-1}] \tag{24}$$

Define $\lambda = \frac{M}{N}$ as the coding rate and $R = N - M$ as the coding redundancy. Better spectral containments can be achieved at the cost of higher coding redundancy. If $R \geq K$, $\boldsymbol{x} = 0$ for any arbitrary data vector $\boldsymbol{d}_l$ due to the fact that $\boldsymbol{s}_l$ is in the null space of $\boldsymbol{P}$. In this paper, we choose $R \leq K$.

The main idea of PAPR precoding is to project the M-dimensional data symbol vector $\boldsymbol{d}_l$ into a N-dimensional linear subspace that ensures the desired PAPR performance. The transmit signal in (1) can be rewritten as:

$$\phi_l(t) = \sum_{m=N_0}^{N_0+M-1} d_{l,m}p'_m(t-lT) \tag{25}$$

where $p'_m(t)$ is the inverse discrete Fourier transform (IDFT) of the mth column of the precoder $\boldsymbol{G}_{\text{PAPR}}$

$$p'_m(t) \equiv \sum_{n=0}^{N-1} G_{n,m}e^{j\frac{2\pi n}{T_s}t}. \tag{26}$$

Note that the mth data $d_{l,m}$ is carried by the waveform $p'_m(t)$. Then, to suppress the PAPR, we can choose $\{G_{n,m}\}$ such that the peaks of different $p'_m(t)$ for different m occurs at different times within an OFDM symbol. The waveform design has been investigated in [20, 21], and goes beyond the scope of this paper. For simplicity, we use the discrete Fourier transform (DFT) precoding matrix $\boldsymbol{G}_{\text{DFT}}$ in [22] with elements:

$$[\boldsymbol{G}_{\text{PAPR}}]_{k,n} = [\boldsymbol{G}_{\text{DFT}}]_{k,n} = \frac{1}{\sqrt{N}}e^{-j\frac{2\pi k}{N}n} \tag{27}$$

which makes different $p'_m(t)$ as sinc functions with their peaks at different times.

Note that, the precoding matrix in (24) can be easily extended to reduce the PAPR by multiplying the spectral precoding matrix $\boldsymbol{G}_S$ by the DFT matrix $\boldsymbol{G}_{\text{DFT}}$:

$$\boldsymbol{G}_{\text{Mix}} = \boldsymbol{G}_S\boldsymbol{G}_{\text{DFT}} \tag{28}$$

2.4 Time-domain approaches: F-OFDM and W-OFDM

Two main time-domain OOBE suppression approaches are summarized here. For F-OFDM, to suppress the OOBE at the transmitter, the CP-OFDM signal is convolved with a transmit filter, $f_{\text{tx}}(t)$, with a length of $\alpha_{\text{tx}}T_s$, where α_{tx} is the normalized transmit filter length, and to reject the ACI at the receiver, the received signal is convolved with a receive filter, $f_{rx}(t)$, with a length of $\alpha_{rx}T_s$, where α_{rx} is the normalized receive filter length. For convenience, an identical filter is used at both the transmitter and receiver, i.e., $f_{tx}(t) = f_{rx}(t)$. An example of the identical filter is a square root raised cosine filter. For W-OFDM, to suppress OOBE at the transmitter, the CP-OFDM signal is point-to-point multiplied with a transmit windowing function, $w_{tx}(t)$, before transmission, and to reject ACI at the receiver, the received signal is point-to-point multiplied with a receive windowing function, $w_{rx}(t)$. An example of $w_{tx}(t)$ is given below:

$$w_{tx}(t) = \begin{cases} 0.5\left(1+\cos\left\{\pi\left(1+\frac{t}{\beta_{tx}T_s}\right)\right\}\right), & 0 \le t < \beta_{tx}T_s \\ 1, & \beta_{tx}T_s \le t < T_s + T_{CP} \\ 0.5\left(1+\cos\left\{\pi\left(\frac{t-T_s}{\beta_{tx}T_s}\right)\right\}\right), & T_s + T_{CP} \le t \le (1+\beta_{tx})T_s + T_{CP} \end{cases} \tag{29}$$

where $\beta_{tx}T_s$ is the roll-off portion of $w_{tx}(t)$. As shown in Fig. 2, the CP-OFDM signal is extended with CP and cyclic suffix (CS), then is point-to-point multiplied with the windowing function $w_{tx}(t)$. In order to keep the symbol duration of OFDM symbol unchanged after windowing, the ending $\beta_{tx}T_s$ portion of the current windowed-OFDM symbol overlaps with the beginning $\beta_{tx}T_s$ portion of the next windowed-OFDM symbol.

At the receiver, as the CP is discarded before FFT to reduce ISI, the windowing process should be carried out on the signal portion that goes through the FFT for rejecting ACI. Thus, an example of the windowing function is:

$$w_{rx}(t) = \begin{cases} 0, & 0 \le t < T_{CP} - \beta_{rx}T_s \\ 0.5\left(1+\cos\left\{\pi\left(1+\frac{t}{\beta_{rx}T_s}\right)\right\}\right), & T_{CP} - \beta_{rx}T_s \le t < T_{CP} \\ 1, & T_{CP} \le t \le (1-\beta_{rx})T_s + T_{CP} \\ 0.5\left(1+\cos\left\{\pi\left(\frac{t-T_s}{\beta_{rx}T_s}\right)\right\}\right), & (1-\beta_{rx})T_s + T_{CP} < t \le T_s + T_{CP} \end{cases} \tag{30}$$

where $\beta_{rx}T_s$ is the roll-off duration (see Fig. 3). Note that, in order to keep the orthogonality of OFDM symbol, the windowing function is chosen to satisfy

$$w_{rx}(t) + w_{rx}(t + T_s) = 1, T_{CP} - \beta_{rx}T_s \le t < T_{CP} \tag{31}$$

so that, as shown in Fig. 3, the roll-off section of OFDM symbol after windowing can be added back to the tail section (the last $\beta_{rx}T_s$) of the OFDM symbol before FFT.

2.5 Hybrid approaches: F-P-OFDM and W-P-OFDM

According to Fig. 1, the F-P-OFDM can be simply constructed by using the filters presented in Section 2.4 and the spectral decoder presented in Section 2.3. Similarly, the W-P-OFDM can be simply constructed by using the window functions presented in Section 2.4 and the spectral decoder presented in Section 2.3.

3 Practical issues and numerical examples: without ACI

Various practical issues including non-contiguous and dynamically changing spectrum, multipath delay spread, radio impairments (FO and TO), implementation complexity, spectral efficiency, PAPR, spectral containment, interference rejection, etc., will be addressed in this section and Section 4. The performances of CP-OFDM (without any OOBE or ACI suppression approach), P-OFDM, F-OFDM, W-OFDM, F-P-OFDM, and W-P-

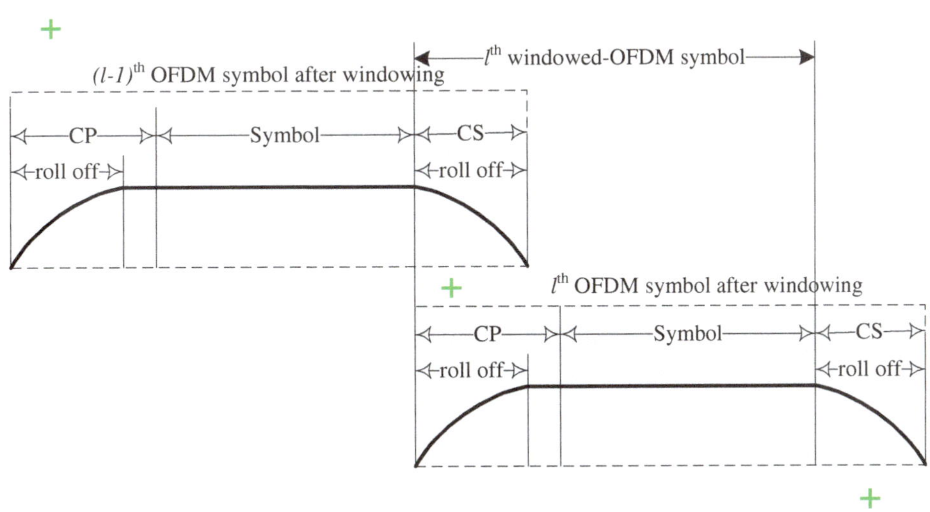

Fig. 2 Illustration of a windowed-OFDM transmitted symbol

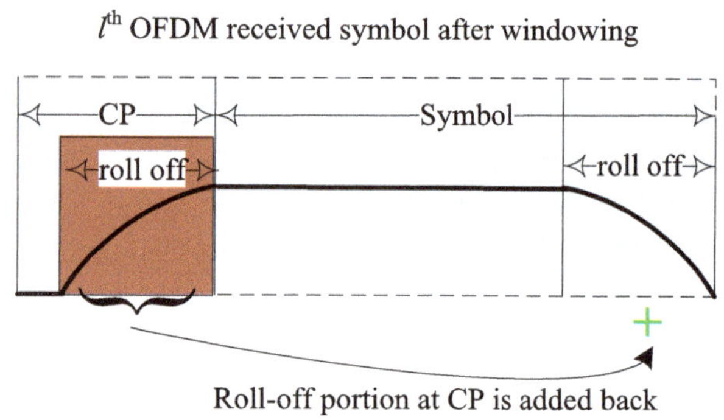

Fig. 3 Illustration of a W-OFDM receiver symbol

OFDM are evaluated in terms of complexity, coding rate, spectral containment, PAPR, and BER based on a LTE-like system. Specifically, results shown in this section are without ACI and results shown in Section 4 are with ACI. The simulation parameters are specified in Table 1, if not otherwise indicated.

3.1 Complexity and coding rate analysis

In this subsection, the coding rates and computational complexities of various OOBE suppression or ACI rejection approaches are presented. The analysis of coding rate λ is straight forward. For approaches involving precoding (P-OFDM, F-P-OFDM, and W-P-OFDM), their coding rates are $1 - R/N$. However, for approaches without precoding (CP-OFDM, F-OFDM, and W-OFDM), their coding rates are 1. The analysis of complexity is much more involved. Here, the computational complexity for each approach is evaluated in terms of the number of real multiplications needed for completing the transmission of one quadrature amplitude modulation (QAM) input symbol vector $\boldsymbol{d}_l$ from the transmitter to the receiver. But the multiplications with ± 1 and $\pm j$ are not included since they are merely flips of sign and/or flips of real and imaginary parts. In addition, since the channel and equalizer coefficients are assumed known, the computational efforts for channel estimation and equalizer design are not included.

Table 1 LTE-like systems

System parameters
Symbol duration: $T_s = 1/\Delta f$
Subcarrier spacing: $\Delta f = 15$ kHz
Number of samples per symbol duration: $L = 1024$
Active subcarrier index = [−300:−1,1:300]
CP parameters
One subframe: 2 slots/subframe
One slot: 7 data symbols ($7T_s$ and half additional symbol duration ($0.5T_s$ is used for CP, with CP duration $T_{cp1} = 80T_s/1024$ $L_{cp1} = 80\,L/1024$ for the first symbol and $T_{cp2} = 72T_s/1024$ $L_{cp2} = 72\,L/1024$ for the other symbols.
Average CP length: $L_{cp} = L/14$

Note that the complexity of an L-point FFT/IFFT (via split radix FFT [23]) with L complex inputs is $L\log_2 L - 3L + 4$. And the complexity of a scalar 1-tap MMSE FDE equalization is $4N$ for N subcarriers. Therefore, the complexity of the CP-OFDM without any OOBE or ACI suppression approach is $2(L\log_2 L - 3L + 4) + 4N$.

For P-OFDM, including SP-OFDM, PP-OFDM, and PMix-OFDM, the complexity of precoding and decoding is $8N(N - R)$. For a matrix MMSE FDE, the complexity is $4N^2$ since a matrix product is required. Thus, including the complexity for IFFT and FFT, the complexity for P-OFDM is $2(L\log_2 L - 3L + 4) + 8N(N - R) + 4N^2$, which is proportional to N^2 if N is large.

For F-OFDM, it is assumed that the transmit filtering and addition of the CP could be combined such that the filtering is only done once for the CP samples [24]. Therefore, the complexity of transmit filtering is $2LL\alpha_{tx}$. But the filtering at the receiver is done for all symbol duration. Therefore, the complexity of receive filtering is $2(L + L_{CP})L\alpha_{rx}$. Thus, including the complexity of IFFT and FFT, the complexity of F-OFDM with a scalar 1-tap MMSE FDE is $2(L\log_2 L - 3L + 4) + 2L^2\alpha_{tx} + 2(L + L_{CP})L\alpha_{rx} + 4N$. Note that, the complexity increases as α_{tx} or α_{rx} increases.

For W-OFDM, the complexity of point-to-point multiplication of CP-OFDM signal and windowing at the transmitter and receiver is $2L\beta_{tx}$ and $2L\beta_{rx}$, respectively, which are relatively small. Including the complexity of IFFT and FFT, the complexity of for W-OFDM with a scalar 1-tap MMSE FDE is $2(L\log_2 L - 3L + 4) + 2L(\beta_{tx} + \beta_{rx}) + 4N$. Therefore, the complexity of W-OFDM is very close to that of CP-OFDM.

For F-P-OFDM, including F-SP-OFDM, F-PP-OFDM, and F-PMix-OFDM, the complexity, including IFFT/FFT, filtering, precoding, decoding and equalization using a matrix MMSE FDE, is $2(L\log_2 L - 3L + 4) + 2LL\alpha_{tx} + 2(L + L_{CP})L\alpha_{rx} + 8N(N - R) + 4N^2$ which is proportional to N^2 if N is large.

For W-P-OFDM, including W-SP-OFDM, W-PP-OFDM and W-PMix-OFDM, the complexity, including IFFT/FFT, windowing, precoding, decoding and equalization using a matrix MMSE FDE, is $2(L\log_2 L - 3L + 4) + 2L(\beta_{tx} + \beta_{rx}) + 8N(N - R) + 4N^2$ which is proportional to N^2 if N is large.

Table 2 shows the per-symbol normalized complexities (with respect to the complexity of CP-OFDM) of various approaches for $N = 60$ and 300. As discussed before, for W-OFDM, the complexity is very close to that of CP-OFDM no matter how many subcarriers are used. For F-OFDM with the fixed subcarrier number, the complexity increases as the filter length increases and are almost independent of the number of subcarriers. Therefore, for low complexity, a filter with a minimum length is

Table 2 Normalized complexity (with respect to CP-OFDM) for $N = 60$ and 300. Note that P-OFDM represents SP-OFDM, PP-OFDM, or PMix-OFDM and W/F-P-OFDM represents W/F-SP-OFDM, W/F-PP-OFDM, or W/F-PMix-OFDM

$L = 1024$		$N = 60$	$N = 300$
CP-OFDM		1	1
P-OFDM	$R = 3$ or $R = 6$	3.85 ($R = 3$)	69.48 ($R = 6$)
	$R = 6$ or $R = 12$	3.75 ($R = 6$)	68.55 ($R = 12$)[1]
			34.74 ($R = 12$)[2]
			23.47 ($R = 12$)[3]
W-OFDM	$\beta_{tx} = \beta_{rx} = 24/L$	1.01	1.01
	$\beta_{tx} = \beta_{rx} = 48/L$	1.01	1.01
F-OFDM	$\alpha_{tx} = \alpha_{rx} = 18/L$	6.24	5.91.
	$\alpha_{tx} = \alpha_{rx} = 35/L$	11.18	10.55
F-P-OFDM	$R = 3$ *or* 6, $\alpha_{tx} = \alpha_{rx} = 18/L$	9.08 ($R = 3$)	74.39 ($R = 6$)
W-P-OFDM	$R = 3$ *or* 6, $\beta_{tx} = \beta_{rx} = 24/L$	3.85 ($R = 3$)	69.48 ($R = 6$)

desired if OOBE and ACI can be kept within the specs. For precoding related approaches, the complexities of P-OFDM, F-P-OFDM, and W-P-OFDM for $N = 300$ are very high because they are proportional to N^2 when N is large. Fortunately, the complexity can be reduced by dividing the big spectrum block into several smaller spectrum blocks. For example, for $N = 300$, the normalized complexity of P-OFDM is 68.55 for one spectrum block (which requires one precoder/decoder pair for 300 subcarriers). The normalized complexity is reduced to 34.74 (or 23.47) for two (or three) spectrum blocks where we use two (or three) precoder/decoder pairs, and each pair is designed to precode/decode 150 (or 100) subcarriers. (The three cases of one, two, and three spectrum blocks are denoted by the superscripts 1, 2, and 3 in Table 2, respectively.) However, the reduction of complexity may cause degradation in spectral containment if the coding rate is kept unchanged (an example will be discussed in the next subsection).

3.2 Implications for non-contiguous and dynamic changing spectrum

In CR, the available bandwidth may be non-contiguous and dynamically changing. Among the approaches discussed in Section 2, only the windowing approach can adapt to the changes or the non-contiguousness of the frequency bands easily. However, the performance of windowing is limited by the constraint on the CP length and is not always satisfactory as shown in the simulation result in Section 4. For the filtering and precoding approaches, the computational complexity is high if a single pair of transmit/receive filters or precoder/decoder is employed to deal with the entire possible band (which includes all possible stopbands and passbands). In addition, the design of filters or precoder/decoder has to change whenever the available spectrum changes. To deal with these problems caused by non-contiguous and dynamically changing spectrum, a resource block (RB) strategy for the filtering approach is proposed in [14]. Similarly, we have also proposed a RB strategy for the precoding approach (see Appendix and [25]) to adapt to the non-contiguous and dynamically changing spectrums.

3.3 Spectral containment and PAPR

In this subsection, PSD and PAPR of various approaches based on the parameters defined in Tables 1 and 2 with $N = 300$ are evaluated and compared. In the legends of Figs. 4 and 5: for F(k), k is the transmitter filter length in samples; for W(k), k is the roll-off length of transmit windowing function in samples; for SP(R), only spectral precoding is used and R is the redundancy length of spectral precoding; and for PMix(R), the combination of spectral precoding and PAPR precoding in (28) is used and R is the redundancy length of the spectral precoding.

For F-OFDM, as shown in Fig. 4, the longer the filter is, the larger the stopband attenuation can be achieved. For W-OFDM, similar to F-OFDM, better spectral containments can be achieved at the cost of a longer roll-off portion of the windowing function. However, W-OFDM has a slower fall-off rate (i.e., larger transition bands) than CP-OFDM (denoted as "None"; representing the case where there is no OOBE suppression process at the transmitter), as shown in Fig. 4. We aim to use a filter with the longest filter length or a windowing function with the longest possible roll-off portion length to reduce inter-user interference as long as the performance loss caused by ISI is below a threshold. For SP-OFDM, the higher the spectral efficiency loss (i.e., the smaller coding rate) is, the larger the stopband attenuation can be achieved. Therefore, better spectral containment is at the cost of higher spectral efficiency loss. For the given examples, comparing with SP, filtering has a slower fall-off rate near the edges of the passband but a smaller PSD in the stopband away from the edges, and windowing has a slower fall-off rate and a larger fall-off PSD in the stopband away from the edges.

Generally, a hybrid approach outperforms its individual constituent approaches. For example, F(18)-SP(6)-OFDM is better than SP(6)-OFDM and F(18)-OFDM in terms of spectral containment. Furthermore, it is less sensitive to ISI compared with F(35)-OFDM. Similar observations can also be made for W(24)-SP(6)-OFDM. Furthermore, comparing PMix with SP, it is remarkable that the spectral containment performances of PMix-OFDM, F-PMix-OFDM, and W-PMix-OFDM are almost the same as (or even better than) those of SP-OFDM, F-SP-OFDM, and W-SP-OFDM, respectively. Thus, we

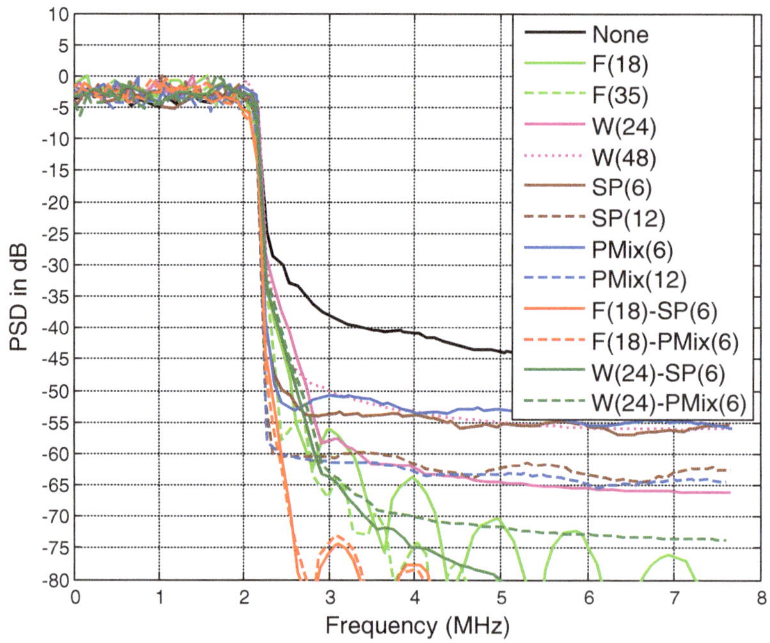

Fig. 4 Spectral containments. In the legends, for F(*k*), *k* is the filter length; for W(*k*), *k* is the windowing roll-off length; and for SP(*R*) and PMix(*R*), *R* is the redundancy length of the spectral precoding

conclude that PMix outperforms SP in spectral containment and PAPR (to be shown in Fig. 5).

The effects of RB on PSD will be briefly discussed below. Note that the PSD curve for SP(12)-OFDM in Fig. 4 is derived by using one precoder to precode the entire spectrum block of 300 subcarriers where the 12 redundancies are employed to nullify the frequency responses at 12 notched frequencies (in which, six in the upper stopband and six in the lower stopband). For the two-precoder case discussed in the previous subsection, the passband is divided into two blocks and each precoder is to precode a spectrum block of 150 subcarriers. In this case, the precoder for the upper spectrum block has six notched frequencies in the upper stopband and zero notched frequencies in the lower stopband and the precoder for the lower frequency block has six notched frequencies in the lower stopband and zero notched frequencies in the upper stopband. The PSD result of

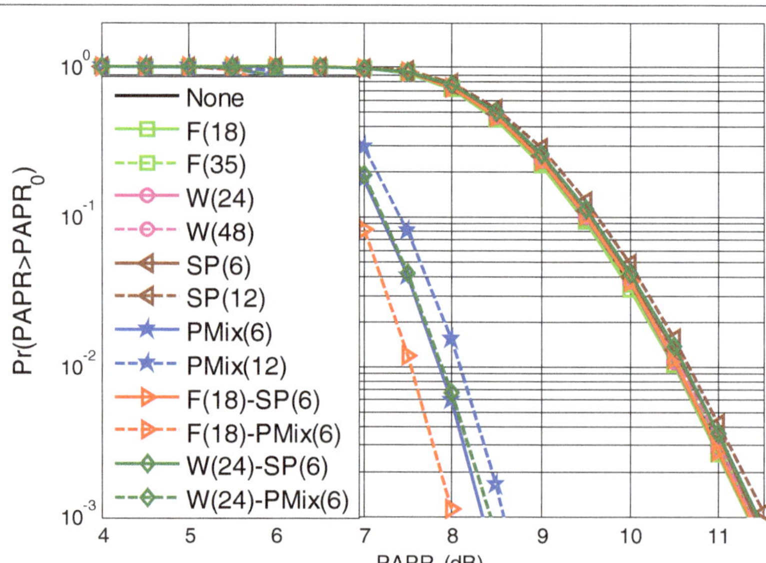

Fig. 5 PAPR performances. The meanings of the legends are the same as those in Fig. 4

this case is essentially the same as that shown in Fig. 4 (i.e., the one precoder for the entire spectrum block of 300 subcarriers). There is no degradation in spectrum containment, but the normalized complexity is reduced from 68.55 to 34.74 as shown in Table 2. For the three-precoder case which is also discussed in the previous subsection, the OOBE increases 5 dB if the redundancy number R is kept at 12. Thus, the reduction of the normalized complexity from 68.55 to 23.47, as shown in Table 2, is at the expense of an OOBE increase of 5 dB. Similar results have been shown in [25] and are omitted here.

The PAPRs of various approaches are plotted in Fig. 5. It is shown that the PAPRs of W-OFDM and F-OFDM are almost the same as that of CP-OFDM. However, for SP-OFDM, W-SP-OFDM, and F-SP-OFDM, their PAPR performances degrade compared with that of CP-OFDM and the degradation increases as R increases. This is due to the fact that the precoded elements in $\mathbf{s}$ become correlated. As mentioned in Section 2.3, SP-OFDM, F-SP-OFDM, and W-SP-OFDM can be easily extended to reduce the PAPR drastically by using (28). As shown in Fig. 5, at Pr (PAPR > PAPR$_0$) = 10^{-3}, the PAPRs of SP-OFDM, W-SP-OFDM, and F-SP-OFDM are reduced by more than 3 dB if PMix-OFDM, F-PMix-OFDM, and W-PMix-OFDM are used respectively. Therefore, concluding from Figs. 4 and 5, the combination of spectral and PAPR precoding in (28) can suppress the OOBE and reduce the PAPR at the same time without adding extra complexity (compared with spectral precoding in (24)).

3.4 BER

In an ideal situation where there is no ACI, FO, or TO, Fig. 6 shows the raw BER versus the E_b/N_0 (the energy per bit to noise PSD ratio) for various approaches. The transmitter is transmitting on subcarriers [−144:−1, 1:156] where QPSK (or 4QAM) modulation is used. The system parameters defined in Tables 1 and 2 ($N = 300$), and a multipath channel with delay spread of 39 samples are used in the simulation. It is assumed that the channel is perfectly known and a 1-tap MMSE FDE is used at the receiver. Each BER point is obtained by averaging over 10,000 subframes.

In Fig. 6, the BERs of F-OFDMs and W-OFDMs are slightly worse than that of CP-OFDM (denoted as None; representing the case where there is no OOBE suppression process at the transmitter and no ACI rejection at the receiver), due to ISI caused by delay spread. The BER performance degradation mainly depends on the filter or windowing roll-off length, i.e., the F/W-OFDM with longer filter or windowing roll-off length has larger ISI and therefore worse BER performance. On the contrary, all approaches using precoding have small but noticeable BER performance improvements over CP-OFDM due to frequency diversity gain, since each data symbol is modulated by all the subcarriers due to the precoding.

It is remarkable that, with the same amounts of spectral coding rate, the performances of approaches with PMix are much better than approaches with SP. For example, PMix-OFDM, F-PMix-OFDM, and W-PMix-OFDM are much better than those of SP-OFDM, F-SP-OFDM, and W-SP-OFDM (the improvement is larger than 8 dB at BER = 10^{-3}) due to the higher diversity gains provided by PMix. This is demonstrated in Fig. 7 which shows the PSDs of several arbitrarily selected modulated data symbols in SP-OFDM and PMix-OFDM. We can see that, in PMix-OFDM, each data symbol $d_{l,i}$ is modulated by all

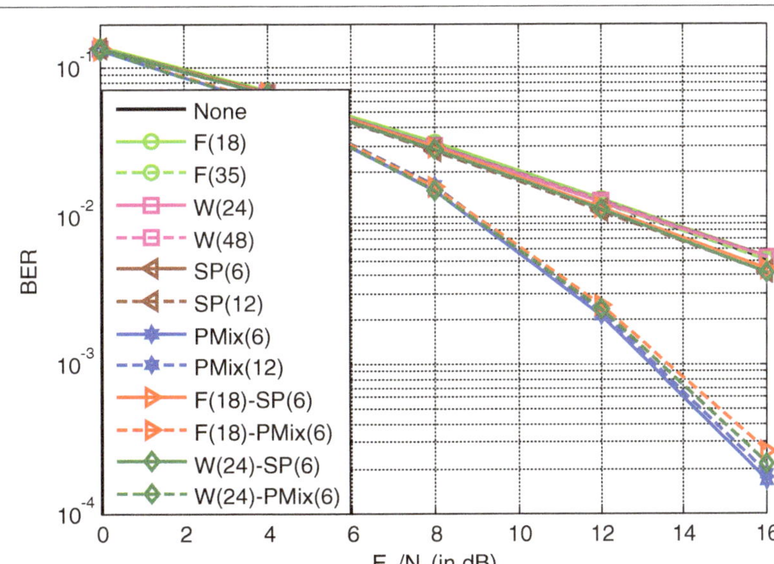

Fig. 6 BER performances when there is no ACI. The meanings of the legends are the same as those in Fig. 4

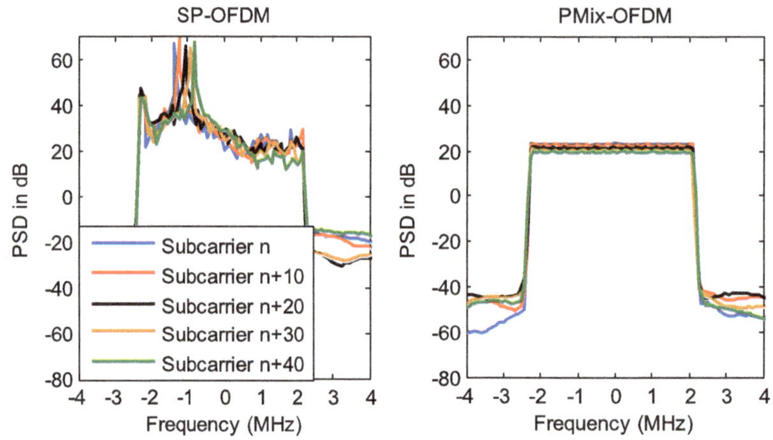

Fig. 7 PSDs of $\sum_{n=0}^{N-1} G_{n,i} d_{l,i} p_n(t-lT), i = k, k+10, k+20, k+30, k+40$ in SP-OFDM ($\boldsymbol{G} = \boldsymbol{G}_S$) and PMix-OFDM ($\boldsymbol{G} = \boldsymbol{G}_S \boldsymbol{G}_{DFT}$)

the subcarriers with almost equal power. However, there is frequency selectivity in SP-OFDM, where the power of each data symbol is spread mainly only on several subcarriers. Thus, in the following simulations where ACI is present, approaches with SP (spectral precoding only) will not be shown since PMix (combined spectral and PAPR precoding) has better BER performance when SP and PMix have the same redundancy length (note that the redundancy length for PAPR precoding is zero).

4 Practical issues and numerical examples: with ACI

Various practical issues are addressed in this section and the previous section (Section 3) based on a LTE-like system. Specifically, results shown in this section are with ACI, and results shown in Section 3 are without ACI. The simulation parameters are specified in Table 1, if not otherwise indicated. To examine the effects of ACI on a given receiver, consider two transmitters: one transmitter is desired and generates the desired signal, and the other transmitter is not desired and generates ACI. Presented in this section are the following three scenarios. In the first scenario, the reference receiver belongs to a PU, the desired transmitter belongs to another PU, and the interfering transmitter belongs to a SU. In the second scenario, the reference receiver belongs to a SU, the desired transmitter belongs to another SU, and the interfering transmitter belongs to a PU. In the third scenario, the reference receiver belongs to a SU, the desired transmitter belongs to another SU, and the interfering transmitter belongs to a third SU.

In our simulations, CP-OFDM is used in all PU and SU transmitter-receiver pairs. Note that a PU transmitter does not need to suppress its OOBE to the SU receivers and a PU receiver does not need to have special design to reject ACI from interfering SUs. However, a SU transmitter needs to suppress its OOBE to the PU or non-designated SU receiver, and a SU receiver needs to have special design to reject ACI from the interfering PU and SU transmitters. Thus, we assume no transmit filtering, windowing, or precoding for suppressing OOBE in the PU transmitter. We also assume that in the PU receiver, there is a standard low-pass filter with the cutoff frequency of 4.5 MHz and the filter length of 35 samples (note that the minimum CP length is 72 samples and the channel impulse response length is 39 samples). For the SU transmitters and receivers, any of the approaches tested in Section 3 can be employed. For simplicity, we assume that the SU transmitters and receivers use the same filtering, windowing, precoding, or hybrid approach.

4.1 OOBE suppression by a SU transmitter

Consider an example where a transmit-receive pair of PUs are operating in the frequency band with subcarriers [−144:−1, 1:156] and a transmit-receive pair of SUs are operating in the adjacent frequency band of the same bandwidth. The frequency separation of the two bands is 1.5 times of subcarrier spacing (i.e., 22.5 kHz) where the 0.5 (in 1.5) implies that there is FO between the PU pair and SU pair. Two cases are investigated. In the first case, TO = 0, where the PU pair and SU pair are perfectly synchronous in time. In the second case, TO = 200/1024 T_s, where there is a 200-sample delay between the PU pair and SU pair.

The OOBE suppression performance of W-OFDM, F-OFDM, PMix-OFDM, F-PMix-OFDM, and W-PMix-OFDM by the SU transmitter will be investigated by evaluating the raw BER performance of the PU receiver. In this simulation, the power of the SU transmitter is 10 dB higher than that of the PU transmitter. Figure 8 shows the BER performance of the PU receiver for TO = 0 and TO = 200 when the SU transmitter uses various

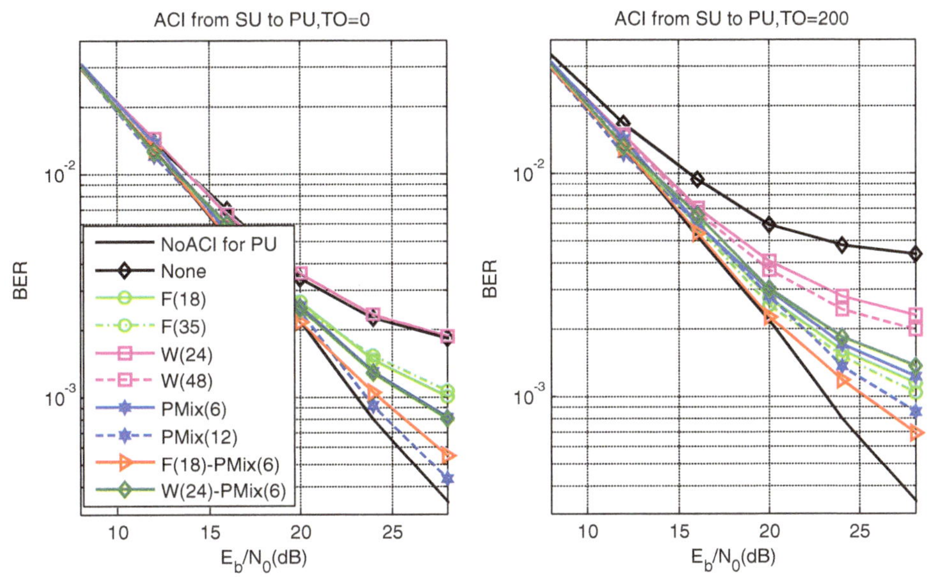

Fig. 8 BER performance of PU accounting for OOBE suppression of a SU transmitter. The meanings of the legends are the same as those in Fig. 4

OOBE suppression approaches. For comparison, the best possible performance (denoted as "No ACI for PU"; representing the case where the interfering SU pair is absent) and the *worst* possible performance (denoted as None; representing the case where there is no OOBE suppression process in the interfering SU transmitter) are included in Fig. 8. It is clear that this is an interference-limited system because error floors exist for all approaches at high signal to noise ratios (SNRs).

When TO = 0, Fig. 8 shows that W-OFDM has almost the same OOBE suppression performance as None (i.e., windowing does very little in suppressing OOBE in this example where the frequency spacing of the desired and interfering signals is 22.5 kHz). This is because W-OFDM has larger stopband attenuation but a slower fall-off rate than None (see Fig. 4). For F-OFDM, we observe that the F-OFDM with longer filter length has a slightly worse BER performance, which seems contrary to the spectral containment result in Fig. 4. In fact, there is no contradiction because the BER performances are affected not only by OOBE but also ISI. A filter with longer filter length has smaller OOBE but causes larger ISI. So, the ISI effects dominate the ACI effects in this example. Comparing F-OFDM with W-OFDM, we observe that F-OFDM outperforms W-OFDM. Comparing F-OFDM and W-OFDM with PMix-OFDM, we observe that PMix-OFDM has the best OOBE suppression results but at the cost of a small spectral efficiency loss and a significant complexity increase. Furthermore, the PMix-OFDM with more redundancy has smaller OOBE and causes less interference to the PU. It is remarkable that F(18)-PMix(6)-OFDM outperforms both F(18)-OFDM and PMix(6)-OFDM but at the cost of a small spectral efficiency loss and a significant complexity increase. As W-OFDM did little to suppress OOBE in this example, W(24)-PMix(6)-OFDM has the same BER performance as PMix(6)-OFDM. All these observations are consistent with the spectral containment results in Fig. 4.

When TO between the desired and interfering signals changes from 0 to 200, the BER performances of all approaches degrade in different degrees. As shown in Fig. 8, the degradation of None is profound because the ability to reject interferences of the PU receiver depends very much on correct sampling of each interfering OFDM symbol. The degradation of "PMix" is also quite significant because, to a certain extent, the OOBE suppression ability of the interfering SU transmitter requires the PU receiver to have correct samples of each interfering OFDM symbols. Windowing has small degradations because its OOBE suppression ability does not depend a lot on the OFDM symbol structure. However, as shown in TO = 0 case, windowing is not a good choice since it did very little to suppress OOBE. Filtering has the least degradation because its OOBE suppression ability depends very little on the OFDM symbol structure. As expected, the hybrid approach F-PMix (or W-PMix) has the combined features of filtering (or windowing) and PMix.

In practice, as PU and SU will generally not be synchronized, filtering is then a good choice for OOBE suppression because of its low complexity and robustness to TO. However, filtering alone is usually not adequate to produce sufficient OOBE suppression because the filter length of F-OFDM is limited by the CP length. In order to satisfy the OOBE suppression requirement and to minimize the sensitivity with respect to TO, the combination of filtering with precoding would be necessary if the CP length is not long enough to allow a sufficient long filtering function. Thus, we conclude from this example that the combination of filtering with precoding would be the most efficient way to suppress OOBE of OFDM signal but at the cost of a small spectral efficiency loss and a significant complexity increase. This can be seen in Fig. 8 where, among all results shown, the hybrid approach F(18)-PMix(6)-OFDM for the SU has the best OOBE suppression performance when TO = 200.

4.2 ACI rejection by a SU receiver

Consider the same example as in the previous subsection (Section 4.1) where a transmit-receive pair of PUs are operating in the frequency band with subcarriers [−144:−1, 1:156] and a transmit-receive pair of SUs are operating in the adjacent frequency band of the same bandwidth. The frequency separation of the two bands is 1.5 times of subcarrier spacing where the 0.5 (in 1.5) implies that there is FO between the PU pair and SU pair. Two TO cases are investigated. In the first case, TO = 0, where the PU pair and SU pair are perfectly synchronous in time. In the second case, TO = 200/1024 T_s, where there is a 200-sample delay between the PU pair and SU pair.

The performances of W-OFDM, F-OFDM, PMix-OFDM, F-PMix-OFDM, and W-PMix-OFDM used by the SU receiver to reject the ACI from the PU transmitter will be investigated by evaluating the BER performance of the SU receiver. For comparison, the worst possible performance (denoted as None; representing the case where there is no OOBE suppression or ACI rejection process in the SU transmitter and receiver) are included in Fig. 9. In this simulation, the power of the PU transmitter is 10 dB higher than that of the SU transmitter.

When TO = 0, the ACI rejection performance of F-OFDM/W-OFDM mainly depends on the filtering/windowing length (the longer the filter length/windowing roll-off length, the better the ACI rejection performance), and the performance of PMix-OFDM depends on the redundancy length (the longer the redundancy length, the better the ACI rejection performance). In addition, PMix outperforms F-OFDM and W-OFDM. These observations are similar to those made for the OOBE suppression results in Fig. 8. However, unlike the OOBE suppression results in Fig. 8, windowing outperforms filtering for ACI rejection in Fig. 9. This is due to the fact that, at the SU receiver, the process of discarding CP generates discontinuities of the received signal. Since discontinuities usually cause spread in the frequency, the ACI caused by the interfering signal before discarding CP is smaller than that after discarding CP. Now that filtering is done before discarding CP, the ACI rejection ability of filtering approaches is greatly reduced by CP removal. On the other hand, windowing process is carried out on the signal portion that goes through the FFT for rejecting ACI, and the ACI rejection ability of windowing approaches is unaffected by CP removal because the discarded portion is compensated (see Fig. 3). For hybrid approaches, F(18)-PMix(6)-OFDM outperforms both F(18)-OFDM and PMix(6)-OFDM but at the cost of a small spectral efficiency loss and a significant complexity increase. Similarly, W(24)-PMix(6)-OFDM outperforms both W(24)-OFDM and PMix(6)-OFDM but at the cost of a small spectral efficiency loss and a significant complexity increase. Remarkably, F(18)-PMix(6)-OFDM outperforms W(24)-PMix(6)-OFDM

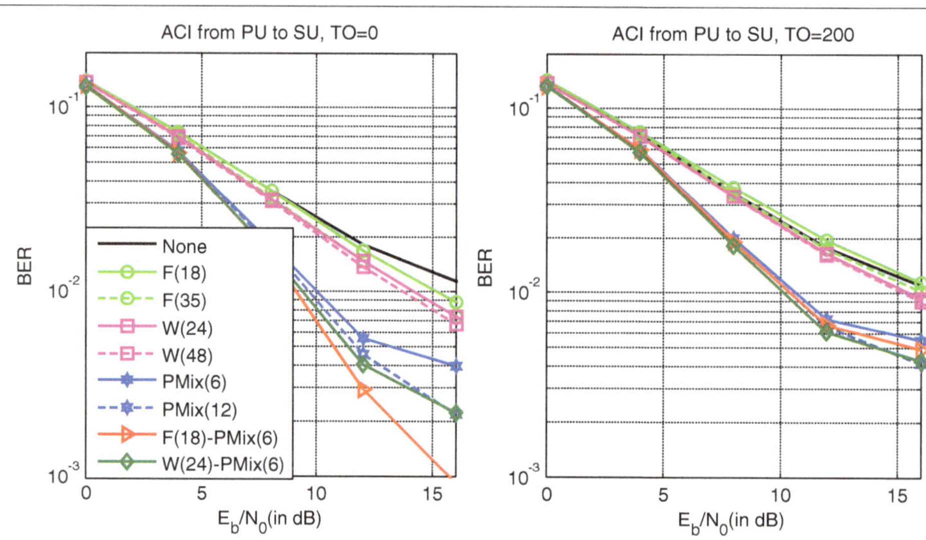

Fig. 9 BER performance of the SU receiver accounting for rejection of the ACI caused by the PU transmitter. The meanings of the legends are the same as those in Fig. 4

even though W(24)-OFDM outperforms F(18)-OFDM. It seems that the combination of filtering and spectral precoding work better than the combination of windowing and spectral precoding at TO = 0 since both filtering and spectral precoding approaches suppress the unwanted signal in the frequency domain.

When TO changes from 0 to 200, the BER performances of all approaches in Fig. 9 degrade in different degrees. Similar to Fig. 8, the degradation of PMix is significant because, to a certain extent, the ACI rejection ability of the SU receiver requires to have correct samples of each interfering OFDM symbols. Windowing and filtering have smaller degradations because their ACI rejection ability does not depend a lot on the OFDM symbol structure. However, as shown in TO = 0 case, filtering is not a good choice since it does not deal with the ACI regeneration due to the CP removal. As expected, the hybrid approach F-PMix (or W-PMix) has the combined features of filtering (or windowing) and PMix. What is remarkably different from the results in Fig. 8 is that BERs for TO = 0 and TO = 200 are similar for None because TO = 200 does not introduce more ACI than TO = 0 when no ACI suppression technique is employed.

In practice, as PU and SU will generally not be synchronized, windowing is then a good choice for ACI rejection because of its low complexity, robustness to TO, and insensitivity to CP removal. However, windowing alone is usually not adequate to produce sufficient ACI rejection because the roll-off portion length of the windowing function in W-OFDM is limited by the CP length. In order to satisfy the ACI rejection requirement and to minimize the sensitivity with respect to TO, the combination of windowing with precoding would be necessary if the CP length is not long enough to allow a sufficient long windowing function. Thus, we conclude from this example that the combination of windowing with precoding would be the most efficient way to reject ACI from interfering OFDM signal but at the cost of a small spectral efficiency loss and a significant complexity increase. This can be seen in Fig. 9 where, among all results shown, the hybrid approach W(24)-PMix(6)-OFDM for the SU has the best ACI rejection performance when TO = 200.

4.3 OOBE suppression and ACI rejection by a SU (with respect to other SU, with FO and TO)

In this scenario, there are two transmit-receive pairs of SUs operating in adjacent (non-overlapping) frequency bands with the same bandwidth of 4.5 MHz. For convenience, we assume that all SUs use the same OOBE/ACI suppression approach with the parameters defined in Table 2. The frequency separation of the two bands is 1.5 times of subcarrier spacing where the 0.5 implies there is FO between the two pairs. Two cases are investigated. In the first case, TO = 0, the two pairs of SUs are perfectly synchronous in time. In the second case, TO = 200, there is a 200-sample delay between the two pairs of SUs.

We will evaluate the BER performances of one of the two SU receivers to study the combined effects of OOBE suppression and ACI rejection for various approaches. Among the two SU transmitters, one is chosen to be as desired and the other as unwanted (i.e., an interferer). We assume that the transmit power of the interferer is 10 dB higher than that of the desired SU transmitter. Any of the approaches discussed in Section 2 can be employed.

Figure 10 shows the BER performances of the desired SU pair with various approaches when TO = 0 and 200.

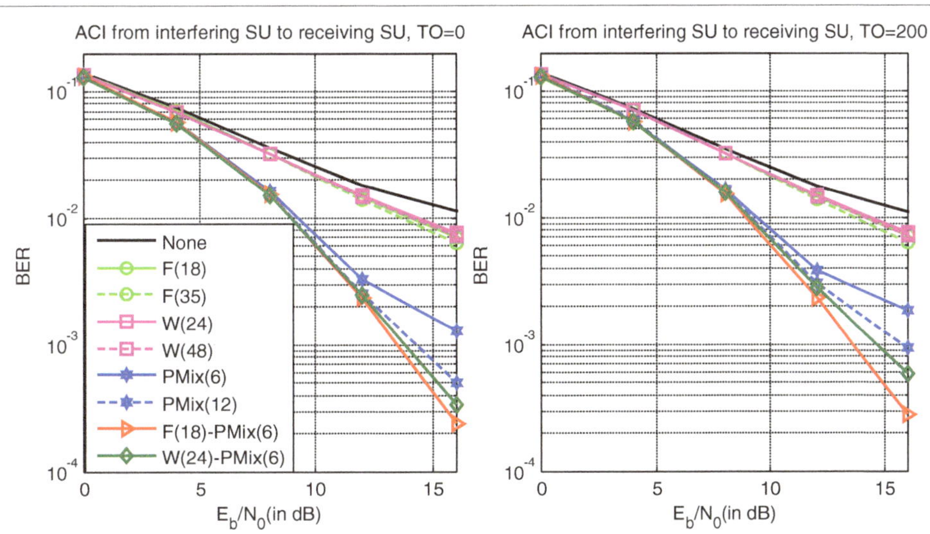

Fig. 10 BER performance of the SU receiver accounting for rejection of the ACI and suppression of OOBE by an interfering SU transmitter. The meanings of the legends are the same as those in Fig. 4

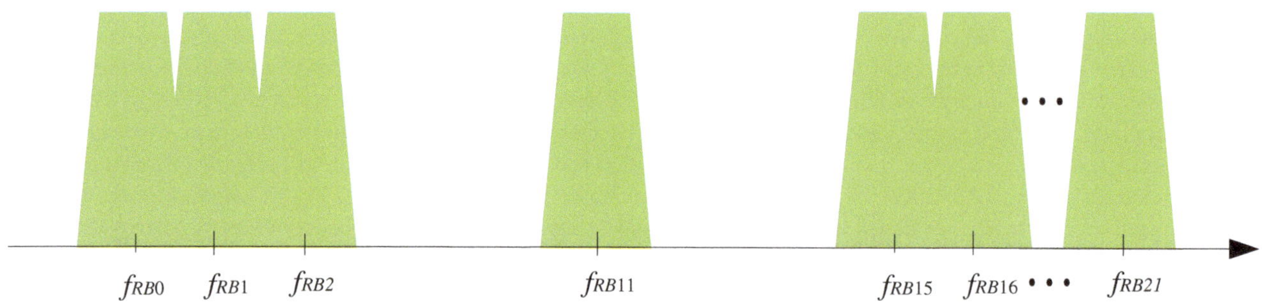

Fig. 11 Non-contiguous spectrum divided into RBs

Among all, None is the worst because nothing is done to suppress OOBE or reject ACI; F(18)-PMix(6)-OFDM is the best but at the cost of spectral efficiency loss. For W-OFDM, W(24) and W(48) have the same BER performance. For F-OFDM, that with a longer filter length has a better performance. As we have designed, W(24)-PMix(6)-OFDM and F(18)-PMix(6)-OFDM both achieve better performance than W(24)-OFDM, W(48)-OFDM, F(18)-OFDM, F(35)-OFDM, PMix(6)-OFDM, and PMix(12)-OFDM.

When TO is changed from 0 to 200, there is very little degradation on the performances of None, F-OFDM, and W-OFDM. On the contrary, the degradations on the performance of PMix-OFDM are significant. For the hybrid approaches, the degradations are visible but not significant.

Comparing the BERs of the corresponding approaches in Figs. 9 and 10, the BERs in Fig. 10 (for two pairs of SUs) are smaller than those in Fig. 9 (for one pair of PU and one pair of SUs). This is due to the fact that the transmit interferer in Fig. 10 (i.e., a SU) performs OOBE suppression but the transmit interferer (i.e., a PU) in Fig. 9 does not perform OOBE suppression.

5 Conclusions

If the signaling structures are not known, windowing and filtering are the two primary designing principles for the conventional OOBE suppression and ACI rejection approaches. However, when the signaling structures are known, new designs can be added into the conventional designs. For OFDM cognitive radio systems, precoding approaches can be employed to exploit the OFDM signaling structures before IFFT at the transmitter and after FFT at the receiver. In addition, a lot of non-trivial considerations need to be factored into the design. For example, the removal of CP at the receiver causes ACI regeneration if the truncation of interference signal due to the CP removal process is not properly taken into consideration by the adopted ACI rejection approach. On the other hand, effective OOBE is much larger than the estimated OOBE if there is a timing offset (TO) between PU and SU. Thus, among the two conventional approaches, windowing is more appropriate for ACI rejection since the windowing process is carried out on the portion that goes through the FFT and proper compensation for the discarded portion has been made, and filtering is more appropriate for OOBE suppression since it is less sensitive to TO.

In this paper, both time-domain (before FFT at the receiver and after IFFT at the transmitter) and frequency-domain (after FFT at the receiver and before IFFT at the transmitter) OOBE suppression and ACI rejection approaches for OFDM cognitive radio systems are examined

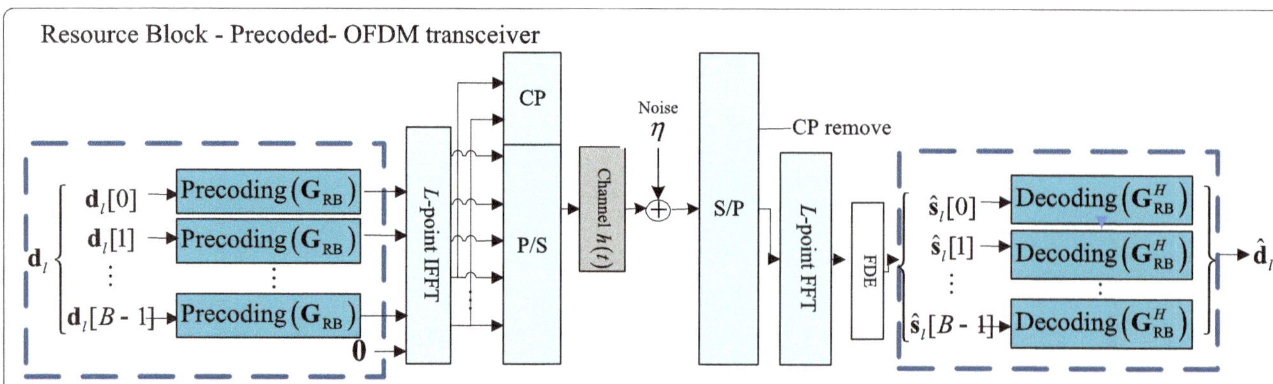

Fig. 12 Precoding and decoding of RB-P-OFDM

in details. Through the evaluation of the OOBE suppression (a SU transmitter to suppress its OOBE) and ACI rejection (a SU receiver to reject ACI caused by an interfering PU or SU transmitter) performances of all these approaches and their possible combinations under practical conditions (broadband channels with radio impairments such as FO and TO), we see that all of them can be used at the SU transmitter to minimize OOBE or at the SU receiver to reject the ACI. However, each of the existing approaches has some merits and some shortcomings, none of them alone can deal with all practical challenges (such as long multipath delay spread, finite CP length, requirements of low PAPR, low BER, low complexity and high spectral efficiency, robustness with respect to FO and TO, ability to adapt to non-contiguous and dynamically changing spectrum, etc.) alone. The two time-domain approaches, filtering and windowing, are proved to be efficient approaches to reduce OOBE at the SU transmitter and reject ACI at the OFDM-based SU receiver, respectively, because of their robustness with respect to TO. However, their performance will be limited by the CP length. The developed hybrid precoding approach PMix, which combines spectral precoding and PAPR precoding in the frequency domain, can provide satisfactory performances on OOBE suppression, ACI rejection, PAPR reduction, low BER and robustness with respect to FO, etc., but with the drawbacks of spectral efficiency loss, high complexity (when the available spectrum is broadband, non-contiguous, or time-varying), and being sensitive to TO. Fortunately, the complexity of PMix can be reduced with the resource block scheme introduced in our previous work. Furthermore, to reduce spectral efficiency loss and TO sensitivity of PMix, the idea of combining the frequency-domain PMix with a time-domain approach (filtering or windowing) is presented in details in this paper with simulation results. The combination of PMix with filtering or windowing can mitigate not only the shortcomings of PMix (i.e., high complexity, spectral efficiency loss, and performance loss due to TO) but also the shortcomings of filtering and windowing (i.e., sensitivity to delay spread). Simulation results show that F-PMix-OFDM provides the best OOBE suppression and W-PMix-OFDM provides the best ACI rejection characteristics in our numerical examples.

6 Appendix

6.1 Resource block scheme

In a multicarrier modulation system, a RB consists of a fixed number of contiguous subcarriers, which is designated to be the smallest dynamically available transmission band. All the assigned transmission bands or opportunistically detected frequency bands are consisting of one or multiple RBs. As depicted in Fig. 11, a non-contiguous spectrum may be divided into RBs of the same size. Based on the concept of RB, it is possible to remove the dependence of precoding approach on the allocated spectrum.

The main idea of RB-P-OFDM is to precode *all* RB by a predetermined precoding, $\boldsymbol{G}_{\mathrm{RB}}$, independently. Once N_{B}, the size of each RB, the spacing between the central frequency of each RB, and the notched frequencies, f_i, $i = 0, 1,.., K-1$, are determined, the matrix $\boldsymbol{P}_{\mathrm{RB}}$ is fixed.

$$\boldsymbol{P}_{\mathrm{RB}} = \begin{bmatrix} P_0(f_0) & \cdots & P_{N_{\mathrm{B}}-1}(f_0) \\ \vdots & \ddots & \vdots \\ P_0(f_{K-1}) & \cdots & P_{N_{\mathrm{B}}-1}(f_{K-1}) \end{bmatrix} \tag{32}$$

Then, a unique precoding matrix $\boldsymbol{G}_{\mathrm{RB}}$ with coding redundancy R_{B} can be obtained by (24). In addition, since the dimension of $\boldsymbol{G}_{\mathrm{RB}}$ is decreased from N to N_{B}, the computational complexity of the precoding and decoding operations is greatly reduced.

A block diagram of RB-P-OFDM is shown in Fig. 12. At the transmitter, the data of each RB, $\boldsymbol{d}_l[k]$, $k \in \mho$, is precoded by $\boldsymbol{G}_{\mathrm{RB}}$ independently to form a per-RB precoded data $\boldsymbol{s}_l[k]$. Then, all the precoded data $\boldsymbol{s}_l[k]$ instead of $\boldsymbol{d}_l[k]$ are modulated by the assigned subcarriers. At the receiver, the demodulated data of each RB, $\hat{\boldsymbol{s}}_l[k]$, is decoded independently by $\boldsymbol{G}_{\mathrm{RB}}^H$ to recover the per-RB transmitted data $\hat{\boldsymbol{d}}_l[k]$. The extension of the RB approach to the W-P-OFDM is straightforward. The extension to the F-P-OFDM is a bit involved but can be done by applying the RB approach to the filtering process as well. However, the RB approach may degrade the spectrum containment unless the redundancy order is increased (which increases the spectral efficiency loss) as discussed in Section 3.2.

Competing interests

The authors declare that they have no competing interests.

References

1. J Mitola, GQ Maguire Jr, Cognitive radio: making software radios more personal. IEEE Personal Comm. **6**(4), 13–18 (1999)
2. TA Weiss, FK Jondral, Spectrum pooling: an innovative strategy for the enhancement of spectrum efficiency. IEEE Comm. Mag. **42**(3), 8–14 (2004)
3. F Meucci, O Cabral, FJ Velez, A Mihovska, NR Prasad, Spectrum aggregation with multi-band user allocation over two frequency bands, in *Proc. IEEE Mobile WiMAX Symp*, 2009, pp. 81–86
4. Y Wu, WY Zou, Orthogonal frequency division multiplexing: a multi-carrier modulation scheme. IEEE Trans. Consum. Electron. **41**(3), 392–399 (1995)
5. S Alamouti, A simple transmit diversity technique for wireless communications. IEEE J. Sel. Areas Comm. **16**(8), 1451–1458 (1998)
6. C Cordeiro, K Challapali, D Birru, N Sai Shankar, IEEE 802.22: an introduction to the first wireless standard based on cognitive radios. J. Commun. **1**, 1 (2006)
7. A Sahin, H Arslan, Edge windowing for OFDM based systems. IEEE Commun. Lett. **15**(11), 1208–1211 (2011)
8. D. Noguet, M. Gautier, V. Berg, Advances in opportunistic radio technologies for TVWS, EURASIP Journal on Wireless Communications and Networking, Nov. 2011, http://jwcn.eurasipjournals.com/content/2011/1/170

9. I Cosovic, S Brandes, M Schnell, Subcarrier weighting: a method for sidelobe suppression in OFDM systems. IEEE Commun. Lett. **10**(6), 444–446 (2006)
10. S Brandes, I Cosovic, M Schnell, *Sidelobe suppression in OFDM systems by insertion of cancellation carriers*. IEEE 62nd Vehicular Technology Conference (VTC), 2005, pp. 152–156
11. X Fu, J Wang, SQ Li, *Sidelobe suppression for OFDM based cognitive radio systems*. 4th International Conference on Communications and Networking in China (ChinaCOM), 2009, pp. 1–5
12. CD Chung, Spectrally precoded OFDM. IEEE Trans. Comm. **54**(12), 2173–2185 (2006)
13. J Van de Beek, Sculpting the multicarrier spectrum: a novel projection precoder. IEEE Comm. Lett **13**(12), 881–883 (2009)
14. J Li, E Bala, R Yang, Resource block filtered-OFDM for future spectrally agile and power efficient systems. Phys. Commun **11**, 36–55 (2014)
15. E Bala, J Li, R Yang, Shaping spectral leakage: a novel low-complexity transceiver architecture for cognitive radio. IEEE Vehicular Magazine **8**(3), 38–46 (2013)
16. J Fang, L I-Tai, Precoder designs for jointly suppressing out-of-band emission and peak-to-average power ratio in an orthogonal frequency division multiplexing system. IET Commun. **8**(10), 1705–1713 (2014)
17. TS 36.211 Evolved Universal Terrestrial Radio Access (E-UTRA); Physical channels and modulation; Release 9, 2010
18. FCC 12–36, Third Memorandum Opinion and Order in the TV white spaces proceeding, ET Docket Nos. 04–186 and 02–380, April 4, 2012.
19. J Van de Beek, Orthogonal multiplexing in a subspace of frequency well-localized signals. IEEE Comm. Lett **14**(10), 882–884 (2010)
20. SB Slimane, Reducing the peak-to-average power ratio of OFDM signals through precoding. IEEE Trans. Veh. Tech. **56**(2), 686–695 (2007)
21. I Baig, V Jeoti, *PAPR reduction in OFDM systems: Zadoff-Chu matrix transform based pre/post-coding techniques* (Second International Conference on Computational Intelligence, Communication Systems and Networks (CICSyN), 2010), pp. 373–377
22. HG Myung, L Junsung, DJ Goodman, Single carrier FDMA for uplink wireless transmission. IEEE Veh. Technol. Mag. **1**, 30–38 (2006)
23. HS Malvar, *Signal processing with lapped transforms*. 357 pages, Artech House, 1992 ISBN 0890064679
24. D Noguet, M Gautier, V Berg, Advances in opportunistic radio technologies for TVWS. EURASIP J. Wireless Commun. Networking **170**, 1–12 (2011)
25. J Fang, Z You, I-T Lu, J Li, R Yang, *Resource block based precoding schemes for suppressing out-of-band emission*. Military Communications Conference (MILCOM), 2013, pp. 1570–1575

15

Network-coded primary-secondary cooperation in OFDM-based cognitive multicast networks

Zhihui Liu, Wenjun Xu*, Shengyu Li and Jiaru Lin

Abstract
This paper investigates resource allocation for network-coded primary-secondary cooperation in orthogonal frequency division multiplexing (OFDM)-based cognitive multicast networks, in which cognitive radio (CR), network coding (NC), multicast, and OFDM are effectively combined toward the spectrum efficient fifth generation (5G) wireless communication systems. Different from the conventional one-to-one/one-to-more primary-secondary cooperation, the paper concentrates on two-to-one primary-secondary cooperation, in which one secondary user (SU) cooperates with two primary users (PUs) to gain more spectrum access opportunities. To accomplish the cooperation, an agreement is established between the SU and PUs. Meanwhile, to alleviate spectrum bands consumed for PUs' data transmission, network coding is employed at the SU transmitter. Subject to per-primary-user minimum rate requirement and the total power budget at the secondary transmitter, the investigated primary-secondary cooperation strategy is formulated as a mixed integer optimization problem with the aim of maximizing the average multicast transmission rate. The formulated problem is non-convex and difficult to solve directly. In this paper, a stepwise optimization algorithm is proposed in which the subcarrier assignment and power allocation are executed separately to reduce the computation complexity. The simulation results show that compared to existing schemes, the achieved secondary multicast transmission rate in the proposed scheme is greatly improved. The presented network-coded primary-secondary cooperation is a promising paradigm to improve the spectrum efficiency for the future 5G systems.

Keywords: Cognitive multicast networks (CMNs); Primary-secondary cooperation; Orthogonal frequency division multiplexing (OFDM); Network coding (NC)

1 Introduction

The dramatic growth of mobile data services driven by wireless Internet and smart devices has triggered the investigation of fifth generation (5G) for the next generation of terrestrial mobile telecommunications [1]. Facing great challenges of future mobile networks, the essential requirements for 5G which mainly include higher traffic volume, spectrum, energy, and cost efficiency are pointed out. Therein, cognitive radio (CR) technology, which provides the authorized spectrum of primary users (PUs) to various unlicensed users also called secondary users (SUs) in an opportunistic (time-limited), interference-limited, or paid way [2], handles flexibly the predicament of spectrum underutilization and spectrum shortage resulting from the surging wireless requirements and applications and, thus, has been regarded as the inevitable option for 5G to improve spectrum efficiency [3,4]. Particularly, cognitive cooperation, not only allowing SUs in cognitive radio networks (CRNs) to share authorized spectrum but also inheriting the unique advantages of cooperative communications that promise significant capacity and multiplexing gain increase through distributed transmission/processing, has been becoming an appealing communication paradigm [5,6].

Meanwhile, due to its high spectrum utilization, multicast transmission has become an indispensable part of mobile communication systems nowadays [7]. In this paper, cognitive cooperation and multicast are joint

*Correspondence: wjxu@bupt.edu.cn
Key Lab of Universal Wireless Communications, Ministry of Education, Beijing University of Posts and Telecommunications, Xitucheng Road 10, Haidian District, 100876 Beijing, China

considered. For the primary-secondary cooperation in cognitive multicast networks (CMNs), the secondary source (SS) with limited transmit power needs to broadcast message to multiple secondary destinations (SDs), and hence, the transmission data rate is confined to the worst channel condition among all SDs. Thus, the quality of service (QoS) of the SU suffers severely, and the spectrum accessed by the SU might not be able to afford satisfactory communication services for the SU. One effective protection countermeasure is that the SU assists simultaneously multiple PUs to gain more spectrum access opportunities. Moreover, the SU turns to spend least power and spectrum on transmission data for PUs and scrambles to save resources any way for multicast members. Wireless network coding (NC), which mixes the data from different sessions before signal forwarding to increase per-transmission information content, has been a promising approach [8,9]. Motivated by all these profits, NC technique is adopted by the SU. Furthermore, to enhance the spectrum efficiency, orthogonal frequency division multiplexing (OFDM) [10] is considered in this paper. Combining these technologies mentioned above, this paper investigates the resource allocation strategy for the one-secondary-user and two-primary-user (1S2P) cooperation with NC in OFDM modulated CMNs.

In this paper, first of all, the essential conditions for 1S2P cooperation is analyzed. Secondly, PU 1 and PU 2 separately commit to listen for signal from each other and SU, in exchange for safeguarding their minimum transmission rate requirements. Meanwhile, the SU can access both PUs' authorized spectrum on the premise that the SU assists PUs in their data transmission with corresponding minimum rate requirements for PUs. Thirdly, the SU switches adaptively between NC mode and decode-and-forward (DF) mode and performs subcarrier assignment and power allocation to maximize its own benefits. Fourthly, the cooperation strategy problem for the SU is formulated as a mixed integer optimization problem, in which the subcarrier assignment and power allocation for NC-based data forwarding, DF-based data forwarding, and SU's data transmission are tightly coupled and hard to search the optimal solution. And finally, we present a stepwise subcarrier and power optimization algorithm with a low computational complexity.

The rest of this paper is organized as follows: in Section 2, we introduce the related work about cognitive cooperation and summarize the current work related to network coding-based cooperation in cognitive radio system. In Section 3, the system model and 1S2P cooperation are introduced. In Section 4, formulation for the profit-driven 1S2P cooperation strategy problem is presented. The stepwise optimization algorithm is derived in Section 5. Section 6 lists evaluation results and Section 7 concludes our study.

2 Related works

Most researches focus on non-NC cognitive cooperation [11], which from the perspective of profit-driven collaborators, can fall into three categories: selfless relay cooperation (SLRC), profit-driven equal-priority user cooperation (PEPC), and profit-driven unequal-priority primary-secondary user cooperation (PUPSC). In the SLRC, one or more relays put all resources into assisting PUs or SUs to forward data [12-15]. This kind of research is established on the assumption that all relays are selfless. However, in the future cooperation networks, the relay nodes deployed probably by different service providers or individuals have self-serving features and heterogeneous resource requirements. Hence, effective profit-driven mechanism becomes an inevitable development trend for cognitive cooperative transmission. In the PEPC, cognitive users with equal priority collaborate with each other to share resources such as spectrum sensing information and antennas to decrease inter-cell interference and the interference to PUs [16-19]. On the other hand, in the PUPSC, the SUs assists PUs to forward data in exchange for obtaining more spectrum access opportunities to achieve the most benefit, such as transmission rate maximization and transmit power minimization [19-21].

In the current literature, the researches on NC in cognitive cooperation spring up for its satisfactory performance gain. Comparatively speaking, there are a few researches on network-coded cognitive cooperation which mainly focus on non-primary-secondary cooperation [22,23]. In [22], Jin et al. present an optimization framework for multicast scheduling in CRNs, where secondary base station opportunistically utilizes 'spectrum holes' to multicast data to SUs and SUs help with each other with network coding in a local neighborhood to reduce overhead and perform error control and recovery. Chun et al. in [23] consider network-coded cooperation for cognitive relay networks, in which SUs communicate to the secondary destination through multiple relay nodes in underlay way while some relay nodes generate the network-encoded packet using linear combination.

At last, researches related to cognitive primary-secondary cooperation can be classified into the following two sides: on one hand, most studies in the literature focus on one-secondary-user and one-primary-user cooperation or multi-secondary-user and one-primary-user cooperation model [24,25], where one PU selects one or multiple SUs to assist it in data communication, while one-secondary-user and multi-primary-user cooperation, in which one SU assists two or more primary transmissions simultaneously, can provide more spectrum access opportunities for the SU. On the other hand, there are only a few researches on NC-based cognitive primary-secondary cooperation. Zou et al. [26,27] study resource allocation problem at the PU side in NC-based cognitive cooperative networks, in which PUs assist in the transmission of SUs, and in exchange for this concession, PUs receive payments from SUs for the spectrum and cooperative transmit power being used in cooperation. In their study, one PT combines together its own data sent in the first phase and the data overheard from the secondary transmitter in the second phase and sends the additive data out with network coding. Then, both PR and secondary receiver extract their desired data from the combined data by subtracting the data they overheard.

Based on the research [28], which demonstrates that the NC noise received at each session's destinations increases along with the number of sessions increasing, it is unfavorable to employ NC blindly in cooperation. In our previous work [29], the power and spectrum allocation at the SU side is researched for network-coded 1S2P cooperation in CRNs. However, to the best of our knowledge, there are few literature concentrating on profit-driven PUPSC strategy problem at the SU side with NC in OFDM-modulated CMNs. Therefore, in this paper, 1S2P cooperation for network-coded OFDM-modulated CMNs will be studied in which one SU assists the data transmission of two PUs and utilizes NC to save the spectrum resource used for PUs' data transmission.

Compared with the recent researches, the main contributions of the paper are summarized as follows.

1) Essential condition and content of agreement for network-coded 1S2P cooperation is analyzed. It contains objective environmental factors of the SU and two PUs, their respective transmission requirements and resource constraints, and their respective responsibilities that should be performed.
2) The resource allocation strategy for network-coded 1S2P cooperation is modeled as a mixed integer optimization problem. The problem mainly covers adaptive cooperative mode switch between NC and DF, the assignment of each subcarrier, and the power allocation on each subcarrier.
3) The stepwise subcarrier and power allocation optimization algorithm is derived for the formulated problem, and extensive simulations verify the effectiveness and superiority of the proposed cooperation scheme compared with existing schemes.

3 System model and analysis for cooperation

3.1 System model

Consider an OFDM-modulated cognitive multicast cooperation network with one SU and two PUs, i.e., PU 1 and PU 2. As shown in Figure 1, the SU in the secondary network is composed of a SS and G secondary multicast groups (SMGs) $\mathcal{K}_g$, $g \in \mathcal{G} = \{1, \cdots, G\}$ with K_g SDs. The PU $i, i = 1, 2$ in the primary network is composed of a primary transmitter (PT) i and a primary receiver (PR) i. The OFDM technology is adopted in the system to facilitate fine-grained spectrum management. The whole spectrum band is divided into multiple subcarriers (SCs), and the occupied SC set of PT i is denoted as Ω_i. The transmit power on the SC l at the PT i is $P^l_{\mathrm{PT}i}, l \in \Omega_i$, and the minimum transmission rate requirement at the PR i is r_i. Assume that SC set 1 and SC set 2 have different elements, i.e., $\Omega_1 \cap \Omega_2 = \emptyset$.

Focusing on network-coded 1S2P cooperation in OFDM-modulated cognitive multicast network, as shown in Figure 1, two PTs that intend to send data to their corresponding PRs cannot communicate directly and ask for assistance from the SU.

Meanwhile, the SS which assists the data transmission of these two PUs can access their licensed spectrum bands and multicasts data to its different SMGs at different rates. To assist the traffic of PT 1 and PT 2 with fewer spectrum bands and save more for its own multicast transmission, the SS acts as a relay to assist PTs' transmissions by the NC and DF relaying protocol. Thus, for the sake of favorable cooperation, PU 1 (PT 1→PR 1), PU 2 (PT 2→PR 2), and SU (SS→SDs) reach a tripartite cooperation agreement: 1) The SD assists both PUs in data transmission. As a reward, SS can access both the spectrum bands of PUs and reallocate it for both the data relaying of PUs and its own multicast data delivery. 2) The PU 1(2) overhears the data of PU 2(1) as auxiliary information for decoding coded data from SS and both PRs are able to operate and receive signals on the whole spectrum bands.

Remark. In the network-coded 1S2P cooperation scheme, for simplicity, we focus on the scenario of no direct link from each PT to its corresponding PR, i.e., both PTs cannot communicate directly with their corresponding PRs and the direct transmission between each PT and its corresponding PR is not available [30-32]. The case with direct links between the PT and its corresponding PR involves many complicated problems such as the maximum ratio combination of network-coded data and direct transmitted data and is beyond the focus of the paper. In addition, due to the long distance or the shielding effect caused by some barrier between each PT and its corresponding PR, the PR is not within the communication range of the PT, and thus, the assumption that there is no direct link between the source and destination is reasonable for some practical applications [33-35]. □

Before analyzing the network-coded 1S2P cooperation and resource allocation for the SU, we first introduce the essential condition for cooperation among two PUs and one SU in OFDM-modulated CMNs.

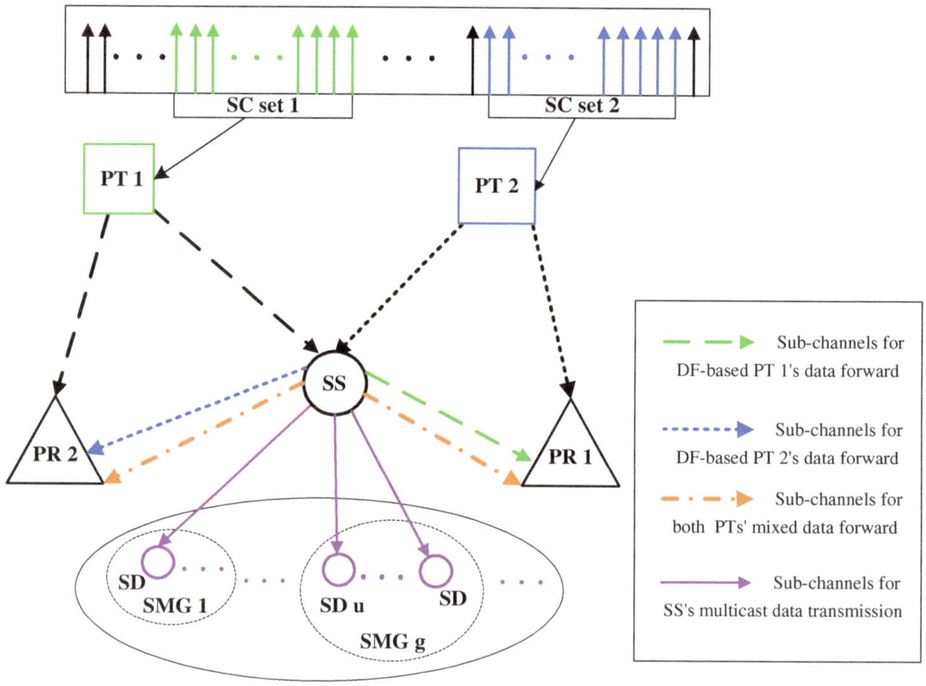

Figure 1 System model for network-coded primary-secondary cooperation in OFDM-based CMNs.

3.2 Essential condition of network-coded 1S2P cooperation

In order to achieve cooperation among PU 1, PU 2, and SU, some essential conditions should be satisfied.

On one hand, the SS must be located within both the transmission ranges of PT 1 and PT 2. Otherwise, as depicted in Figure 2a, if the location of SS is not within PT 2's transmission range, then the SS cannot receive the signal of PT 2 and assist the data transmission of PT 2. On the other hand, each PR should be located within the transmission range of its cooperative PU member for the decoding of its own data from the coded data transmitted by the SU. If one PR, for example, PR 2 shown in Figure 2b, is not located within the transmission range of PT 1, PR 2 cannot listen the data from PT 1 and decode the needed data after it receives the mixed data of PT 1 and PT 2 transmitted by the SS.

Only when both these sides are satisfied, as shown in Figure 2c, their own benefits of the three members for network-coded 1S2P cooperation can be guaranteed, and the network-coded 1S2P cooperation among them can be meaningful. Therefore, the essential conditions for the network-coded 1S2P cooperation are summarized as follows:

$\langle 1 \rangle$ $\mathbb{L}(\text{SS}) \in \Upsilon(\text{PT } 1) \bigcap \Upsilon(\text{PT } 2)$;
$\langle 2 \rangle$ $\mathbb{L}(\text{PR } 2) \in \Upsilon(\text{PT } 1)$, and $\mathbb{L}(\text{PR } 1) \in \Upsilon(\text{PT } 2)$.

where $\Upsilon(\text{PT } i)$ $(i = 1, 2)$ denotes the transmission range of PT i and $\mathbb{L}(\text{X})$ denotes the location of X.

4 Problem formulation for network-coded 1S2P cooperation

Concentrating on the resource allocation problem at the SS in network-coded 1S2P cooperation, the SS reallocates SCs in the set $\mathcal{S} = \Omega_1 \cup \Omega_2$ into $G + 3$ SC sets $\mathcal{S}_k$, $k \in \Pi = \{1, \cdots, G+3\}$, as depicted in Figure 3. The SCs in set $\mathcal{S}_k$, $k = 1, \cdots, G$ are used for broadcasting SU's data to $\mathcal{K}_g$, $g = 1, \cdots, G$. The SCs in set $\mathcal{S}_{G+i}$, $i = 1, 2$ are used for assisting transmission of PT i's data using DF protocol. The SCs in set $\mathcal{S}_{G+3}$ are used for forwarding transmission of the XOR-ed data from PT 1 and PT 2. Next, we start with the cooperation transmission analysis for network-coded 1S2P cooperation.

4.1 Cooperation transmission analysis

As sketched in Figure 3, two time slots are applied in each cooperation transmission. In the first time slot, PTs broadcast their data, respectively, and hence, the time slot is called as primary user broadcast phase. In the second time slot, the SS communicates with PRs and multicasts data to SMGs, and hence, the time slot is also referred as secondary user relay and multicast phase.

4.1.1 Primary user broadcast phase

In the broadcast phase, two PTs broadcasts their symbols simultaneously. Specially, PT $i, i = 1, 2$ sends signals on subcarrier l in SC set Ω_i. The received signals at the SS and PR i can be expressed, respectively, as

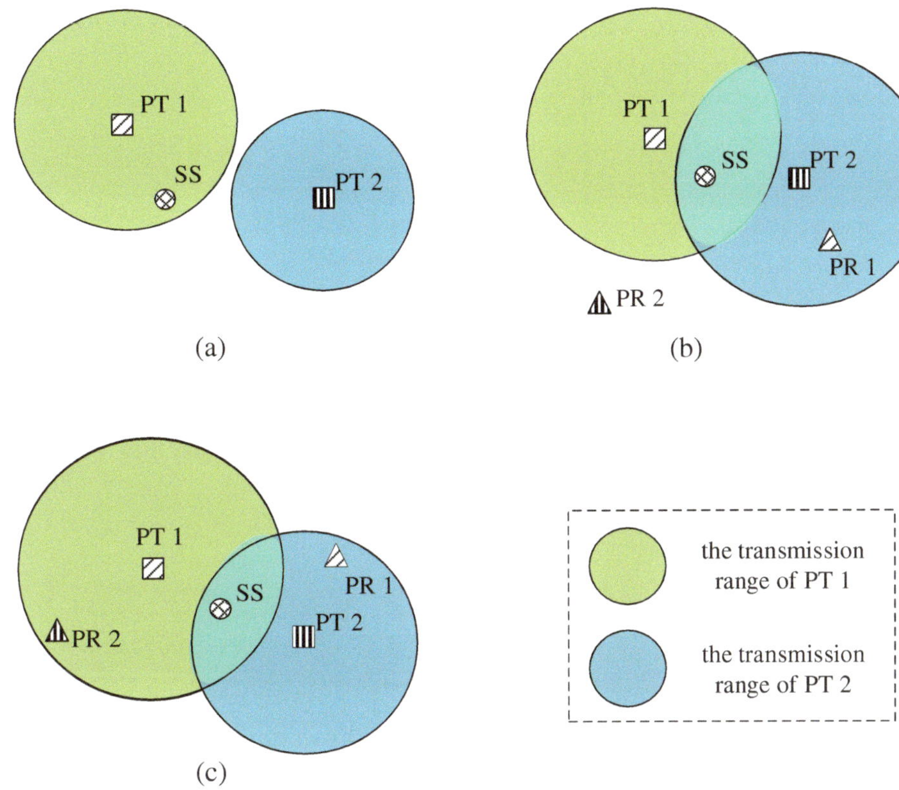

Figure 2 Location-based necessary condition sketch for tripartite cooperation. **(a)** Node SS is not located within the intersection transmission range of PT 1 and PT 2. **(b)** One primary receiver is not located within the transmission range of its cooperative primary member. **(c)** Both conditions are satisfied and necessary condition of 1S2P cooperation for further progress is guaranteed.

$$\begin{aligned} y^l_{\mathrm{PT}i\to\mathrm{SS}} &= \sqrt{p^l_{\mathrm{PT}i}}h^l_{\mathrm{PT}i,\mathrm{SS}}x^l_{\mathrm{PT}i} + n_{\mathrm{SS}}, \\ \forall\, i &= 1,2, l \in \Omega_i \end{aligned} \tag{1}$$

$$\begin{aligned} y^l_{\mathrm{PT}i\to\mathrm{PR}i'} &= \sqrt{p^l_{\mathrm{PT}i}}h^l_{\mathrm{PT}i,\mathrm{PR}i'}x^l_{\mathrm{PT}i} + n_{\mathrm{PR}i'}, \\ \forall\, i,i' &= 1,2, i \neq i', l \in \Omega_i \end{aligned} \tag{2}$$

where $x^l_{\mathrm{PT}i}$ is the transmitted symbol with unit power from PT i on subcarrier l, $p^l_{\mathrm{PT}i}$ is the transmit power for PT i on subcarrier l, and n_{SS} and $n_{\mathrm{PR}i'}$ are the additive white Gaussian noise at the SS, PR i', respectively. We assume that $n_{\mathrm{SS}}, n_{\mathrm{PR}i} \sim \mathcal{CN}(0, N_0)$. $h^l_{\mathrm{PT}i,\mathrm{SS}}$ $\left(h^l_{\mathrm{PT}i,\mathrm{PR}i'}\right)$ denotes the channel coefficients from PT i to the SS (PR i') on the subcarrier l.

The transmission rate of the first hop on the subcarrier l for PT i, $i = 1, 2$, $l \in \Omega_i$ is

$$\begin{aligned} C^l_{\mathrm{p},i} = \min &\left\{ \frac{1}{2}\Delta f \log_2\left(1 + \frac{p^l_{\mathrm{PT}i}h^l_{\mathrm{PT}i,\mathrm{SS}}}{N_0\Delta f}\right), \right. \\ &\left. \frac{1}{2}\Delta f \log_2\left(1 + \frac{p^l_{\mathrm{PT}i}h^l_{\mathrm{PT}i,\mathrm{PR}i'}}{N_0\Delta f}\right) \right\}, \\ \forall\, i' &= 1,2,\ i' \neq i \end{aligned} \tag{3}$$

where the spanning bandwidth of each subcarrier is Δf Hz, and coefficient $\frac{1}{2}$ is derived from the time-division transmission as depicted in Figure 3.

4.1.2 *Secondary user relay and multicast phase*

In the secondary user relay and multicast phase, the SU reallocates SCs in $\mathcal{S}$ into SC sets $\mathcal{S}_k$ $(k \in \Pi)$. Therein, the SC sets $\mathcal{S}_k$ $(k \in \{1, \cdots, G\})$ are used for transmitting its own multicast data to SDs belonging to SMGs $\mathcal{K}_k$, respectively. The SC sets $\mathcal{S}_k$ $(k = G + i, i \in \{1, 2\})$ are allocated for transmitting the received signal from PT i using DF protocol, respectively. The SC set $\mathcal{S}_k$ $(k = G+3)$ is assigned for transmitting the XOR-ed data of received data from both PTs. Define the indicator functions δ^l_k that

$$\text{if } l \in \mathcal{S}_k, \text{ then } \delta^l_k = 1, \text{ and } \delta^l_{k'} = 0, \forall k' \in \Pi, k' \neq k \tag{4}$$

Accordingly, the received signals at SDs and PRs on each SC can be expressed, respectively, as

$$\begin{aligned} y^l_{\mathrm{SS},\mathrm{SD}u} &= \sqrt{P^l_{\mathrm{s}}}h^l_{\mathrm{SS},\mathrm{SD}u}x^l_{\mathrm{SS}} + n_{\mathrm{SD}u}, \\ &\forall\, l \in \mathcal{S}, u \in \mathcal{K}_g, g \in \mathcal{G}, \delta^l_g = 1 \end{aligned} \tag{5}$$

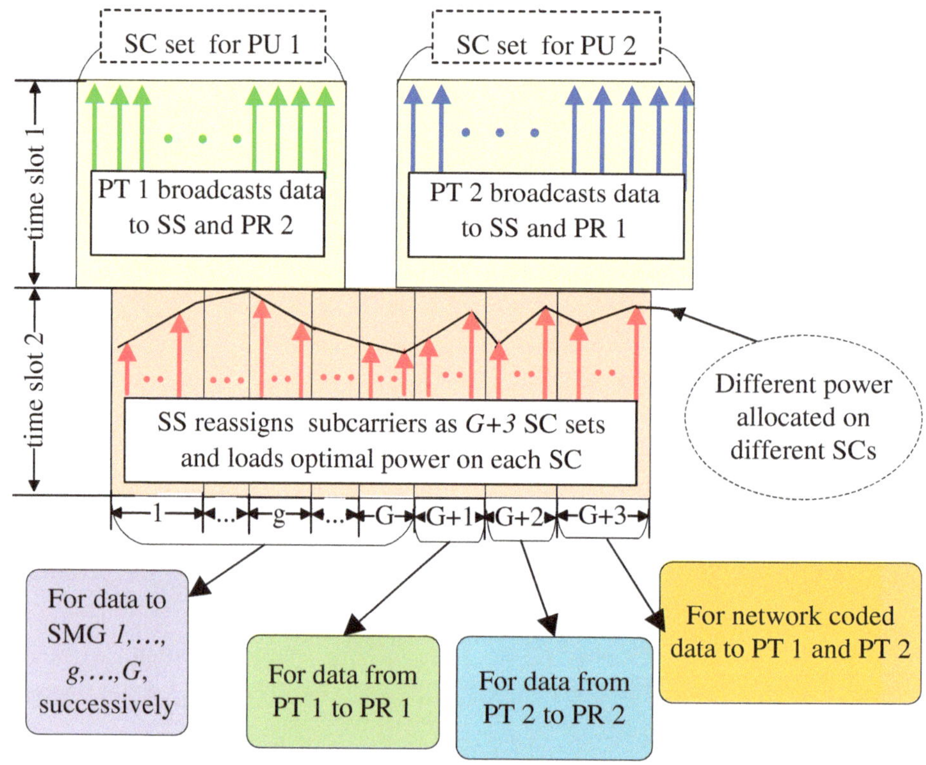

Figure 3 Resource allocation schematic for the OFDM-based network-coded 1S2P cooperation.

$$y^l_{\mathrm{SS,PR}i} = \begin{cases} \sqrt{P^l_{\mathrm{s}}}\, h^l_{\mathrm{SS,PR}i} \tilde{x}^l_{\mathrm{PT}i} + n_{\mathrm{PR}i}, \\ \quad \forall\, l \in \mathcal{S}, i \in \{1,2\}, \delta^l_{G+i} = 1 \\ \sqrt{P^l_{\mathrm{s}}}\, h^l_{\mathrm{SS,PR3}} \left(\tilde{x}^l_{\mathrm{PT1}} \oplus \tilde{x}^l_{\mathrm{PT2}}\right) + n_{\mathrm{PR}i}, \\ \quad \forall\, l \in \mathcal{S}, i \in \{1,2\}, \delta^l_{G+3} = 1 \end{cases} \tag{6}$$

where x^l_{SS} is the transmitted symbol with unit power by the SS on the SC l for SDs, $\tilde{x}^l_{\mathrm{PT}i}$ is the symbol from PT i transmitted by the SS with unit power on the SC l using DF protocol for PR i, P^l_{s} denotes the power loaded on the SC l by the SS, $h^l_{\mathrm{SS,PR}i}$ denotes the channel coefficients from the SS to the PR i on the SC l, $h^l_{\mathrm{SS,PR3}} = \min\left\{h^l_{\mathrm{SS,PR1}}, h^l_{\mathrm{SS,PR2}}\right\}$, $h^l_{\mathrm{SS,SD}u}$ is the channel coefficient from the SS to the SD u belonging to the SMG $\mathcal{K}_g$, we denote that $h^l_{\mathrm{SS,SG}g} = \min\left\{h^l_{\mathrm{SS,SD}u} | u \in \mathcal{K}_g\right\}$ which means the received signal at the SD of multicast group g is limited by the minimum channel gain among members of the group g, $n_{\mathrm{SD}u}$ is the additive white Gaussian noise at the SD u, and we also assume that $n_{\mathrm{SD}u} \sim \mathcal{CN}(0, N_0)$.

Denote that

$$\gamma^l_{\mathrm{ss},g} = \frac{h^l_{\mathrm{SS,SG}g}}{N_0 \Delta f},\ \forall\, g \in \mathcal{G} \text{ and } \gamma^l_{\mathrm{sp},j} = \frac{h^l_{\mathrm{SS,PR}j}}{N_0 \Delta f},\ \forall\, j \in \{1,2,3\} \tag{7}$$

Owing to signal transmission uniqueness for each subcarrier, mathematically, we have the following constraints $\mathcal{C}1$ and $\mathcal{C}2$.

$$\mathcal{C}1: \quad \sum\nolimits_{k \in \Pi} \delta^l_k = 1,\ \ \forall l \in \mathcal{S} \tag{8}$$

$$\mathcal{C}2: \ \delta^l_k \in \{0,1\},\ \ \forall l \in \mathcal{S}, k \in \Pi \tag{9}$$

4.2 Formulation for network-coded 1S2P cooperative resource allocation

According to the description above and network coding constraint that the receiver terminal needs enough information to decode its desired data by XORing the received data with its own data or auxiliary data [8,33,36], the achievable transmission rate for coded data at the second hop is limited by the minimum transmission rate R_3 of data transmitted by both PTs from the first hop, i.e.,

$$R_3 = \min\left\{\sum\nolimits_{l \in \Omega_1} C^l_{\mathrm{p},1}, \sum\nolimits_{l \in \Omega_2} C^l_{\mathrm{p},2}\right\} \tag{10}$$

and we have

$$\mathcal{C}3: \ \sum\nolimits_{l \in \mathcal{S}} \delta^l_{G+3} C^l_{\mathrm{sp},3}\left(\gamma^l_{\mathrm{sp},3}, P^l_{\mathrm{s}}\right) \le R_3 \tag{11}$$

where $C^l_{\mathrm{sp},3}\left(\gamma^l_{\mathrm{sp},3}, P^l_{\mathrm{s}}\right) = \frac{1}{2}\Delta f \log_2\left(1 + \gamma^l_{\mathrm{sp},3} P^l_{\mathrm{s}}\right)$ denotes the achievable transmission rate of network coded data transmitted by the SS on the subcarrier l if $l \in \mathcal{S}_{G+3}$.

If $l \in \mathcal{S}_{G+i}, i = 1,2$, the achievable transmission rate of data to the PU i transmitted by the SS according to DF protocol on the subcarrier l can be denoted as $C^l_{\text{sp},i}\left(\gamma^l_{\text{sp},i}, P^l_{\text{s}}\right) = \frac{1}{2}\Delta f \log_2\left(1+\gamma^l_{\text{sp},i}P^l_{\text{s}}\right)$. To achieve the cooperation agreement, the SS need to guarantee both PUs' minimum transmission rate requirements $r_i, i = 1,2$. Mathematically, the following constraint should be satisfied

$$\mathcal{C}4: \sum\nolimits_{l\in\mathcal{S}}\left(\delta^l_{G+3}C^l_{\text{sp},3}\left(\gamma^l_{\text{sp},3},P^l_{\text{s}}\right)+\delta^l_{G+i}C^l_{\text{sp},i}\left(\gamma^l_{\text{sp},i},P^l_{\text{s}}\right)\right) \geq r_i, \forall\, i=1,2 \tag{12}$$

It should be noticed that the total achieved transmission rate for PU i is limited by the transmission data rate R_i from PU i during the first hop, i.e., $\sum_{l\in\mathcal{S}}\left(\delta^l_{G+3}C^l_{\text{sp},3}\left(\gamma^l_{\text{sp},3},P^l_{\text{s}}\right)+\delta^l_{G+i}C^l_{\text{sp},i}\left(\gamma^l_{\text{sp},i},P^l_{\text{s}}\right)\right) \leq R_i$, $\forall\, i = 1,2$, where $R_i = \sum_{l\in\Omega_i} C^l_{\text{p},i}$. If $R_i < r_i$, then the conditions (12) cannot be satisfied, and hence the network-coded 1S2P cooperation cannot guarantee the minimum rate requirements of PRs, and the cooperation agreement cannot be achieved. Thus, cooperation members for the network-coded 1S2P cooperation should be reselected based on the essential conditions mentioned in Section 3.2 and the criterion $R_i \geq r_i$ for PU i involved in the collaboration. Considering the power budget P_{th} at the SS and the maximum power P_{high} that can be loaded on each SC, the power loaded on the SCs should be constrained as

$$\mathcal{C}5: \quad \sum\nolimits_{l\in\mathcal{S}} P^l_{\text{s}} \leq P_{\text{th}} \tag{13}$$

$$\mathcal{C}6: \; 0 \leq P^l_{\text{s}} \leq P_{\text{high}}, \forall\, l \in \mathcal{S} \tag{14}$$

The achievable transmission rate of data for the SMG $\mathcal{K}_g, g \in \mathcal{G}$ on the subcarrier l, if $l \in \mathcal{S}_g$, is $C^l_{\text{ss},g}\left(\gamma^l_{\text{ss},g}, P^l_{\text{s}}\right) = \frac{1}{2}\Delta f \log_2\left(1+\gamma^l_g P^l_{\text{s}}\right)$. To maximize the achieved transmission rate of the SS to all SDs, the network-coded 1S2P cooperative resource allocation problem can be formulated as the following model $\mathcal{OP}_1$.

$$\mathcal{OP}_1: \quad \max_{P^l_{\text{s}},\delta^l_k, l\in\mathcal{S}, k\in\Pi} \sum\nolimits_{g\in\mathcal{G}, l\in\mathcal{S}} K_g \delta^l_g C^l_{\text{ss},g}\left(\gamma^l_{\text{ss},g}, P^l_{\text{s}}\right) \quad s.t. \quad \mathcal{C}1 \sim \mathcal{C}6. \tag{15}$$

5 Subcarrier assignment and power allocation for network-coded 1S2P cooperation problem

In the section, the solution for problem $\mathcal{OP}_1$ is investigated. First of all, in order to facilitate problem solving, the reformulation of problem $\mathcal{OP}_1$ is implemented and the transformation method has been mentioned in our previous work [29]. It is found that in order to save power and SCs resource and to maximize the achievable transmission rate of the SS to all SDs, the SS would give priority to network-coded data forward transmission for satisfying the minimum rate requirements of both PRs. Therefore, the power and SCs allocated for network-coded data forward transmission must make full use of the given resource R_3, i.e.,

$$\mathcal{C}3': \sum\nolimits_{l\in\mathcal{S}} \delta^l_{G+3}C^l_{\text{sp},3}\left(\gamma^l_{\text{sp},3},P^l_{\text{s}}\right) \geq W_3, \tag{16}$$

where $W_3 = \min\{R_3, r_1, r_2\}$, which is to say that if $R_3 \geq \min\{r_i\}$, then $W_3 = \min\{r_i\}$; otherwise, $W_3 = R_3$.

Meanwhile, due to the different rate requirements and channel conditions for two PUs, the surplus rate requirements of PRs should be provided by the SS with respective cooperation with each PU. Thus, the constraints $\mathcal{C}4$ can be reformulated as

$$\mathcal{C}4': \sum\nolimits_{l\in\mathcal{S}} \delta^l_{G+i}C^l_{\text{sp},i}\left(\gamma^l_{\text{sp},i},P^l_{\text{s}}\right) \geq r_i - W_3,\; i=1,2. \tag{17}$$

Denote that $W_i = r_i - W_3, \forall\, i = 1,2$, the constrains $\mathcal{C}3' \sim \mathcal{C}4'$ can be jointly expressed as

$$\mathcal{C}3'': \sum\nolimits_{l\in\mathcal{S}} \delta^l_{G+j}C^l_{\text{sp},j}\left(\gamma^l_{\text{sp},j},P^l_{\text{s}}\right) \geq W_j,\; \forall\, j=1,2,3 \tag{18}$$

Therefore, problem $\mathcal{OP}_1$ can be reformulated as $\mathcal{OP}_2$ described below

$$\mathcal{OP}_2: \quad \max_{P^l_{\text{s}},\delta^l_k, l\in\mathcal{S}, k\in\Pi} \sum\nolimits_{g\in\mathcal{G}, l\in\mathcal{S}} K_g \delta^l_g C^l_{\text{ss},g}\left(\gamma^l_{\text{ss},g}, P^l_{\text{s}}\right) \quad s.t. \quad \mathcal{C}1, \mathcal{C}2, \mathcal{C}3'', \mathcal{C}5, \mathcal{C}6. \tag{19}$$

Specifically, we present some basic analysis for the equivalence of problem $\mathcal{OP}_1$ and $\mathcal{OP}_2$: If $R_3 \geq \min\{r_i\}$, without loss of generality, assume $r_1 \geq r_2$, which means that the resource R_3 can afford the QoS of the PT 2 and the constraint $\mathcal{C}4$ for the PT 2 becomes loose, then the constraints $\mathcal{C}3$ and $\mathcal{C}4$ in the problem $\mathcal{OP}_1$ can be converted as $\sum_{l\in\mathcal{S}} \delta^l_{G+3}C^l_{\text{sp},3}\left(\gamma^l_{\text{sp},3},P^l_{\text{s}}\right) \geq r_2$ and $\sum_{l\in\mathcal{S}} \delta^l_{G+1}C^l_{\text{sp},1}\left(\gamma^l_{\text{sp},1},P^l_{\text{s}}\right) \geq r_1 - r_2$ with $\delta^l_{G+2} = 0$, which are in accordance with problem $\mathcal{OP}_2$. Other situations suffer the similar analysis omitted to avoid wordiness.

Next, the solution algorithm for problem $\mathcal{OP}_2$ is derived.

5.1 Stepwise optimization problem presentation

The allocation issue of joint subcarrier and power has high computational complexity, and it is even intractable when subcarrier number and secondary multicast group number are very large. In the section, we try to present a low-complexity and efficient solution by dividing the problem $\mathcal{OP}_2$ into two separate subproblems $\mathcal{OP}_{2a}$ and $\mathcal{OP}_{2b}$ as follows.

$\mathcal{OP}_{2a}$: Optimize the sum rate by varying SC assignment δ_k^l when fixing power allocation P_s^l denoted as $\tilde{P}_s^l$.

$$\max_{\delta_k^l, l\in\mathcal{S}, k\in\Pi} \sum\nolimits_{g\in\mathcal{G}, l\in\mathcal{S}} K_g \tilde{C}_{ss,g}^l \delta_g^l$$

$$\mathcal{C}1 : \sum\nolimits_{k\in\Pi} \delta_k^l = 1, \forall\, l \in \mathcal{S}$$

$$\mathcal{C}2 : \delta_k^l \in \{0,1\}, \forall\, l \in \mathcal{S}, k \in \Pi$$

$$\mathcal{C}3'' : \sum\nolimits_{l\in\mathcal{S}} \delta_{G+j}^l \tilde{C}_{sp,j}^l \geq W_j,\ \forall\, j = 1,2,3$$

where $\tilde{C}_{ss,g}^l = C_{ss,g}^l\left(\gamma_{ss,g}^l, \tilde{P}_s^l\right)$ and $\tilde{C}_{sp,j}^l = C_{sp,j}^l\left(\gamma_{sp,j}^l, \tilde{P}_s^l\right)$.

$\mathcal{OP}_{2b}$: Optimize the sum rate by varying power allocation P_s^l when fixing SC assignment δ_k^l denoted as $\left(\delta_k^l\right)^*$, respectively.

$$\max_{P_s^l, l\in\mathcal{S}} \sum\nolimits_{g\in\mathcal{G}, l\in\mathcal{S}} K_g \left(\delta_g^l\right)^* C_{ss,g}^l\left(\gamma_{ss,g}^l, P_s^l\right)$$

$$\mathcal{C}3'' : \sum\nolimits_{l\in\mathcal{S}} \left(\delta_{G+j}^l\right)^* C_{sp,j}^l\left(\gamma_{sp,j}^l, P_s^l\right) \geq W_j,\ \forall\, j = 1,2,3$$

$$\mathcal{C}5 : \sum\nolimits_{l\in\mathcal{S}} P_s^l \leq P_{th}$$

$$\mathcal{C}6 : 0 \leq P_s^l \leq P_{high}, \forall\, l \in \mathcal{S}$$

Through the division of problem $\mathcal{OP}_2$, suboptimal power and subcarrier allocation is obtained. However, it has been shown that the stepwise low-complexity suboptimal allocation scheme can achieve the similar performance with the optimal one [33,37].

5.2 The heuristic solving method for $\mathcal{OP}_{2a}$

To solve the subproblem $\mathcal{OP}_{2a}$, a heuristic solving method is applied. The main idea of the heuristic solving method is elaborated as follows.

On one hand, let set $J = \{1,2,3\}$ denote the current unsatisfied minimum constraint item set for the $\mathcal{C}3''$ in the subproblem $\mathcal{OP}_{2a}$. One top priority is to assign the SCs for $\mathcal{S}_{G+j}, j \in J$ to guarantee the minimum rate requirements.

Specifically, for rate transmission targets $W_j, j \in J$, find the maximum W_j and the corresponding j, and select the first best channel, the second best one, etc. from current unassigned subcarrier set $\mathcal{S}_0$ which is initialized to $\mathcal{S}$ in turn until the constraint $\sum_{l\in\mathcal{S}} \delta_{G+j}^l \tilde{C}_{sp,j}^l \geq W_j$ is satisfied. The point is that the SC l in set $\mathcal{S}_0$ is removed from $\mathcal{S}_0$ once it is selected, and l is assigned to $\mathcal{S}_{G+j}$. Another point is that once the target W_j is guaranteed, the J is updated as $J\backslash\{j\}$.

On the other hand, each remaining unassigned SC $l \in \mathcal{S}_0$ is assigned in turn to the SMG $\mathcal{K}_g$ that maximizes the value of $K_g \tilde{C}_{ss,g}^l$, where $g \in \mathcal{G}$.

The heuristic SC assignment (HSCA) sub-algorithm is provided as shown in Algorithm 1.

Algorithm 1. HSCA sub-algorithm

Input:
1: for each $l \in \mathcal{S}_0$, $\tilde{P}_s^l = \min\{P_{th}/|\mathcal{S}|, P_{high}\}$
Initialization:
2: $J = \{1,2,3\}$, $\mathcal{S}_0 = \mathcal{S}$; for each $k \in \Pi, l \in \mathcal{S}_0$, $\delta_k^l = 0$; for each $\forall\, j \in J$, $\hat{W}_j = W_j$
Iteration:
3: **while** $J \neq \emptyset$ **do**
4: Compute $W_{max} = \max_{j\in J} \hat{W}_j$ and $j_{max} = \arg\max_{j\in J} \hat{W}_j$
5: Select the best channel l^* in the set $\left\{h_{sp,j_{max}}^l, l \in \mathcal{S}_0\right\}$
6: Assign the SC element l^* to the member j_{max}, i.e., $\delta_{G+j_{max}}^{l^*} = 1$, and update $\mathcal{S}_0 \leftarrow \mathcal{S}_0\backslash\{l^*\}$
7: Calculate $C_p = \sum_l \delta_{p,j_{max}}^l \tilde{C}_{sp,j_{max}}^l$
8: **if** $C_p \geq W_{j_{max}}$
9: update $J \leftarrow J\backslash\{j_{max}\}$
10: go to step 17
11: **else** update $\hat{W}_{j_{max}} \leftarrow W_{j_{max}} - C_p$
12: **if** $\mathcal{S}_0 = \emptyset$
13: terminate the iteration stage
14: **end if**
15: **end if**
16: **end while**
17: **while** $\mathcal{S}_0 \neq \emptyset$ **do**
18: Find the maximal element in $\left\{K_g \tilde{C}_{ss,g}^l, \forall l \in \mathcal{S}_0, g \in \mathcal{G}\right\}$ and the corresponding l^* and g^*
19: Assign the SC l^* in $\mathcal{S}_0$ to the SMG $\mathcal{K}_{g^*}$, i.e., $\delta_{g^*}^{l^*} = 1$, and update $\mathcal{S}_0 \leftarrow \mathcal{S}_0\backslash\{l^*\}$
20: **end while**

5.3 The dual method for $\mathcal{OP}_{2b}$

Based on the SC assignment $\left(\delta_k^l\right)^*$ determined through solving the subproblem $\mathcal{OP}_{2a}$, the subproblem $\mathcal{OP}_{2b}$ can be solved by the dual method.

By introducing lagrangian multipliers $\lambda_j, j = 1,2,3$ and μ for the constraints $C3''$ and $C4$ of the subproblem $\mathcal{OP}_{2b}$ respectively, the lagrangian function $\mathcal{L}\left(\left\{P_s^l\right\}, \{\lambda_j\}, \mu\right)$ can be expressed as

$$\mathcal{L}\left(\left\{P_s^l\right\}, \{\lambda_j\}, \mu\right) = \sum\nolimits_{g\in\mathcal{G}, l\in\mathcal{S}} K_g \left(\delta_g^l\right)^* C_{ss,g}^l\left(\gamma_{ss,g}^l, P_s^l\right) - \mu\left(\sum\nolimits_{l\in\mathcal{S}} P_s^l - P_{th}\right) - \sum\nolimits_{j=1}^{3} \lambda_j\left(W_j - \sum\nolimits_{l\in\mathcal{S}} \left(\delta_{G+j}^l\right)^* C_{sp,j}^l\left(\gamma_{sp,j}^l, P_s^l\right)\right) \tag{20}$$

To obtain necessary conditions for optimal solution, differentiating $\mathcal{L}\left(P_s^l, \lambda_j, \mu\right)$ with respect to P_s^l provides the following Karush-Kuhn-Tucker (KKT) conditions [38] as

$$\sum_{g\in\mathcal{G}} K_g \left(\delta_g^l\right)^* \frac{\Delta f}{2\ln 2} \frac{\gamma_{ss,g}^l}{1+\gamma_{ss,g}^l P_s^l} + \sum_{j=1,2,3} \lambda_j \left(\delta_{G+j}^l\right)^* \frac{\Delta f}{2\ln 2} \frac{\gamma_{sp,j}^l}{1+\gamma_{sp,j}^l P_s^l} - \mu = 0 \tag{21}$$

Since for each l, only one $(\delta_k^l)^*, k \in \Pi$ is non-zero. Associating with the constraints $\mathcal{C}6$ of subproblem $\mathcal{OP}_{2b}$, the power loaded on each SC l can be provided as the following equation

$$P_s^l = \sum_{g\in\mathcal{G}} \left(\delta_g^l\right)^* \left[K_g \frac{\Delta f}{2\ln 2\mu} - \frac{1}{\gamma_{ss,g}^l}\right]_0^{P_{\text{high}}} + \sum_{j=1}^{3} \left(\delta_{G+j}^l\right)^* \left[\lambda_j \frac{\Delta f}{2\ln 2\mu} - \frac{1}{\gamma_{sp,j}^l}\right]_0^{P_{\text{high}}} \tag{22}$$

where $[z]_a^b$ is the projection on $[a, b]$ of z.

The lagrangian multipliers λ_j and μ can be obtained by solving the dual problem of $\mathcal{OP}_{2b}$ which is presented as Equation 23.

$$\min_{\{\lambda_j\},\, \mu>0} \mathcal{D}(\{\lambda_j\}, \mu) = \min_{\{\lambda_j\},\, \mu>0} \max_{0\le P_s^l \le P_{\text{high}}} \mathcal{L}\left(\left\{P_s^l\right\}, \{\lambda_j\}, \mu\right) \tag{23}$$

The dual problem for subproblem $\mathcal{OP}_{2b}$ can be solved using the sub-gradient method [39] presented as Equations 24 and 25.

$$\lambda_j = \left[\lambda_j + \alpha_j \left(W_j - \sum_{l\in\mathcal{S}} \left(\delta_{G+j}^l\right)^* C_{sp,j}^l \left(\gamma_{sp,j}^l, P_s^l\right)\right)\right]^+, \quad \forall j = 1,2,3 \tag{24}$$

$$\mu = \left[\mu + \beta \left(\sum_{l\in\mathcal{S}} P_s^l - P_{\text{th}}\right)\right]^+ \tag{25}$$

where $[z]^+$ is the projection on $[0, \infty)$ of z, and α_j and β are diminishing step sizes for guaranteeing the convergence of the sub-gradient method [39].

Therefore, the stepwise subcarrier-power allocation (SSPA) algorithm is summarized in Algorithm 2.

Algorithm 2. SSPA algorithm

Input:

1: Implement the HSCA sub-algorithm to obtain the SC assignment $\delta_k^l, l \in \mathcal{S}, k \in \Pi$

Initialization:

2: $t \leftarrow 0$, set the lagrangian multipliers $\lambda_j, j = 1,2,3$ and μ to some nonnegative values, and set tolerance error ε_0, $\Delta > \varepsilon_0$ and tune $t \leftarrow 1$

Iteration:

3: **while** $\Delta > \varepsilon_0$ **do**

4: Determine the power P_s^l loaded on each $l \in \mathcal{S}$ as Equation 22

5: Update $\lambda_j, j = 1,2,3$ and μ according to relations (24) and (25)

6: Calculate $\Delta = \sum_{j=1}^3 |\lambda_j(t) - \lambda_j(t-1)| + |\mu(t) - \mu(t-1)|$, and set $t \leftarrow t+1$

7: **end while**

6 Simulation results

In this section, simulation results are presented to demonstrate the performance of the proposed network-coded 1S2P cooperation scheme in terms of average achieved transmission rate of SUs and success probability of both PT 1 and PT 2 using the MATLAB R2010b software on a PC equipped with an Intel(R) Core(TM) i3-2130 CPU (3.40 GHz).

6.1 Simulation parameters

The channel gains between each transmitter and receiver is modeled as $h = d^{-\alpha/2}\xi$ with Rayleigh fading ξ, where d is the normalized distance, and α is the path loss exponent, chosen as 3 [40]. Each point in the simulation curves is the average of 2,000 channel realizations. Moreover, the success probability is calculated by the probability that both the minimum rate requirements of PTs are satisfied. It should be noted that during 2,000 channel realizations, some channel states may be so severely bad that neither minimum rate requirements of PUs can be satisfied and the cooperation agreement cannot be achieved, and thus, the success probability is not always 1.

Some other simulation parameters are listed as follows:

1) Parameters about subcarriers: the subcarrier numbers owned by PT 1 and PT 2 are both equal to 32, i.e., $K = M = 32$; the bandwidth of each subcarrier $\Delta f = 0.3125$MHz; the maximum power constraint on each SC $P_{\text{high}} = 0.5W$.
2) Parameters about PUs: unless noted otherwise, the transmission target rates of PU 1 and PU 2 are $r_1 = 30$ Mbps and $r_2 = 20$ Mbps, respectively; the transmission power of PU 1 and PU 2 equal to 1 W; equal power allocation on SCs is adopted for each PT.
3) Parameters about SUs: the total power for the SS $P_{\text{th}} = 3W$; the SD number U equals to 5; the SMG number G equals to 2.

4) Parameters about scene: the distances among cooperation members are listed as follows. For each $i, i' = 1, 2, i \neq i'$, $d(\text{PT } i, \text{PR } i) = 3$, $d(\text{PT } i, \text{PR } i') = 1$, and $d(\text{PT } i, \text{SS}) = 1.2$, $d(\text{SS}, \text{PR } i) = 2$. For each $u \in \mathcal{K}_g, g \in \mathcal{G}$, $d(\text{SS}, \text{SD } u)$ is distributed uniformly over range [1, 2].

6.2 The compared schemes

The proposed network-coded 1S2P cooperation scheme is compared with these three schemes:

- No-code 2PUs: the cooperation scheme with two primary users assisted by the SS with decode-and-forward protocol but without network code.
- No-code 1PU: the cooperation scheme with one primary user assisted by the SS with decode-and-forward protocol. Specifically, SS selects the PT→PR pair whose target transmission rate requirement is the most unsatisfied to assist.
- Direct: non-cooperation scheme. Specifically, the PT 1→PR1 pair and PT 2→PR 2 pair transmit directly their data by themselves without the cooperation of the SS.

6.3 Simulation results with different noise power

In this subsection, we compare the average transmission rate of SUs provided by the proposed scheme and the schemes mentioned in the above subsection. Meanwhile, the effects of these schemes on the success transmission probability of PUs under the different channel noise power density N_0 are analyzed.

Figure 4 depicts the average transmission rate of SUs under different noise power density N_0 with different schemes. The curves of Figure 4 accord with the fact that the average transmission rates of SUs for all schemes get smaller as the channel conditions get worse. It is found that the proposed scheme can provide the largest average transmission rate of SUs. When the noise power density N_0 is lower than a certain value Θ, e.g., about 32 dBW/MHz in Figure 4, the 'No-Code 2PUs' scheme can supply higher average transmission rate of SUs than 'No-Code 1PU' supplies. When the N_0 gets higher than Θ, the average transmission rate of SUs obtained by 'No-Code 2PUs' scheme is a litter lower than that obtained by 'No-Code 1PU' scheme. It stems from the fact that the 'No-Code 2PUs' scheme needs to satisfy the minimum transmission rate requirements of both PUs, and when the channel conditions get worse, it is a little hard to realize for 'No-Code 2PUs' scheme. Therefore, we can discover that the proposed network-coded 1S2P cooperation scheme provides the largest profits for SUs.

On the other hand, success transmission probabilities for PUs are displayed under different noise power density N_0 in Figure 5. The success transmission probabilities become smaller with noise power gradually louder for all schemes, and this is mostly the result of channel conditions influence and is within our expectation. We find that the proposed scheme can guarantee the minimum transmission rate requirements of PUs for network-coded 1S2P cooperation with very similar

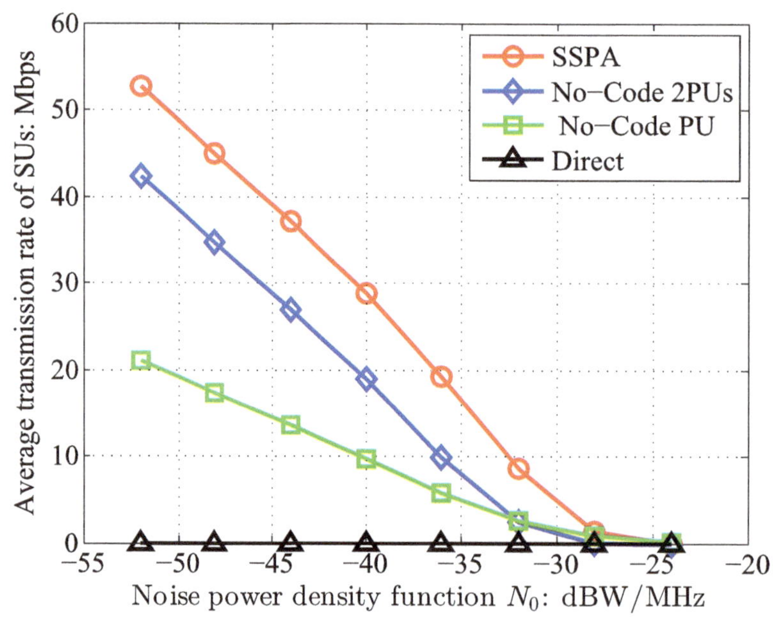

Figure 4 Average transmission rate of SUs with different noise power density N_0. Where the target transmission requirements of PU 1 and PU 2 are 30 and 20 Mbps.

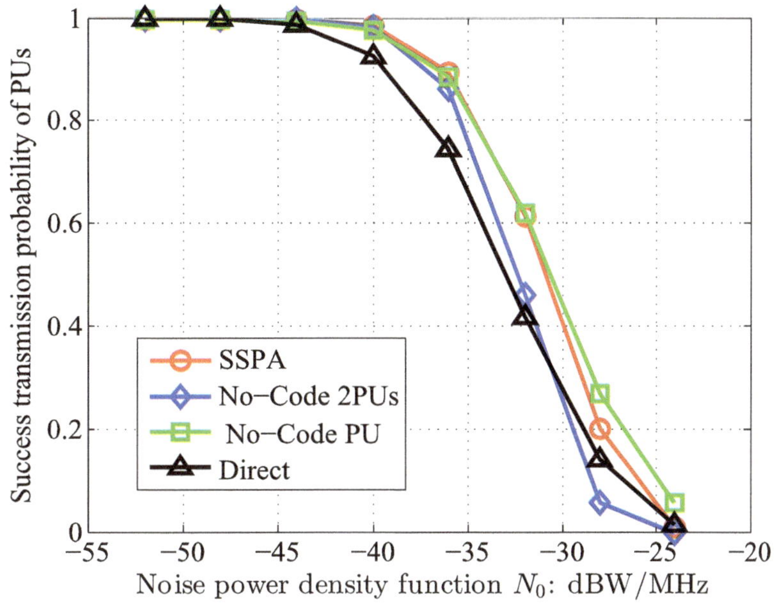

Figure 5 Success probability for transmissions of PUs with different noise power N_0. Where the target transmission requirements of PU 1 and PU 2 are 30 and 20 Mbps.

probability as that of the 'No-Code 1PU' scheme. Moreover, with channel conditions becoming severely worse, the success transmission probability of the network-coded 1S2P cooperation suffers a little higher degradation than that of the 'No-Code 1PU' scheme for guaranteeing two PUs' minimum transmission rate requirements. Moreover, the 'No-Code 2PUs' scheme can guarantee the similar success transmission probability with the proposed scheme when the channel condition is very good, e.g., with the noise power density lower than -40 dBW/MHz in Figure 5. However, when the channels get worse, the success transmission probability of the 'No-Code 2PUs' scheme becomes much lower than that of the proposed scheme and even lower than that of the 'Direct' scheme. Thus, it can be concluded that the proposed scheme can provide high average

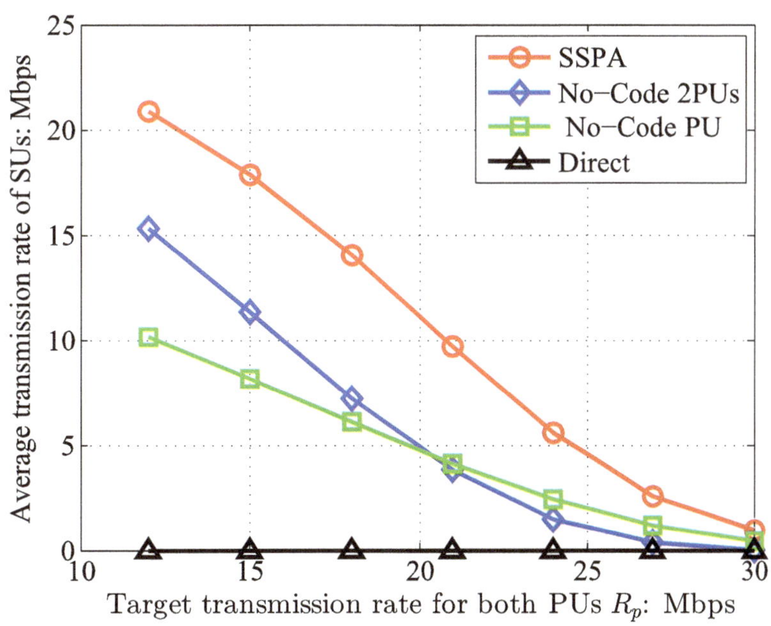

Figure 6 Average transmission rate of SUs with different target transmission rate $r_p = r_1 = r_2$. Where $N_0 = -30$ dBW/MHz.

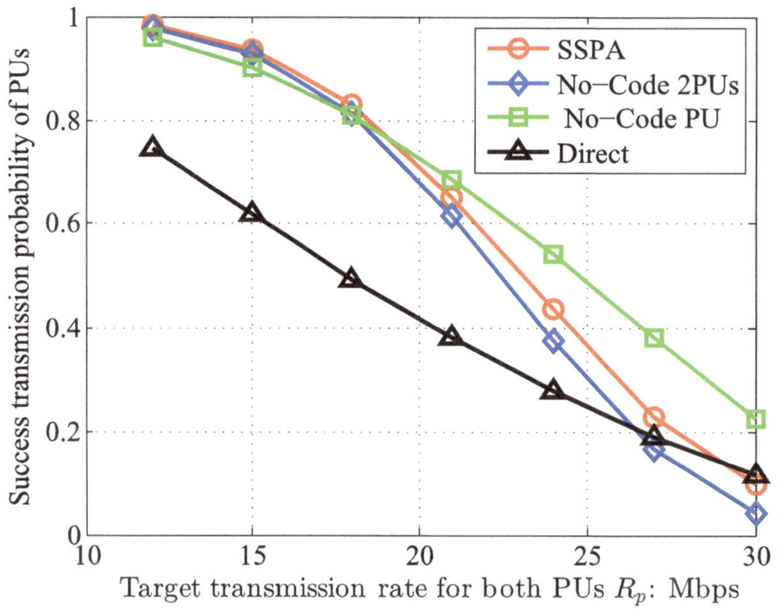

Figure 7 Success probability for transmissions of PUs with different target transmission rate $r_p = r_1 = r_2$. Where $N_0 = -30$ dBW/MHz.

transmission rate of SUs with satisfactory success transmission probability.

6.4 Simulation results with different transmission rate requirements

In this subsection, we examine the effects of these schemes on the average transmission rate of SUs and success transmission probability of PUs with the different transmission rate requirements of PUs under the channel noise density N_0 equalling to -30 dBW/MHz. For ease of analysis, both the minimum transmission rates of PUs r_1 and r_2 equal to r_p.

From Figure 6, we find that the average transmission rate of SUs versus different target transmission rate R_p demonstrates the similar downtrend and intersection of curves shown in Figure 4. It verifies the superiority and the satisfactory performance of the proposed network-coded 1S2P cooperation scheme.

Success probabilities of all schemes for transmissions of PUs under different target transmission rate requirement r_p are described in Figure 7. Within a certain target transmission rate range, the proposed scheme gains much higher transmission rate at the expense of lowering a little success transmission probability of PRs. We have to be alert that when the cooperation members PUs make claims for excessive minimum rate requirements, the SU may not afford higher success transmission probability of PUs than that SU assists one PU or even that no SU helps the data forward of PUs. That is because the SU has to guarantee both the transmission rate requirements of PUs, and it also reserves a portion of resources to multicast data for SMGs.

7 Conclusions

In this paper, the 1S2P cooperation based on network coding in OFDM-modulated cognitive multicast networks is investigated. Concentrating on the essential conditions and cooperation context, the network-coded 1S2P cooperation scheme is presented. Meanwhile, the corresponding resource allocation problem is formulated and adopting the heuristic method and dual method, a stepwise subcarrier-power allocation algorithm is proposed with lower computation complexity. The simulation shows that the proposed cooperation scheme guarantees not only the minimum transmission rate requirements of PUs but also provides much higher transmission rate for SUs than that provided by traditional non-NC primary-secondary cooperation schemes. The network-coded 1S2P cooperation provides the inspiration and substance to enhance the spectrum efficiency for 5G systems. In the future research, network-coded multi-secondary-user and multi-primary-user cooperation will be discussed based on the cooperation agreement mentioned in the paper.

Competing interests

The authors declare that they have no competing interests.

Acknowledgements

This work is supported by National Natural Science Foundation of China (61362008), the 863 Project (2014AA01A701), Fundamental Research Funds for the Central Universities (2014ZD03-01), Special Youth Science Foundation of Jiangxi (20133ACB21007), and ZTE's Industry-Academy-Research Cooperation Forum Program (A2014172).

References

1. S Chen, J Zhao, The requirements, challenges, and technologies for 5G of terrestrial mobile telecommunication. IEEE Commun. Mag. **52**(5), 36–43 (2014)
2. J Mitola, GQ Maguire, Cognitive radio: making software radios more personal. IEEE Pers. Commun. **6**, 13–18 (1999)
3. FCC, Spectrum policy task force. ET Docket, 02–135 (2002)
4. J Mitola III, J Guerci, J Reed, Y-D Yao, Y Chen, TC Clancy, J Dwyer, H Li, H Man, R McGwier, G Yi, Accelerating 5G QoE via public-private spectrum sharing. IEEE Commun. Mag. **52**(5), 77–85 (2014)
5. K Letaief, W Zhang, Cooperative communications for cognitive radio networks. Proc. IEEE. **97**(5), 878–893 (2009)
6. A Nosratinia, TE Hunter, A Hedayat, Cooperative communication in wireless networks. IEEE Commun. Mag. **42**(10), 74–80 (2004)
7. Y Zhou, H Liu, Z Pan, L Tian, J Shi, G Yang, Two-stage cooperative multicast transmission with optimized power consumption and guaranteed coverage. IEEE J. Selected Areas Commun. **32**(2), 274–284 (2014)
8. S Katti, H Rahul, W Hu, D Katabi, M Médard, J Crowcroft, XORs in the air: practical wireless network coding. IEEE/ACM Trans. Netw. **16**(3), 497–510 (2008)
9. X Wang, J Li, Network coding aware cooperative MAC protocol for wireless ad hoc networks. IEEE Trans. Parallel Distributed Syst. **25**(1), 167-179 (2013)
10. PH Tan, CK Ho, S Sun, OFDM modulated cooperative multiple-access channel with network-channel coding. IEEE Trans. Wireless Commun. **11**(2), 604–613 (2012)
11. X Chen, H Chen, W Meng, Cooperative communications for cognitive radio networks-from theory to applications. EEE COmmun. Surveys Tuts. **16**(3), 1180-1192 (2014)
12. O Simeone, J Gambini, Y Bar-Ness, U Spagnolini, in *2007 IEEE International Conference on Communications (ICC)*. Cooperation and cognitive radio (IEEE Glasgow, 2007), pp. 6511–6519
13. Y Zou, Y-D Yao, B Zheng, Diversity-multiplexing tradeoff in selective cooperation for cognitive radio. IEEE Trans. Commun. **60**(9), 2467–2481 (2012)
14. W Jaafar, W Ajib, D Haccoun, in *2012 IEEE International Conference on Communications (ICC): 10–15 June 2012*. Opportunistic adaptive relaying in cognitive radio networks (Ottawa, ON, 2012), pp. 1811-1815
15. X Jia, M Zhou, X Dang, L Yang, H Zhu, Adaptive power allocation and outage performance of cognitive best relay cooperation systems with multiple primary transceiver pairs and direct path between cognitive source and destination. EURASIP J. Wireless Commun. Netw. **2014**(1), 122 (2014)
16. Q Zhang, J Jia, J Zhang, Cooperative relay to improve diversity in cognitive radio networks. IEEE Commun. Mag. **47**(2), 111–117 (2009)
17. W Lee, D-H Cho, in *2011 Sixth International ICST Conference on Cognitive Radio Oriented Wireless Networks and Communications (CROWNCOM)*. Cooperation based spectrum management scheme in ad-hoc cognitive radio system (IEEE Osaka, 2011), pp. 36–40
18. R Ramamonjison, A Haghnegahdar, V Bhargava, Joint optimization of clustering and cooperative beamforming in green cognitive wireless networks. IEEE Trans. Wireless Commun. **13**(2), 982-997 (2014)
19. T-Y Wu, K-W Lin, P-H Huang, W Liao, in *2013 IEEE 24th International Symposium on Personal Indoor and Mobile Radio Communications (PIMRC)*. A distributed cooperation strategy in cognitive radio networks (IEEE London, 2013, pp. 2517–2521
20. G Zheng, S Song, K-K Wong, B Ottersten, Cooperative cognitive networks: optimal, distributed and low-complexity algorithms. IEEE Trans. Signal Process. **61**(11), 2778–2790 (2013)
21. F Gabry, N Li, N Schrammar, M Girnyk, L Rasmussen, M Skoglund, On the optimization of the secondary transmitter's strategy in cognitive radio channels with secrecy. IEEE J. Sel. Area Commun. **32**(23), 451-463 (2014)
22. J Jin, H Xu, B Li, in *2010 Proceedings IEEE INFOCOM: 14–19 March 2010*. Multicast scheduling with cooperation and network coding in cognitive radio networks San Diego, CA, 2010), pp. 1-9
23. YJ Chun, NO Hasna, A Ghrayeb, in *2013 IEEE 24th International Symposium on Personal Indoor and Mobile Radio Communications (PIMRC): 8–11 Sept. 2013*. Adaptive network coding over cognitive relay networks (London, United Kingdom, 2013), pp. 105-110
24. W Lu, J Wang, Opportunistic spectrum sharing based on full-duplex cooperative OFDM relaying. IEEE Commu. Lett. **18**(2), 241–244 (2014)
25. T Jing, S Zhu, H Li, X Xing, X Cheng, Y Huo, R Bie, T Znati, Two-stage cooperative multicast transmission with optimized power consumption and guaranteed coverage. IEEE Trans. Vehicular Technol. **PP**(99), 1 (2014)
26. Q Wu, J Zou, Z Wu, in *2014 IEEE International Conference on Communications (ICC)*. Joint spectrum and power allocation in coded cooperative cognitive radio networks (IEEE Sydney, 2014), pp. 1596-1601
27. J Zou, Q Wu, H Xiong, CW Chen, Joint spectrum and power auction with multiauctioneer and multibidder in coded cooperative cognitive radio networks. IEEE Trans. Wireless Commun. **13**(10), 5768–5780 (2014)
28. S Sharma, Y Shi, J Liu, YT Hou, S Kompella, SF Midkiff, Network coding in cooperative communications: friend or foe? IEEE Trans. Mobile Comput. **11**(7), 1073–1085 (2012)
29. Z Liu, W Xu, S Li, J Lin, in *Proceedings of the IEEE Global Telecommunications Conference Workshop (Globecom Workshop2014)*. Power and spectrum allocation for network coded primary-secondary cooperation in cognitive radio networks (IEEE Austin, 2014), pp. 845-880
30. A Omri, MO Hasna, KB Letaief, Inter-relay interference management schemes for wireless multi-user decode-and-forward relay networks. IEEE Trans. Wireless Commun. **PP**(99), 1 (2014)
31. O Taghizadeh, R Mathar, in *International Conference on Wireless for Space and Extreme Environments (WiSEE): 30–31 Oct. 2014*. Cooperative strategies for distributed full-duplex relay networks with limited dynamic range (Noordwijk, 2014), pp. 1-7
32. Q Li, W-K Ma, M Lin, J Ge, J Lin, Robust cooperative beamforming and artificial noise design for physical-layer secrecy in AF multi-antenna multi-relay networks. IEEE Trans. Signal Process. **63**(1), 206–220 (2015)
33. X Gao, Adaptive relay techniques for OFDM-based cooperative communication systems. PhD thesis, The University of Western Ontario, Electrical and Computer Engineering Department. (2014)
34. K Xiong, PY Fan, M Lei, Y Su, Network coding-aware cooperative relaying for downlink cellular relay networks. China Commun. **10**(7), 44–56 (2013)
35. L Fan, X Lei, TQ Duong, M Elkashlan, GK Karagiannidis, Secure multiuser communications in multiple amplify-and-forward relay networks. IEEE Trans. Commun. **62**(9), 3299–3310 (2014)
36. LJ Chen, T Ho, M Chiang, SH Low, JC Doyle, Congestion control for multicast flows with network coding. IEEE Trans. Inf. Theory. **58**(9), 5908–5921 (2012)
37. B Han, M Peng, Z Zhao, W Wang, A multidimensional resource-allocation optimization algorithm for the network-coding-based multiple-access relay channels in OFDM systems. IEEE Trans. Vehicular Technol. **62**(8), 4069–4078 (2013)
38. S Boyd, L Vandenberghe, *Convex Optimization*. (Cambridge university press, Cambridge, 2009)
39. DP Bertsekas, *Nonlinear Programming*. (Athena Scientific, Nashua, 1999)
40. B Rankov, A Wittneben, Spectral efficient protocols for half-duplex fading relay channels. IEEE J. Selected Areas Commun. **25**(2), 379–389 (2007)

In-vehicle channel sounding in the 5.8-GHz band

Pavel Kukolev[1], Aniruddha Chandra[1*], Tomáš Mikulášek[1], Aleš Prokeš[1], Thomas Zemen[2,3] and Christoph F Mecklenbräuker[4]

Abstract

The article reports vehicular channel measurements in the frequency band of 5.8 GHz for IEEE 802.11p standard. Experiments for both intra-vehicle and out-of-vehicle environments were carried out. It was observed that the large-scale variations (LSVs) of the power delay profiles (PDPs) can be best described through a two-term exponential decay model, in contrast to the linear models which are suitable for popular ultra-wideband (UWB) systems operating in the 3- to 11-GHz band. The small-scale variations (SSVs) are separated from the PDP by subtracting the LSV and characterized utilizing logistic, generalized extreme value (GEV), and normal distributions. Two sample Kolmogorov-Smirnov (K-S) tests validated that the logistic distribution is optimal for in-car, whereas the GEV distribution serves better for out-of-car measurements. For each measurement, the LSV trend was used to construct the respective channel impulse response (CIR), i.e., tap gains at different delays. Next, the CIR information is fed to an 802.11p simulation testbed to evaluate the bit error rate (BER) performance, following a Rician model. The BER results strongly vouch for the suitability of the protocol for in-car as well as out-of-car wireless applications in stationary environments.

Keywords: Intra-vehicle channel; IEEE 802.11p; Channel sounding; Ultra-wideband; Power delay profile; Bit error rate

1 Introduction

1.1 Motivation

Vehicle area networks (VANs) and vehicular *ad hoc* networks (VANETs) generally refer to networks between cars and infrastructure points located along the road side [1,2]. These networks are aimed to deliver information about the traffic, to ensure safety of the passengers, and to provide driver assistance and passenger entertainment [3]. The majority of future vehicular networks are envisaged to conform with the IEEE 802.11p standard [4] described in wireless access in vehicular environments (WAVE) [5] and governed by a non-profit organization, the car-2-car communication consortium (C2C-CC) [6]. IEEE 802.11p standardizes signal transmission for vehicle-to-vehicle (V2V) or vehicle-to-infrastructure (V2I) communications, operated at a frequency band of 5.8 GHz [7].

Presently, there had been an upsurge in the interest of the automobile industries regarding wider applications of intra-vehicular wireless transmission. The use of wireless communication in the vehicle cuts down the wiring harness, which will reduce the weight and cost of the car. It also eliminates the need of drilling to pass the cables, thereby improving robustness of the vehicle, and will simplify the design as well as production processes. As mentioned earlier, the IEEE 802.11p protocol was specified primarily for V2V/V2I scenarios where both transmitter and receiver are placed outside the vehicle [8]. In order to realize fully integrated intelligent transportation systems (ITSs), it is important to evaluate the efficacy of the protocol for wireless connections in in-car (both transmitter and receiver are inside the vehicle) and out-of-car (transmitter is inside, and receiver is outside, or vice-versa) scenarios as well.

1.2 Literature survey

There is a good number of open literature available that deal with vehicular channel measurement and modelling. We begin the list with [9], which provides a general review on advances and challenges in V2V channel measurement campaigns. The following papers, namely [10-12], describe measurement-based narrowband channel models for different situations: on highways, in cities, and on

*Correspondence: chandra@feec.vutbr.cz
[1] Department of Radio Electronics, Brno University of Technology, Technicka 12, 61600 Brno, Czech Republic
Full list of author information is available at the end of the article

country roads. Different aspects of non-stationary vehicular channel measurements in the frequency band of our interest are addressed in [13,14]. Reports from some inter-vehicular measurement campaigns in the 5.9-GHz band are also made available in [15] and [16]. In particular, Cheng et al. [15] study the suitability of IEEE 802.11a for vehicular wireless links. Sen and Matolak [16] provide a wide set of power delay profile (PDP) measurements and proposed tapped delay line models based on Markov chains.

As far as intra-vehicular channel measurements are concerned, most of the articles are focused on ultra-wideband (UWB) operating over a broad frequency spectrum of 3 to 11 GHz. For example, the performance of multiband orthogonal frequency division multiplexing (MB-OFDM) in intra-car communication is studied by Maehara et al. [17]. In [18], Schack et al. report a comparison of broadband channel sounding experiments over a different range of cars. Another interesting paper is by Niu et al. [19], where they describe a detailed UWB channel modelling campaign for intra-vehicular environments. The characterization of the 5-GHz intra-vehicle communication channel is only attempted recently in [20], where the authors present power delay profiles, delay spread, and statistical channel models for a minivan and a bus.

Finally, in [21] and [22], the throughput and packet delivery performance of the IEEE 802.11p protocol is studied in V2V and V2I scenarios with different speeds of movement. However, to the best of our knowledge, none of the authors utilize the measured intra-vehicle channel models to evaluate the bit error rate (BER) for the IEEE 802.11p standard.

1.3 Contributions of the paper

In this paper, we present the results of intra-vehicle channel model measurements for the 5.8-GHz band for a typical passenger car. We provide measurements for a vehicle parked in an underground garage. The results were collected in the frequency domain by a vector network analyzer (VNA), then they were transformed into the time domain using the inverse fast Fourier transform (IFFT), and the large-scale variations (LSVs) from the measured PDPs were separated. The small-scale variations (SSVs), which were obtained after subtracting the LSV from the PDP, were characterized statistically thereafter. Finally, simulated BER results were obtained using a 802.11p simulation model. Specifically, the contributions of the paper are as follows:

- Results for intra-vehicle and out-of-vehicle single input single output (SISO) channel measurements performed in the 5.875- to 5.885-GHz frequency band.
- Characterization of the LSV trends via a two-term exponential decay model and the SSV random variations via logistic, normal, and generalized extreme value (GEV) distributions with the two-sample Kolmogorov-Smirnov (K-S) test.
- Comparison of PDP decay trends of UWB and 802.11p for an identical measurement setup.
- BER simulation for 802.11p using the measured channel data.

1.4 Organization of the paper

The paper reads as follows: Section 2 describes the measurement setup realized through a VNA. In Section , we provide a decay model to characterize the PDP, characterize the LSV trends, and find the optimal statistical distribution for the SSV. This section also includes a comparison between 802.11p and vehicular UWB channel characteristics. Next, the BER performance for the 802.11p standard is presented in Section 4. Finally, some conclusions are presented in Section 5.

2 Measurement setup

Measurements were performed with one transmitter (Tx) and three receiver (Rx) antennas in a right-hand, four-door Skoda Octavia car with dimensions: 4.659 × 1.814 × 1.462 m. The vehicle was parked on the sixth floor in an underground garage of the Faculty of Electrical Engineering (FEKT), Brno University of Technology (VUT). Walls and floors of the garage premises are made of reinforced concrete, and they provide an environment that is free from narrowband interference. There were no parked cars on neighboring parking lots.

Figure 1a shows the underground garage and the parking lot with the car and equipments. The receive/ transmit antennas inside the car are zoomed in Figure 1b. The four omni-directional conical monopole antennas that were used are identical. The vertical polarization of the antennas can be confirmed from the radiation patterns presented in Figure 2 which depicts that the azimuth plane pattern is circular.

Equipments for recording data were placed outside of the vehicle. All doors and windows were closed, except the driver's window, which was slightly open to pass the cables between the antennas and the recording equipment.

For measuring and recording the transmission coefficient between transmitter and receiver, a four-port VNA from Agilent Technologies, E5071C (Agilent Technologies, Inc., Santa Clara, CA, USA) [7] was used (Figure 3 [bottom]). The transmit power was set at 5 dBm, and phase stable coaxial cables were used for interconnections. The VNA measures the complex valued channel transfer function in the frequency domain through the s-parameter, S_{x1}, which is the transmission coefficient between port 1 (Tx) and port $x \in \{2,3,4\}$ (Rx). The

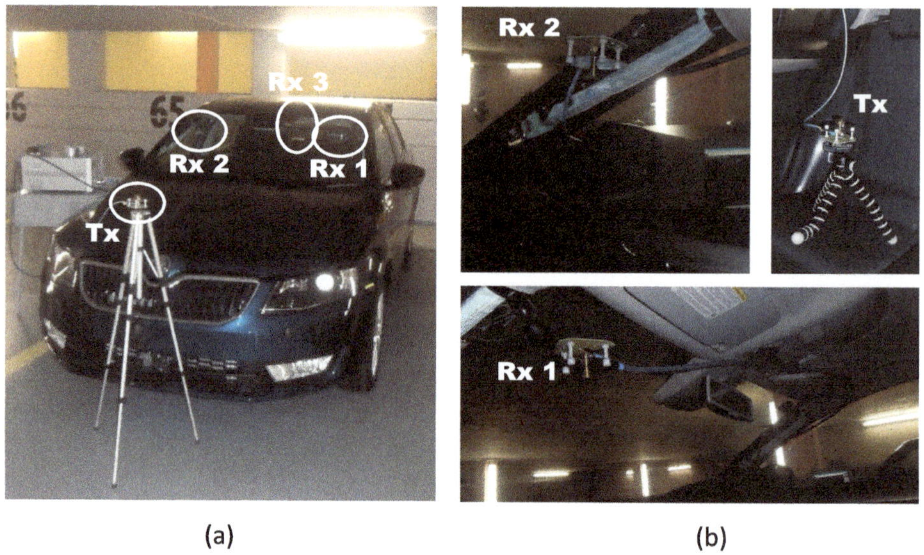

Figure 1 The underground parking lot and the receive/transmit antennas inside the car. (a) The vehicle under test. **(b)** Close up of the antennas inside the car.

measured frequency range (5.775 to 5.875 GHz) was swept with a frequency step size of $f_s = 0.125$ MHz. Due to the fine resolution in the frequency domain, we were able to construct the delay domain up to $t_d = 1/f_s = 8$ µs without aliasing effect of the VNA. The information, when translated into the spatial domain, gives us a propagation space resolution $P_{sr} = c/\text{BW} = 3$ m and a maximum propagation distance of $L_{d\,\max} = c/f_s = 2.4$ km. The parameters, c and BW, denote propagation speed of electromagnetic waves in free space (3×10^8 m/s) and the bandwidth of the measured band, respectively.

The schematic diagram of the setup for IEEE 802.11p frequency domain channel sounding is presented in Figure 3 [top] [23]. Real and imaginary parts of the transfer function (S_{x1}) were exported to MATLAB. The frequency domain data over the entire bandwidth BW = 100 MHz were partitioned into 10 MHz bins, where each bin corresponds to a sub-channel of 802.11p. All results were transformed from the frequency domain into the time domain, $h(\tau) = \mathcal{F}^{-1}H(f)$, utilizing the IFFT with a typical rectangular window. The PDP was calculated by averaging over ten sub-channel time-domain data.

Figure 4 shows transmitter and receiver antenna positions. A total of 15 different Tx-Rx combinations were tested with separations ranging from 0.53 to 3.38 m. The Tx antenna was placed at three different locations inside the vehicle and at two locations outside the vehicle. In-car Tx antenna positions were set at the right rear seat, armrest in the middle of the car, driver seat, while for locations outside the car, two positions at a height of 1.09 m were chosen: one in front of the car and the other one near the right headlamp. The Rx antennas were installed

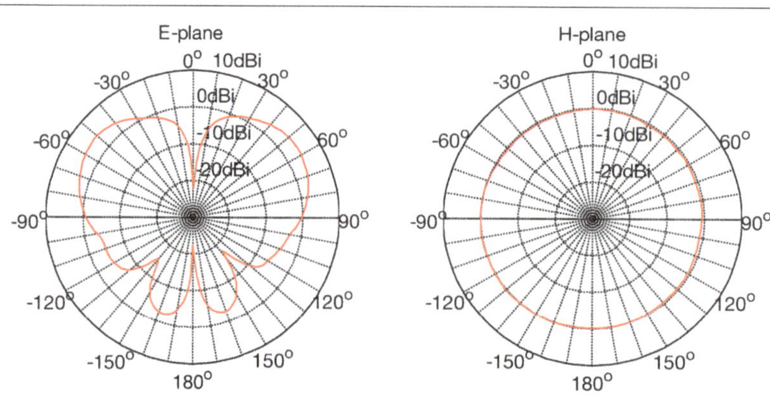

Figure 2 Simulated gain pattern of the conical antennas for central frequency of 5.8 GHz.

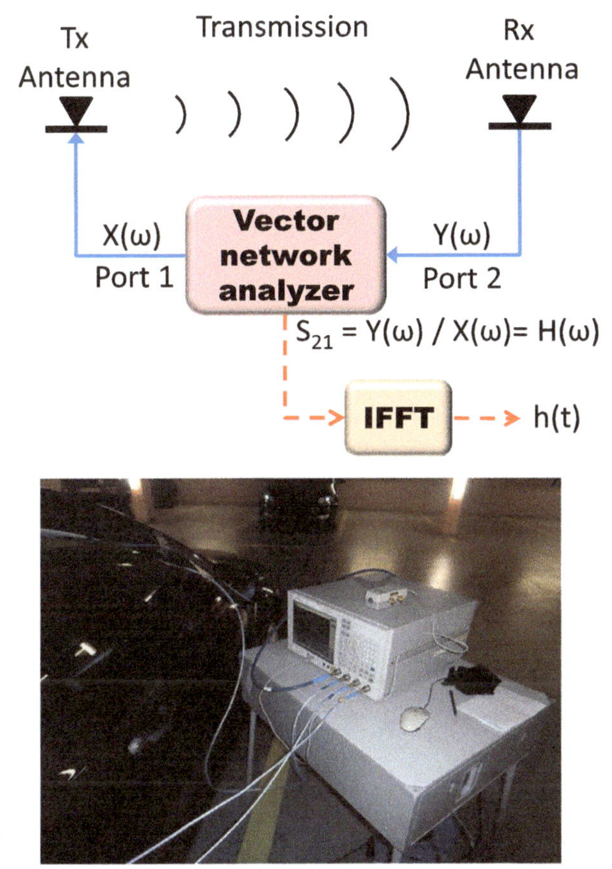

Figure 3 Block diagram of the measurement setup [top] and Vector Network Analyzer Agilent Technologies E5071C with four ports [bottom].

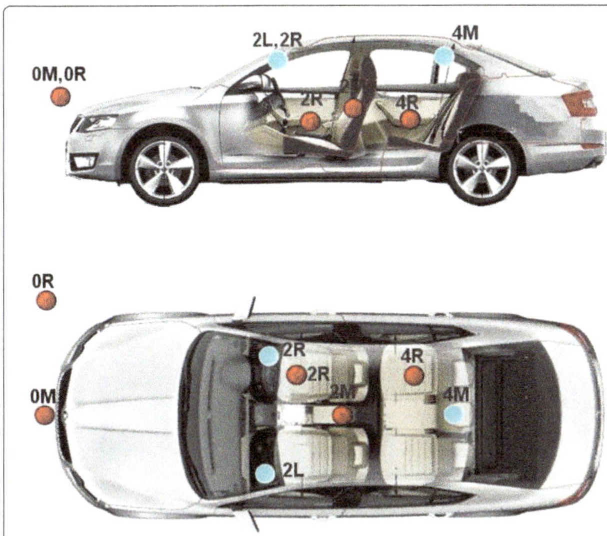

Figure 4 Schematic of antenna positions, RED - transmitter antenna, BLUE - receiver antenna.

on the left and right upper edges of the windshield and on the roof in the rear part of the vehicle. The positions of antennas were chosen for realizing both line-of-sight (LOS) and non-line-of-sight (NLOS) scenarios. All these combinations are described in detail in Table 1.

3 Channel description

The measured channel impulse response (CIR) is $h(\tau) = h_{ch}(\tau) * h_{fil}(\tau)$ where $h_{ch}(\tau)$ refers to the actual channel response, $*$ denotes time-domain convolution, and $H_{Fil}(f) = \mathcal{F}\{h_{fil}(\tau)\}$ is the transfer function of the windowing operation. The band-limiting post-processing, however, does not lead to any serious causality issues [24]. The generic multipath CIR is described by:

$$h_{ch}(\tau) = \sum_{i=0}^{N-1} a_i \exp(j\theta_i)\delta(\tau - \tau_i) \qquad (1)$$

where τ_i is the propagation delay, $a_i \exp(j\theta_i)$ is the complex amplitude coefficient of the ith multipath component, $\delta(.)$ is the Dirac delta function, and N is the total number of multipath components.

The PDP is simply the squared channel impulse response $P(\tau) = |h(\tau)|^2$, and it includes LSV and SSV, which can be designated mathematically in the following manner: $P(\tau) = \gamma(\tau) + \xi(\tau)$, where $\gamma(\tau)$ denotes LSV and $\xi(\tau)$ is the SSV.

Table 1 Detailed settings for measurements (Tx and Rx legends are shown in Figure 4)

Measurement number	Tx-Rx separation (m)	Tx position	Rx position	Remarks
1	0.53	2R	2R	LOS
2	0.69	4R	4M	LOS
3	0.74	2M	2L	LOS
4	0.85	2M	2R	LOS
5	0.87	2M	4M	LOS
6	0.94	2R	2L	LOS
7	1.26	4R	2R	NLOS
8	1.28	2R	4M	NLOS
9	1.41	4R	2L	NLOS
10	1.86	0R	2L	Tx at an angle
11	2.03	0M	2L	Tx in front
12	2.08	0M	2R	Tx in front
13	2.74	0R	2R	Tx at an angle
14	3.15	0R	4M	Tx at an angle
15	3.38	0M	4M	Tx in front

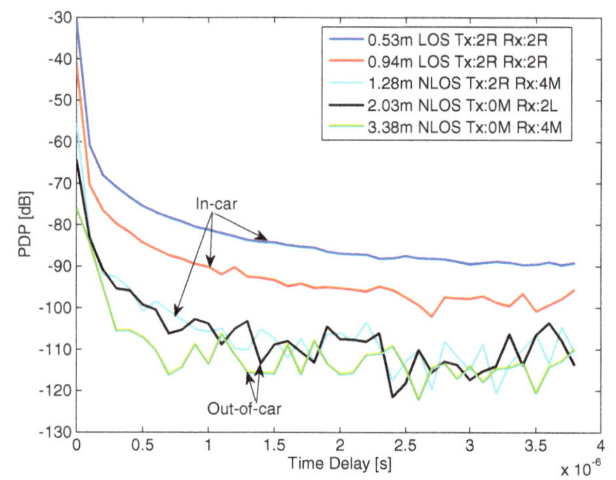

Figure 5 **Power delay profile for LOS/ NLOS measurements carried out for different Tx-Rx separations.**

Figure 5 shows the measured PDP for different LOS and NLOS conditions for both inside and outside of the vehicle. As expected, the power levels are significantly higher for the LOS measurements compared to the NLOS situations. The power levels for the NLOS case show rapid variations near the noise floor (−110 dB).

3.1 Large-scale variations

Analytical calculations performed for closed spaces [25,26] demonstrated an exponential decay of the PDP. As intra-vehicular channels appear to have similar signal propagation environments, we express the LSV with a two-term exponential model:

$$\gamma(\tau) = A\exp(B\tau) + C\exp(D\tau) \quad ; 0 < \tau < t_d/2. \quad (2)$$

Table 2 Parameter values for LSV

Tx-Rx separation (m)	Parameter values A	B ($\times 10^4$)	C	D ($\times 10^5$)
0.53	−49.81	0.63	45.98	−7.05
0.94	−48.56	0.64	43.89	−6.47
1.28	−50.66	0.46	45.37	−5.20
2.03	−47.36	0.34	38.82	−3.68
3.38	−45.32	0.14	38.25	−4.28

The first term includes power from direct and major reflected rays, and the second term, with a very low slope (close to linear), reflects the power from diffused multipath components. This exponential model (2) offers more flexibility compared to single-term exponential or linear models. Further, it also avoids discontinuities in the extracted LSVs, which are present when one attempts to divide the PDP into two (or more) delay parts and tries to fit individual expressions for each of these parts [27].

In order to evaluate how well the LSV model (2) corresponds to the actual PDP, we evaluated the mean square error (MSE) between the PDP and the LSV:

$$\text{MSE} = \frac{1}{M}\sum_{i=1}^{M}[P_i(\tau) - \gamma_i(\tau)]^2 \quad ; 0 < \tau < t_d/2. \quad (3)$$

The MSE is 6.59×10^{-5}, averaged over all the $M = 15$ measurements.

The LSVs for the two-term exponential model for different distance between Tx and Rx antennas of 0.53, 0.87, 1.28, 2.74, and 3.15 m are shown in Figure 6. It is observed that LSV for the LOS case drops more rapidly than for the NLOS scenario. This is in perfect harmony with earlier observations [28].

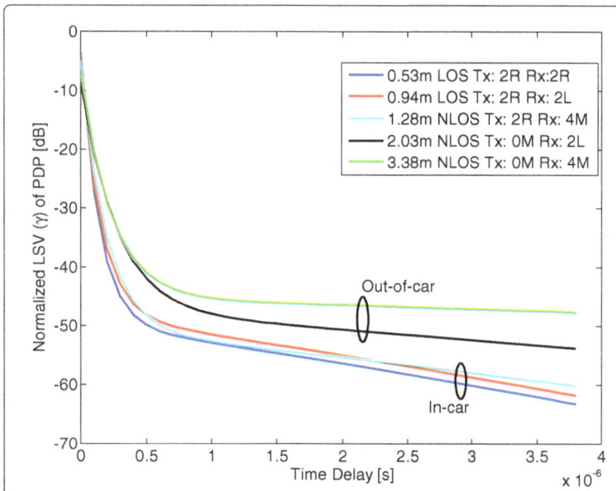

Figure 6 **Normalized large-scale variation (γ) of the PDP for LOS/ NLOS measurements carried out for different Tx-Rx separations.**

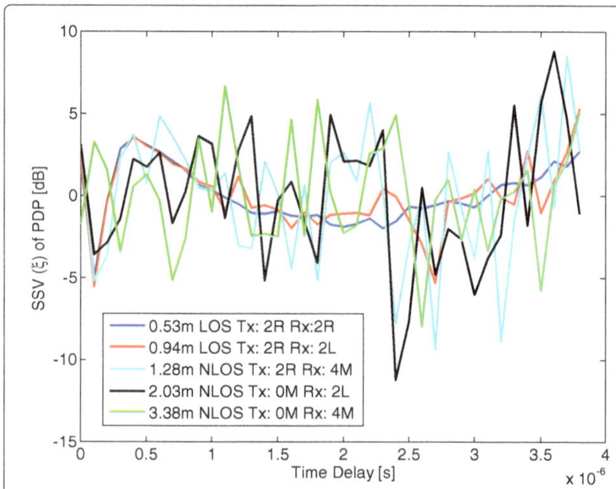

Figure 7 **Small-scale variation (ξ) of the PDP for LOS/ NLOS measurements carried out for different Tx-Rx separations.**

Table 3 Two sample K-S test *p* values for fitting SSV with different continuous distributions

Tx	Tx-Rx	Distributions		
position	separation (m)	GEV	Logistic	Normal
	0.53	0.3456	0.8838	0.7085
In-car	0.94	0.8838	0.8838	0.7085
	1.28	0.7085	0.7085	0.8838
Out-of-car	2.03	0.9997	0.6297	0.5142
	3.38	0.8838	0.7085	0.8838

The parameters of the model for these five distances are given in Table 2. Both the coefficients, A and C, show similar values for in-car (or out-of-car) situations. On the other hand, the coefficient B decreases steadily with increasing Tx-Rx separation whereas D increases.

3.2 Small-scale variations

The SSV was obtained through subtracting the LSV from the measured PDP, i.e., $\xi(\tau) = P(\tau) - \gamma(\tau)$. Considering that the LSV follows (2), we separated SSVs for five different distances, and the results are depicted in Figure 7.

The characterization of the SSV is based on finding the appropriate random process using the normal, the logistic, and the GEV distributions. The GEV distribution [29] appeared to be a good fit for both UWB [30] and milimeter wave [31] small-scale variations. On the other hand, the logistic family of distributions [32] has been used successfully in ray-cluster-based modelling at 5 GHz [33] and modelling the probability density function (PDF) of the rms delay spread at 2.35 GHz [34]. We have also included the normal distribution as a reference. The PDF of these three distributions are given in (9), (10), and (11), respectively.

$$f(x|\mu,\sigma) = \frac{1}{\mu\sqrt{2\pi}} \exp\left[\frac{-(x-\mu)^2}{2\sigma^2}\right] \quad (4)$$

$$f(x|\mu,\sigma) = \frac{\exp\left(\frac{x-\mu}{\sigma}\right)}{\sigma\left[1+\exp\left(\frac{x-\mu}{\sigma}\right)\right]^2} \quad (5)$$

$$f(x|k,\mu,\sigma) = \frac{1}{\sigma}\exp\left(-\beta^{-k^{-1}}\right)\beta^{-1-k^{-1}} \quad (6)$$

where $\beta = 1 + k\frac{x-\mu}{\sigma}$. The parameters, μ, σ, and k, are the location, scale, and shape parameters, respectively.

To compare distributions and SSV, we have performed the two-sample K-S test [35]. The p values are presented in Table 3. Higher p values for the logistic distribution inside the car and for the GEV distribution outside the car prove that they are the best candidates among the three for describing the SSV in in-car and out-of-car situations, respectively. The same can be observed from Figure 8, where the cumulative distribution functions (CDFs) of measured SSVs are fitted with these three distributions for results measured at different distances.

3.3 Comparison with UWB

UWB possesses a great potential for high-speed data communication and precise localization operations in cluttered closed-space hostile towards radio frequency (RF) signal propagation. Traditionally, the PDP for UWB channels are described with a modified Saleh-Valenzuela (S-V) model [19]. However, in a recent work by Demir et al. [36], the authors found that the PDP for UWB transmission in vehicular networks may be characterized by segmenting the PDP (in dB scale) into several linear slopes. Inspired by these facts, we tried to compare our results for 802.11p protocol with measurements for UWB (3 to 11 GHz) performed using the same VNA-based setup. The goal was to find if one can attain similar LSV trends in wider bandwidths as well.

The number of measured points was the same; however, due to the larger BW (8 GHz) and a frequency step size of $f_s = 100$ MHz, we have a smaller time range $t_d = 1/f_s = 10$ ns. Thus, we cannot compare the PDPs in an one-to-one basis. For analyzing LSV models, we use

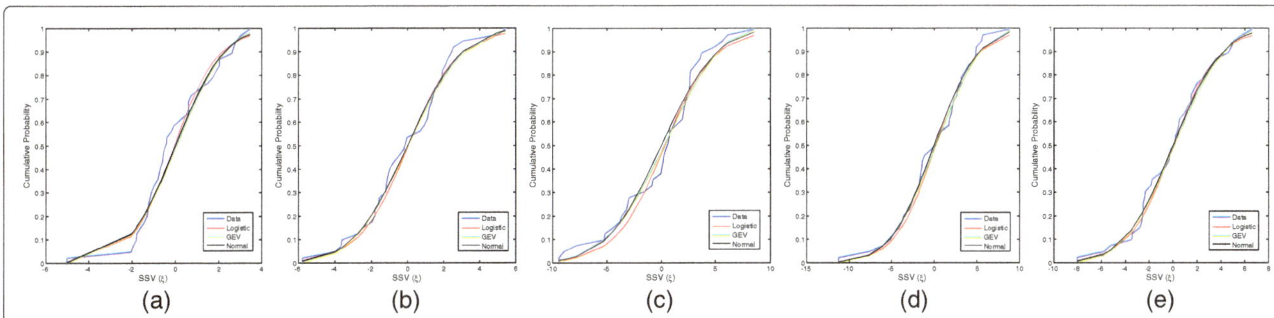

Figure 8 Empirical CDF of small-scale variation fitted with three statistical distributions (logistic, GEV, and normal) for different Tx-Rx separations. (a) $d = 0.53$ m, **(b)** $d = 0.94$ m, **(c)** $d = 1.28$ m, **(d)** $d = 2.03$ m, and **(e)** $d = 3.38$ m.

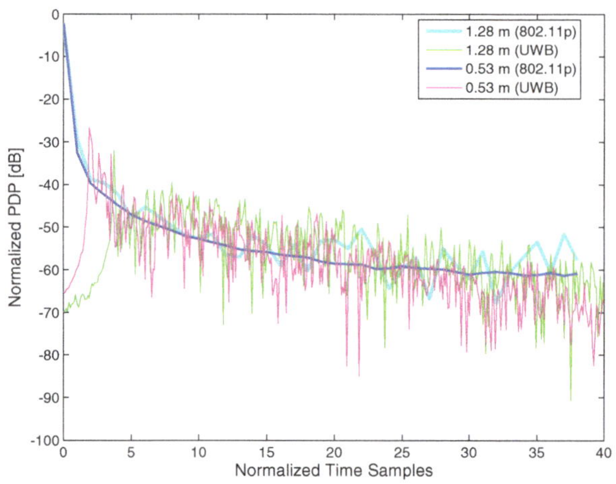

Figure 9 Comparison of 802.11p and UWB PDP trends in normalized time scale.

a scaled comparison instead. For the UWB measurement, the propagation space resolution is $P_{sr} = 3$ cm and the maximum propagation distance is $L_{d\,max} = 30$ m.

Figure 9 shows the UWB and the 802.11p normalized PDP vs normalized time samples. For UWB, it was found that the PDP constitutes one (or a few) major peaks followed by somewhat linear decreasing slope. The reader may also note the delay and lower power for the first peak of the UWB PDP. In addition, the delay and the maximum value of the peak is more for larger distances between the transmitting and receiving antennas. However, this phenomenon is not observed in narrowband 5.8 GHz measurements for 802.11p.

4 Simulation results

For BER simulation, we used MATLAB. A detailed description of the simulation methodology can be found in [37]. Parameters for generating the signal are available in [38]. A block-scheme diagram that corresponds to the SISO transmission model is presented in Figure 10.

The Tx generates 10^6 data sub-frames for a given energy per bit to noise power spectral density ratio (E_b/N_0). For signal transmission, we use rate 1/2 convolutional coding and BPSK modulation. On the receiver side, we use simple least square (LS) estimation and a hard Viterbi decoder. The 802.11p standard is based on OFDM with $N_{\mathrm{IFFT}} = 64$ for a bandwidth of 10 MHz. In the time domain, each OFDM symbol contains 80 chips including a cyclic prefix of length 16. The final transmitted signal has a duration of 8 μs.

Rician distribution had been successfully utilized for non-geometrical stochastic modelling of vehicular channels [11]. A Rician model also encompasses the Rayleigh model as a special case. For the channel block in Figure 10, the tap gains were calculated from the LSV based on a simple Rician model. The respective CIRs are shown in Table 4. It is interesting to find that real-life measurements exhibit a very fast decrease of the tap gains, similar to some earlier reports [39].

The BER results for different distances between Tx and Rx, for LOS and NLOS cases, and for intra-vehicle and out-of-vehicle situations are shown in Figure 11. Results can be classified quite unambiguously into two groups, one for intra-vehicle and the other for out-of-vehicle measurements. The BER worsens for higher Tx-Rx distance

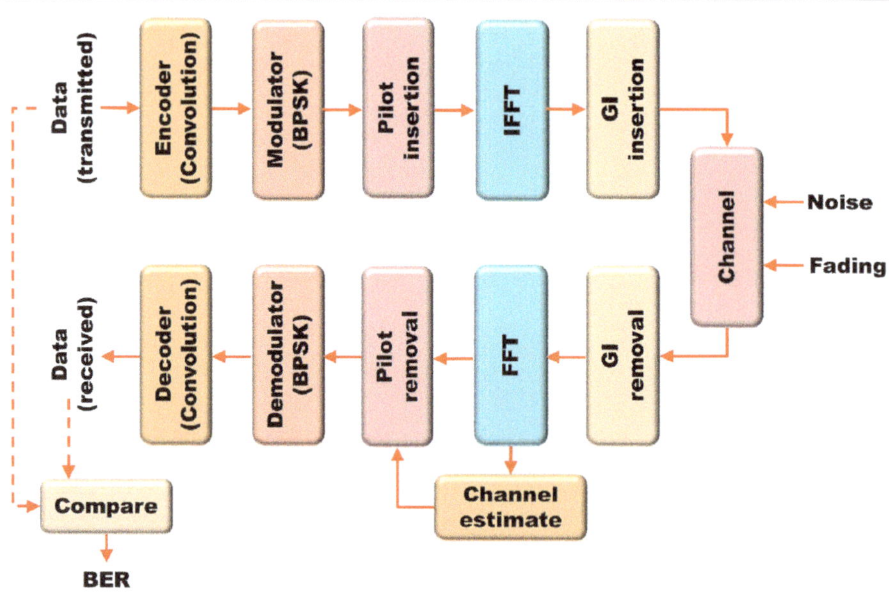

Figure 10 Block diagram of the simulation setup.

Table 4 Characteristics of the channel models

Tap delay (ns)	Tap gain (dB)				
	0.53 m LOS	0.94 m LOS	1.28 m NLOS	2.03 m NLOS	3.38 m NLOS
0	0	0	0	0	0
10	−11.78	−10.61	−9.61	−6.06	−6.69
20	−17.69	−16.24	−14.90	−10.28	−11.06
30	−20.69	−19.27	−18.27	−13.22	−13.93
40	−22.26	−20.93	−20.32	−15.28	−15.80
50	−23.11	−21.87	−21.58	−16.74	−17.04
60	−23.62	−22.45	−22.39	−17.76	−17.85
70	−23.95	−22.83	−22.91	−18.50	−18.40
80	−24.20	−23.10	−23.27	−19.04	−18.76
90	−21.41	−23.33	−23.53	−19.43	−19.01
K factor (dB)	21.30	20.07	19.61	15.15	15.40

inside the car, but outside the vehicle, the BER values almost remain unaffected when the Tx-Rx separation is altered. The BER results for intra-vehicle simulations are slightly better than the results for the out-of-vehicle channel. For a target BER of 10^{-5}, the required E_b/N_0 differs by 6 dB.

5 Conclusions

The aim of the text was to assess the suitability of protocol 802.11p for in-car wireless applications. For the purpose, extensive vehicular channel measurements in the 5.8-GHz band were carried out. A double exponential decay model was used for describing the basic trend of the measured PDP, and we utilized it for IEEE 802.11p BER simulation.

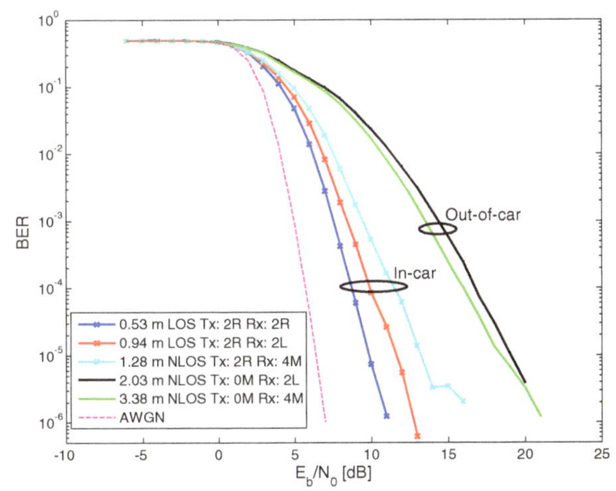

Figure 11 BER performance of IEEE 802.11p standard over the measured channels.

The BER achieves the recommended values for all variants of the channel, and it can be concluded that 802.11p standard (in particular, the PHY protocol stack) can be adopted for intra-vehicular communication systems.

Competing interests

The authors declare that they have no competing interests.

Acknowledgements

This work was supported by the SoMoPro II programme, Project No. 3SGA5720 *Localization via UWB*, co-financed by the People Programme (Marie Curie action) of the Seventh Framework Programme of EU according to the REA Grant Agreement No. 291782 and by the South-Moravian Region. The research is further co-financed by the Czech Science Foundation, Project No. 13-38735S *Research into wireless channels for intra-vehicle communication and positioning*, and by Czech Ministry of Education in frame of National Sustainability Program under grant LO1401. For research, infrastructure of the SIX Center was used. The generous support from Skoda a.s. Mlada Boleslav are also gratefully acknowledged.

Author details

[1]Department of Radio Electronics, Brno University of Technology, Technicka 12, 61600 Brno, Czech Republic. [2]AIT Austrian Institute of Technology GmbH, 1220 Vienna, Austria. [3]Telecommunications Research Center Vienna (FTW), Donau-City-Straße 1, 1220 Vienna, Austria. [4]Institute of Telecommunications, Vienna University of Technology, Gusshausstraße 25, 1040 Vienna, Austria.

References

1. J Gozalvez, M Sepulcre, R Bauza, IEEE 802.11p vehicle to infrastructure communications in urban environments. IEEE Commun. Mag. **50**(5), 176–183 (2012)
2. M Faezipour, M Nourani, A Saeed, S Addepalli, Progress and challenges in intelligent vehicle area networks. Commun. ACM. **55**(2), 90–100 (2012)
3. H Hartenstein, KP Laberteaux, A tutorial survey on vehicular ad hoc networks. IEEE Commun. Mag. **46**(6), 164–171 (2008)
4. IEEE Computer Society, IEEE Standard for Information Technology - telecommunications and information exchange between systems - local and metropolitan area networks - specific requirements - part 11: wireless LAN medium access control (MAC) and physical layer (PHY) specifications amendment 6: wireless access in vehicular environments. IEEE Std. **802**, 1–51 (Jul. 2010). doi: 10.1109/IEEESTD.2010.5514475

5. Society IEEE Vehicular Technology, IEEE guide for wireless access in vehicular environments (WAVE) - architecture. IEEE Std. **1069**, P1609.0/D7.0, , 1–77 (Mar. 2014). http://ieeexplore.ieee.org/servlet/opac?punumber=6576803
6. CAR 2 CAR communication consortium: overview of the C2C-CC system. Manifesto, 1–94 (Aug. 2007). http://elib.dlr.de/48380/1/C2C-CC_manifesto_v1.1.pdf
7. M Muller, WLAN 802.11p Measurements for Vehicle to Vehicle (V2V) DSRC. Application note Rohde & Schwarz: 1 MA152_Oe, 1–25 (Sep. 2009). www.rohde-schwarz.cz/file/1MA152_2e.pdf
8. CF Mecklenbrauker, AF Molisch, J Karedal, F Tufvesson, A Paier, L Bernado, T Zemen, O Klemp, N Czink, Vehicular channel characterization and its implications for wireless system design and performance. Proc. IEEE. **99**(7), 1189–1212 (2011)
9. W Cheng-Xiang, C Xiang, DI Laurenson, Vehicle-to-vehicle channel modeling and measurements: recent advances and future challenges. IEEE Commun. Mag. **47**, 96–103 (2009)
10. A Paier, J Karedal, N Czink, H Hofstetter, C Dumard, T Zemen, F Tufvesson, CF Mecklenbrauker, AF Molisch, in *Proc. IEEE PIMRC*. First results from car-to-car and car-to-infrastructure radio channel measurements at 5.2 GHz (Athens, 2007), pp. 1–5
11. G Acosta-Marum, MA Ingram, Six time- and frequency-selective empirical channel models for vehicular wireless LANs. IEEE Veh. Tech. Mag. **2**, 4–11 (2007)
12. S Biswas, R Tatchikou, F Dion, Vehicle-to-vehicle wireless communication protocols for enhancing highway traffic safety. IEEE Commun. Mag. **44**, 74–82 (2006)
13. L Bernado, T Zemen, F Tufvesson, AF Molisch, CF Mecklenbrauker, Delay and Doppler spreads of non-stationary vehicular channels for safety relevant scenarios. IEEE Trans. Veh. Tech. **63**(1), 82–93 (2014)
14. L Bernado, T Zemen, F Tufvesson, AF Molisch, CF Mecklenbrauker, Time- and frequency-varying K-factor of non-stationary vehicular channels for safety-relevant scenarios. IEEE Trans. Intell. Transport. Syst, 1–11 (2015). doi: 10.1109/TITS.2014.2349364
15. L Cheng, BE Henty, R Cooper, DD Stancil, F Bai, A measurement study of time-scaled 802.11a waveforms over the mobile-to-mobile vehicular channel at 5.9 GHz. IEEE Commun. Mag. **46**(5), 84–91 (2008)
16. I Sen, DW Matolak, Vehicle-vehicle channel models for the 5-GHz band. IEEE Trans. Intell. Transport. Syst. **9**(2), 235–245 (2008)
17. F Maehara, R Kaneko, A Yamakita, S Goto, in *Proc. IEEE CSNDSP*. Coverage performance of MB-OFDM UWB in-car wireless communication (Newcastle upon Tyne UK, 2010), pp. 417–421
18. M Schack, R Geise, I Schmidt, R Piesiewiczk, T Kurner, in *Proc. IEEE EuCAP*. UWB channel measurements inside different car types (Berlin, Germany, 2009), pp. 640–644
19. W Niu, J Li, T Talty, Ultra-wideband channel modeling for intravehicle environment. EURASIP J. Wirel. Commun. Net. **2009**(806209), 1–12 (2009)
20. DW Matolak, A Chandrasekaran, in *Proc. IEEE VTC*. 5 GHz intra-vehicle channel characterization (Quebec City, QC, 2012), pp. 1–5
21. M Wellens, B Westphal, P Mahonen, in *Proc. IEEE VTC*. Performance evaluation of IEEE 802.11-based WLANs in vehicular scenarios (Dublin, 2007), pp. 1167–1171
22. FA Teixeira, VF e Silva, JL Leoni, DF Macedo, JMS Nogueira, Vehicular networks using the IEEE 802.11p standard: an experimental analysis. Vehic. Commun. **1**(2), 91–96 (2014)
23. M Jankiraman, *Space-Time Codes and MIMO Systems*. (Artech House, Boston, London, 2004)
24. P Pupalaikis, in *Proc. DesignCon*. The relationship between discrete frequency s-parameters and continuous frequency responses (UBM, 2012)
25. J Hansen, M Nold, in *Proc. IEEE GLOBECOM*. Analytic calculation of the power delay profile for single room wireless LAN environments, vol. 1 (San Francisco, CA, USA, 2000), pp. 98–102
26. J Hansen, An analytical calculation of power delay profile and delay spread with experimental verification. IEEE Commun. Lett. **7**(6), 257–259 (2003)
27. H Yang, P Smulders, M Herben, Channel characteristics and transmission performance for various channel configurations at 60 GHz. EURASIP J. Wirel. Commun. Net. **2007**(19613), 1–15 (2007)
28. S Guerin, YJ Guo, SK Barton, in *Proc. IEEE ICAP*. Indoor propagation measurements at 5 GHz for HIPERLAN, vol. 2 (Edinburgh, UK, 1997), pp. 306–310
29. S Kotz, S Nadarajah, *Extreme Value Distributions: Theory and Applications*. (Imperial College Press, London, UK, 2000)
30. J Blumenstein, T Mikulasek, R Marsalek, A Chandra, A Prokes, T Zemen, C Mecklenbrauker, in *IEEE VNC*. In-vehicle UWB channel measurement, model and spatial consistency (Paderborn, Germany, 2014), pp. 77–80
31. J Blumenstein, T Mikulasek, R Marsalek, A Prokes, T Zemen, C Mecklenbrauker, in *Proc. IEEE VTC*. In-vehicle mm-wave channel model and measurement (Vancouver, Canada, 2014), pp. 1–5
32. N Balakrishnan, VB Nebzorov, *A Primer on Statistical Distributions*. (John Wiley and Sons, New York, USA, 2003)
33. H El-Sallabi, DS Baum, P Zetterberg, P Kyosti, T Rautiainen, C Schneider, in *Proc. IEEE VTC*. Wideband spatial channel model for MIMO systems at 5 GHz in indoor and outdoor environments, vol. 6 (Melbourne, Australia, 2006), pp. 2916–2921
34. Y Guo, J Zhang, C Tao, L Liu, L Tian, Propagation characteristics of wideband high-speed railway channel in viaduct scenario at 2.35 GHz. J Modern Transportation. **20**(4), 206–212 (2012)
35. FJ Massey, The Kolmogorov-Smirnov test for goodness of fit. J. Am. Statist. Assoc. **46**(253), 66–78 (1951)
36. U Demir, CU Bas, SC Ergen, Engine compartment UWB channel model for intravehicular wireless sensor networks. IEEE Trans. Veh. Tech. **63**(6), 2497–2505 (2014)
37. P Kukolev, Comparison of 802.11a and 802. 11p over fading channels. Elektrorevue. **4**(1), 7–11 (2013)
38. P Kukolev, in *Proc. Radioelektronika*. BER performance of 802.11p standard over an ITU- R multipath channel (Pardubice, Czech Republic, 2013), pp. 391–396
39. A Paier, *The vehicular radio channel in the 5 GHz band*. PhD Thesis. (Technischen Universitat Wien, Vienna, Austria, 2010)

17

EASO-TDMA: enhanced *ad hoc* self-organizing TDMA MAC protocol for shipborne *ad hoc* networks (SANETs)

Changho Yun* and Yong-kon Lim

Abstract

As very high frequency (VHF) data links are overloaded due to the rise of Automatic Identification System (AIS) applications, additional VHF channels and the development of corresponding communication technologies are necessary. For this reason, shipborne *ad hoc* networks (SANETs) and the medium access (MAC) protocol for SANETs, *ad hoc* self-organizing-time division multiple access (ASO-TDMA), have been proposed, which provide ocean-going ships with diverse data services cost-effectively. In this paper, a new MAC protocol for SANETs, enhanced ASO-TDMA (EASO-TDMA), is proposed in order to manage the bottleneck and under-utilization problems resulting from ASO-TDMA and thus enhances the overall performance of SANETs. To accomplish this, in EASO-TDMA, a ship adaptively allocates its time slots according to the number of subscribing ships, which make their communication route to a base station at the shore via the ship, by referring to the routing table. The performance of EASO-TDMA is investigated through simulations and compared with that of ASO-TDMA. The simulation results indicate that EASO-TDMA outperforms ASO-TDMA, regardless of the network conditions, with 25% higher reception success rate, 25% lower collision rate, and 21% higher channel utilization at maximum. In particular, the end-to-end delay of EASO-TDMA is on average 1/12 of that of ASO-TDMA, and this implies that EASO-TDMA can be effective for delay-sensitive data services.

Keywords: *Ad hoc*; Communication; MAC; Maritime; Network; Simulation; TDMA; VDE; VHF

1 Introduction

Automatic Identification System (AIS) for the safety of navigation is now used in diverse applications (e.g., aids to navigation (AtoN), application-specific message (ASM), and man overboard (MOB) unit), and this causes the overload of the very high frequency (VHF) data links [1]. Simultaneously, the demand for maritime radio spectra for digital data exchange that requires a higher data rate and longer payload size than those of AIS is increasing [2]. To handle these situations, several VHF channels were allocated for the purpose of digital VHF data exchange (VDE) at the World Radiocommunication Conference 2012 (WRC-12) [3]. The VDE system which integrates the function of AIS, ASM, and VDE was also introduced by the International Association of Lighthouse Authorities (IALA) [4]. In addition, the characteristics of VHF radio systems and equipment for use in the VDE have been recommended by the International Telecommunication Union Radiocommunication sector (ITU-R) [5].

Under these circumstances, shipborne *ad hoc* networks (SANETs), counterparts of mobile *ad hoc* networks (MANETs) on land, were conceptualized in [6] and introduced in the International Marine Organization (IMO) Facilitation sessions (i.e., IMO FAL-38 and FAL-39), as well as the IMO Navigation, Communications and Search and Rescue session (i.e., IMO NCSR-2) [7-9]. In SANETs, a ship located far away from the base station at the shore can transmit its data to the base station with the help of *ad hoc* communication, provided that it can locate its neighboring ships. Thus, SANETs can cost-effectively provide ships with VDE services at a high data rate in place of satellite communications [2,6]. The communication system for use in SANETs can be implemented by referring to the system characteristics recommended in

*Correspondence: sgn0178@kriso.re.kr
Ocean System Engineering Research Division, KRISO, 32, Yuseong-daero 1312 beon-gil, Yuseong-gu, Daejeon 305-343, Republic of Korea

[10], which include the modulation technique, the range of transmitting power and receiver sensitivity, the radiation pattern, and so on. The medium access control (MAC) protocol for the VDE recommended in [10] is carrier sense-time division multiple access (CS-TDMA) that was originally designed for AIS class B traffic [11]. CS-TDMA can be inefficient for SANETs because it is targeted for peer-to-peer communication. For this reason, a new MAC protocol for SANETs, *ad hoc* self-organizing TDMA (ASO-TDMA), has been designed in [2], which is based on the MAC protocol for AIS class A traffic, SO-TDMA, where a ship allocates a fixed number of time slots per minute in a decentralized manner.

ASO-TDMA provides a ship with the means to reliably transmit its data and other ships' data by piggybacking them from the source to the destination with the help of one-hop acknowledgement. In particular, the channel access of ASO-TDMA is designed to reduce the possibility of receiver collisions as much as possible in order to guarantee end-to-end data transfer without failures [2]. To accomplish this, the frame used in ASO-TDMA is fragmented into several sub-frames and the network is divided into several hops, where a hop is defined as a zone determined by its distance from the base station. A ship located in an arbitrary hop can only reserve a fixed number of time slots in the available sub-frames in order to avoid receiver collisions. The simulation results indicate that ASO-TDMA outperforms SO-TDMA and CS-TDMA in terms of the receiver collision rate and the end-to-end delay, regardless of the network conditions. Moreover, the performance margin increases as the volume of ship traffic increases [2].

On the other hand, the effort required to reduce the receiver collision rate of ASO-TDMA (i.e., allocating time slots only at available sub-frames) can result in poorer channel utilization as the volume of ship traffic decreases. This can also be ineffective for delay-sensitive VDE services, because the long time difference between two consecutively allocated time slots belonging to a different sub-frame can cause time-outs during the establishment of a connection. In addition, the fact that all of the ships take the same number of time slots at each sub-frame can cause a bottleneck problem at a specific ship, where the volume of incoming data exceeds the number of allocated time slots due to the excessive relaying of other ships' data. Accordingly, a new MAC protocol for SANETs, enhanced ASO-TDMA (EASO-TDMA), is proposed in this paper, in order to overcome the weaknesses of ASO-TDMA and improve the overall performance.

Let us consider two ocean-going ships, ships A and B, to explain EASO-TDMA. Ship A is located at a position where it is able to receive the base station's signal, while ship B is not directly connected to the base station and transfers its data to the base station via ship A in an *ad hoc* manner. Then, ship B is defined as ship A's subscribing ship. Let us also consider a hyperframe and a superframe in EASO-TDMA, which correspond to the frame and subframe in ASO-TDMA, respectively. In EASO-TDMA, a ship determines adaptively the number of allocated time slots according to the number of subscribing ships, in order to ease the bottleneck problem. To do this, the ship checks the routing table to obtain the information of the number of its subscribing ships at the end of the current hyperframe and determines the number of allocated time slots for the next hyperframe. In addition, a ship can allocate time slots at all superframes in EASO-TDMA, whereas it can do so only at the available sub-frames in ASO-TDMA. This enables data transfer between the source and the destination to be done during one hyperframe, and this can reduce the overall end-to-end delay. Accordingly, it is expected that EASO-TDMA can outperform ASO-TDMA in terms of the reception success rate, collision rate, end-to-end delay, and channel utilization.

The rest of this paper is structured as follows. Section 2 describes the characteristics of the SANET, and EASO-TDMA is explained in detail in Section 3. In Section 4, the performance of EASO-TDMA is investigated via simulations and the results are compared to those of ASO-TDMA. In Section 5, this paper is concluded.

2 The SANET

Diverse ships of different usage and tonnage, in different positions at sea, can be nodes for the SANET. This is comparable to AIS where ships with a gross tonnage of 300 or more (engaged in international voyages), cargo ships with a gross tonnage of 500 or more (not engaged on international voyages), and passenger ships irrespective of their size are involved [12]. As illustrated in Figure 1, the SANET consists of a base station at the shore and ocean-going ships, and the topology of the SANET is infrastructureless, due to the uncertainty of the ships' location and navigational direction. The base station is connected to the terrestrial backbone network and plays a role as an edge router, in order to provide ships with various data services such as general electronic mail, safety-related information, weather maps, electronic maps, ship position and movement reporting, and so on [10]. A ship attempts to make IP connectivity to the base station in order to obtain uplink IP-level interoperability (i.e., ship-to-shore) with the terrestrial backbone network and access these data services effectively through the SANET.

In terms of obtaining IP connectivity, a ship can be directly connected to a base station, provided that the base station is within its communication range by guaranteeing line-of-sight (LOS). Otherwise, the ship can be indirectly connected to the base station with the help of neighboring ships in an *ad hoc* manner. The VHF communication range of a ship depends on the length of its mast,

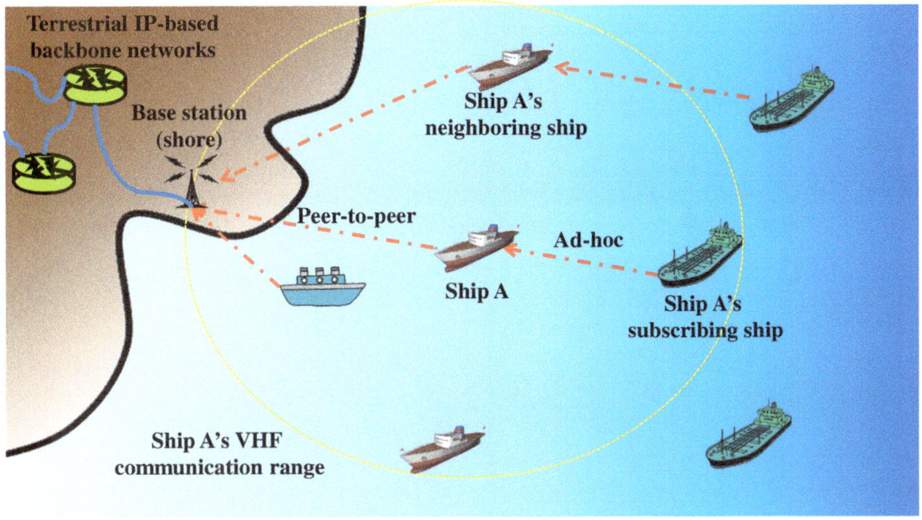

Figure 1 SANET architecture.

which is determined by the length overall (LOA) [13]. In the case where the length of the mast is less than 5 m, the VHF communication range is from 20 to 30 km. In the case where the length of the mast is from 12 to 30 m, the VHF communication range is from 55 to 120 km. As an example, a ship with a mast of 30 m that makes a voyage along a sea route which is normally 100 nautical miles (i.e., 185 km) from the shore, it can achieve IP connectivity to a base station by going through three or four hops, provided that it locates its neighboring ships.

The procedure to join the SANET is as follows:

1. A ship primarily checks the availability of the SANET by listening to the neighboring ships' signals.
2. If the ship can locate its neighbors, it tries to allocate time slots on the basis of the given MAC protocol.
3. After allocating time slots, the ship can achieve IP connectivity to the base station by determining a communication route based on the given routing protocol.
4. The base station provides ships with IP connectivity by relaying data from the backbone network to a ship, or vice versa.
5. If a ship can neither reach a base station nor locate its neighboring ships, it should find another communication method to connect to the base station.

3 EASO-TDMA

It is hard for a base station in the SANET to control the channel access of ocean-going ships, due to the uncertainty of their location and navigational direction. Thus, the ships need to allocate time slots for themselves in a decentralized manner by using EASO-TDMA. In this section, EASO-TDMA is described, including the time structure, the main algorithm that consists of the first phase (FP) and the normal phase (NP), as well as the channel access algorithm.

3.1 Time structure

A universal timing reference is necessary because all ships and base stations in the SANET allocate their time slots in a decentralized way. The time structure for the VDE has not been determined yet, and both IALA and ITU-R SG5 WP5B are still working to propose the time structure by referring to that of Terrestrial Trunked Radio (TETRA) [14,15]. In this paper, we refer to the time structure specified in [14] in order to define that of EASO-TDMA.

As illustrated in Figure 2, the time structure is based on one EASO-TDMA cycle that takes an hour. One EASO-TDMA cycle is also subdivided into several shorter time units. The longest time unit is a hyperframe, and the shortest one is a time slot. The time units for use in EASO-TDMA are explained in descending order of their length.

1. Hyperframe: A hyperframe is the counterpart of a frame in ASO-TDMA, and the hyperframe index belonging to one EASO-TDMA cycle is defined as HN ($HN = 1:1:60$). The length of one hyperframe is 1 min and consists of five superframes.
2. Superframe: A superframe is the counterpart of a sub-frame in ASO-TDMA, and the superframe index belonging to one hyperframe is given as SN ($SN = 1:1:5$). The length of a superframe is 12 s and consists of 15 multiframes.
3. Multiframe: The multiframe index belonging to one superframe is defined as MN ($MN = 1:1:15$). The

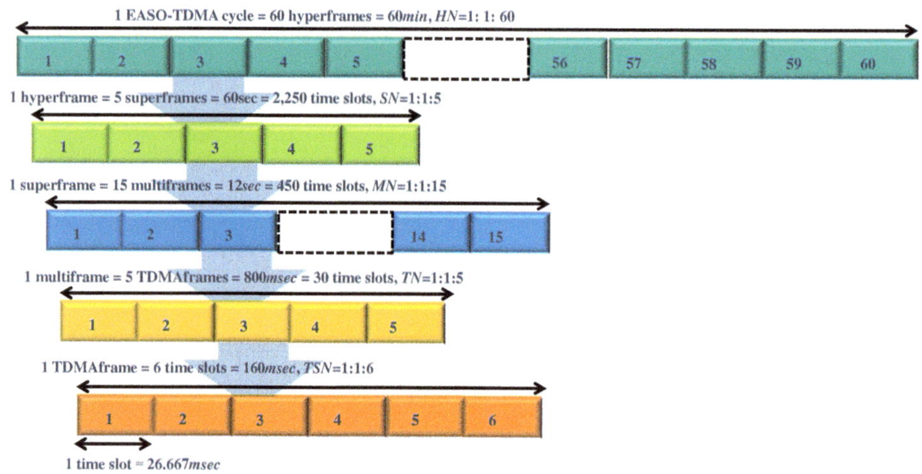

Figure 2 The time structure of EASO-TDMA.

length of one multiframe is 800 ms and consists of five TDMAframes.

4. TDMAframe: The TDMAframe index belonging to one multiframe is TN ($TN = 1:1:5$). The length of one TDMA frame is 160 ms and consists of six time slots.
5. Time slot: The length of one time slot is approximately 26.667 ms, and the time slot index belonging to one TDMAframe is given as TSN ($TSN = 1:1:6$).

If a ship tries to join the SANET, it can synchronize the time by checking the current time point and converting the time point into EASO-TDMA time units which can be expressed in terms of *HN*, *SN*, *MN*, *TN*, and *TSN*.

3.2 Main algorithm

Figure 3 illustrates the overall flowchart of EASO-TDMA, and Table 1 defines the parameters to describe EASO-TDMA. EASO-TDMA broadly consists of the FP and the NP. A ship joins the network in the FP by allocating its time slots and making a connection to the base station. Then, the ship goes to the NP and stays in this phase unless it terminates the connection.

3.2.1 *First phase*

Let us consider ship i that is not currently a member of the network. If ship i tries to join the network, it primarily executes the scan one hyperframe (SOHF) process in the FP (see the square block (1) in Figure 3). The SOHF is a process where ship i enters reception mode by receiving the packets from its neighboring ships and the base station during one hyperframe. The details of the SOHF are given as follows:

1. Ship i primarily synchronizes the time structure of EASO-TDMA by using GPS.
2. Ship i receives packets time slot by time slot (i.e., for 2,250 time slots).
3. At each time slot, ship i checks the MAC headers of the received packet and updates the channel status table that is an important reference for allocating its time slots. The channel status table includes the availability of the time slot (busy or idle), the identification of neighboring ship j that allocates the time slot, the number of ship j's subscribing ships, and the occupancy number of ship j.
4. If the received packet is found to be corrupted after error checking, it is just dropped.
5. At the end of each superframe (i.e., $TSN = 6, TN = 5, MN = 15$), ship i determines its time slots with the help of the channel access algorithm, based on the channel status table.

After finishing the SOHF, ship i checks the existence of neighboring ships (see the diamond block (2) in Figure 3). If there are no neighboring ships, the number of failures, N_{Fi}, decreases by one. The N_{Fi} parameter is used in order to avoid the formation of an infinite loop of ships trying to join the network, as well as to try to find another communication mode instead. The value of N_{Fi} is determined by considering the communication environments of the SANET. If N_{Fi} is greater than one, ship i goes to the SOHF again, in order to detect its neighboring ships again. $N_{Fi} = 0$ implies that the SANET is unavailable due to the absence of neighboring ships. Then, ship i determines whether it stops joining the network (see the diamond block (6) in Figure 3). If ship i wants to continue to join the network, it goes to the SOHF. Otherwise, ship i stops joining the network and needs to use another communication system. If there are any neighboring ships, it goes to the transmit and receive one hyperframe (TROHF) (see the square block (3) in Figure 3). The TROHF is a process

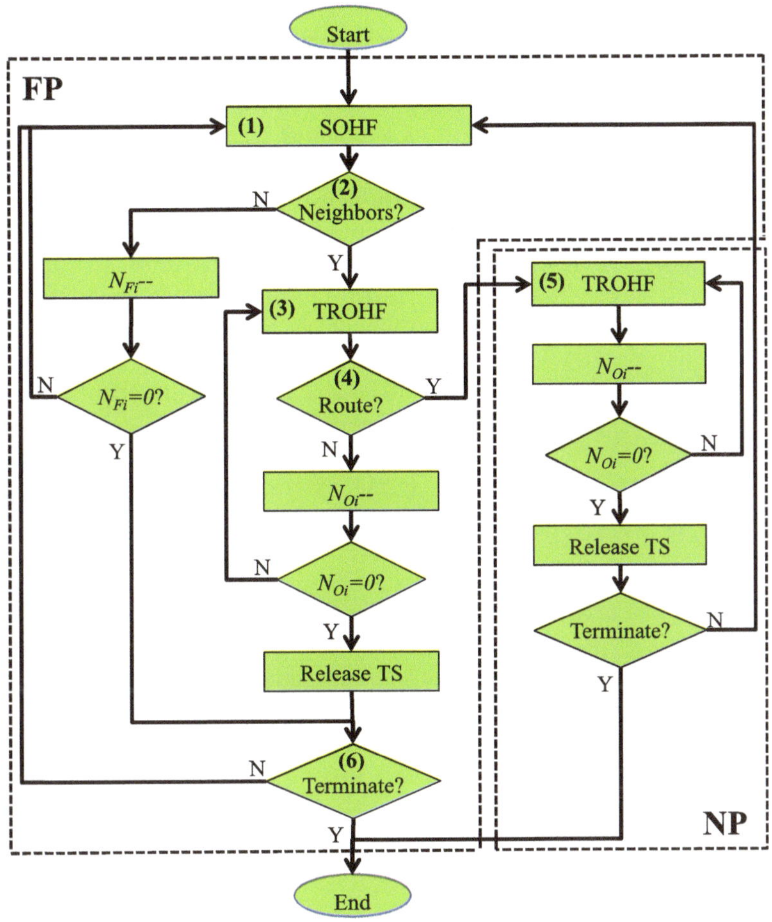

Figure 3 The flowchart of EASO-TDMA.

where ship i is in the transmission and reception modes interchangeably and can be described as follows:

1. The TROHF is also executed time slot by time slot during one hyperframe.
2. In the transmission mode, ship i transmits its packet at its allocated time slots, which have been determined in the SOHF, in order to make a route to the base station.
3. In the reception mode, ship i receives the other ships' packets and updates the channel status table at the rest of the time slots in the same manner as in the SOHF. Ship i does not relay the received packet because it is not connected to the base station. If the received packet is found to be corrupted after error checking, it is just dropped.

After finishing the TROHF, ship i checks whether a route between ship i and the base station is available (see the diamond block (4) in Figure 3). If a route is set up, ship i updates the routing table and goes to the TROHF in the NP (see the square block (5) in Figure 3). Otherwise, the occupancy number of ship i, N_{Oi}, decreases by one. The use of the N_{Oi} parameter can avoid the possibility that a specific ship monopolizes the time slots consistently. The value of N_{Oi} is determined by considering the communication environments of the SANET. If N_{Oi} is greater than one, ship i goes to the TROHF again in order to make a route to the base station. If N_{Oi} is zero, ship i releases its time slots and determines whether it should continue to try to join the network (see the diamond block (6) in Figure 3). If ship i wants to continue to make a route to the base station, it goes to the SOHF again in order to allocate new time slots. Otherwise, ship i stops trying to join the network.

3.2.2 *Normal phase*

In the NP, ship i continuously executes the TROHF (see the square block (5) in Figure 3) unless it terminates its connection to the network. The characteristics of the TROHF in the NP are given as follows:

1. In the transmission mode, ship i transmits its own packet or relays the received packet from its subscribing ship at its allocated time slots.

Table 1 Parameters for EASO-TDMA

Parameters	Description
N	The number of ships
i	The index of a ship ($1 \leq i \leq N$)
j	The index of ship i's neighboring ships
k	The index of ship i's subscribing ships
TN	The index of a TDMAframe
MN	The index of a multiframe
SN	The index of a superframe
HN	The index of a hyperframe
TSN	The index of a time slot
TSN_i	The index of time slot reserved by ship i
N_{BSMF}	The number of multiframes per superframe for use in a base station
N_{Oi}	The occupancy number of ship i
N_{Fi}	The number of failures of ship i
$n_R(i, HN)$	The number of ship i's subscribing ships at hyperframe HF
$Set_R(i, HN)$	The set of ship i's subscribing ships at hyperframe HF
$N_{SF}(i, HN, SN)$	The number of time slots allocated by ship i at hyperframe HF and superframe SN
$N_{SFidle}(i, HN, SN)$	The number of idle time slots scanned by ship i at hyperframe HF and superframe SN
$N_{MFidle}(i, HN, SN, MN)$	The number of idle time slots scanned by ship i at hyperframe HF, superframe SN, and multiframe MN
$N_{TFidle}(i, HN, SN, MN, TN)$	The number of idle time slots scanned by ship i at hyperframe HF, superframe SN, multiframe MN, and TDMAframe TN

2. In the reception mode, ship i receives other ships' packets and updates the channel status table. If the received packet is found to be from ship i's subscribing ship when checking the routing table, it is sent to the transmitting queue. Otherwise, the received packet is just dropped. If a new route is added or an existing route is disconnected, ship i updates the routing table. In addition, if the received packet is found to be corrupted after error checking, it is just dropped.
3. The channel access algorithm is executed at the end of the current hyperframe because ship i needs to add more time slots or release allocated time slots according to the number of subscribing ships.

As in the FP, ship i decreases N_{Oi} by one after finishing the TROHF. If N_{Oi} is greater than one, ship i goes to the TROHF again. If N_{Oi} is zero, ship i releases its time slots and determines whether it should stay in the network. If ship i wants to stay in the network, it goes to the SOHF in the FP again in order to allocate new time slots. Otherwise, ship i terminates its connection to the network.

3.3 Channel access algorithm

Before explaining the channel access algorithm, it is assumed that a base station can allocate a fixed portion of time slots per hyperframe in order to guarantee downlink data services. To do this, the base station can allocate N_{BSMF} multiframes per superframe, and it informs the ships of N_{BSMF} by broadcasting BEACON packets regularly.

The channel access algorithm for the ships is executed in the SOHF of the FP and the TROHF of the NP in order to allocate time slots for use in the next hyperframe. The way to allocate time slots in the SOHF is as follows:

1. Ship i can basically allocate ten time slots which is the number of allocated time slots during one report interval in CS-TDMA [11]. Thus, ship i allocates two time slots per superframe. To do this, ship i checks the number of idle time slots, $N_{SFidle}(i, HN, SN)$, at the end of the current superframe ($TSN = 6, TN = 5, MN = 15$), and it can allocate time slots in the case of $N_{SFidle}(i, HN, SN) \geq 4$ (see the diamond block (1) in Figure 4) [11]. Then, the number of time slots for the same numbered superframe in the next hyperframe, $N_{SF}(i, HN + 1, SN)$, can be determined case by case according to the value of $N_{SFidle}(i, HN, SN)$, two in the case of $N_{SFidle}(i, HN, SN) \geq 4$ and zero in the case of $N_{SFidle}(i, HN, SN) > 4$.
2. Figure 5 shows an example showing that ship i allocates its time slots per superframe. At one superframe, ship i primarily excludes N_{BSMF} multiframes, that are initially assigned for the base station, among the 15 multiframes of one superframe (see the dotted box (1) in Figure 5). Then, ship i selects two multiframes in descending order of $N_{MFidle}(HN, SN, MN)$ (see the dotted box (2) in Figure 5). If there are more than two multiframes whose $N_{MFidle}(HN, SN, MN)$ values are equal, ship i randomly selects one multiframe among them.
3. Ship i selects one TDMAframe of the highest $N_{TFidle}(HN, SN, MN, TN)$ out of the five TDMAframes at the selected multiframe (see the dotted box (3) in Figure 5). If there are more than two TDMAframes whose $N_{TFidle}(HN, SN, MN, TN)$ values are equal, ship i randomly selects one TDMAframe among them.
4. Ship i selects one time slot among the idle time slots at the selected TDMAframe (see the dotted box (4) in Figure 5).

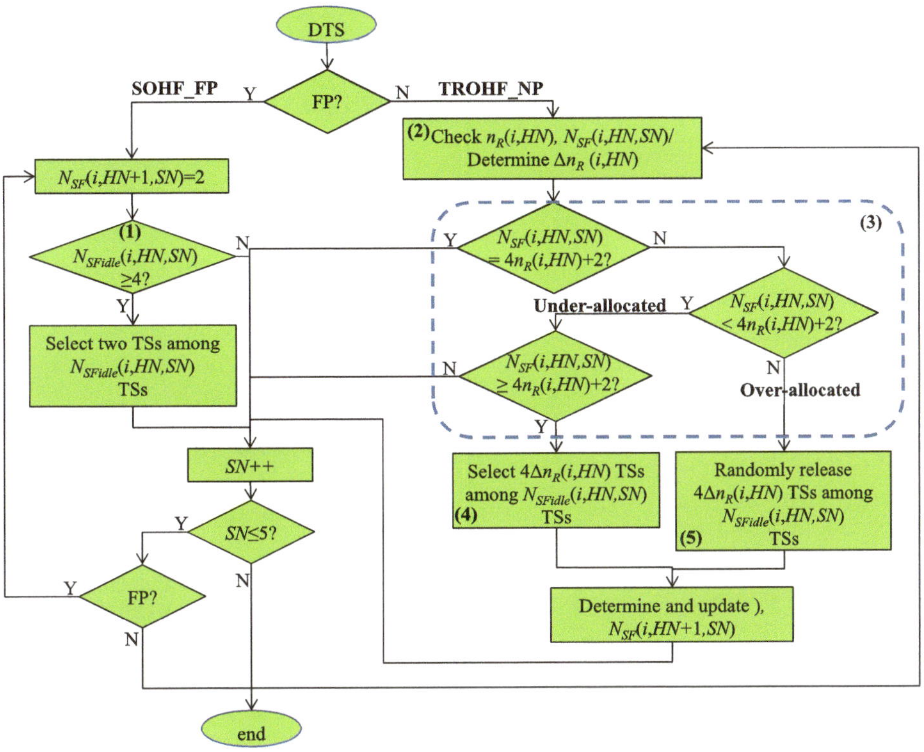

Figure 4 The flowchart of the channel access algorithm.

5. Ship i can allocate time slots hyperframe by hyperframe using methods 2 to 4.

In the TROHF of the NP, ship i can allocate more than two time slots according to the number of subscribing ships. Ship i basically needs two time slots for its transmission and four time slots for its subscribing ship's uplink and downlink transmission per superframe. The number of ship i's subscribing ships is determined at the end of the current hyperframe HN, in order to determine the number of time slots for the next hyperframe $HN + 1$ at that time. The procedures employed to allocate time slots are as follows:

1. Ship i determines the number of its subscribing ships, $n_R(i, HN)$, by referring to the routing table at

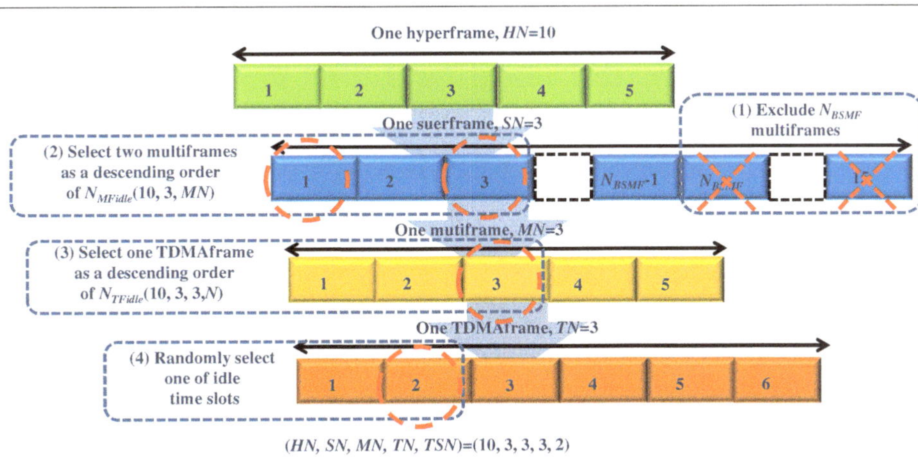

Figure 5 An example of allocating time slots in the SOHF.

the end of the current hyperframe HN (see the square box (2) in Figure 4). Let us consider ship k as ship i's subscribing ship. Then, ship k as well as ship k's subscribing ships become ship i's subscribing ships. Thus, $n_R(i, HN)$ can be represented as

$$n_R(i, HN) = \sum_{k \in Set_R(i,HN)} [n_R(k, HN) + 1]. \quad (1)$$

Thus, the number of time slots allocated per superframe for the next hyperframe $HN + 1$, $N_{SF}(i, HN + 1, SN)$, can be obtained as $4n_R(i, HN) + 2$. Let us define the variation of the number of ship i's subscribing ships at the end of hyperframe HN as $\Delta n_R(i, HN)$ which can be expressed as $|n_R(i, HN - 1) - n_R(i, HN)|$.

2. Ship i checks the status of each superframe SN at hyperframe HN in order to determine $N_{SF}(i, HN + 1, SN)$. There are three cases as follows (see the dotted box (3) in Figure 4):

   (a) Under-allocated: $N_{SF}(i, HN, SN) < 4n_R(i, HN) + 2$.
   (b) Over-allocated: $N_{SF}(i, HN, SN) > 4n_R(i, HN) + 2$.
   (c) Unchanged: $N_{SF}(i, HN, SN) = 4n_R(i, HN) + 2$.

   In the under-allocated case, $N_{SF}(i, HN + 1, SN)$ can be determined case by case according to the values of $N_{SFidle}(i, HN, SN)$ and $\Delta n_R(i, HN)$; $N_{SF}(i, HN, SN) + 4\Delta n_R(i, HN)$ in the case of $N_{SFidle}(i, HN, SN) \geq 4$ and $\Delta n_R(i, HN) \geq 1$, and $N_{SFidle}(i, HN, SN)$ in the case of $N_{SFidle}(i, HN, SN) < 4$. In the over-allocated case, $N_{SF}(i, HN + 1, SN)$ can be obtained as

$$N_{SF}(i, HN+1, SN) = N_{SF}(i, HN, SN) - 4\Delta n_R(i, HN). \quad (2)$$

   In the unchanged case, $N_{SF}(i, HN + 1, SN)$ is the same as $N_{SF}(i, HN, SN)$.

3. In the under-allocated case, $4\Delta n_R(i, HN)$ time slots are additionally assigned (see the square box (4) in Figure 4). The way to allocate additional time slots per superframe depends on ΔN_{SF} which is defined as $15 - N_{BSMF} - 4\Delta n_R(i, HN)$.

   (a) In the case of $\Delta N_{SF} \geq 0$, ship i selects $4\Delta n_R(i, HN)$ multiframes in a similar manner to the allocation of time slots in the SOHF of the FP.
   (b) In the case of $\Delta N_{SF} < 0$, ship i periodically selects $15 - N_{BSMF}$ multiframes as similarly as allocating time slots in the SOHF of the FP until the number of allocated time slots becomes $4\Delta n_R(i, HN)$ at each superframe.

4. In the over-allocated case, $4\Delta n_R(i, HN)$ time slots need to be released (see the square box (5) in Figure 4). The way to release time slots per superframe also depends on ΔN_{SF}.

   (a) Figure 6 provides an example showing that ship i releases its time slots per superframe. In the case of $\Delta N_{SF} \geq 0$, ship i selects $4\Delta n_R(i, HN)$ multiframes in ascending order of $N_{MFidle}(HN, SN, MN)$ (see the dotted box (1) in Figure 6). If there are more than two multiframes whose $N_{MFidle}(HN, SN, MN)$

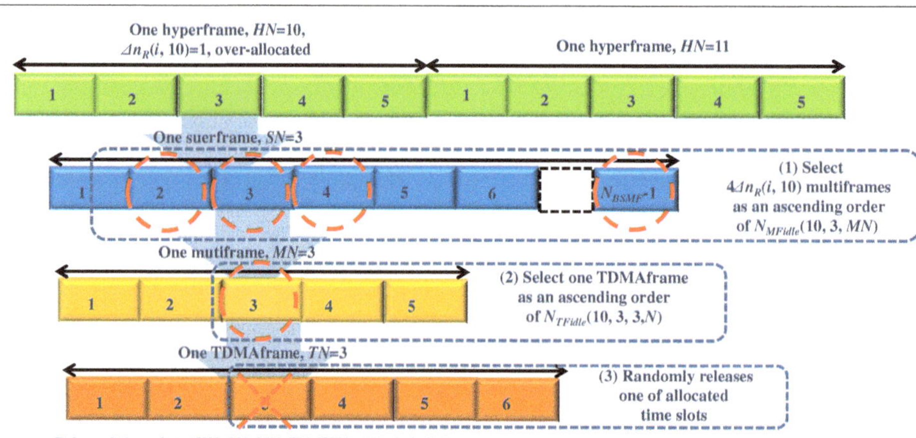

Figure 6 An example of releasing time slots in the TROHF in the NP.

values are equal, ship i randomly selects one multiframe among them. Ship i selects one TDMAframe of the lowest $N_{TFidle}(HN, SN, MN, TN)$ out of the five TDMAframes at the selected multiframe (see the dotted box (2) in Figure 6). If there are more than two TDMAframes whose $N_{TFidle}(HN, SN, MN, TN)$ values are equal, ship i randomly selects one TDMAframe among them. Finally, ship i releases one of the allocated time slots at the selected TDMA frame (see the dotted box (3) in Figure 6).

(b) In the case of $\Delta N_{SF} < 0$, ship i periodically selects $15 - N_{BSMF}$ multiframes in a similar manner to the release of the time slots in method 4(a) until the number of time slots released becomes $4\Delta n_R(i, HN)$ at each superframe.

4 Simulations

In this section, the performance of EASO-TDMA is investigated via simulations and compared to that of ASO-TDMA by considering various network conditions.

4.1 Conditions

The network conditions for simulations are given as follows:

1. As illustrated in Figure 7, a square-shaped network, which consists of one base station and N ships, is considered. The size of the network is 120 km by 120 km.
2. The position of the base station is stationary and given as $(0, 60)$ as shown in Figure 7. N ships are randomly deployed, and their initial positions are randomly determined within the network. The maximum of X- and Y-coordinates for all ships, defined as $X_{\max}$ and $Y_{\max}$, are respectively given as 120. The minimum of X- and Y-coordinates for all ships, defined as $X_{\min}$ and $Y_{\min}$, are respectively given as 0.
3. All ships move with a velocity in the range of the uniform random variable, v, within the network. The mobility of a ship is modeled as follows:
   - At each simulation time, two uniform random numbers C_x and C_y are generated, where $0 \leq C_x, C_y \leq 1$.
   - In the case of $C_x \geq 0.5$, the X-coordinate increases as much as $v \cdot \Delta t$, where Δt is the inter-simulation time, provided that the newly increased X-coordinate is less than $X_{\max}$. Otherwise, the X-coordinate decreases as much as $v \cdot \Delta t$ in order to make the ship stay in the given network area during simulation. In the case of $C_x < 0.5$, the X-coordinate decreases as much as $v \cdot \Delta t$, provided that the newly decreased X-coordinate is greater than $X_{\min}$. Otherwise, the X-coordinate increases as much as $v \cdot \Delta t$ by the same reason.
   - The Y-axis coordinate also increases or decreases as much as $v \cdot \Delta t$ with respect to the value of C_y as similarly as the way to determine the X-axis coordinate.
   - Finally, the X- and Y- coordinate increment or decrement results in the direction and the distance of a ship's movement.
4. As assumed in [2], all ships and the base station have a maximum communication distance, L, in order to easily achieve *ad hoc* communication.
5. Among the N ships, there are N_a active ships which allocate their time slots, make a route to the base station, and transmit their packet as well as their subscribing ships' packets.
6. The number of time slots that are allocated by the base station is given as 300 per hyperframe (or frame) for the two MAC protocols. In addition, the number of allocated time slots per sub-frame for ASO-TDMA is given as ten in order to make it equal to the number of allocated time slots per superframe in the SOHF for EASO-TDMA.
7. Both uplink and downlink traffic are considered. All active ships generate n uplink packets per hyperframe which are destined for the base station. The base station also generates n downlink packets per hyperframe, and each packet is destined for one of the active ships that is randomly selected.
8. The length of ingress and egress queues for all ships is fixed as K, and the queueing discipline is simply given as first-in first-out (FIFO). If the queue is found to be full when a packet arrives, the packet is just dropped out.
9. The values for the simulations are given in Table 2.

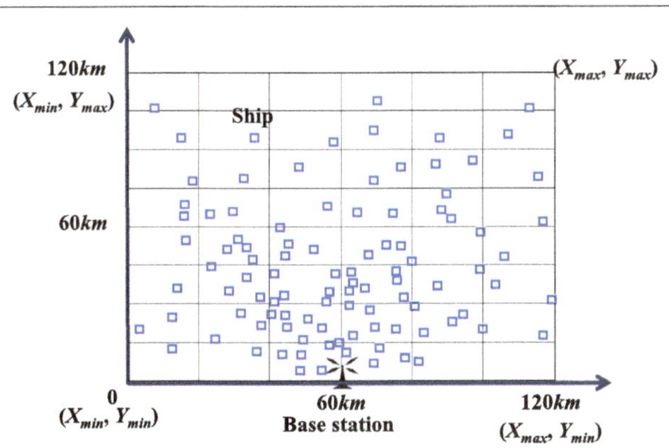

Figure 7 The network topology for the simulation.

Table 2 Simulation conditions

Parameters	Values
N	100
N_a	20 : 20 : 100
n	{1, 5, 10, 15, 20}
v (knots)	Uniformly distributed random variable in the range of [10, 30]
K	1,000
L (km)	30
Simulation time (min)	60
Simulation frequency	100

The performance parameters considered are the reception success rate (RSR), the collision rate (CR), the end-to-end delay (E2ED), and the channel utilization (CU). First, the reception success rate is defined as the ratio of the number of received packets to the number of transmitted packets and expressed as

$$\mathrm{RSR} = \frac{\sum_{a=1}^{N_a} n_{RXa} + n_{RXbs}}{\sum_{a=1}^{N_a} n_{TXa} + n_{TXbs}} \tag{3}$$

where a is the index of active ships ($1 \leq a \leq N_a$), n_{RXa} is the number of received downlink packets at ship a, n_{RXbs} is the number of received uplink packets, n_{TXa} is the number of the uplink packets transmitted by ship a, and n_{TXbs} is the number of the transmitted downlink packets during the simulation. Second, the collision rate is defined as the ratio of the number of collided packets to the number of transmitted packets and represented as

$$\mathrm{CR} = \frac{\sum_{a=1}^{N_a} n_{Ca} + n_{Cbs}}{\sum_{a=1}^{N_a} n_{TXa} + n_{TXbs}} \tag{4}$$

where n_{Ca} is the number of collided downlink packets at ship a, and n_{Cbs} is the number of collided uplink packets. Third, the end-to-end delay is defined as the multi-hop delay resulting from packet generation at the source to packet reception at the destination, and it is expressed as

$$\mathrm{E2ED} = \frac{1}{N_a+1}\left[\sum_{a=1}^{N_a}\left(\frac{1}{n_{RXa}}\sum_{x=1}^{n_{RXa}}\mathrm{E2ED}_{ax}\right) + \frac{1}{n_{RXbs}}\sum_{y=1}^{n_{RXbs}}\mathrm{E2ED}_{y}\right] \tag{5}$$

where x is the index of successfully received uplink packets from ship a, y is the index of successfully received downlink packets, $\mathrm{E2ED}_{ax}$ is the multi-hop delay of the x_{th} downlink packet received by ship a, and $\mathrm{E2ED}_y$ is the multi-hop delay of the y_{th} uplink packet received by the base station. Fourth, the channel utilization is defined as the ratio of the number of time slots used to the number of time slots per hyperframe, *viz* 2,250.

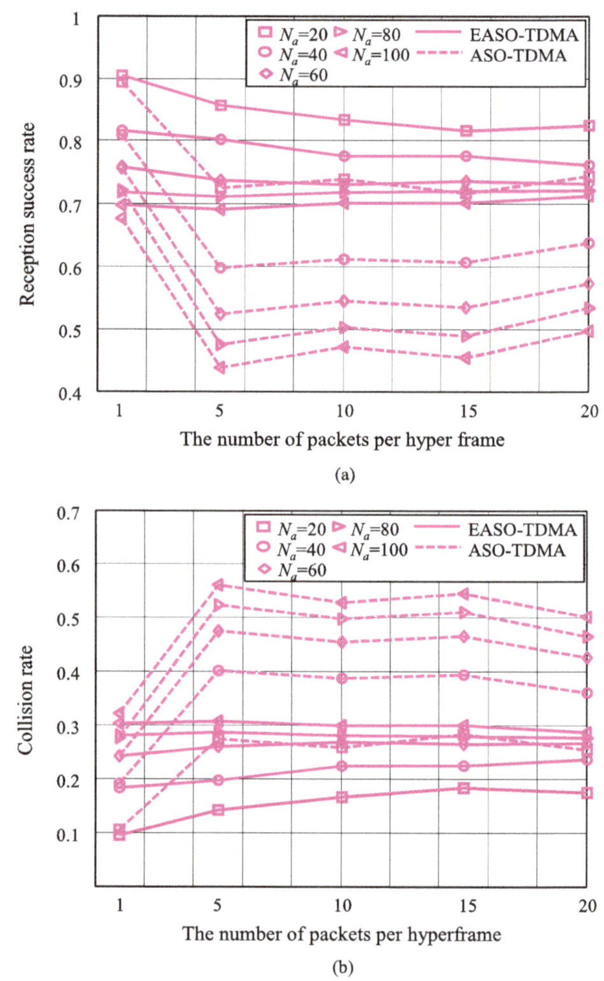

Figure 8 The performance of EASO-TDMA and ASO-TDMA. (a) Reception success rate. **(b)** Collision rate.

4.2 Results

Figures 8 and 9 illustrate the overall performance of EASO-TDMA and ASO-TDMA and demonstrate that EASO-TDMA outperforms ASO-TDMA, regardless of the network conditions. The confidence level of all simulation results is 95%. The simulation results are described as follows:

1. The reception success rate: EASO-TDMA has a 15% higher reception success rate than ASO-TDMA on average. The maximum margin is around 25%, as shown in Figure 8a. The reception success rate of the two MAC protocols decreases as the number of active ships increases. On the other hand, the reception success rate first decreases and then bottoms out with increasing number of packets generated per hyperframe. This implies that the reception success rate mainly depends on the number of active nodes, which increases the possibility of receiver collisions among hidden ships by allocating the same time slots.

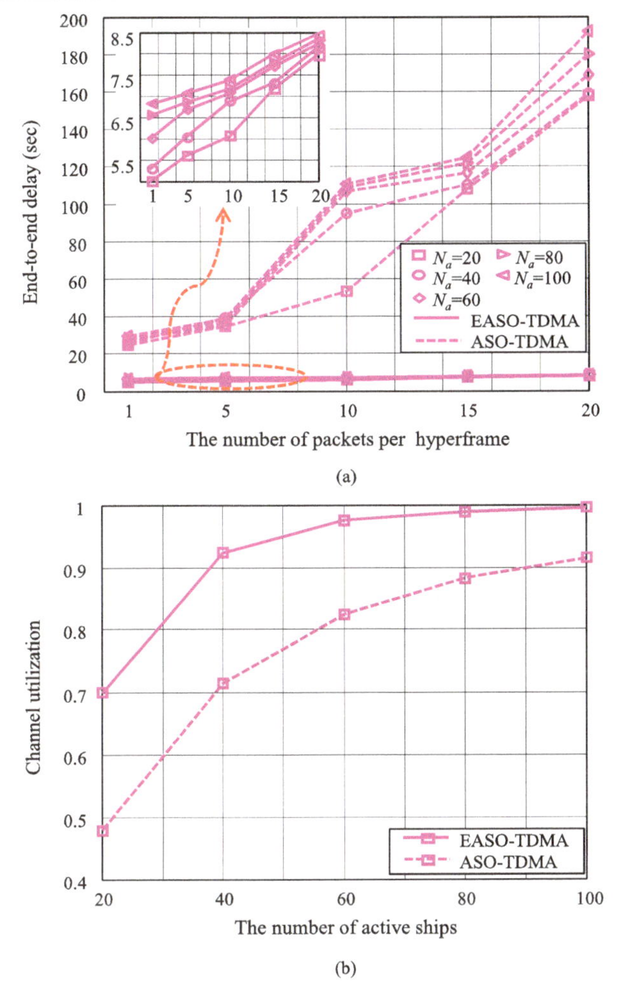

Figure 9 The performance of EASO-TDMA and ASO-TDMA. (a) End-to-end delay. **(b)** Channel utilt!zation.

2. The collision rate: The performance pattern of the collision rate is opposite to that of the reception success rate. That is, the collision rate of the two MAC protocols increases with the increasing number of active ships because the greater the number of ships engaged in accessing the channel, the higher the chance that the same time slots will be selected. The collision rate of EASO-TDMA is 15% lower on average than that of ASO-TDMA. In particular, ASO-TDMA can suffer from severe receiver collision when the number of active ships is over 80, because the collision rate is over 50%, as shown in Figure 8b. This is comparable to the result in [2] in that the collision rate of ASO-TDMA is just 10% in the case that the number of allocated time slots per sub-frame is one. Thus, it is shown that ASO-TDMA is insufficient to manage receiver collisions in the case where the number of time slots allocated per sub-frame is ten.
3. The end-to-end delay: In contrast to the performance pattern of the reception success rate and collision rate, that of the end-to-end delay is affected more by the number of packets than by the number of active ships, as illustrated in Figure 9a. This is because the queueing delay per ship increases as the number of packets increases, and thus, the multihop delay from the source to the destination also increases. ASO-TDMA results in a 12 times longer end-to-end delay than EASO-TDMA on average. ASO-TDMA results at most in a roughly 22 times longer end-to-end delay than EASO-TDMA. This remarkable performance enhancement results from the fact that each ship allocates as many time slots as there are subscribing ships in order to ease the bottleneck problem.
4. The channel utilization: The channel utilization of EASO-TDMA is 15% higher on average than that of ASO-TDMA, as shown in Figure 9b. The maximum margin is around 21%. This result indicates that EASO-TDMA can provide better channel utilization than ASO-TDMA with a lower collision rate because it guides the ships to access the channels more efficiently by avoiding receiver collisions.

5 Conclusions

The SANET was introduced as a technology providing ocean-going ships with VDE services cost-effectively. This paper proposed the MAC protocol for SANETs, EASO-TDMA, which overcomes the weaknesses of the previous approach, ASO-TDMA, and thus enhances the performance of SANETs. The virtue of EASO-TDMA is that it eases the bottleneck problem that occurs at a specific ship by managing the network traffic in a decentralized way. Namely, each ship periodically checks the number of subscribing ships by referring to the routing table and then allocates as many time slots as necessary to manage the traffic from its subscribing ships.

The simulation results showed that EASO-TDMA outperforms ASO-TDMA in terms of the reception rate, collision rate, end-to-end delay, and channel utilization, regardless of the network conditions. In particular, the end-to-end delay of EASO-TDMA is less than 20 s, which is almost 1/12 of that of ASO-TDMA on average. This implies that EASO-TDMA can be efficient for the delay-sensitive VDE services by handling the bottleneck problem. Accordingly, it can be concluded that EASO-TDMA would be an adequate MAC protocol for SANETs. This is an ongoing work. EASO-TDMA will be implemented in SANET systems, and its functionality and performance will be verified through a maritime test for future use.

Competing interests

The authors declare that they have no competing interests.

Acknowledgements

This work was conducted as a part of the research project 'Development of marine RF based ad-hoc network for ship' and was financially supported by the Ministry of Land, Transport and Maritime Affairs (MLTM) of Korea.

References

1. S Lee, J Jeong, M Kim, G Park, A study on real-time message analysis for AIS VDL load management. J. KIIS. **23**(3), 256–261 (2013)
2. C Yun, Y Lim, ASO-TDMA: ad-hoc self-organizing TDMA protocol for shipborne ad-hoc networks. EURASIP WCN. **12**(10), 1–16 (2012)
3. Y Lee, World Radio-communication Conference 2012 (WRC-12). TTA J. **140**, 113–117 (2012)
4. CPG PTC(13) INFO 16, Information paper on VHF data exchange system (VDES). http://www.cept.org/ (Accessed 01-10-2014)
5. C Yun, S Kim, A Cho, Y Lim, Performance analysis of multiple access protocol for maritime VHF data exchange system (VDES). J. JKIICE. **18**(12), 2839–2846 (2014)
6. C Yun, A Cho, S Kim, J Park, Y Lim, Design of multiband maritime network for ships and its applications. Int. J. KIMICS. **7**(3), 314–322 (2009)
7. IMO FAL-38 INF.2, A study of a data communication network at sea for efficient maritime logistics. https://docs.imo.org/ (Accessed 01-28-2013)
8. IMO FAL-39 INF.5, A study on enhancing maritime logistics efficiency utilizing maritime VHF digital communication technology and facilitation method. https://docs.imo.org/ (Accessed 08-01-2014)
9. IMO NCSR-2 INF.9, A study on enhancing maritime logistics efficiency utilizing maritime VHF digital communication technology and facilitation method. https://docs.imo.org/ (Accessed 23-12-2014)
10. Rec. ITU-R M. 1842-1, Characteristics of VHF radio systems and equipment for the exchange of data and electronic mail in the maritime mobile service RR Appendix 18 channels. http://www.itu.int/rec/R-REC-M.1842/en (Accessed 06-09-2009)
11. Rec. ITU-R M.1371-5, Technical characteristics for a universal ship borne automatic identification system using time division multiple access in the VHF maritime mobile band. http://www.itu.int/rec/R-REC-M.1371/en (Accessed 01-02-2014)
12. SOLAS, International Convention for the Safety of Life at Sea (SOLAS). https://docs.imo.org/ (Accessed 18-02-2015)
13. COLREGS, International Regulations for Preventing Collisions at Sea. https://docs.imo.org/ (Accessed 15-05-2014)
14. ITU-R SG5 WP5B working doc, VHF data link protocols for data communication systems in the maritime mobile service band. https://www.itu.int/md/R07-WP5B-C-0601/en (Accessed 04-11-2010)
15. ETSI EN 300 392-2, Terrestrial Trunked Radio (TETRA): voice plus data (V+D); Part2: Air Interface (AI). http://www.itu.int/itu-t/workprog/wp-a5-out.aspxisn=2972 (Accessed 01-08-2010)

18

Distributed multi-hop clustering algorithm for VANETs based on neighborhood follow

Yuzhong Chen[1,2], Mingyue Fang[1,2], Song Shi[1,2], Wenzhong Guo[1,2] and Xianghan Zheng[1,2*]

Abstract

Vehicular *ad hoc* networks (VANETs) have become important components of metropolitan area networks, and clustering for VANETS provides many advantages. However, the stability of current clustering algorithms exhibits poor robustness because a VANET is a highly dynamic scenario. In this study, a novel multi-hop clustering scheme for VANETs, which generates cluster heads (CHs) via neighborhood follow relationship between vehicles, is proposed. The scheme is based on a reasonable assumption that a vehicle cannot certainly identify which vehicle in its multi-hop neighbors is the most suitable to be its CH, but it can easily grasp which vehicle in one-hop distance is the most stable and similar with it, and thus, they most likely belong to the same cluster. Consequently, a vehicle can choose its CH by following the most stable vehicle. The relative mobility between two vehicles combining the gains based on the followed number and the historical following information enables a vehicle to select which target to follow. Extensive simulation experiments are conducted to validate the performance of the proposed clustering scheme.

Keywords: VANETs; Clustering; Multi-hop; Neighborhood follow

1 Introduction

As a new form of mobile *ad hoc* networks (MANETs), the vehicular *ad hoc* network (VANET) has emerged with the rapid development of radio technology that allows vehicle-to-vehicle communication [1]. VANET is one of the important components of an intelligent transport system because it holds great potential in traffic accident warning, traffic flow control, as well as in providing information services and extra serviceability.

Clustering in VANET exhibits good scalability, because clustering can provide a simple information management mechanism and improve communication efficiency. Therefore, clustering algorithms for VANETs are attracting increasing attention.

Unlike traditional MANETs, VANETs exhibit new features such as rapid movement of vehicles, frequent changing of network topology, and limited driving directions; moreover, it does not consider energy problems [2]. These particular features are the reason why traditional clustering algorithms for MANETs can hardly be applied in VANETs. Cluster stability is an important requirement in VANETs. Vehicles move fast which makes clusters broken easily and further affects the routing efficiency. Moreover, unstable clusters are prone to generate more control packets in VANETs and make the networks overload.

Considerable research has explored clustering algorithms for VANETs to satisfy the requirements of their new features [3-10]. Most of these studies are based on one-hop clustering, which only allows communication between a cluster member (CM) and its cluster head (CH) with one-hop distance. The coverage of clusters is small in one-hop clustering, which leads to excess CHs and high-maintenance overhead. Consequently, several multi-hop clustering algorithms have been proposed in the past years [11-13]. These algorithms can extend the coverage of clusters, reduce the number of CHs, and improve cluster stability. However, some issues remain in multi-hop clustering for VANETs. For example, cluster stability must be further improved and maintenance cost must be reduced. Thus, comprehensive solutions must be developed.

This study proposes a distributed multi-hop clustering algorithm for VANETs based on neighborhood follow (DMCNF) to enable fast and stable network setup. The main idea of this study is explained as follows. In large-scale and complex VANETs, a vehicle can hardly acquire

* Correspondence: xianghan.zheng@fzu.edu.cn
[1]College of Mathematics and Computer Science, Fuzhou University, No. 2 Xue Yuan Road, 350108 Fuzhou, China
[2]Fujian Key Laboratory of Network Computing and Intelligent Information Processing, No. 2 Xue Yuan Road, 350108 Fuzhou, China

precise details of multi-hop distanced vehicles and can hardly decide which CH to choose among multi-hop neighbors. However, a vehicle can easily master local information and determine which vehicle within a one-hop distance is the most stable/similar to it, and thus, they most likely belong to the same cluster. Consequently, a vehicle can choose its CH by following the most similar vehicle. This mechanism is identified as the neighborhood follow relationship. In turn, selecting CHs becomes a feedback from the neighborhood follow relationship. Accordingly, this study attempts to quickly obtain the aforementioned relationship chain from large-scale VANETs. CHs are then chosen according to the obtained relationship among vehicles, and other vehicles directly or indirectly follow CHs to form clusters. The main contributions of this work are as follows.

(1) A novel cluster model based on one-hop neighborhood follow is proposed. In the model, a cluster has a CH, which is directly or indirectly followed by other vehicles. The structure of DMCNF can steadily evolve in highly dynamic VANETs.
(2) A neighborhood follow strategy is introduced for vehicles to choose and follow stable target vehicles from one-hop neighbors. Through this strategy, vehicles can adaptively update neighborhood follow information. Consequently, clusters are formed and maintained.
(3) Through the neighborhood follow strategy, clusters are formed and maintained in a distributed manner. Vehicles are only required to regularly communicate with its one-hop neighbors for updating the neighborhood following information and maintaining clusters.
(4) DMCNF does not depend on location service but still provides grouping of related vehicles and fast response to topology changes.

The rest of this paper is organized as follows. Section 2 discusses the review of related literature, and Section 3 provides the preliminaries of the study. Section 4 describes the DMCNF algorithm, and Section 5 presents the experimental results. Finally, Section 6 concludes the paper and suggests a potential subject for future work.

2 Related work

Clustering is a well-known means of organizing networks in MANETs. Many clustering solutions, including identifier neighbor-based clustering, topology-based clustering, mobility-based clustering, energy-based clustering, and weight-based clustering, have been proposed [14-21]. However, these clustering solutions significantly differ from vehicular clustering. MANETs are primarily limited because of their energy [22] and processing power; hence, their clustering is optimized for low-resource usage. However, vehicles are not only rich in resources, but they are also highly mobile. Consequently, clustering algorithms for MANETs are not effective in VANETs, and new solutions must be developed.

Mobility-based clustering algorithm (MOBIC) [16] is a popular clustering algorithm mentioned in various studies. This approach is based on the lowest-ID algorithm; yet, it uses a signal power level mobility metric that is derived from successive receptions. MOBIC does not scale well in VANETs because it is a simple algorithm designed for MANETs; nevertheless, it is frequently compared with other VANET clustering algorithms.

Hafeez et al. [9] proposed a novel clustering algorithm for VANETs by considering speed as the main influential factor to form clusters. These researchers also attempted to improve cluster stability via the fuzzy processing of speed. The algorithm introduced by Hafeez et al. chooses the second optimal vehicle as the temporary CH when the original one becomes unavailable. The algorithm is applied to high-mobility scenarios, but CHs frequently change when they move fast. The rapid change in network topology induces the unstable performance of temporary CHs, which results in unstable clusters.

The affinity propagation (AP) algorithm is one of the most stable clustering algorithms that have been recently proposed. Applying AP to VANETs [10] remarkably improves cluster stability. Nevertheless, AP is a distance-based clustering algorithm, which results in the frequent changing of CHs when speed changes dramatically. Moreover, AP requires several iterative loops that increase the delay time of cluster formation.

The majority of the other known clustering solutions for VANETs are studied in [8]. Goonewardene et al. [4] presented a novel algorithm with the rare feature of cluster overlapping. In particular, this algorithm considers speed, location, and direction via the Global Positioning System (GPS) or other similar services to form clusters. The aggregate local mobility (ALM) algorithm [5] is a new beacon-based clustering scheme that aims to extend the lifetime of clusters by using ALM to decide cluster reorganization. Meanwhile, Rawashdeh and Mahmud [7] considered speed and relative direction to present a novel speed-overlapped clustering algorithm for highways. This system also depends on location services.

The majority of the proposed clustering algorithms for VANET depend on GPS, which may not be the ideal option [23-25]. Positioning services are not available everywhere, and even if they are, their accuracy can significantly vary. Inaccuracy, even within a few meters, endangers position service-based clustering algorithms because it can severely affect cluster stability and leads to communication failure, which are both unacceptable. The complexity and reliability of the entire clustering system are also influenced by its dependence on positioning services. Nonetheless, vehicles should be able to communicate even when positioning services are unavailable or unreliable.

Moreover, most of the aforementioned algorithms are based on one-hop clustering, which only allows communication among one-hop neighbors. Consequently, the coverage rank is small and many clusters are formed, which decrease cluster stability. Cluster formation algorithms should be designed to guarantee cluster stability which is crucial to reduce the maintenance cost of clusters and increase throughput and routing efficiency.

Several scholars have recently analyzed multi-hop clustering algorithms, which are rare but exhibit good achievements.

Wolny [26] presented a modified distributed mobility-adaptive clustering algorithm [27] to adapt the new features of VANET. This modified algorithm is distributed and mobility-adaptive, as well as traffic direction-dependent, and thus, reclustering is avoided when the clusters of vehicles move in different directions. However, this algorithm requires GPS to acquire direction data.

Zhang et al. [11] set packet transfer delay as the relative mobility between multi-hop distanced vehicles and selected vehicles with the smallest aggregate mobility within multi-hop neighbors as CHs. In this scheme, vehicles must identify the aggregate mobility of all N-Hops distance neighbors. Consequently, numerous extra control messages are generated and broadcasted within the network, which eventually reduces the efficiency of cluster formation.

Ucar et al. [12] introduced a vehicular multi-hop algorithm for stable clustering (VMaSC) based on choosing the vehicle with the least mobility. Mobility is calculated with the difference in speed among neighboring vehicles in multi-hop. However, VMaSC requires the support of GPS or similar location services to obtain mobility data.

Hierarchical cluster analysis (HCA) [13] is a fast, randomized clustering algorithm. Instead of attentively selecting the initial CHs and constructing the most stable clusters, HCA attempts to assemble clusters as fast as possible, which leaves cluster optimization in the maintenance phase. The only limiting factor in cluster size is radio propagation, which is avoided by implementing two-hop clustering. HCA is simple and fast, but it does not support optimization related to vehicle movement pattern that can improve cluster stability and CH duration.

Compared with one-hop clusters, multi-hop clusters can extend the range of cluster coverage and gain additional advantages. Hence, multi-hop cluster is a promising direction in VANETs. However, there still remain some open issues to be solved, which can be divided into two main aspects in general: 1) Multi-hop clustering scheme should reduce the control overhead in the processes of forming and maintaining the cluster for VANETs. 2) More complex routing protocols should be designed deliberately to satisfy the requirement of multi-hop communication among vehicles in different clusters.

3 Preliminaries

3.1 Cluster structure

Clusters are virtual groups formed by using clustering algorithms. Each cluster consists of a CH and several CMs. The architecture of a cluster can be divided into two categories, namely, *one-hop* and *multi-hop* clusters, based on routing hops.

In one-hop cluster, each CM can directly communicate with its CH in a one-hop cluster and can directly or indirectly communicate (via CH) with other CMs.

A one-hop cluster structure requires the CH to be directly available to its CMs, which causes many isolated vehicles to emerge after clustering. As shown in Figure 1a, two isolated vehicles (V1 and V2), which are located between two clusters, cannot directly communicate with either CH. Hence, V1 and V2 are divided into two isolated clusters. Isolated clusters increase the number of CHs in a network, which raises communication cost over the network and decreases routing efficiency. In highly dynamic scenarios, a one-hop cluster structure can hardly ensure interconnection between CH and CMs.

In multi-hop cluster, vehicles can communicate with each other in a multi-hop manner.

A multi-hop cluster scheme can decrease the number of CHs to reduce communication cost. As shown in Figure 1b, V1 and V2 can be divided into cluster up

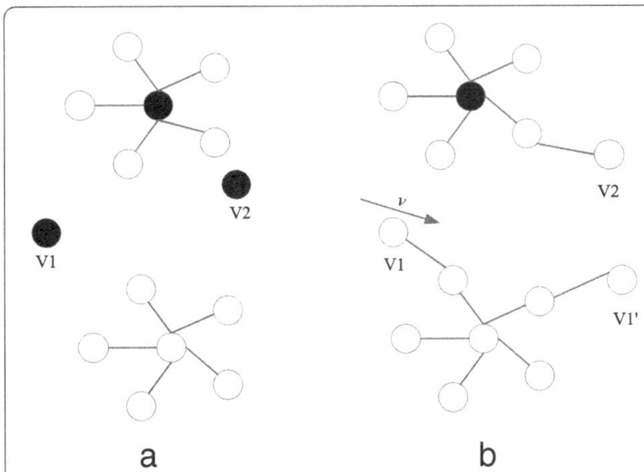

Figure 1 Schematic diagrams of one-hop cluster structure and multi-hop cluster structure. The black and white nodes denote CHs and CMs, respectively. **(a)** The one-hop cluster structure. **(b)** The multi-hop cluster structure.

and down, respectively, because a multi-hop structure allows CHs to indirectly communicate with CMs. Moreover, CMs can communicate with their CHs through other intermediate CMs. These conditions allow CMs and CHs to move in a flexible manner.

However, the cost of forming clusters and maintaining a multi-hop cluster structure may be high. A vehicle must determine the aggregate mobility metric of all vehicles in N-Hop distance [11]. This situation generates and broadcasts numerous control messages within the network and reduces the efficiency of cluster formation. Thus, multi-hop clustering algorithms must be designed to improve the efficiency of cluster formation. This objective is realized in this study by presenting a new multi-hop cluster model that depends on local one-hop topology information. The proposed cluster model is presented in the following section.

3.2 Clustering performance metrics

The clustering of MANETs is primarily limited by their energy [22] and processing power. Hence, existing clustering algorithms are optimized for low-resource usage. In VANETs, the scenario is completely different because vehicles are rich in energy resource and highly mobile. Thus, clustering in VANETs mainly aims to improve cluster stability. Improving cluster stability is helpful to reduce the maintenance cost of clusters and increase throughput and routing efficiency.

Cluster stability can be defined through different mechanisms, but the most frequently used are *CH duration, CM duration, CH change number,* and *cluster number* [10,11].

CH duration is the interval from the period during which a vehicle is selected as a CH to the period it assumes other roles. Similarly, CM duration is the interval between the periods during which a vehicle joins and leaves a cluster. Maximizing the duration of CHs and CMs is useful to improve the stability and minimize the overhead cost of cluster formation [12].

CH change number is the number of vehicles that shift from being CH to other roles. The analysis shows that the cost of selecting CHs and modifying cluster structure is expensive [28]. Consequently, clustering algorithms must attempt to reduce the changing of CHs to decrease the reselection of CHs and the reassociation of CMs. Thus, CH change number is a significant measurable indicator.

The routing efficiency of a VANET exerts the greatest influence. A few clusters accelerate routing in a VANET. Therefore, cluster number is also an important indicator in measuring cluster structure.

3.3 DMCNF algorithm

This section provides a detailed description of DMCNF. First, a novel multi-hop cluster structure model based on the one-hop neighborhood follow strategy is proposed. A neighborhood follow strategy is then introduced to enable vehicles to independently and adaptively choose which target to follow. Clusters are eventually formed and maintained in a distributed manner based on the strategy. The details of the clustering scheme are also described in this section. Table 1 provides the notations used in this paper.

3.4 Neighborhood follow cluster model

Compared with the traditional cluster structure, a more stable structure is constructed and labeled as the neighborhood follow cluster model, which includes the following properties.

Property 1 Multi-hop. Each cluster comprises a CH and CMs. Each CM connects to its CH directly or indirectly via multi-hop.

As shown in Figure 2a, the vehicles in a VANET are divided into two clusters, in which vehicles 5 and 11 are the CHs of the two clusters. The CMs for the left cluster include vehicles 1 to 9, with a hop distance of 2 between vehicles 1, 9, and their CH (vehicle 5).

Property 2 A CM passively selects its CH. A following relation exists between two vehicles in a one-hop distance, in which a CM only chooses and follows a stable neighbor and then owns and shares the same CH with its neighbor.

In a traditional multi-hop structure, a CM directly chooses its CH according to the smallest relative mobility when it receives several head messages. This traditional structure requires each CM to be familiar with the relative mobility among all possible CHs. Consequently, extra control messages are widely broadcasted within the network,

Table 1 Notations and description

Notation	Description
$Nblist_x$	One-hop neighbor list of vehicle x
$PktDelay_{x,y}$	The packet delay of a packet sent from vehicle y to vehicle x
$RelM_{x,y}$	The relative mobility metric between x and y
CH	Cluster head
CM	Cluster member
CH_x	The CH of vehicle x
F_x	The one-hop neighbor follow target of vehicle x
Hello message	The control message sent periodically to maintain the information of neighbors
HI	The time interval of two continuous periods for a vehicle to send hello messages
Follow message	The notification message sent from a follower to its follow target
Follow reply message	The reply message from a vehicle to its followers caused by a follow message
Head message	The message broadcasted from a new CH

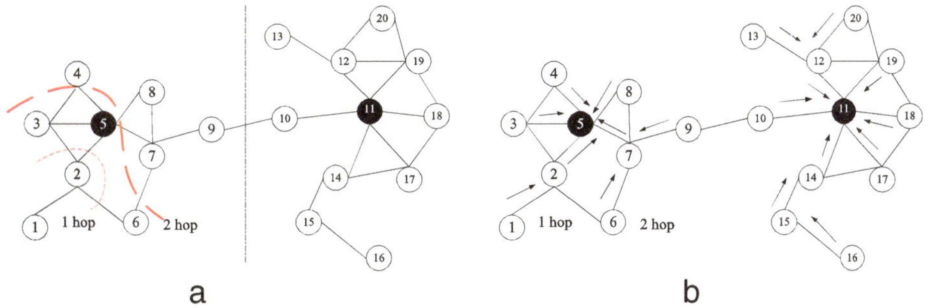

Figure 2 Schematic diagram of neighborhood follow structure. The black nodes denote CHs, and the white nodes denote CMs. The arrows denote the follow relationship between two nodes. **(a)** shows that vehicles 1 to 9 and vehicles 10 to 20 are divided into two clusters, respectively. The hop distance between vehicle 1 and its CH vehicle 5 is 2, and the hop distance between vehicle 10 and its CH vehicle 11 is 3. **(b)** describes the neighborhood follow relationship for each CM, which directly follows one of its neighborhood and directly or indirectly follows its CH. For example, there is a follow chain 16 → 15 → 14 → 11, from vehicle 16 to vehicle 11.

and each vehicle must store and update the relative information of all multi-hop neighbor vehicles. This situation critically increases the overhead.

In large and complex VANETs, a vehicle can hardly obtain the precise details of multi-hop distanced vehicles and decide which CH to choose among multi-hop neighbors. By contrast, a vehicle can quickly obtain the most stable vehicle among its one-hop neighbors. Then, they probably belong to a same cluster. In this study, a vehicle is not required to actively detect its relative mobility with all CHs in multi-hop but must select its CH by following the most stable one-hop neighbor. This mechanism is called the neighborhood follow relationship of belonging to a cluster. Related definitions are provided as follows.

Definition 1 (direct following relationship) Vehicle x exhibits a direct following relationship f, which is defined as follows:

$$f : x \rightarrow y \wedge y \in NBHD(x) \tag{1}$$

where $NBHD(x)$ denotes the one-hop neighbors of vehicle x. A vehicle chooses to follow a target from one-hop neighbors rather than from multi-hop neighbors because obtaining relative mobility with one-hop neighbors is easier and more helpful in reducing extra control messages. As shown in Figure 2, each vehicle selects which target to follow from its one-hop neighbors.

Definition 2 (indirect following relationship) If vehicle y does not belong to $NBHD(x)$, but a following chain exists from x to y, then $x \rightarrow \ldots \rightarrow i \rightarrow \ldots \rightarrow y$. Accordingly, x indirectly follows y; such condition is denoted as $x \mapsto y$.

Consequently, a neighborhood follow cluster is defined as follows.

Definition 3 (neighborhood follow cluster)

$$\mathrm{FC}_c = \{x | x \rightarrow c \vee x \mapsto c \vee x = c\} \tag{2}$$

where FC_c denotes a cluster whose CH is vehicle c, which is directly or indirectly followed by other vehicles. As shown in Figure 2b, among the CM vehicles in the left cluster, vehicles 2, 3, 4, 7, and 8 directly follow vehicle 5, which is also indirectly followed by vehicles 1, 6, and 9. The proposed cluster structure, which is obtained based on the neighborhood follow relationship, can improve stability in highly dynamic VANETs.

Property 3 Follow uniqueness. A CM directly follows only one one-hop neighbor and directly or indirectly follows only one CH.

Similar to the case of the longest follow chain $16 \rightarrow 15 \rightarrow 14 \rightarrow 11$ in the right cluster (Figure 2), each vehicle in the chain only has one target to follow. That is, no other follow chain exists from vehicles 16 to 11. Moreover, vehicle 16 only indirectly follows vehicle 11.

A cluster structure can be constructed by solving two problems, which are as follows: (1) 'How does a vehicle decide which vehicle to follow?' and (2) 'How are CHs selected according to the neighborhood follow relationship?' The solutions to these problems are respectively introduced in Sections 4.2 and 4.3

3.5 Neighborhood follow strategy

3.5.1 Relative mobility

Relative mobility is important for a vehicle to decide which target to follow. This study adopts the relative mobility mentioned in [11], which can be calculated based on the beacon (radio propagation) delay on each vehicle.

$$RelM_{x,y} = 10 \log_{10} \frac{PktDelay_{x,y}^{\mathrm{new}}}{PktDelay_{x,y}^{\mathrm{old}}} \tag{3}$$

where $PktDelay_{x,y}^{\text{old}}$ and $PktDelay_{x,y}^{\text{new}}$ are the transport delays of two continuous messages and are calculated as follows.

$$PktDelay_{x,y} = Now - TT_y \tag{4}$$

where TT_y denotes the time when vehicle y sends a message to vehicle x.

3.6 Neighborhood follow strategy

Vehicles move fast in a VANET. Consequently, a vehicle may easily change its target if it simply chose it based on instant relative mobility. Thus, when deciding whether it should follow vehicle $PktDelay_{x,y} = Now - TT_y$, vehicle x considers three factors, namely, the relative mobility with y ($RelM_{x,y}$), the current number of followers of y (fc_y), and the historical cluster belonging information of y.

The neighborhood follow strategy is presented after defining two following gains, which are based on the number of followers and the historical cluster belonging information, respectively.

Definition 4 (gain based on the number of followers) A vehicle is inclined to follow a target that is already followed by many vehicles. Thus, when choosing which target to follow, vehicle x considers the numbers of followers of possible targets. Gain is defined by the following expression based on the number of followers.

$$gain\_fc_{x,y} = \frac{fc_y}{\sum_{i \in NBHD(x)} fc_i} \tag{5}$$

where $y \in NBHD(x)$. If a vehicle is followed by many vehicles, then these vehicles are relatively stable with it. Consequently, other vehicles are likely to follow vehicles with a stable local topologic structure. It is helpful to improve cluster stability in the dynamic evolution process of a network.

Definition 5 (gain based on the historical cluster belonging information) Vehicle x is likely to follow one target that belongs to the same cluster in which its previous target also belong to. Thus, it is helpful to maintain cluster structure stable. Based on the historical cluster belonging information, gain is defined as follows.

$$gain\_his_{x,y} = \frac{\sum_{i \in NBHD(x)} \|RelM_{x,i}\| \cdot \delta(CH_i, CH_y)}{\sum_{i \in NBHD(x)} \|RelM_{x,i}\|} \tag{6}$$

$$\delta(a,b) = \begin{cases} 1 & a = b \wedge a \neq \text{null} \\ 0 & a \neq b \end{cases} \tag{7}$$

where CH_x denotes the CH of x whose initial value is null. Formula 6 indicates that gain is greater than 0 when a vehicle and its possible target stay in the same cluster; otherwise, gain is equal to 0. This condition verifies the aforementioned proposition.

The gain based on the followed number and the historical cluster belonging information is helpful in improving cluster stability.

According to the two gains defined above, the neighborhood follow strategy based on relative mobility and the two kinds of gain is defined as follows.

$$F_x = \arg\min_{y \in NBHD(x)} RelM_{x,y} \cdot (1 - \text{sgn}(RelM_{x,y}) \cdot gain\_fc_{x,y}) \times (1 - \text{sgn}(RelM_{x,y}) \cdot gain\_his_{x,y}) \tag{8}$$

where F_x denotes the follow target of x. Formula 8 adjusts relative mobility though the two kinds of gains, and then, a vehicle selects its target as the one with the smallest adjusted relative mobility. $RelM_{x,y}$ may be negative, which indicates that x and y are approaching and are likely to be in a same cluster. Thus, the symbol factor sgn($RelM_{x,y}$) is used to reduce the value of the adjusted relative mobility.

Cluster structure should be adjusted with network evolution. When forming clusters, CHs remain unidentified. Hence, the neighborhood follow strategy only considers relative mobility and gain based on the number of followers. Meanwhile, when maintaining cluster structure, the neighborhood follow strategy also considers the gain based on the historical cluster belonging information. In such case, a vehicle does not frequently change its target.

3.7 CH decision rule

Direct and indirect neighborhood follow relation between vehicles can be obtained with the strategy. To obtain the final cluster structure, CHs should be selected.

CHs typically forward packets for communication among vehicles. Selecting stable CHs is beneficial to promote routing efficiency and reduce packet loss probability. According to the neighborhood follow strategy, vehicles with more followers and smaller average relative mobility with their neighbors are more stable. Consequently, these vehicles are suitable to be selected as CHs. The rule in selecting a vehicle to be a CH is described as follows.

A vehicle can be considered as a CH according to the following decision rule.

$$bCh_x = \begin{cases} \text{true} & x \rightarrow y \wedge fc_x > fc_y \wedge AvgRelM_x > AvgRelM_y \\ \text{false} & \text{else} \end{cases} \tag{9}$$

$$AvgRelM_x = \frac{\sum_{y \in NBHD(x)} RelM_{x,y}}{|NBHD(x)|} \tag{10}$$

As indicated in Formula 9, if vehicle x can be considered as a CH, then the number of its followers (fc_x) should be greater than that of its target vehicle y, and its average relative mobility ($AvgRelM_x$) should be less than that of vehicle y. The basis of this rule is obvious. In a following chain, downstream vehicles are prone to be stable, i.e., they have more followers and are more stable than their followers. However, CHs are exceptions. The follow uniqueness property of the proposed cluster model causes CHs to have their own targets to follow from neighbors in one-hop. Nevertheless, these CHs have evidently greater *fc* and less *AvgRelM* than their neighbors.

3.8 Cluster formation and maintenance

This section discusses the detailed processes of forming and maintaining the cluster structure of VANETs based on the neighborhood follow strategy and CH decision rule. DMCNF is a distributed clustering method. Meanwhile, a VANET exhibits a dynamic evolution that presents a dynamic cluster structure, in which each vehicle updates its follow information and dynamically changes its state. Vehicles have three kinds of state, namely, CLUSTER_HEAD, CLUSTER_MEMBER, and CLUSTER_UNDECIDED.

When vehicle x intends to join a VANET, it initializes its state as CLUSTER_UNDECIDED. Then, this vehicle broadcasts a hello message to its one-hop neighbors. Each hello message consists of the *id* of a vehicle, the *id* of its CH, the number of its followers, and a forwarding history record list. The element of this list contains the *id* of a forwarding vehicle and its forwarding time. The time is used to calculate the delay of a packet sent from a vehicle to its one-hop neighbor vehicles. $RelM_{x,y}$, where y is one of the neighbors of vehicle x, can be calculated after a same packet is sent twice between the two vehicles. After receiving a hello message, vehicle x updates its *Nblist*, including the *RelM* values with its neighbors. The detailed fields of *Nblist* are shown in Table 2. The process of receiving a hello message is presented in Algorithm 1.

Table 2 Fields of element *y* of the *Nblist* of vehicle *x*

Fields	Description
id	The sequence number of *y*
pktDelay	The packet transmission delay between *x* and *y*
state	The current state of *y*
ch	The cluster head of *y*
fc	The number of followers of *y*
relm	The relative mobility between *x* and *y*
avgRelM	The average relative mobility between y and each of its neighbors
followFlag	FOLLOWER: *y* follows *x*
	FOLLOWEE: *x* follows *y*
	NULL: no follow relation between them

Algorithm 1. Reception of hello message

```
source_id: the source id of the packet
reply_id: the packet is received from vehicle reply_id
recv_id: this vehicle's id
history: the forward history of the packet
nblist: the neighbor list of this vehicle
state: this vehicle's state
Get the nearest transmission time tt from vehicle recv_id
pktdelay = now - tt
if pktdelay appears in nblist[reply_id] then
  Calculate the relative mobility metric and set the value to
  nblist[reply_id].relm
end if
//Whether drop the packet
forward_cnt = 0
for each ele in history do
  if ele.id == recv_id then
    forward_cnt ++
  end if
end for
if forward_cnt == 2 then
  Drop the packet
else
  //Record the forwarding time
  history.insert(recv_id, now)
  Send the packet to vehicle reply_id
end if
  if state == CLUSTER_ UNDECIDED then
  state = CLUSTER_MEMBER
end if
```

The current state of vehicle x, i.e., CLUSTER_UNDECIDED, transforms into CLUSTER_MEMBER after it sends a hello message. Then, the vehicle starts a timer. After a certain period, referred to as the local learn interval, the vehicle stops receiving hello messages and selects which target to follow. The vehicle chooses a target from the *Nblist* according to the neighborhood follow strategy and sends a follow message to the target vehicle to notify it about the selection. After receiving the follow message, the target vehicle sends back a follow

Table 3 Simulation parameters

Parameters	Value
Simulation time	400 s
Area range	1,000 m × 1,000 m
Maximum velocity	10 to 35 m/s
Number of vehicles	100
Transmission range	100 to 300 m
Propagation model	Two-way ground model
Channel	Mac/802_11
LLI	200 ms
CI	5 s

reply message to vehicle *x*. The follow reply message contains the *id, state, ch,* and *fc* of the target vehicle. After receiving the follow reply message, vehicle *x* updates its *Nblist* with the information carried in the message.

CHs are passively selected in DMCNF. After sending the reply message to the follower, the target vehicle, whose current state is CLUSTER_MEMBER or CLUSTER_HEAD, triggers a process to determine whether to be a CH by adopting the predefined CH decision rule. If the vehicle is determined to be a CH, it changes its state to CLUSTER_HEAD and broadcasts a head message to its followers and the target. Algorithm 2 demonstrates the process of receiving a follow message.

Algorithm 2. Reception of follow message

```
follower_id: the follower's id
cur_id: the id of this vehicle
state: the current state of this vehicle
nblist: the neighbor list of this vehicle
curAvgRelM: the average of all the RelM values between this vehicle and its neighborhood
flwAvgRelM: the AvgRelM of this vehicle's follow target vehicle
cFc: the number of the followers of this vehicle
fFc: the number of the followers of the following vehicle
//Update the nblist
if the follower does not appear in the nblist then
  Add the follower into nblist
end if
nblist[follower_id].bFollower = true
Send a follow reply to the follower
bCh=false
if state != CLUSTER_UNDECIDED then
  //Then this vehicle might be a cluster head
  if cFc > fFc and curAvgRelM > flwAvgRelM then
    bCh = true
  end if
end if
if bCh == true then
  Broadcast a head message to followers and the follow target of this vehicle
end if
```

Upon receiving a head message, the followers with the state of CLUSTER_MEMBER update their CH information and continue to broadcast the head message to their followers (if any). Some followers may be the CHs of other clusters, and these followers directly drop the head messages. If the target vehicle of the CH is also a CH, then the target vehicle is no longer fit to be a CH. Subsequently, it changes its state to CLUSTER_MEMBER, triggers the process of choosing a target, and directly drops the head message. Algorithm 3 presents the process of receiving a head message.

Algorithm 3. Reception of a head message

```
ch_id: the cluster head's id
ch: the ch id of this vehicle
state: the current state of this vehicle
nblist: the neighbor list of this vehicle
for each v in nblist do
  if nblist[v].followFlag == FOLLOWER and state == CLUSTER_HEAD then
    state = CLUSTER_MEMBER
    Boot the follow processing and drop the received packet
    return
  end if
end for
if state == CLUSTER_MEMBER then
  Drop the packet
else
  ch= ch_id
  Broadcast a header message to followers
end if
```

Once each vehicle selects a target, its state turns to CLUSTER_HEAD or CLUSTER_MEMBER, which indicates that a transient cluster structure is formed.

Each vehicle triggers the process of choosing a target in every hello interval (HI). The maintenance work is then completed in the time interval during which a vehicle finishes its state updating until the next HI occurs. Each vehicle must execute a check job in every check interval (CI) during maintenance. A CM vehicle is tasked to ensure its connection with its target. If such connection is lost, then a CM triggers the process of choosing a new target. For a CH vehicle, the process of selecting a new target is triggered if other CH vehicles exist in its one-hop neighbor.

4 Simulation results

The algorithms are run on NS2 [29] (release 2.35) platform. VanetMobiSim [30] is used to generate realistic vehicle track data. Several detailed simulation parameters are listed in Table 3. Maximum velocity varies from 10 to 35 m/s, and the transmission range varies from 100 to 300 m. DMCNF is compared with N-Hop [11]. N-Hop is a novel multi-hop clustering scheme for VANETs. N-Hop proposes a mobility metric to evaluate the multi-hop relative mobility between two vehicles. Then, N-Hop uses the relative mobility to decide whether a vehicle can be selected as a CH or not. As DMCNF applies a relative mobility measure similar to that of N-Hop, and both schemes are to discover multi-hop clusters in VANETs, we choose N-Hop for performance comparison. The parameter MAX_HOP of N-Hop is set to 3, which is the maximum hop distance between a CM and a CH. The analysis of the stability performance of the clustering mechanisms is based on four indicators, namely, CH duration, CM duration, CH change number, and cluster number. Then, the distribution of the number of vehicles on different hops from CHs and the overhead of two algorithms are analyzed.

5 CH duration

In this study, CH duration refers to the interval during which the vehicle is in a CLUSTER_HEAD state until it shifts into a CLUSTER_MEMBER or CLUSTER_UNDECIDED state. The average CH duration is computed by dividing the total CH duration with the total number of state changes from CLUSTER_HEAD to CLUSTER_MEMBER or CLUSTER_UNDECIDED. Figure 3 specifies the average CH duration of the proposed scheme and the N-Hop under different transmission ranges and velocity scenarios.

In particular, Figure 3 indicates that CH duration decreases with increasing vehicle velocity. Increased vehicle velocity makes it difficult for CHs to maintain a relatively stable condition with their neighbor vehicles for a long period. If the CHs do not satisfy the conditions for being a CH, then they are demoted into CMs. Nevertheless, with the increase in vehicle velocity, CH duration is moderately reduced in DMCNF than in N-Hop. Meanwhile, CHs in N-Hop must exhibit the least average relative mobility among their N-Hop neighbors. Such condition increases the difficulty of maintaining stable CHs for a long period under high vehicle velocity. In DMCNF, the stability of CHs only considers the relative mobility with their followers and targets, thereby enhancing the robustness of vehicle velocity.

The transmission range factor influences the stability of CH vehicles, and CH duration increases with increasing transmission range. With a wide transmission range, vehicles do not lose connection with their neighbors. Similar to the case in DMCNF, the vehicle reboots the process of choosing a new target once it loses its target. This situation may reduce the number of followers of CHs, which may be demoted to CMs, and consequently, increases the number of state changes from CLUSTER_HEAD to CLUSTER_MEMBER. Therefore, increasing transmission range is beneficial to cluster stability.

In DMCNF, a vehicle triggers the process of choosing its target every HI to maintain a dynamic cluster structure for a VANET. Figure 3 reveals that CH duration increases with increasing HI. During each HI, a vehicle checks its connection with its target in every CI. The stability of a vehicle may deteriorate after a while; yet, the vehicle will not reboot the process of choosing a new target as long as it does not lose its connection with its target. However, when HI is too small, the vehicle will reselect a target based on its new local topology structure, which may vary from the old one. Consequently, some CHs lose their followers and become CMs. DMCNF is superior to N-Hop under different scenarios.

5.1 CM duration

CM duration is the interval from the time during which a vehicle joins a specified cluster to the time when it leaves the cluster. As shown in Figure 4, vehicle velocity moderately affects DMCNF in terms of CM duration. In DMCNF, a cluster is developed based on the neighborhood follow relation among vehicles. The cohesiveness of a cluster is strong, i.e., most neighbors of a vehicle are from the same cluster. A vehicle may change its target during its transition. Nevertheless, the new target is prone to have the same CH as the old one, which implies that the vehicle still follows the same CH indirectly. Such situation helps increase CM duration in the cluster. In N-Hop, a vehicle easily

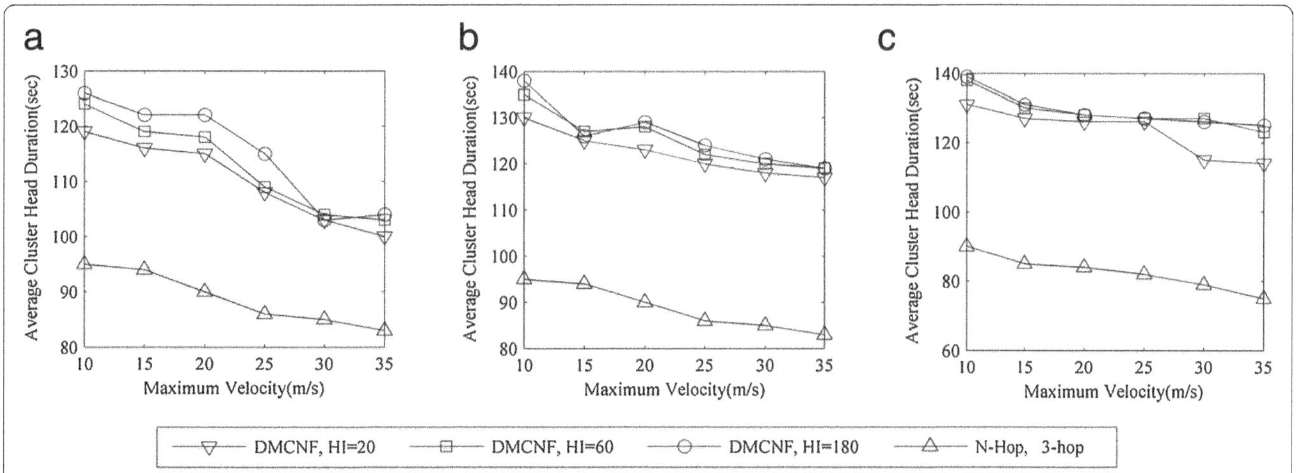

Figure 3 Average CH durations under different transmission ranges and velocity scenarios. (a) 100 m. **(b)** 200 m. **(c)** 300 m.

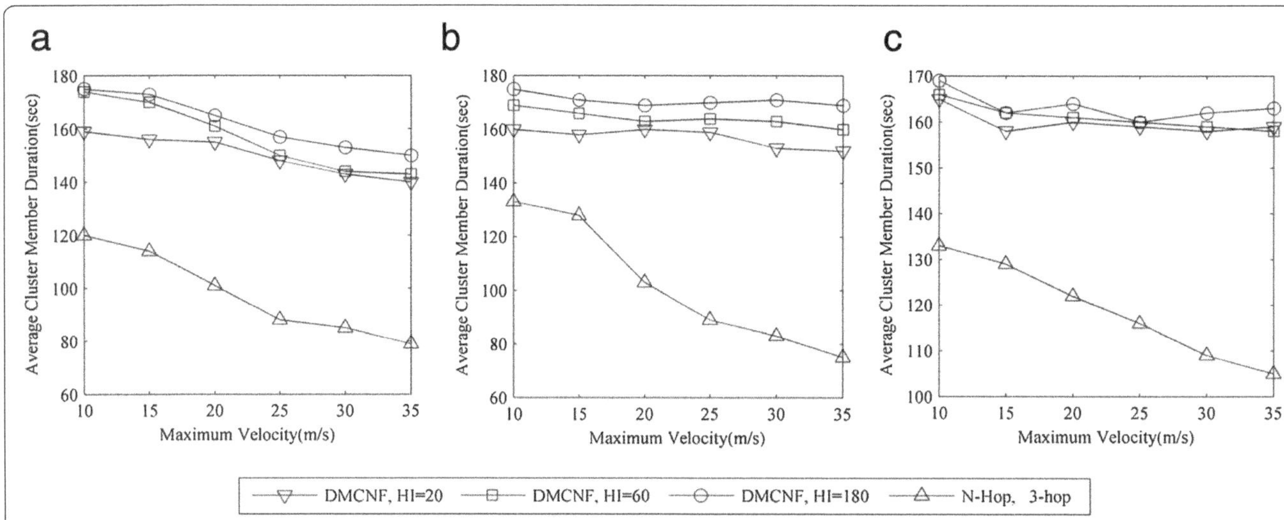

Figure 4 Average CM durations under different transmission ranges and velocity scenarios. (a) 100 m. **(b)** 200 m. **(c)** 300 m.

breaks away from its old cluster once it increases relative mobility with its old CH.

CM duration increases with increasing propagation range because vehicles are stable with their neighbors under a high propagation range. Moreover, the increase in HI allows CM vehicles to reduce the frequency of rebooting the process of choosing a target, which allows them to leave their old cluster.

5.2 CH change number

CH change number is the number of vehicles whose state changes from CLUSTER_HEAD to CLUSTER_-MEMBER or CLUSTER_UNDECIDED during a simulation process. A low CH change number leads to a stable cluster structure. Figure 5 shows the results of the average CH change number under different transmission ranges and velocity scenarios. The figure demonstrates that the indicator is thrice more in N-Hop than in DMCNF. Moreover, by using DMCNF, the indicator moderately increases with increasing velocity because a vehicle satisfies the conditions for a CH, which generally has more direct and stable followers (one-hop followers) than CMs. CH is passively reselected when its one-hop neighbors restart the follow process. The target changes of multi-hop neighbors minimally influence CH selection. Consequently, a CH is not easily weakened to other states.

5.3 Number of clusters

A few clusters can improve routing efficiency in a VANET. Figure 6 indicates the average number of clusters under different transmission ranges and velocity scenarios, in

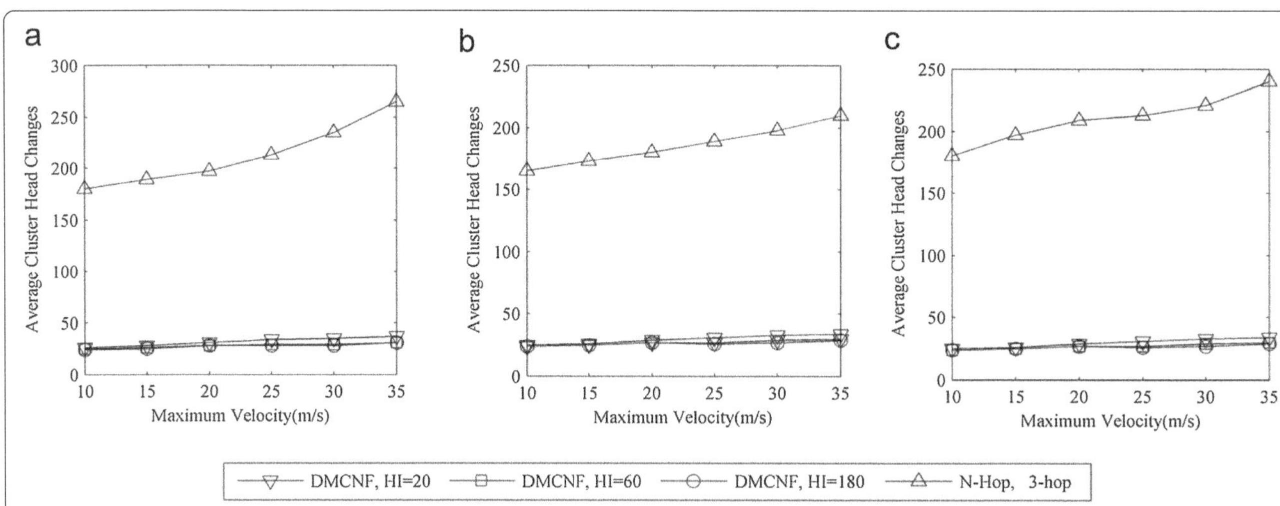

Figure 5 Average CH change number under different transmission ranges and velocity scenarios. (a) 100 m. **(b)** 200 m. **(c)** 300 m.

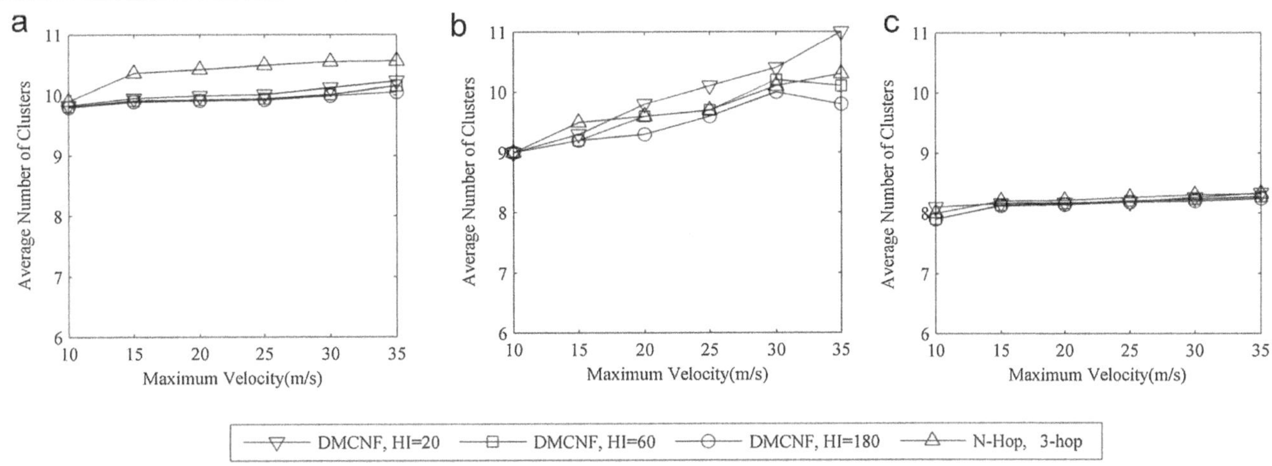

Figure 6 Average number of clusters under different transmission ranges and velocity scenarios. (a) 100 m. **(b)** 200 m. **(c)** 300 m.

which the average number of clusters is over 50 simulations. The number of clusters in each simulation is the average number of instant clusters that is periodically obtained. As shown in Figure 6, the number of clusters changes minimally with increasing velocity. A vehicle can still identify the cluster to which it belongs to via the multi-hop scheme even if velocity increases and the number of isolated vehicles is reduced. Accordingly, DMCNF can effectively reduce the number of clusters and isolated vehicles, thereby decreasing the number of clusters with increasing propagation range.

5.4 Analysis of follow situation in cluster structure

In DMCNF, CMs directly or indirectly follow the CH. Clusters are stable if the cohesiveness of the cluster structure is good, and numerous CMs follow the CH with few hops. Figure 7 illustrates the distribution of vehicles under different hops of the cluster structure obtained by DMCNF. The figure demonstrates that more than half of the vehicles directly follow their CHs and that the number of vehicles decreases with increasing hops, which satisfies the requirement of high cohesion. When the maximum velocity is set to 25 and 15 m/s, the maximum hop is 3. Meanwhile, when the maximum velocity is set to 35 m/s, the relevant maximum hop is 4. These results imply that DMCNF can obtain a cluster structure with high cohesion.

5.5 Analysis of overhead

Figure 8 shows the average control overhead of DMCNF and N-Hop at different velocities. The overhead counts all the control message received by each vehicle in the

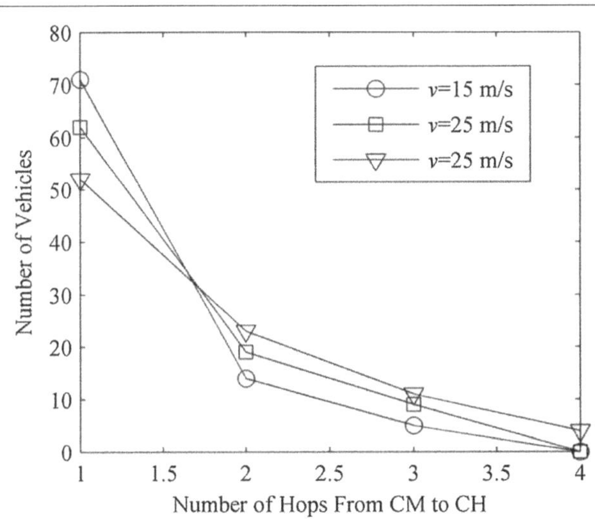

Figure 7 **Distribution of the number of vehicles on different hops.**

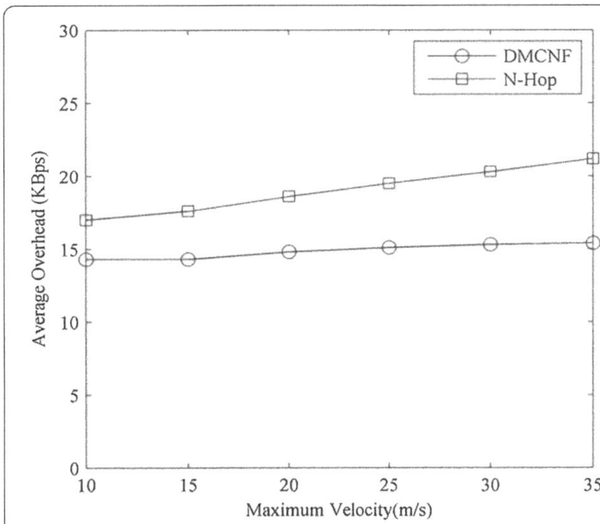

Figure 8 **Average overhead of DMCNF and N-Hop at different velocities.**

entire network and is averaged over the transmission range from 100 to 300 m. Compared with N-Hop, DMCNF performs better in terms of overhead. In DMCNF, each vehicle creates a hello message, a follow message, and a head message during every HI. For the hello message, a vehicle only needs to broadcast a hello message to its one-hop neighbors to calculate the relative mobility between the vehicle and its neighbors. On the contrary, N-Hop needs to broadcast hello messages in multi-hop distance and the size of a hello message increases through the forwarding path, which in turn increases the overhead of VANETs.

6 Conclusions

In this study, a multi-hop clustering scheme with improved stability is obtained. First, the cluster model is presented based on the neighborhood follow strategy. Then, a novel multi-hop clustering algorithm, called DMCNF, is proposed. DMCNF allows vehicles to periodically choose their targets from one-hop neighbors in a distributed manner. The neighborhood follow strategy considers relative mobility, gain based on the number of followers, and gain based on the historical cluster belonging information. Moreover, this strategy improves the stability of clusters during network evolution. The distributed manner of choosing a target leads to easy maintenance of the cluster structure. The efficient routing protocols based on the neighborhood follow cluster structure for VANETs will be explored in the future.

Competing interests

The authors declare that they have no competing interests.

Acknowledgements

The authors would like to thank the support of the Technology Innovation Platform Project of Fujian Province under Grant No. 2009 J1007, the Key Project of Fujian Education Committee under Grant No. JK2012003, the Program of National Natural Science Foundation of China under Grant No. 61300104, 61103175, and 61370210, and the Natural Science Foundation of Fujian Province under Grant No. 2013 J01232.

References

1. H Hartenstein, K Laberteaux, A tutorial survey on vehicular ad hoc networks. IEEE Commun Mag **46**(6), 164–171 (2008)
2. S Tayal, M Triphathi, VANET-challenges in selection of vehicular mobility model, in *Proc. of the Second International Conference on Advanced Computing & Communication Technologies* (IEEE, Rohtak-Haryana India, 2012), pp. 231–235
3. Z Wang, L Liu, M Zhou, A position-based clustering technique for ad hoc intervehicle communication. Appl Rev IEEE Transactions Systems, Man, and Cybernetics, Part C **38**(2), 201–208 (2008)
4. RT Goonewardene, F Ali, E Stipidis, Robust mobility adaptive clustering scheme with support for geographic routing for vehicular ad hoc networks. IET Intell Transp SY **3**(2), 148–158 (2009)
5. E Souza, I Nikolaidis, P Gburzynski, A new aggregate local mobility (ALM) clustering algorithm for VANETs, in *Proc. of Communications (ICC)* (IEEE, Cape Town- South Africa, 2010), pp. 1–5
6. A Koulakezian, *ASPIRE: adaptive service provider infrastructure for VANETS (University of Toronto*, 2011
7. Z Rawashdeh, SM Mahmud, A novel algorithm to form stable clusters in vehicular ad hoc networks on highways. EURASIP J Wirel Commun Netw **1**, 1–13 (2012)
8. S Vodopivec, J Bester, A Kos, A survey on clustering algorithms for vehicular ad-hoc networks, in *Proc. of Telecommunications and Signal Processing (TSP), 2012 35th International Conference on* (IEEE, Brague-Czech Republic, 2012), pp. 52–56
9. K Hafeez, L Zhao, Z Liao, B Ma, A fuzzy-logic-based cluster head selection algorithm in VANETs, in *Proc. of Communications (ICC)* (IEEE, Orrawa-Canada, 2012), pp. 203–207. 3-27 11–15 June 2012
10. B Hassanabadi, C Shea, L Zhang, S Valaee, Clustering in vehicular ad hoc networks using affinity propagation. Ad Hoc Netw **13**, 535–548 (2014)
11. Z Zhang, A Boukerche, R Pazzi, A novel multi-hop clustering scheme for vehicular ad-hoc networks, in *Proc. of the 9th ACM international symposium on Mobility management and wireless access* (ACM, Paris-France, 2011), pp. 19–26
12. S Ucar, S Ergen, O Ozkasap, Vehicular multi-hop algorithm for stable clustering in vehicular ad hoc networks, in *Wireless Communications and Networking Conference (WCNC)* (IEEE, Shanghai-China, 2013), pp. 2381–2386
13. E Dror, C Avin, Z Lotkerk, Fast randomized algorithm for hierarchical clustering in vehicular ad-hoc networks, in *Ad Hoc Networking Workshop (Med-Hoc-Net), 2011 The 10th IFIP Annual Mediterranean. IEEE*, 2011, pp. 1–8
14. C Lin, M Gerla, Adaptive clustering for mobile wireless networks. IEEE J Sel Areas Commun **15**(7), 1265–1275 (1997)
15. J Yu, P Chong, 3hbac (3-hop between adjacent clusterheads): a novel non-overlapping clustering algorithm for mobile ad hoc networks, in *Proc. of 4th IEEE Pacific Rim Conference on Communications, Computers and signal Processing* (IEEE, Singapore, 2003), pp. 318–321
16. P Basu, N Khan, T Little, A mobility based metric for clustering in mobile ad hoc networks, in *Proc. of 21st International Conference on Distributed Computing Systems Workshops* (IEEE, Arizona-USA, 2001), pp. 413–418
17. M Ni, Z Zhong, D Zhao, MPBC: a mobility prediction-based clustering scheme for ad hoc networks. IEEE Trans Veh Technol **60**(9), 4549–4559 (2011)
18. Y Xu, S Bien, Y Mori, J Heidemann, D Estrin, Topology control protocols to conserve energy in wireless ad hoc networks. Center for Embedded Network Sensing, 2003.
19. R Selvam, V Palanisamy, Stable and flexible weight based clustering algorithm in mobile ad hoc networks. Int J Com Sci Inf Technol **2**(2), 824–828 (2011)
20. R Agarwal, D Motwani, *Survey of clustering algorithms for MANET. arXiv preprint arXiv:0912.2303*, 2009
21. A Bentaleb, A Boubetra, S Harous, Survey of clustering schemes in mobile ad hoc networks. Commun Netwo **5**(02), 8 (2013)
22. I Humar, X Ge, L Xiang, M Chen, J Zhang, Rethinking energy efficiency models of cellular networks with embodied energy. IEEE Netw **25**(2), 40–49 (2011)
23. J Volpe, *Vulnerability assessment of the transportation infrastructure relying on the global positioning system*, 2001
24. M Thomas, J Norton, A Jones, *Global navigation space systems: reliance and vulnerabilities* (The Royal Academy of, Engineering, 2011)
25. S Vodopivec, J Bešter, A Kos, A multihoming clustering algorithm for vehicular ad hoc networks. Int J Distributed Sensor Networks **2014**(2014), 8 (2014). Article ID 107085
26. G Wolny, Modified DMAC clustering algorithm for VANETs, in *Proc. of 3rd International Conference on Systems and Networks Communications* (IEEE, San Francisco-USA, 2008), pp. 268–273
27. S Basagni, Distributed clustering for ad hoc networks, in *Proc. of 4th International Symposium on Parallel Architectures, Algorithms, and Networks (PAAP)*. 1999, pp. 310–315
28. M Ni, Z Zhong, K Wu, D Zhao, A new stable clustering scheme for highly mobile ad hoc networks, in *Proc. of 2010 IEEE Wireless Communications and Networking Conference* (WCNC 2010, IEEE, Sydney-Australia, 2010), pp. 1–6
29. K Fall, K Varadhan, The network simulator (ns-2). http://nsnam.isi.edu/nsnam/index.php/Main_Page
30. M Fiore, J Harri, F Filali, C Bonnet, Vehicular mobility simulation for VANETs, in *Proc. of 40th Annual Simulation Symposium* (ANSS'07, Norfolk-England, 2007), pp. 301–309

19

Interference management for moving networks in ultra-dense urban scenarios

Yutao Sui[1*], Ismail Guvenc[2] and Tommy Svensson[1]

Abstract

The number of users relying on broadband wireless connectivity while riding public transportation vehicles is increasing significantly. One of the promising solutions is to deploy moving base stations on public transportation vehicles to form moving networks (MNs) that serve these vehicular users inside the vehicles. In this study, we investigated the benefits and challenges in deploying MNs in ultra-dense urban scenarios. We identified that the key challenge limiting the performance of MNs in ultra-dense urban scenarios is inter-cell interference, which is exacerbated by the urban canyon effects. To address this challenge, we evaluated different inter-cell interference coordination and multi-antenna interference suppression techniques for MNs. We showed that in using MNs together with effective interference management approaches, the quality of service for users in vehicles can be significantly improved, with negligible impacts on the performance of regular outdoor users.

Keywords: 5G; HetSNet; Interference management; Moving networks; Ultra-dense urban networks; Vehicular users

1 Introduction

The demand for ubiquitous availability of broadband mobile data is exploding. The industry has predicted a 1,000-fold increase in mobile data traffic within the next decade [1-3]. To meet the capacity demand of mobile communication systems, academia and industry are working together closely. The Third Generation Partnership Project (3GPP) has recently standardized the Long Term Evolution-Advanced (LTE-A) networks and has been working on developing its evolution by adding more features to support higher system throughput. Several large scale projects such as EU FP7 Mobile and wireless communications Enablers for the Twenty-twenty Information Society (METIS) and 5GNOW[a] [2,4], have also recently started to investigate 5G mobile communication systems, ranging from new radio access interfaces to new system architectures.

With the increasing capabilities of mobile devices and cloud computing, user experiences are not only based on the capacity a mobile communication system can offer but are also determined by the latency of the communication, reliability of the services, and pervasive availability of the networks. The METIS project describes five features of the next-generation mobile communication systems as '1) amazingly fast, 2) great service in a crowd, 3) ubiquitous things communicating, 4) best experience follows you, and 5) super real-time and reliable connections' [4]. The number of users relying on wireless data connections while riding public transportation vehicles, e.g., buses, trams, and trains, is increasing significantly, because of the high penetration of smart phones, tablets, and increasingly portable notebooks. Users are expecting similar experiences at home, in the office, when stationery or traveling, or commuting. Therefore, in this study, we target a particular user group, i.e., vehicular users located in well-isolated public transportation vehicles.

Serving these vehicular users efficiently has been a problem that has recently attracted significant attention [5-8]. In a coverage-limited scenario, e.g., high-speed trains or inter-city coaches in suburban areas, the vehicular penetration loss (VPL)[b] is identified as one of the biggest factors that limits the performance for vehicular users [5]. According to the measurements in [9], VPL can be as high as 25 dB in a minivan at the frequency of 2.4-GHz, and higher VPLs are foreseeable in well-isolated vehicles and at higher frequency bands, e.g., the 3.6-GHz band allocated to future mobile communication

*Correspondence: sui.yutao@chalmers.se
[1] Department of Signals and Systems, Chalmers University of Technology, EDIT-huset, Hörsalsvägen 9-11 Gothenburg, SE-41296, Sweden
Full list of author information is available at the end of the article

systems [10]. One of the most promising solutions to better serve vehicular users is to use moving relay nodes (MRNs) mounted on top of public transportation vehicles [6,7,11]. An important advantage of using MRNs is that by the proper placement of the antennas on a vehicle, the VPL can be circumvented, and thereby, the quality of service (QoS) for the users can be significantly improved. Moreover, compared with regular user equipment (UE) devices, an MRN has fewer constraints in terms of transmit power and transceiver complexity, and therefore, more antennas and advanced signal processing algorithms can be used in MRNs to further boost performance.

To the best of our knowledge, all existing studies related to MRNs either deal with system architecture issues or study the performance of MRNs in simplified scenarios (see Section 2). There has been no thorough investigation on using MRNs in typical outdoor urban scenarios. In typical outdoor urban deployments, because of the densely deployed small cells and street canyon effects, complicated inter-cell interference becomes one of the key factors limiting the performance of mobile communication systems. To better understand the practical challenges of using MRNs in ultra-dense urban scenarios, in this study, we extend our earlier investigations on MRNs in [7,11,12] in several aspects. In this study, we extend the studies of half-duplex MRNs to full-duplex moving networks (MNs). Both MRNs and MNs form their own cells inside public transportation vehicles to serve the vehicular UE (VUE) devices on board, and communicate with the macro BSs through the radio interface. However, the MRNs can only work in a half-duplex fashion, as the backhaul links and access links of MRNs are assumed to operate in the same frequency band. In the vision of METIS, more bandwidth can be freed up for small cell deployment in the 5G system. Therefore, it becomes feasible in the 5G systems to have vehicles constitutes a full-duplex MN with backhaul and access links on different frequency bands.

We focus on the deployment of MNs on public transportation vehicles in an ultra-dense urban area with a practical densely deployed heterogeneous and small cell networks (HetSNets) framework defined by the test case 'dense urban information society' in the EU 5G project METIS [4][c]. More specifically, an MN communicates with a macro BS via its backhaul link and forms a small cell inside a public transportation vehicle. Instead of being served directly by the macro BS, the VUEs located inside a public transportation vehicle are offloaded to the vehicle's MN. In such a setup, the complicated interference situation becomes a significant problem, where MNs are subject to inter-cell interference from both macro and micro cells. As an example, when MNs are passing through the small cells, the access links of MNs experience dynamic fluctuations of interference power. Moreover, MNs may also potentially bring unexpected interference challenges into the system.

Contributions: In this study, we first identify the challenges in deploying MNs in ultra-dense urban HetSNets scenarios in terms of interference situations. To exploit the benefits of MNs, based on the availability of channel state information (CSI) and multi-antenna deployments, we then apply and compare various inter-cell interference coordination (ICIC) and interference cancellation schemes to alleviate the impact of inter-cell interference on MNs. Furthermore, the fairness between VUEs and regular outdoor UEs is also evaluated. We show that compared with serving the VUEs directly from macro BSs, by using MNs, the QoS at VUEs can be significantly improved without sacrificing the performance of regular outdoor UEs. Therefore, MNs can bring noticeable benefits to future mobile communication systems.

The paper is organized as follows. In Section 2, we introduce previous studies related to MNs and various ICIC solutions used in HetSNets. Section 3 presents the detailed system model that we used to evaluate the performance of MNs. In Sections 4 and 5, we discuss both the time domain and multi-antenna solutions considered in this study to cope with inter-cell interference challenges for backhaul links of MNs and macro UEs. The ICIC scheme used for micro cells is discussed in Section 6, and the detailed evaluation results using the METIS 5G simulation framework are presented in Section 7. In Section 8, we discuss the complexity of the schemes considered in this paper, the constraints in the current LTE-A systems that limit further performance improvement of MNs and some important future research directions using MNs. Section 9 concludes the paper.

2 Related studies

Because the demand for mobile data capacity is growing, especially in urban areas, network densification is inevitable. HetSNets are identified as one of the key solutions to address the capacity demand of next-generation mobile communication systems, as they maximize the reuse of existing spectra by deploying more nodes in a given area [1,3,13]. In HetSNets, various low-power nodes, i.e., remote radio heads (RRHs), micro and femto BSs, and fixed-relay nodes (FRNs), are deployed underlying traditional high power macro BSs. In this way, the low-power nodes can meet the capacity needs in certain hot spot areas while high-power macro BSs are used to provide basic coverage. In the study of the 5G mobile communication system, a HetSNets deployment has been considered from the very beginning [3,4]. For example, in the test case 'dense urban information society' of the METIS project, a dedicated HetSNets deployment called the *Madrid grid model* was developed based on the characteristics of typical European cities [14].

In [15], it is shown that in an ideal HetSNets setup, even though interference becomes severe as transmitters get closer, network densification can still bring significant gains to a wireless communication system. In practical deployment of HetSNets, to improve the small cell coverage and achieve more offloading gain, cell range expansion (CRE) is introduced in 3GPP LTE Rel-10 ([16,17] Ch. 24 and 31). In CRE, a positive bias is added to the reference signal received power (RSRP) of small cells, and thereby more UEs can be offloaded to small cells. Using CRE, the coverage of small cells can be expanded to offload more macro users to small cells, and the overall uplink performance of the network can be improved [1,17,18]. However, because of severe inter-cell interference, CRE causes downlink performance degradation, especially for the cell-edge UEs of the small cells.

Therefore, to realize the CRE gain and protect victim UEs, it is critical to use ICIC schemes in the deployment of HetSNets [1,13,19]. Both time and frequency domain solutions have been extensively studied to cope with the inter-cell interference [20,21]. The use of almost blank subframes (ABSs) is standardized in 3GPP LTE Rel-10 as a part of the enhanced inter-cell interference coordination (eICIC) framework to protect UEs that are subject to excessive co-channel interference ([16] Ch. 31.2.2) [18,22-24]. Moreover, the use of multi-antennas to suppress interference is also studied for practical deployments ([20] Ch. 5). For example, network-assisted interference cancellation has been studied for 3GPP LTE Rel-12 [25]. In later sections, we discuss the ICIC schemes, used in this study, in further detail.

Providing ubiquitous service experience for users on well-isolated high-speed vehicles is very challenging because of the speed of the vehicles and the severe attenuation of the received signal caused by the VPL. Several solutions using elements of existing systems, such as deploying dedicated macro BSs, or using layer-1 repeaters and WiFi access points, have been discussed in 3GPP [5]. These solutions, however, are temporary and mainly aim at offering a basic coverage for VUEs rather than capacity improvement.

To better serve the VUEs, using MRNs has been proposed and studied in [5-7,12,26-29]. The idea of using mobile devices to relay information to UEs outside the coverage of a BS is not new. Mobile devices can form a mobile *ad hoc* network (MANET) to enable communications between mobile devices directly without (or with limited support of) network infrastructure [30]. Nowadays, MANET attracts more attention because of the investigation into device-to-device (D2D) communications in the next-generation mobile communication systems [31-33]. The mobile communication system can benefit from mobile-device-enabled relaying in several different ways. For example, with the approach in [34], using a nomadic network formed by randomly parked vehicles to relay data from BSs to UEs, significant energy savings can be achieved. Other relevant studies show that in using D2D-enabled relaying in mobile communication systems, the coverage of a cellular network can be extended, the network capacity can be improved, and the outage probability of the communication can be reduced [32,33,35-37]. Interference management in MANET is very challenging. In a MANET without central control nodes, interference management based on power control or coordination between different nodes can only be implemented in a distributed manner [38], which usually leads to local or suboptimal solutions. If a central control node is available, more effective interference and resource management algorithms can be implemented, and therefore, MANET can bring more gains to the current mobile communication systems. For example, studies in [36,39,40] show that with effective interference management and resource allocation methods, a mobile communication system can benefit in various ways by using D2D-enabled communication underlying the coverage of a BS. However, MANET provides limited support to high-mobility users. Because the topology of a MANET may change very fast, it may not always be possible to find available nodes to relay the information. As pointed out in Section 3, complicated inter-cell interference is one of the biggest factors limiting the performance gains that can be obtained using MNs. Therefore, in an interference-limited scenario with high-mobility users, e.g., to serve VUEs in a densely deployed urban scenario, the MANET solutions cannot be directly applied.

Previous studies show that by using MRNs, in a coverage-limited scenario [6,7,11] or scenarios with insignificant interference [12,27,28], both the QoS of VUEs and the system throughput can be significantly improved. From a system perspective, several architectures have been proposed and compared for deploying MRNs to current mobile systems in [5,8]. Moreover, it has been identified that the backhaul links are the bottlenecks to further improve the performance of MRNs [11,28]. Compared with regular UEs, MRNs have fewer constraints in terms of transmit power and transceiver complexity, and thereby, more antennas and advanced signal processing algorithms can be used in MRNs to further boost performance. In [41,42], it is shown that by using the so-called predictor antennas mounted on top of a vehicle, reliable CSI for the backhaul links of MRNs can be obtained to support various multi-antenna operations. Therefore, the throughput of the backhaul links of MRNs can be significantly improved. In [29], the energy efficiency aspects of using MRNs are studied, showing that the energy consumption of the network can be reduced by using MRNs, when compared with serving the VUEs directly from the macro BSs. The mobility management aspects of MRNs have been studied in [8,12,43]. In [12], a generalized framework to optimize the parameters related to the handovers with MRNs was proposed. In [43], it is shown that by using MRNs on high-speed trains, the number of ping-pong handovers and handover failures can be significantly reduced.

In the EU 5G project METIS, based on the study of MRNs, the concept of MNs has been developed and integrated as part of the METIS 5G system concept [44]. In METIS, the concept of MNs not only covers mobility-robust high data-rate communication links for VUEs but also includes flexible demand-driven network deployment. The MNs are also envisioned to provide low-latency services for road safety and traffic efficiency [44]. To understand the impact of deploying MNs to an ultra-dense urban HetSNets scenario, in this study, we evaluated the performance of MNs within the HetSNets framework defined by METIS. In the next section, we describe the detailed METIS setup to provide a background for the performance investigation of MNs.

3 System model for moving networks

We considered the downlink of a densely deployed mobile network in a typical urban setup, which was developed and used in the METIS project [14]. To facilitate the comparison of different schemes, based on the Madrid grid model, an abstract and simplified deployment model is suggested by METIS, as shown in Figure 1, where heterogeneously deployed macro and micro BSs are considered. In this study, we consider the calibration deployment given in Table 3.7 in [14], where macro BSs offer basic coverage, and micro BSs are used for capacity-demanding applications and local coverage. A macro BS has three sectors and transmits at the 800-MHz band with a 20-MHz bandwidth. Each micro BS has two sectors and transmits at the 2.6-GHz band with an 80-MHz bandwidth. We assume the general scenario that an MN is deployed on a public transportation vehicle, e.g., a bus or a tram. Following the 3GPP convention, we name the link between an MN and a macro BS as the backhaul link and the link between an MN and its VUEs as the access link, respectively. The backhaul of the MN communicates with a macro BS sector at 800-MHz, and the access link operates at 2.6-GHz to serve the users inside the vehicle. Public transportation vehicles enter each street through a Poisson process with an inter-arrival time of 2 min and a fixed velocity of 50 km/h ([14] Sec. 9.2). Since the backhaul links and access links operate at different frequency bands, in this study, we assume that MNs work in a full-duplex mode.

The inter-cell interference experienced by the backhaul links of MNs is very similar to the inter-cell interference faced by the VUEs, if the backhaul links are served directly by macro cells. Therefore, some of the traditional ICIC schemes, e.g., using ABS (see Section 4), can be directly applied. Introducing MNs to a mobile communication system brings in more opportunities to better serve the VUEs but also creates more challenges. Preliminary studies in METIS show that without using any ICIC schemes or interference cancellation, there are limited gains in using MNs to serve VUEs in densely deployed urban scenarios ([45] pp. 33), because the MNs are operating mostly at the same interference-limited scenario as regular outdoor UEs. This is discussed in more detail in Section 3.1. Therefore, techniques for alleviating the impact of inter-cell interference become important when using MNs in a densely deployed urban scenario. Unlike the regular UEs, MNs are less constrained by size, power, and transceiver complexities, and therefore more antenna elements can be deployed and more advanced signal processing schemes can be implemented at the MNs to combat the interference. In Section 5, we present multi-antenna techniques for the backhaul receivers of MNs to combat inter-cell interference.

Furthermore, if we simply treat an MN as a regular UE with high data-rate demand and do not consider the QoS of each of the individual VUEs served by the MN, the scheduler at the BSs may not allocate enough resources to the MN. This is because the scheduler at the BSs needs to ensure fairness between different UEs sharing the same time and frequency resources. Therefore, an unfair situation may be faced by the VUEs. Hence, in this study, we regard an MN as a super user that aggregates the traffic from all the VUEs it serves. Thus, when we design the scheduling algorithm, the number of VUEs served by an MN has to be taken into account. We analyze this problem further in Section 3.2.

In this study, we consider an orthogonal frequency-division multiple access (OFDMA)-based system which is the multiple access technology for the existing LTE-A systems. Corresponding parameters such as the sub-carrier spacing, the physical resource block (PRB) size, the subframe length, the modulation and coding schemes, and the link adaptation constraints that are defined in the frequency division duplex (FDD) LTE-A specifications are adopted throughout the paper. As the current FDD LTE-A system supports a maximum of 20 MHz bandwidth for a specific carrier frequency, we consider an intra-carrier aggregation configuration for the 80-MHz bandwidth used by the micro cells. A PRB, which consists of 12 subcarriers, with a duration of 0.5 ms, is the basic unit of resource allocation for a user in an LTE-A system. In most of the cases, the channel coherence bandwidth and time are much larger than the size of a PRB. Thus, we can assume a frequency flat-fading channel within a PRB.

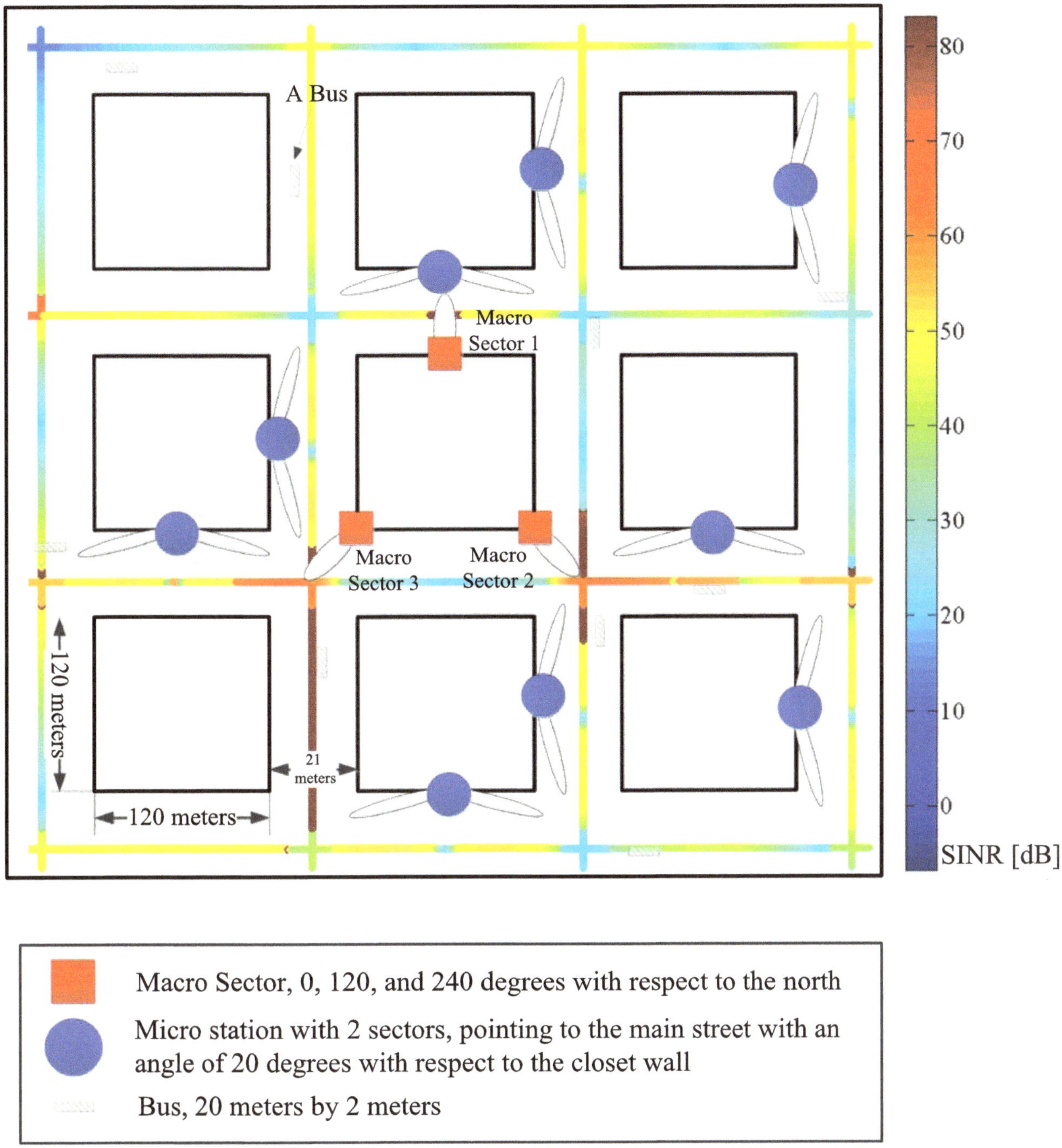

Figure 1 Simplified Madrid grid model with a heterogeneous deployment of macro and micro BSs. In the baseline scenario, macro and micro BSs operate at different frequency bands. The SINR levels from the strongest serving macro sector or micro station are also shown on the streets. The detailed parameters used for this plot are given in Section 7.

3.1 Interference identification for MNs

As demonstrated in [7,11], in a noise-limited system, MNs can improve the QoS of vehicular UEs, because the backhaul links of MNs have better propagation conditions and an MN can circumvent the VPL. In an ultra-dense urban scenario, however, situations are very different for both the backhaul and access links of MNs, particularly because of different central frequency propagation environments and bandwidths. Figure 2 illustrates the interference situations for different users. There is no interference between macro and micro cells, as they operate on different frequency bands. Macro users and the backhaul links of MNs

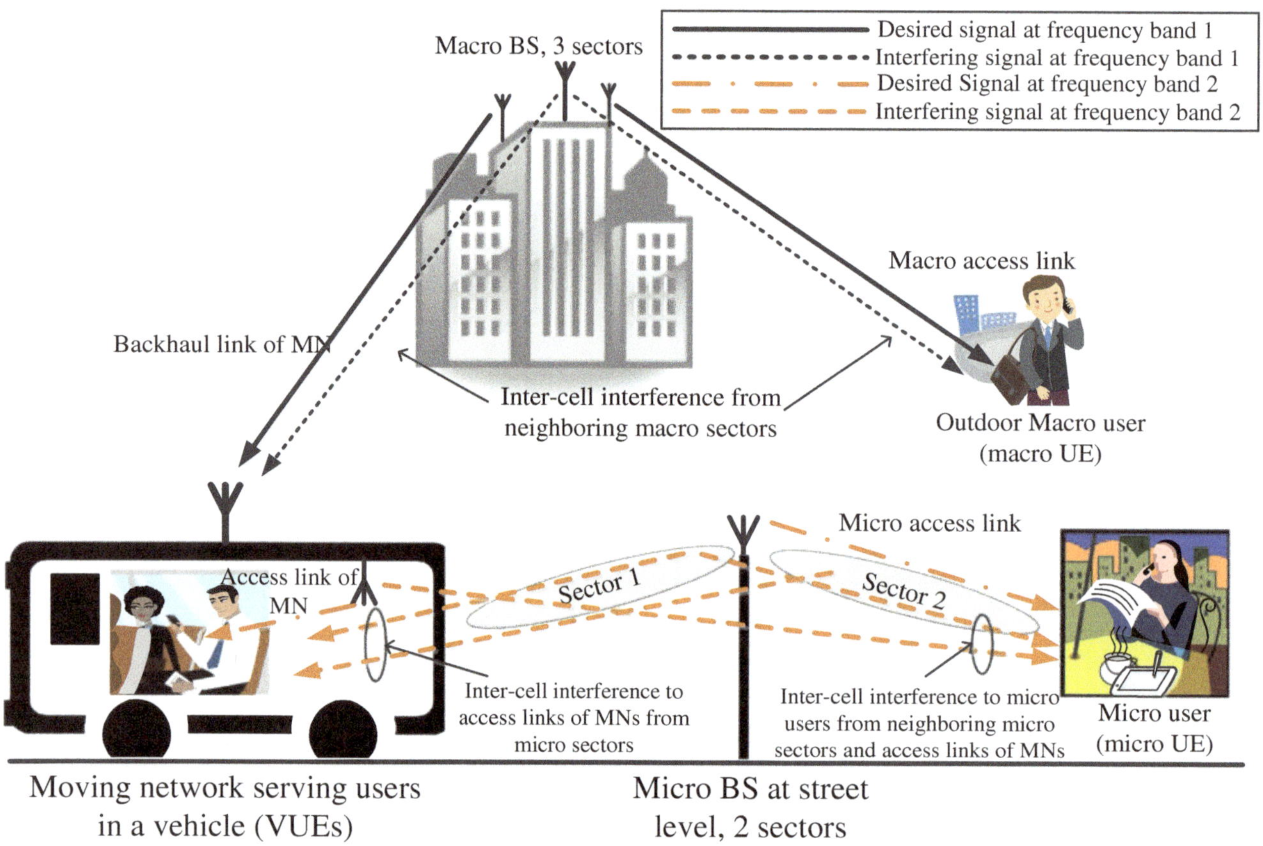

Figure 2 Illustration of desired and interfering signals. Macro and small cells are operating at different frequency bands. Macro UEs and backhaul links of MNs are subject to the inter-cell interference from neighboring macro sectors. Micro UEs are subject to inter-cell interference from neighboring micro sectors and potentially the interference from the access links of nearby MNs. Access links of MNs suffer from interference from nearby micro sectors.

experience interference from neighboring macro sectors. Micro users are subject to interference from micro sectors close by and the access links of MNs. A vehicular user can experience interference from micro cells and potentially by the access links of other close-by MNs.

Figure 1 plots the signal-to-interference-plus-noise ratio (SINR) that an outdoor user experiences in the considered Madrid grid set up, assuming the user always associates with the strongest cell (5 dB CRE is considered for the micro cells, and detailed parameters used for the plot are given in Section 7). As we can see from the SINR plot, in an ultra-dense urban scenario, because of inter-cell interference, the SINR can still be low, even when a user is close to a BS, e.g., between Macro sectors 2 and 3. The same situation applies for the MNs. Although, MNs can circumvent VPL and improve the signal-to-noise ratio (SNR) at the VUEs, in a densely deployed scenario, the severe inter-cell interference has a noticeable impact on the received SINR. To illustrate this, for the deployment scenario shown in Figure 1, Figure 3 shows an example of the SNR and SINR cumulative distribution function (cdf) at the backhaul links of MNs and the VUEs when the VPL is 15 dB. Compared with when VUEs are directly served by macro cells, the SNR is much improved via MNs because of better propagation conditions and the circumvention of VPL. However, SINR is only slightly improved by using MNs. Although the desired signal strength can be improved at the backhaul antennas, the desired signal subject to significant inter-cell interference. Furthermore, different from the traditional hexagonal cell deployment model, in a Madrid grid model, interference situations experienced by the receivers are more complicated because of urban canyon effects and may degrade the SINR observed at an MN.

The access links of MNs are assumed to transmit at the same frequency bands as the micro BSs. Although micro BSs transmit at a much lower power than the macro BSs, the interference they bring to the access links of MNs is still significant. This is because micro BSs are densely deployed at street level (around 80 m apart from each other in the Madrid grid baseline case), and, in the worst case scenario, the access link of an MN can be subject

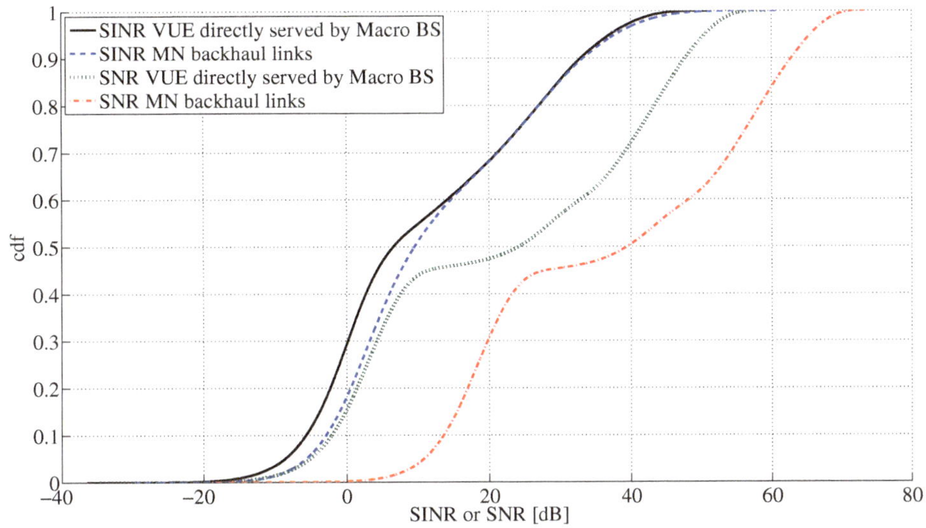

Figure 3 SNR and SINR plot of VUEs at 15 dB VPL. The detailed parameters used for this plot are given in Section 7. The solid black line is the SINR at VUEs when the VUEs are served directly by the macro BS. The dashed blue line is the SINR of the backhaul links of MNs. The dotted dark green line is the SNR at VUEs when the VUEs are served directly by the macro BS. The dash-dot red line is the SNR of the backhaul links of MNs.

to interference from several micro BS sectors at the same time. As shown in Figure 4, without interference coordination methods, even though the available bandwidth of the access links of MNs is large, the access links of MNs may not always accommodate the traffic from their backhaul links.

Therefore, to improve the performance of VUEs served via MNs, we have to alleviate the impact of inter-cell interference, both for the backhaul and access links of MNs. The methods used to improve the performance of backhaul links are discussed in Sections 4 and 5, while the ICIC scheme used to protect access links is outlined in Section 6.

The access links of the MNs might also cause interference with each other. However, in the scenario that we consider in this study, the interference between the access links of the MNs has negligible impact on the throughput of the VUEs. There are two main reasons for this. First, the access links of MNs only need to accommodate the traffic from their backhaul links. Because of the relatively large bandwidth compared with the backhaul links, and short distance (around 5 to 10 m) between the access

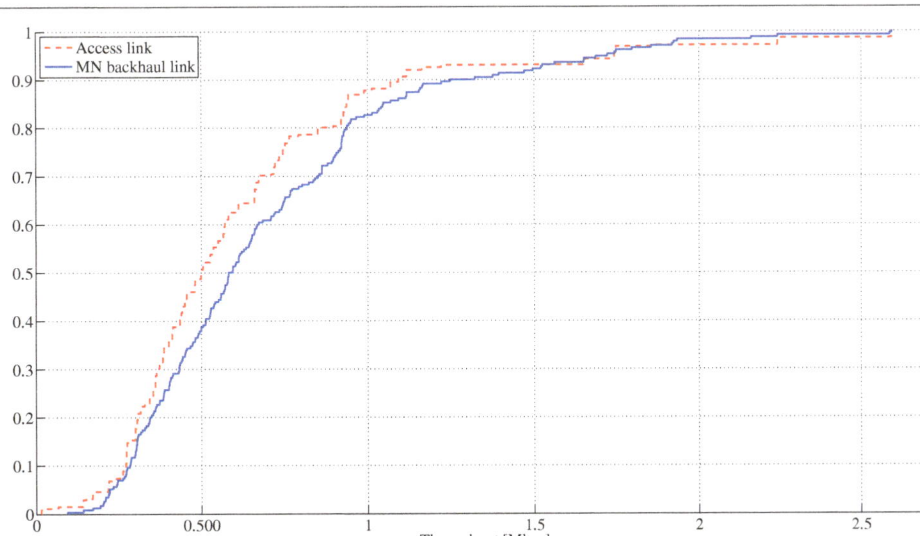

Figure 4 An example for the throughput cdf of an MN without ICIC in micro cells. The schemes used for MN backhaul is MRC combining with four antennas, and the VPL is at 15 dB. The detailed parameters used for this plot are given in Section 7. The dashed red line is the throughput of the access links. The solid blue line is the throughput of the backhaul links of MNs.

link antennas and the VUEs, a high-transmit power is not needed for the access links of MNs. Second, the interference generated by the access link of an MN is attenuated twice by the VPL, from its own vehicle and from the other vehicles. We observed similar effects in our previous studies in [12,28], which showed that the vehicle itself can help to reduce the impact of co-channel interference. This is very beneficial from a system perspective, because the frequency can be reused more efficiently. Therefore, in this study, we do not consider any interference management scheme between the access links of the MNs.

3.2 Fairness between regular UEs and VUEs

Besides the study of different ICIC schemes that can be used for MNs, our study also aims at improving the performance of VUEs without sacrificing the fairness between VUEs and regular outdoor UEs. We use proportional fairness (PF)-based resource allocation to ensure a balance between system throughput and fairness among all the users [46]. We assume a full buffer traffic situation as well as the same priorities for all the UEs, i.e., VUEs, macro UEs, and micro UEs. Furthermore, in this study, each MN is regarded as a super user that aggregates the data traffic of all its VUEs. In this section, we briefly describe how to modify the PF scheduler to integrate the MNs and their VUEs into the scheduling decisions.

In a network without MNs, the throughput of N_u users is proportionally fair [46], if it maximizes the following objective function:

$$f_1 = \sum_{u=1}^{N_\mathrm{u}} \log\left(s(u)\right), \tag{1}$$

where $s(u)$, $(u = 1, 2, \ldots, N_\mathrm{u})$, is the average throughput of the uth user. Note that in an OFDMA system, the minimum scheduling granularity is given by a PRB. Without loss of generality, we use $b\,(l, m, u)$ to denote the scheduling decision for the uth user in mth PRB in the lth subframe, i.e.,

$$b\,(l, m, u) = \begin{cases} 1, & \text{if user } u \text{ is assigned at } (l, m) \\ 0, & \text{if user } u \text{ is not assigned at } (l, m) \end{cases}. \tag{2}$$

In a general case, $b\,(l, m, u)$ is determined by a PF scheduler as follows [47,48]. First, for each user u, its throughput at mth PRB in the lth subframe is calculated as $r\,(l, m, u)$. Based on the SINR information, there are several ways to calculate the throughput $r\,(l, m, u)$, e.g., look-up tables or variations of the Shannon capacity equation [49,50]. In this study, we use the SINR look-up table method [50]. Then, a user u^* is selected at (l, m) as:

$$u^*\,(l, m) = \arg \max_{u\in\{1,2,\ldots,N_\mathrm{u}\}} \left(\frac{r\,(l, m\, u)}{\mu\,(l, u)}\right), \tag{3}$$

where $\mu\,(l, u)$ is the average throughput of user u, which is updated in each subframe as:

$$\mu\,(l+1, u) = (1-\varepsilon)\,\mu\,(l, u) + \frac{\varepsilon}{M}\sum_{m=1}^{M} r\,(l, m, u)\; b\,(l, m, u). \tag{4}$$

In Equation 4, M is the total number of PRBs in each subframe, and $0 \leq \varepsilon \leq 1$ is an averaging constant, which is usually set to 0.5 [47]. Finally, the resource assignment is determined as:

$$b\,(l, m, u) = \begin{cases} 1, & \text{if } u = u^*\,(l, m) \\ 0, & \text{otherwise} \end{cases}. \tag{5}$$

When the MNs are introduced into the system, to ensure fairness between VUEs and macro UEs, the PF scheduler needs to be modified in order to take the aggregate traffic carried by the MNs into account. As mentioned before, we treat an MN as a super user that aggregates the traffic of all the VUEs it serves. The MNs whose backhaul links are served by macro cells share the same resources as macro UEs. Since several VUEs are served by one MN, the throughput of the MN backhaul link needs to be scaled by the number of VUEs that it serves. Otherwise, the backhaul link of an MN will not get a fair scheduling assignment. Let there be a total of N_m MNs in the system, and let i_ν denote the number of active VUEs on the νth MN, where $\nu = 1, 2, \ldots, N_\mathrm{m}$. Then, as opposed to using Equation 1, the objective function of the PF scheduler is modified as:

$$f_2 = \sum_{u=1}^{N_\mathrm{u}} \log\left(s(u)\right) + \sum_{\nu=1}^{N_\mathrm{m}} \log\left(\frac{s(\nu)}{i_\nu}\right), \tag{6}$$

where $s(\nu)$, $(\nu = 1, 2, \ldots, N_\mathrm{m})$ is the average throughput of the backhaul link of the νth MN. In this study, the number of active VUEs served by an MN is a uniformly distributed discrete random variable, with $i_\nu \sim \mathcal{U}(1, 50)$, where $\mathcal{U}$ denotes uniform distribution. The difference between Equation 1 and Equation 6 is that the traffic of the backhaul link of each MN is averaged by the number of its VUEs. The scheduling algorithm involving MNs and the time domain ICIC scheme is described in Section 4.2.

As mentioned before, in an interference-limited scenario, certain users at the cell edges are subject to strong interference from neighboring cells. Although PF scheduling is a way to achieve a balance between user fairness and system throughput, it cannot alleviate the impact of interference. In the following sections, we discuss the use of both time domain methods and multi-antenna solutions to alleviate the impact of interference, and the PF scheduling is used together with these schemes to ensure fair resource allocation among all the UEs.

4 Time domain interference coordination for the backhaul links of MNs

As mentioned in Section 2, using ABSs is standardized in 3GPP to protect victim UEs that are subject to excessive co-channel interference. The ABSs can be configured by either macro or micro cells, and different ABSs patterns and density can be configured by each individual cell to offer various levels of protection to UEs influenced by co-channel interference ([16] Ch. 31)[22-24]. The benefits of using ABSs are that it is standard compliant and can be combined with other existing ICIC methods, e.g., frequency reuse, power control, or various multi-antenna solutions.

As mentioned in Section 3, we consider a macro BS that has three sectors. Three types of macro UEs are identified in the system. The first group of UEs has high SINR values, which means these UEs are in the cell center, and are less affected by inter-cell interference. We label this group of UEs as type 1. The second group of UEs experience higher interference from one of the macro BS sectors, i.e., the interference power from one sector is much larger than the other interfering sector. We label this group of UEs as type 2. The third group of UEs experience similar interference power from both interfering sectors. We label this group of UEs as type 3. Obviously, type 2 and 3 UEs need to be protected from interference.

Figure 5 shows an example of the ABS configurations of the three sectors of a macro BS. Scheduling restrictions can be applied to achieve different levels of protection for the UEs. The idea of the scheduling restriction is to restrict the UEs that are in good channel conditions to the non-protected sub-frames, so that more resources can be made available for the victim UEs. To be more specific, the protected resources are reserved for victim UEs that are subject to strong inter-cell interference. However, these victim UEs can also be scheduled in the non-protected sub-frames, if there are available resources and/or the throughput of one of the protected resources is too low to maintain good fairness between the UEs. Therefore, type 1 UEs are not subject to interference, so they are strictly scheduled in the unprotected sub-frames. Type 2 UEs are restricted to sub-frames where their dominant interferer is muted (recall that one significant interferer is observed by a type 2 UE) and where the sub-frames are unprotected. No scheduling restrictions are applied to type 3 UEs.

Therefore, two questions need to be answered to use the time domain ICIC method: 1) how to configure the number of ABSs for each macro BS sector, and 2) how to classify the UEs and schedule them to achieve a balance between maximum protection for the victim UEs and the system throughput. Previous studies show that even in simplified setups, a joint optimization of these two problems is very difficult [23,47,51]. Next, we briefly review and discuss the possible ways to address these two problems.

4.1 Configuration of ABSs

The ABSs patterns of each cell may be dynamically changed. For example, in LTE-A systems, the duty cycle of ABSs can be as short as 40 ms [22] ([16] Ch. 31). However, for practical reasons, e.g., for obtaining traffic distribution statistics and considering the reconfiguration overhead,[d] the time scales of changing ABSs patterns could be at lest in the order of a few minutes [22,23].

When considering different scheduling algorithms, QoS constraints, and other factors, it is challenging to obtain an optimal ABS pattern across multiple cells within a reasonable time. The ABS pattern optimization problem is shown to be NP-hard in a multi-cell full buffer traffic scenario [23]. Previous studies either consider fixed ABS patterns based on traffic distribution statics or use sub-optimal solutions based on simplified assumptions [47,51]. In this study, we consider the use of fixed ABSs patterns (described in the appendix). We consider two setups, i.e., 12 ABSs out of 40 frames (30%) and 15 ABSs out of 40 frames (37.5%). In the next section, we discuss how to classify the UEs and the criteria of identifying their strong interferers, as well as how to incorporate the PF scheduling with the scheduling restrictions.

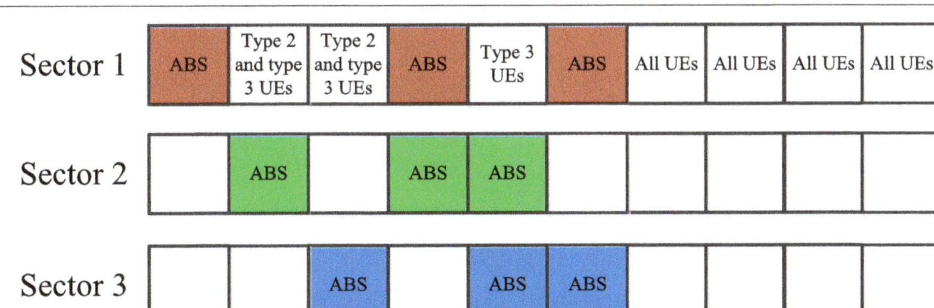

Figure 5 An example of frames with ABS configurations and scheduling restrictions at different macro sectors. Frames with solid colors are ABSs. We use sector 1 as an example to demonstrate scheduling restrictions.

4.2 Scheduling restriction schemes

After determining the ABS patterns of each cell, the next step is to determine how to schedule the three types of UEs. Several criteria have been proposed in determining scheduling restrictions. In [52,53], cell edge UEs and cell center UEs are scheduled in protected and non-protected resources, respectively. In [47], scheduling restrictions are applied based on the channel quality differences between protected and non-protected resources. In our study, we apply the generalized method in [47] to classify the three types of UEs. The framework proposed in [47] takes several factors into account, e.g., user distribution and the ratio between protected and non-protected resources, to decide how to apply the scheduling restrictions. While in [47] only two cells are considered, the method can easily be extended to our scenario by applying pairwise comparisons. Next, we briefly introduce the method used in [47] and discuss how to use it to determine the three different types of UEs in this study.

We begin with a two-cell setup, and we use macro BS sector 1 as an example. The user sets for macro UEs and MNs are $\mathcal{M} \triangleq \{1, 2 \ldots, N_u\}$ and $\mathcal{V} \triangleq \{1, 2 \ldots, N_m\}$, respectively. In total, $N = N_u + N_m$ devices connect to macro BS sector 1, and the interference is from macro BS sector 2. The macro BS sector 2 configures ABSs patterns, and macro BS sector 1 makes the scheduling decisions accordingly. From the perspective of BS sector 1, we define two sets of PRBs, i.e., protected PRBs $\mathcal{R}_p$ and non-protected PRBs $\mathcal{R}_{np}$. Denote the total number of protected PRBs as $B_p = |\mathcal{R}_p|$, and the total number of non-protected PRBs as $B_{np} = |\mathcal{R}_{np}|$, where $|\cdot|$ is the cardinality of a set. With scheduling restrictions, users served by BS sector 1 can be partitioned into three sets: $\mathcal{S}_1 = \{\text{users only scheduled in } R_p\}$, $\mathcal{S}_2 = \{\text{users only scheduled in } R_{np}\}$, and $\mathcal{S}_3 = \{\text{users can be scheduled both in } \mathcal{R}_p \text{and } \mathcal{R}_{np}\}$.

As mentioned in Section 3.2, the throughput of the backhaul links of MNs needs to be scaled by the number of VUEs that it serves. Therefore, for device n, its throughput at mth PRB in the lth subframe is calculated as:

$$r(l, m, n) = \begin{cases} r(l, m, u), & n \in \mathcal{M} \\ \frac{r(l, m, v)}{i_v}, & n \in \mathcal{V} \end{cases}, \tag{7}$$

where $r(l, m, u)$, $(u = 1, 2, \ldots, N_u)$ and $r(l, m, v)$, $(v = 1, 2, \ldots, N_m)$ are the throughput of the uth macro UE and the backhaul link of the vth MN at mth PRB in the lth subframe, respectively. The number of VUEs served by the vth MN is i_v. The PF scheduler at BS sector 1 is restricted to the subset of users, i.e., $(\mathcal{S}_1 \cup \mathcal{S}_3)$ and $(\mathcal{S}_2 \cup \mathcal{S}_3)$. The PF scheduling algorithm can be applied with the following modification to the user selection method:

$$n^*(l, m) = \begin{cases} \arg \max\limits_{n \in (\mathcal{S}_1 \cup \mathcal{S}_3)} \left(\frac{r(l,m,n)}{\mu(l,n)}\right), & \text{if } (l, m) \in \mathcal{R}_p \\ \arg \max\limits_{n \in (\mathcal{S}_2 \cup \mathcal{S}_3)} \left(\frac{r(l,m,n)}{\mu(l,n)}\right) & \text{if } (l, m) \in \mathcal{R}_{np} \end{cases}, \tag{8}$$

where $\mu(l, m)$ is updated as given in Equation 4 with the instantaneous rate calculated by Equation 7.

In order to apply the scheduling restrictions for the PF scheduling, we have to properly partitioned the users. It is shown in [47] that at most one user can be in set $\mathcal{S}_3$. Let $s(n)$ denote the average throughput of device n, with scheduling restriction, our objective function of a PF scheduler becomes:

$$\begin{aligned} g &= \sum_{n=1}^{N} \log(s(n)) \\ &= \sum_{n=1}^{N} \log\left(\frac{1}{LM}\left(r_p(n)\, b_p(n) + r_{np}(n)\, b_{np}(n)\right)\right) \end{aligned} \tag{9}$$

where L is the total number of subframes in which $r_p(n)$ and $r_{np}(n)$ are observed. In this study, we set L as the number of subframes in one ABS duty cycle. Moreover, $r_p(n)$ and $r_{np}(n)$ are average rate of the device n over $\mathcal{R}_p$ and $\mathcal{R}_{np}$, respectively. In addition,

$$b_p(n) = \sum_{(l,m)\in R_p} b(l, m, n), \tag{10}$$

$$b_{np}(n) = \sum_{(l,m)\in R_{np}} b(l, m, n). \tag{11}$$

The detailed solution of how to maximize the proportional fairness metric in Equation 9 in a two cell setup is given in [47]. Next, we shortly revisit the solution to facilitate our discussion. The main idea of the solution is first to identify the users with large quality difference between protected and non-protected resources, and then schedule the users accordingly.

In [47], in order to solve the problem, a function $d(n)$ is defined as:

$$d(n) = \begin{cases} \infty, & \text{if } n = 0 \\ \frac{B_p}{B_{np}} \frac{r_p(n)}{r_{np}(n)} & \text{if } n = 1, 2, \ldots, N \end{cases}. \tag{12}$$

Recall that B_p and B_{np} are the total number of protected and non-protected PRBs, and $r_p(n)$ and $r_{np}(n)$ are the average rates of device n over the protected and non-protected PRBs, respectively. Therefore, $d(n)$ measures channel quality difference for the nth device over the protected and non-protected PRBs. When $d(n)$ is large, it indicates that the nth device is sensitive to the interference in the unprotected resources and thus it benefits largely from being scheduled to the protected resources. One assumption made in Equation 12 is that by appropriate network planning and UE association, there is no user n such that $r_p(n) = r_{np}(n) = 0$, i.e., no user has zero throughput. Furthermore, a dummy user $n = 0$ with $d(0) = \infty$ is defined for mathematical correctness of the problem formulation. The solution that maximizes Equation 9 is given in [47], and we summarize the solution in the following paragraph. The step by step derivations are given in the appendix of [47].

The N devices, including both macro UEs and MNs, are sorted descendingly according to Equation 12, i.e., $d(n-1) \geq d(n)$, for $n = 1, 2 \ldots, N$, and:

$$\begin{cases} b_{\mathrm{p}}(n) = \frac{B_{\mathrm{p}}}{\lambda_{\mathrm{p}}},\ b_{\mathrm{np}}(n) = 0, & \text{if } n = 1, \ldots K \\ b_{\mathrm{p}}(n) = B_{\mathrm{p}} \frac{\alpha}{\lambda_{\mathrm{p}}},\ b_{\mathrm{np}}(n) = B_{\mathrm{np}} \frac{1-\alpha}{\lambda_{\mathrm{np}}}, & \text{if } n = K+1 \\ b_{\mathrm{p}}(n) = 0,\ b_{\mathrm{np}}(n) = \frac{B_{\mathrm{np}}}{\lambda_{\mathrm{np}}}, & \text{if } n = K+2, \ldots N \end{cases} \tag{13}$$

where $K \in \{0, 1, \ldots, N-1\}$. Parameters K, α, λ_{p} and λ_{np} are determined as follows.

$$G(d(K+1)) \leq K \leq G(d(K)) \tag{14}$$

with $G(\nu) = N \frac{\nu}{(1+\nu)}$ for $\nu \geq 0$, and:

$$\lambda_{\mathrm{p}} = \max\left(G(d(K+1)),\ K\right), \tag{15}$$

$$\lambda_{\mathrm{np}} = N - \lambda_{\mathrm{p}}, \tag{16}$$

$$\alpha = \lambda_{\mathrm{p}} - K. \tag{17}$$

Recall that from Equation 5, we always have $b(l, m, n) \geq 0$. Thus, Equation 13 indicates that when $b_{\mathrm{np}}(n) = 0$, the first K users are scheduled in the protected resources; likewise, when $b_{\mathrm{p}}(n) = 0$, the users $n = K+2, \ldots, N$ are scheduled in the non-protected resources. Only the $(K+1)$th user can be scheduled in both the protected and non-protected resources if $0 < \alpha < 1$, or the $(K+1)$th user can be placed to either of the two types of resources by setting $\alpha = 0$ or $\alpha = 1$.

As mentioned in Section 3, we consider a BS with 3 sectors. Recall that three different types of users need to be identified in order to apply the scheduling restrictions. In our study, we apply pairwise comparison between BS sectors by using the methods briefed above to identify the three types of users. As an example, let us consider sector 1, and suppose N_1 devices are served by it. For an arbitrary device q_{n_1}, $n_1 \in \{1, 2, \ldots, N_1\}$ served by sector 1, we apply the aforementioned methods twice and compare sector 1 with sector 2 and sector 3 alternatively. If in both cases q_{n_1} is identified to be scheduled in non-protected resources of sector 1, then q_{n_1} is classified as a type 1 user. If in both cases q_{n_1} is identified to be scheduled in protected resources of sector 2 and sector 3, then q_{n_1} is a type 3 user. The rest of the users (if any) are type 2 users that have either sector 2 or sector 3 as their dominant interferer.

5 Multi-antenna solutions for backhaul links of MNs

Beside the time domain methods, we can also use multi-antenna solutions to improve the received signal quality at the backhaul links of MNs. One of the benefits of using MNs is that public transportation vehicles are less constrained by antenna array size or power compared to regular UEs. Thus, more antenna elements and more advanced signal processing algorithms can be employed. In this study, we consider a single antenna at each sector of the macro BS, and multiple antennas are used for the backhaul link reception at the MNs. The idea of using multiple receiving antennas in wireless communication systems has a long history ([54] Ch. 20). In practice, depending on whether CSI from the interferers can be obtained, different schemes can be applied to improve the received SINR ([54] Ch. 20). In this study, we consider two cases, i.e., maximum ratio combining (MRC), and SINR maximization by interference suppression.

5.1 Maximum ratio combining

If only the CSI of the desired signal can be estimated at each of the antennas at the receiver, MRC can be used to combine the desired signal from each antenna at the receiver constructively to maximize the desired signal power. An MRC receiver is optimal in terms of SNR, i.e., it maximizes the output power of the received desired signal [55]. The received signal consists of the desired signal, interference signal, and thermal noise. Suppose the receiver is equipped with p antennas. In a frequency flat-fading case, e.g., within a PRB, at time k, if we have N_{L} interferers, the received signal can be expressed as:

$$\mathbf{y}_k = \mathbf{h}_k x_k + \sum_{j=1}^{N_{\mathrm{L}}} \mathbf{h}_{j,k} x_{j,k} + \mathbf{n}_k, \tag{18}$$

where $\mathbf{y}$ is a $p \times 1$ column vector that contains the received signal at each antenna, x_k is the desired signal, and $x_{j,k}$ is the interfering signal from the jth interferer. Moreover, $\mathbf{h}_k$ and $\mathbf{h}_{j,k}$ represent the desired and jth interfering signal propagation vectors, respectively, and $\mathbf{n}_k$ is the thermal noise. We assume that the desired signal and interfering signals are uncorrelated.

At the receiver, let $\mathbf{w}_k$ be a vector that contains the weight applied at each antenna at time k. Then, the combined signal at the MN is given by:

$$\mathbf{z}_k = \mathbf{w}_k^{\mathrm{H}} \mathbf{y}_k = \mathbf{w}_k^{\mathrm{H}} \mathbf{h}_k x_k + \sum_{j=1}^{N_{\mathrm{L}}} \mathbf{w}_k^{\mathrm{H}} \mathbf{h}_{j,k} x_{j,k} + \mathbf{w}_k^{\mathrm{H}} \mathbf{n}_k, \tag{19}$$

where $(\cdot)^{\mathrm{H}}$ denotes the complex conjugate transpose. The MRC solution that maximizes the desired signal power uses the combining weights as $\mathbf{w}_k^* = \mathbf{h}_k$, which yields the following instantaneous output SINR at the MN:

$$\gamma_k = \frac{\left|\mathbf{h}_k^{\mathrm{H}}\mathbf{h}_k\right|^2}{\left|\sum_{j=1}^{L}\mathbf{h}_k^{\mathrm{H}}\mathbf{h}_{j,k}x_{j,k} + \mathbf{h}_k^{\mathrm{H}}\mathbf{n}_k\right|^2}. \quad (20)$$

From Equation 20, we can observe that, although at a given time the output power of the desired signal is maximized, the SINR is not maximized due to the presence of the interfering signal. However, if no CSI from the interferers can be obtained, MRC is the optimal solution to maximize the received desired signal power.

5.2 SINR maximization by interference suppression

As mentioned before, MRC combining is optimal in terms of SNR; however, we can do better if the phase and amplitude of the interferers can also be estimated. If CSI from the interferers can be obtained, we can exploit the multiple antennas at the receiver to suppress the interference signals, and thereby maximize the output SINR. This scheme falls into the category of interference rejection combining (IRC) [56]([57] Ch. 10).

From Equation 19, we can obtain the received SINR with some arbitrary weight vector $\mathbf{w}_k$ as:

$$\gamma_k = \frac{\mathbf{w}_k^{\mathrm{H}}\mathbf{h}_k\mathbf{h}_k^{\mathrm{H}}\mathbf{w}_k}{\mathbf{w}_k^{\mathrm{H}}\left(\sum_{j=1}^{L}\mathbf{h}_{j,k}^{\mathrm{H}}\mathbf{h}_{j,k} + \mathbf{R}\right)\mathbf{w}_k}, \quad (21)$$

where $\mathbf{R}$ is the noise covariance for the receiver antennas, same for all k. Thus, if we know $\mathbf{h}_{j,k}$, we can obtain the optimal weight that maximizes the SINR by solving:

$$\mathbf{w}_k^* = \arg\max_{\mathbf{w}_k}\frac{\mathbf{w}_k^{\mathrm{H}}\mathbf{h}_k\mathbf{h}_k^{\mathrm{H}}\mathbf{w}_k}{\mathbf{w}_k^{\mathrm{H}}\left(\sum_{j=1}^{L}\mathbf{h}_{j,k}^{\mathrm{H}}\mathbf{h}_{j,k} + \mathbf{R}\right)\mathbf{w}_k}. \quad (22)$$

The optimal weight problem (22) is a generalized eigenvalue problem, and the optimal weight is given as [56]:

$$\mathbf{w}_k^* = \mathrm{EIG}_{\max}\left(\left(\sum_{j=1}^{L}\mathbf{h}_{j,k}^{\mathrm{H}}\mathbf{h}_{j,k} + \mathbf{R}\right)^{-1}\mathbf{h}_k\mathbf{h}_k^{\mathrm{H}}\right), \quad (23)$$

where $\mathrm{EIG}_{\max}(\cdot)$ denotes the dominant eigenvector of a matrix [e], and we assume that antenna noise is independent among the antenna elements, i.e., $\mathbf{R} = \sigma_n^2 \mathrm{I}_{p\times p}$, where σ_n^2 is the noise power at each antenna, and $\mathrm{I}_{p\times p}$ is the identity matrix of size p.

As an example, Figure 6 compares the cdf of SINR at the backhaul links of MNs. As we can see from the figure, the IRC gives the best result as expected, since it maximizes the SINR. Regarding the use of ABSs and MRC, compared to the case without using any ICIC scheme, the SINR at the receiver is also improved. The MRC scheme improves the performance in the low SINR region, but since MRC cannot remove the interference, in an interference limited scenario, the general performance of it is not as good as the other two schemes. Section 7 provides more detailed system level evaluation results.

6 Interference coordination for access links of MNs

Since the access links of MNs are operating at the same frequency bands as the micro cells, as discussed in Section 3.1, the access links need to be protected in order to accommodate the traffic from their backhaul links. We

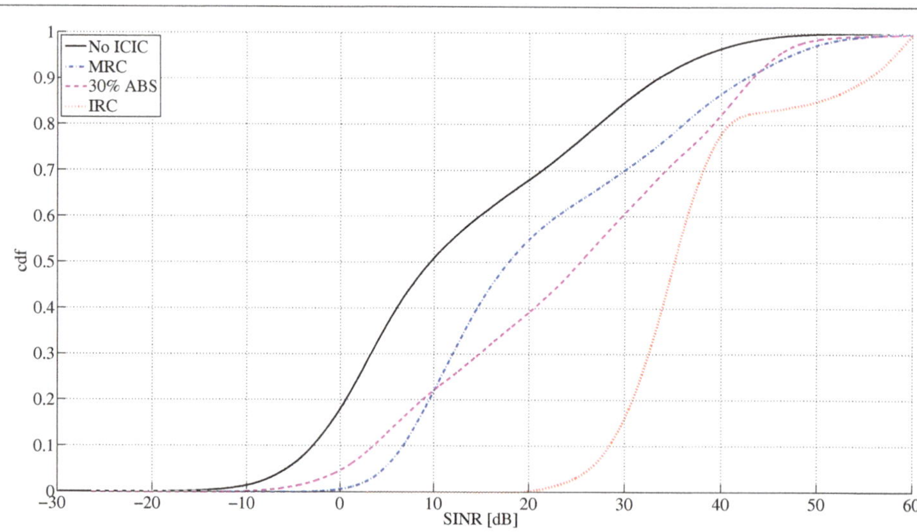

Figure 6 The cdf of SINR per PRB of the backhaul links of MNs. For MRC and IRC, four antennas are used at the receiver. The detailed parameters used for this plot are given in Section 7. The solid black line is the SINR when no ICIC schemes are used. The dash-dot blue line is the SINR when MRC is used. The dashed purple line is the SINR when the number of ABSs is 30% at the macro sectors. The dotted red link is the SINR when IRC is used.

consider to configure ABSs at the micro cells to alleviate the interference problem that they cause to the access links of MNs. There are two ways to configure the ABS patterns for the micro cells: 1) each micro cell configures its own ABS pattern, or 2) the micro cells in a given area use the same ABS pattern. Recall that rather than maximizing the throughput of the access links, our goal is to protect the access links of MNs, and thereby the traffic from their backhaul links can be accommodated. Therefore, we just need to configure a minimum number of ABSs to ensure that the access links of MNs can successfully convey the traffic from their backhaul links. However, as mentioned in Section 4.1, the ABS patterns can only be updated in a semi-static pattern. The criteria of configuring ABS patterns should be based on an average throughput (e.g., one or several ABS duty cycles), rather than the instantaneous throughput of each of the links. This is reasonable, as we can assume an MN can buffer the data from its backhaul link. As long as the buffer is big enough, the access link does not need to instantly accommodate the traffic from its backhaul link. Moreover, we can also optimize the density and ABS patterns to minimize the end-to-end transmission delays.

Furthermore, while a public transportation vehicle is moving, it can pass through several micro cells in a fairly short time. As discussed in [22,23], the typical updating time of ABS configurations is in the order of minutes, and therefore it is reasonable to assume that the same ABS configuration will be used by several micro cells in a given area. In this way, not only the signaling overhead to communicate the ABS configurations between different nodes can be reduced,but also the measurement and feedback from the MNs to the network can be less frequent. Hence, in this study, we use the same ABSs pattern for all the micro cells in the area given in Figure 1.

We consider the average throughput over N_{sf} subframes, where N_{sf} is the number of total subframes in one or several ABSs duty cycles. Our object is to configure a minimum number of ABSs in these N_{sf} subframes, such that the access links of MNs can successfully convey the traffic from their backhaul links. Let $\mathcal{I}_i \in \{0, 1\}$, $i = 1, 2, 3 \ldots N_{sf}$ indicate the status of the ABSs, where $\mathcal{I}_i = 0$ indicates the ith subframe is configured to be an ABS, and $\mathcal{I}_i = 1$ indicates the ith subframe is not configured to be an ABS. Therefore, for each subframe, depending whether it is configured as one of the ABSs by the micro cells, two different rates can be observed at each MN. Let $r_{0,i}(\nu)$ be the throughput of the access link of the νth MN when the ith subframe is configured as ABS, and $r_{1,i}(\nu)$ be the throughput of the access link of the νth MN when the ith subframe is not configured as ABS. Then, we can formulate the optimization problem as:

$$\max \sum_{i=1}^{N_{sf}} \mathcal{I}_i$$

subject to

$$\sum_{i=1}^{N_{sf}} (1 - \mathcal{I}_i)\, r_{0,i}(\nu) + \mathcal{I}_i\, r_{1,i}(\nu) \geq \bar{R}(\nu), \quad (24)$$

where $\bar{R}(\nu)$ is the average throughput of the backhaul link of the νth MN during N_{sf} subframes. Problem (24) is an integer linear programming problem, and if $r_0(\nu)_i$, $r_1(\nu)_i$ and $\bar{R}(\nu)$ are known, the solution of problem 24 can be easily found ([58] Ch. 11).

However, in practice $\bar{R}(\nu)$ cannot be obtained beforehand, and $r_{0,i}(\nu)$, $r_{1,i}(\nu)$ cannot even be observed due to the randomness of wireless channels, measurement restrictions, and feedback delays. Thus, in this study, a reinforcement-learning approach [59] is adopted to configure the ABS patterns for the micro cells. The main idea is that we assume VUEs are primary users, and a set of ABS patterns is pre-defined for the micro cells. Our goal is to minimize the number of ABSs used by the micro BSs in a given area and satisfy the average throughput requirement for the access link of the worst MN in that area.

First, the micro BSs in a given area choose a configuration from the pre-defined set of ABSs patterns[f], and collect feedback from the MNs in that area according to this setting. If during the measurement time, e.g., one or several ABSs duty cycles, this setting cannot satisfy the average throughput of the access link of the worst MN in that area, then the micro BSs choose a new ABSs pattern that offers more protection. Otherwise, the micro BSs can lower the number of ABSs until it reaches the minimum number that satisfy the average throughput of the access link of the worst MN in that area. Recall that we assume MNs can buffer the data from their backhaul links. Therefore, as long as the buffer is big enough, the initial value of ABSs pattern is reasonable, and the measurement time is relatively short, there are no concerns about unnecessary drop of packets and excessive end-to-end delays.

The algorithm used in this evaluation is summarized in Algorithm 1. In practice, step 3 in Algorithm 1 may be optional, as it only ensures that the number of ABSs is not set too high at the micro cells. For example, due to the sudden increase in the number of MNs during rush hours, more ABSs may be allocated by the micro cells, and after the rush hours when the number of MNs drops in a given area, we may need to lower the number of ABSs in order not to waste the resources at the micro cells. This can also be achieved by using a timer at the micro cells, which will lower the number of ABSs after a given time. We emphasize that the configuration of ABSs should not be frequent, at least not in the order of several seconds.

Thus, in practice, a small margin may be added to the minimum number of ABSs to accommodate the sudden change of the number of vehicles or the propagation conditions. In a larger time scale, i.e, in an order of several minutes or even hours, the micro BSs in the given area can update the configuration of ABS patterns according to the feedback from the MNs.

Algorithm 1 Configuration of the number of ABSs for the micro BSs

Step 1 Initialize the ABS configuration

1: Index all the possible ABSs configurations in ascending orders according to the number of ABSs.
2: Set all the micro BSs to use the configuration with the minimum number of ABSs.

Step 2 Update the ABS configurations

1: All the MNs in the given area feed back the measurement of their average access link throughput to the network.
2: **if** one of the MNs reports is insufficient in capacity, **then**
3: The network identifies the index of the ABS configuration with minimum increase of the ABSs, and inform the micro BSs to update their ABSs settings.
4: **else**
5: The micro BSs keep using the previous ABS configuration.
6: **end if**

Step 3 Re-configure the ABSs configurations (optional)

1: **if** none of the MNs report is insufficient in number of ABSs after some minutes, **then**
2: Set all the micro BSs to use the configuration with the minimum number of ABSs.
3: Repeat step 2.
4: **end if**

7 Performance evaluation

In this section, we compare and evaluate the MN interference techniques and discussed in earlier sections, through system level simulations. A semi-static drop based approach is adopted. The simulation area is shown in Figure 1. According to calibration setups given in METIS simulation assumptions ([14] Table 3.7), we consider that macro and micro BSs are operating on different frequency bands, with different bandwidths. To help compare different technical solutions, METIS provides implementations of the path-loss and shadowing channel models for the macro and micro sectors [14].[g] The METIS path-loss and shadowing modeling is based on ray tracing and takes into account the positions of the BSs and the geometry of the Madrid grid to ensure correct spatial property of the channels. Therefore, as long as the position and the antenna height of a receiver are determined, its path-loss and shadowing are also determined. As suggested by METIS, the small scale fading coefficients are generated according to the ITU-R urban macro cell (UMa) and urban micro cell (UMi) channel models for the macro and micro sectors, respectively [14].

To ensure the correct statistical properties of the small scale fading coefficients, all the channel coefficients are generated and normalized at the beginning of each drop. The coherence time of the channel is based on the Doppler spectrum of the receivers according to their velocities. A grid-based geometry is set in the ITU-R channel models to ensure the proper coherence bandwidth. Detailed descriptions of the ITU-R UMa and UMi channel models are given in [60].[h] A non-line-of-sight (NLOS) propagation condition is considered for the macro sectors, and for the micro sectors line-of-sight (LOS) propagation is considered if the receivers are on the same street as the transmitters, otherwise the channel is considered to be NLOS. The propagation of the interference signals between the access links of MNs are modeled in the same way as the micro cells, but with an antenna height of 3.5 m. Because of the size of the public transportation vehicles, we assume that a fairly large space between antenna elements can be achieved. Therefore, at the backhaul receivers of the MNs, independent fading is assumed between desired signals and between the desired and interfering signals. Because a semi-static drop based simulation approach is used, the number of MNs on each street is first calculated based on the mobility model (see Table 1), and the MNs are then dropped uniformly on each street.

The VUEs are either served by macro BS sectors directly (baseline case) or via MNs that are formed by their own vehicles. Because of the estimated speed of the vehicle, and the size of the micro cells, VUEs are assumed not to be offloaded to micro cells. MNs are assumed to be using full-duplex communication with macro BS sectors and their VUEs on different frequency bands. To understand the impact of the number of VUEs on board, in addition to the case of dropping VUEs uniformly on each vehicle with $i_v \sim \mathcal{U}(1, 50)$, we also consider a case when each vehicle has a fixed number of 25 VUEs on board. The expected total number of users, i.e., VUEs, macro UEs, and micro UEs is kept the same for all cases. In the simulation, for each case, 850 drops are conducted with 40 sub-frames in each drop. The channel realizations and the positions of MNs and UEs are updated on a drop basis. The detailed parameters are summarized in Table 1 [14], and the detailed ABS configurations are given in the appendix.

Table 1 Simulation parameters

Component	Configuration parameters
Buildings and streets	Nine buildings 120 by 120 m with six floors (3.5 m height of each floor), see Figure 1.
	Road width 21 m (including sidewalks and parking lanes)
Macro BS	Height: 5 m above the top of the middle building (see Figure 1)
	Maximum transmit power (per 10 MHz): 43 dBm
	Carrier: 800 MHz
	Bandwidth: 20 MHz
	Antenna configuration: 17 dBi gain, three sectors (one antenna per sector), 0°, 120°, and 240° with respect to the north
Micro BS	Height: 10 m above the ground close to middle point of south and east walls
	Positions: see Figure 1
	Maximum transmit power (per 10 MHz): 30 dBm
	Carrier: 2.6 GHz
	Bandwidth: 80 MHz
	Cell range expansion bias: 5 dB
	Antenna configuration: 17 dBi gain, two sectors (one antenna per sector), pointing to the main street with an angle of 20° with respect to the closest wall
Moving network	Full-duplex
	Speed: 50 km/h
	Height: 3.5 m above the ground
	Mobility model: public transportation vehicles enter each street through a Poisson process with an inter-arrival time of 2 min
	Maximum transmit power indoor (per 10 MHz): 10 dBm
	Carrier: 800 MHz for backhaul links, and 2.6 GHz for access links
	Bandwidth: 20 MHz for backhaul links and 80 MHz for access links
	Antenna configuration both indoor and outdoor: single antenna,
	0 dBi gain omnidirectional antenna
	Receiver noise figure: 5 dB
Outdoor UEs (macro UE, micro UE)	Speed: 0 to 3 km/h
	Height: 1.5 m above the ground
	Positions: uniformly randomly dropped, 50 UEs per road
	Cell selection: based on received power with 5 dB CRE bias for micro cells
	Receiver noise figure: 9 dB
VUEs	Height: 1.5 m above the ground
	Position: uniformly randomly dropped inside a vehicle
	Number of VUEs in each vehicle: 1) uniformly from the interval
	[1, 50]; 2) 25 VUEs per vehicle
	Cell selection: 1) same as macro UEs (baseline case); 2) always connect to the MN of their own vehicles (other cases)
	Receiver noise figure: 9 dB

A full-buffer traffic model is considered, and the end user throughput is our metric. The throughput, for the backhaul links of MNs and end users, is calculated based on a look-up table method according to the instantaneous received SINR [50] assuming all the information bits are correctly decoded. Moreover, we also follow the constraints of the current LTE-A systems in terms of scheduling, link adaptation, and modulation and coding schemes.[i] Details of the considered simulation implementations can be found in, for example, [50]. In the following, first, throughput performance of macro UEs and VUEs will be investigated, followed by the throughput simulations of micro UEs.

7.1 Performance analysis for macro UEs and VUEs

VUEs are served directly by the macro BSs in the baseline case, and via MNs for the other cases. An MN is regarded as a super user that communicates with macro BS sectors and aggregates traffic for its VUEs. In the cases when ABSs are configured at the macro sectors, the same scheduling restrictions and protections are applied for both the backhaul links of MNs and macro UEs, and a single antenna is assumed for all transmitters and receivers. For other cases, only the receivers of MN backhaul links are assumed to be equipped with multiple antennas, and no combination using ABSs together with multi-antennas is considered. The detailed scheduling and resource allocation methods of the macro BSs are presented in Section 4.

Figures 7, 8, 9, and 10 plot the throughput cdf of VUEs when the VPL is at 15 and 30 dB. When the VPL is 15 dB (Figures 7 and 8), we observe that compared with the baseline case, the use of MNs does not improve the throughput of VUEs, unless the IRC is used for the backhaul links of MNs. This is very sensible, because the low throughput for VUEs is mostly caused by strong interference, and the IRC scheme can effectively maximize the received SINR by suppressing the interference. The other cases cannot outperform the baseline case. There are two reasons for this: 1) by serving VUEs directly from the macro BSs, we can achieve multi-user scheduling gain (not all the VUEs are necessarily scheduled together), in addition to gains from frequency adaptive modulations. These gains overshadow the gain of circumventing VPL using MNs (see Figure 3); 2). To minimize the signaling overhead, the current LTE-A system requires that all the PRBs allocated to a given link in one sub-frame have to use the same coding and modulation scheme ([16] Ch. 9 and 10). Under this constraint, even though the PF scheduler allocates more PRBs to the backhaul link of an MN than to a regular UE, the gain from using frequency adaptive modulation is limited. However, when the VPL is high, the benefits of using MNs are more obvious. In Figures 9 and 10, we see that all schemes outperform the baseline case for the cell edge VUEs, and the VUEs benefit the most from MNs when IRC is used for backhaul links of MNs.

Moreover, the number of VUEs on board also has an impact on the system performance. When comparing Figure 7 with Figures 8, 9, and 10, we notice some obvious differences in the throughput of the VUEs. For example, where a fixed number of 25 VUEs are on board, the 50th percentile of the VUE throughput is slightly higher than when dropping a random number of VUEs on board. This is mostly because of the scheduling decisions, because when an MN is scheduled, all its VUEs need to be scheduled at the same time. We present detailed discussion on this later in this section.

Figures 11, 12, 13, and 14 plot the throughput cdf of macro UEs when the VPL is at 15 and 30 dB. We emphasize that, by introducing MNs to the system, the performances of the macro UEs are not obviously impacted unless ABSs are used as the ICIC scheme. As expected, the macro UEs have a lower throughput when ABSs are configured, because certain sub-frames from macro BSs are

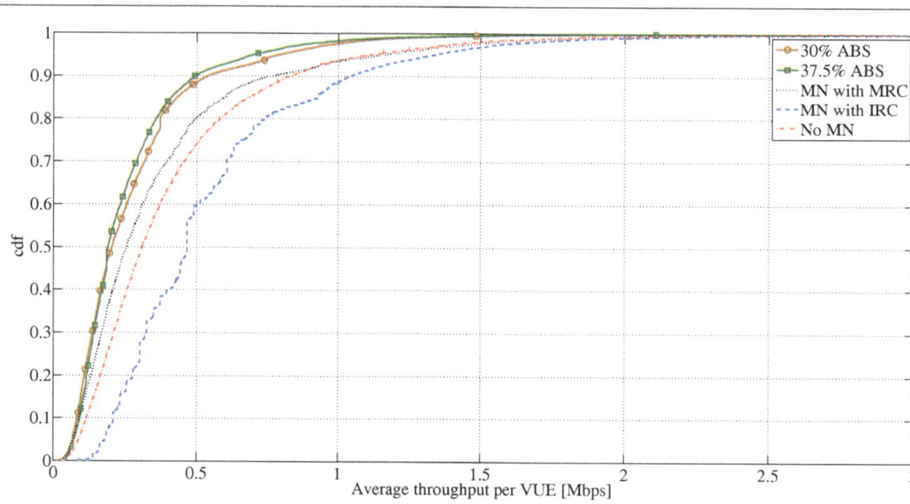

Figure 7 The cdf of VUE throughput, uniformly distributed VUEs, VPL at 15 dB. The solid green line with square markers is the average throughput per VUE when the number of ABSs is 37.5% at the macro sectors. The solid brown line with circle markers is the average throughput per VUE when the number of ABSs is 30% at the macro sectors. The dotted black line is the average throughput per VUE when MRC is used at backhaul links of MNs. The dash-dot red line is the average throughput per VUE when no MNs are used. The dashed blue line is the average throughput per VUE when IRC is used at backhaul links of MNs.

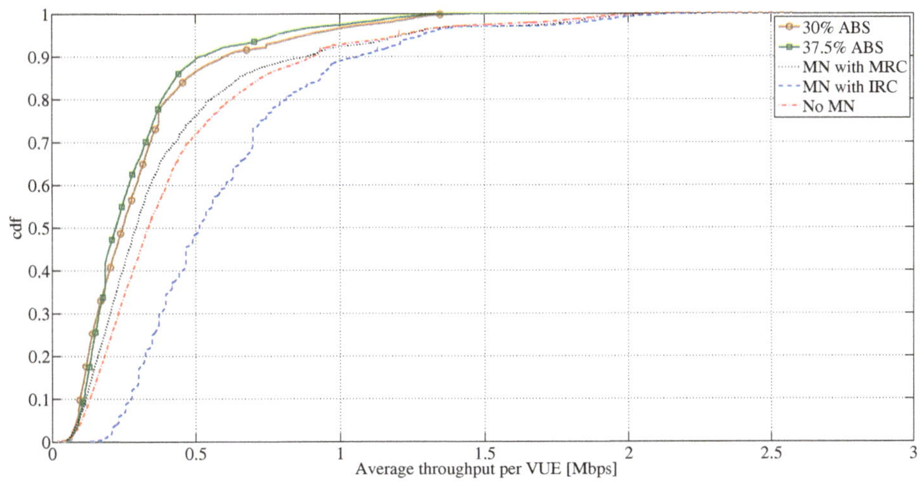

Figure 8 The cdf of VUE throughput, 25 VUEs per vehicle, VPL at 15 dB. The solid green line with square markers is the average throughput per VUE when the number of ABSs is 37.5% at the macro sectors. The solid brown line with circle markers is the average throughput per VUE when the number of ABSs is 30% at the macro sectors. The dotted black line is the average throughput per VUE when MRC is used at backhaul links of MNs. The dash-dot red line is the average throughput per VUE when no MNs are used. The dashed blue line is the average throughput per VUE when IRC is used at backhaul links of MNs.

not used. Moreover, from the figures, we observe that different VPL values have no impact on the performance of macro UEs. This is because the PF scheduler ensures the same scheduling opportunities for all the users. The number of VUEs per bus, however, has some impact on the throughput of macro UEs.

To gain a better insight into the different schemes, in Figures 15, 16, 17, and 18, we plot the 5th percentile user throughput versus the 90th percentile user throughput for the VUEs and macro UEs in all the cases. Different schemes are differentiated by the shapes of the markers. Hollow markers represent the VUEs, and solid markers are the macro UEs. Figures 15 and 17 plot the 5th percentile versus 90th percentile user throughput for a random number of VUEs when the VPL is 15 and 30 dB, respectively. Figures 16 and 18 plot the 5th percentile versus 90th percentile user throughput for a fixed number of 25 VUEs on each vehicle, when VPL is 15 and 30 dB, respectively. In Figures 15, 16, 17, and 18, we observe that the IRC scheme gives the best performance for VUEs

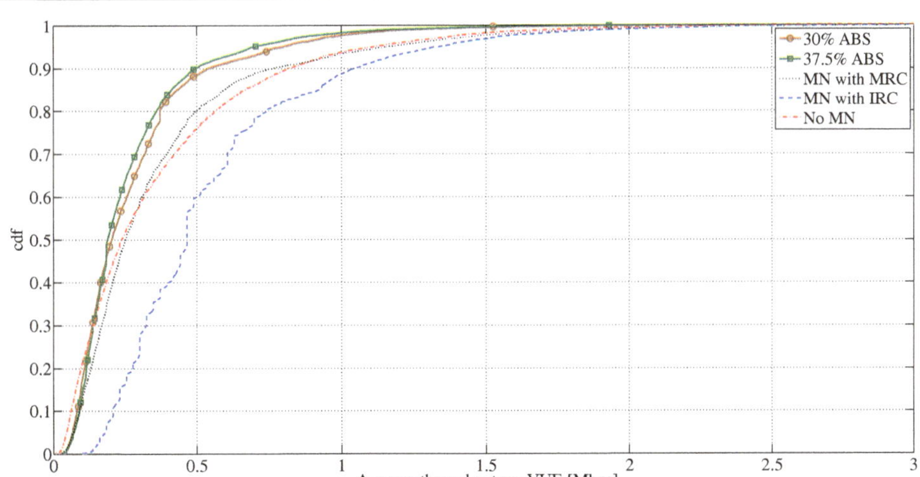

Figure 9 The cdf of VUE throughput, uniformly distributed VUEs, VPL at 30 dB. The solid green line with square markers is the average throughput per VUE when the number of ABSs is 37.5% at the macro sectors. The solid brown line with circle markers is the average throughput per VUE when the number of ABSs is 30% at the macro sectors. The dotted black line is the average throughput per VUE when MRC is used at backhaul links of MNs. The dash-dot red line is the average throughput per VUE when no MNs are used. The dashed blue line is the average throughput per VUE when IRC is used at backhaul links of MNs.

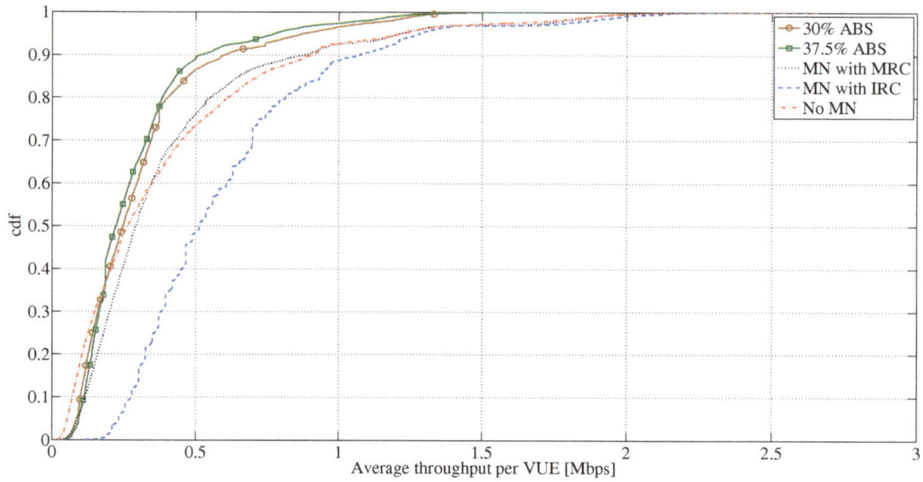

Figure 10 The cdf of VUE throughput, 25 VUEs per vehicle, VPL at 30 dB. The number of VUEs is fixed at 25 for each vehicle and VPL is at 30 dB. The solid green line with square markers is the average throughput per VUE when the number of ABSs is 37.5% at the macro sectors. The solid brown line with circle markers is the average throughput per VUE when the number of ABSs is 30% at the macro sectors. The dotted black line is the average throughput per VUE when MRC is used at backhaul links of MNs. The dash-dot red line is the average throughput per VUE when no MNs are used. The dashed blue line is the average throughput per VUE when IRC is used at backhaul links of MNs.

for all cases. Both the 5th percentile and 90th percentile throughputs have improved significantly compared with the baseline case. However, at low VPL, i.e., VPL at 15 dB, compared with the baseline case, there is no advantage for VUEs when MNs use schemes other than IRC for their backhaul links.

When the VPL increases to 30 dB, the advantage of using MNs to serve VUEs becomes obvious, because the VPL can be circumvented, and the impact of interference can be mitigated. When using MNs to serve VUEs, depending on the ICIC schemes, compared with the baseline setup, the performance of VUEs, for either the 5th percentile or both the 5th percentile and 90th percentile throughput, is improved. The IRC gives the best performance as expected. More specifically, in the low SINR region, both the use of MRC and ABSs can improve the 5th percentile throughput of VUEs. Because ABSs are used to mute certain sub-frames from the macro cells to protect the victim UEs, compared with the baseline case, there is an obvious improvement in the 5th percentile

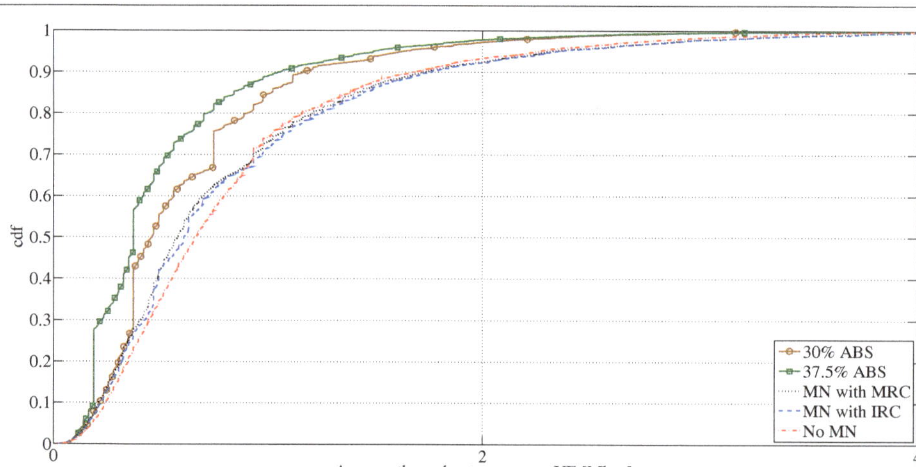

Figure 11 The cdf of macro UE throughput, uniformly distributed VUEs, VPL at 15 dB. The solid green line with square markers is the average throughput per VUE when the number of ABSs is 37.5% at the macro sectors. The solid brown line with circle markers is the average throughput per VUE when the number of ABSs is 30% at the macro sectors. The dotted black line is the average throughput per VUE when MRC is used at backhaul links of MNs. The dash-dot red line is the average throughput per VUE when no MNs are used. The dashed blue line is the average throughput per VUE when IRC is used at backhaul links of MNs.

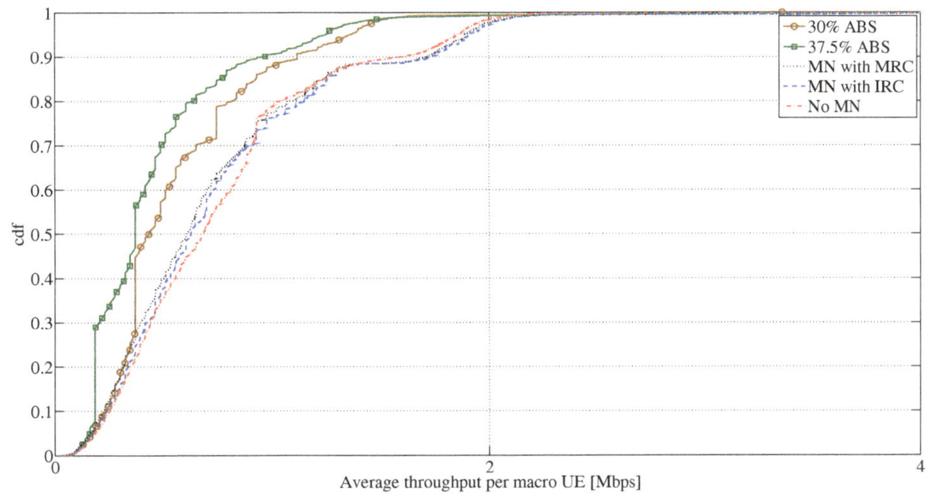

Figure 12 The cdf of macro UE throughput, 25 VUEs per vehicle, VPL at 15 dB. The solid green line with square markers is the average throughput per VUE when the number of ABSs is 37.5% at the macro sectors. The solid brown line with circle markers is the average throughput per VUE when the number of ABSs is 30% at the macro sectors. The dotted black line is the average throughput per VUE when MRC is used at backhaul links of MNs. The dash-dot red line is the average throughput per VUE when no MNs are used. The dashed blue line is the average throughput per VUE when IRC is used at backhaul links of MNs.

throughput of the VUEs, but a lower throughput at the high SINR regions, as expected, and, as also expected, the IRC gives the best performance for both the 5th percentile and 90th percentile throughput of the VUEs.

Furthermore, the number of VUEs on board affects the user throughput. Because an MN is regarded as a super user that aggregates the traffic of its VUEs, if an MN is scheduled, the PRBs allocated to its backhaul link are proportional to the number of VUEs it serves. Therefore, the PF scheduler has less scheduling flexibility. The inflexibility can be easily observed from the cdf plots of the throughput of macro UEs. When comparing Figure 11 with Figures 12, 13, and 14, we observe that the cdf curves are less smooth when there is a fixed number of 25 VUEs on board. From Figures 15, 16, 17, and 18, we observe that even though the total expected number of VUEs in the system is the same, compared with the random VUEs per vehicle, if we have a fixed number of 25 VUEs per vehicle, the 5th percentile of the VUE throughput increases for all the cases when the VUEs are served by the MNs. But

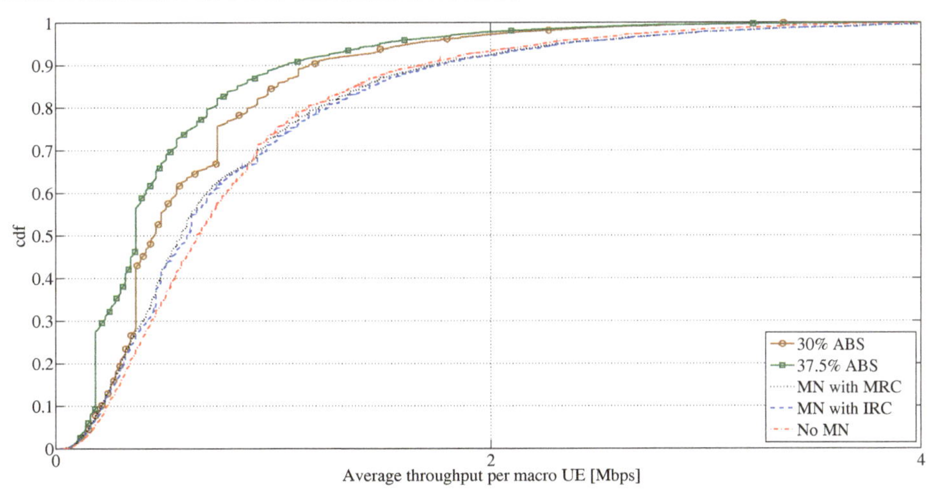

Figure 13 The cdf of macro UE throughput, uniformly distributed VUEs, VPL at 30 dB. The solid green line with square markers is the average throughput per VUE when the number of ABSs is 37.5% at the macro sectors. The solid brown line with circle markers is the average throughput per VUE when the number of ABSs is 30% at the macro sectors. The dotted black line is the average throughput per VUE when MRC is used at backhaul links of MNs. The dash-dot red line is the average throughput per VUE when no MNs are used. The dashed blue line is the average throughput per VUE when IRC is used at backhaul links of MNs.

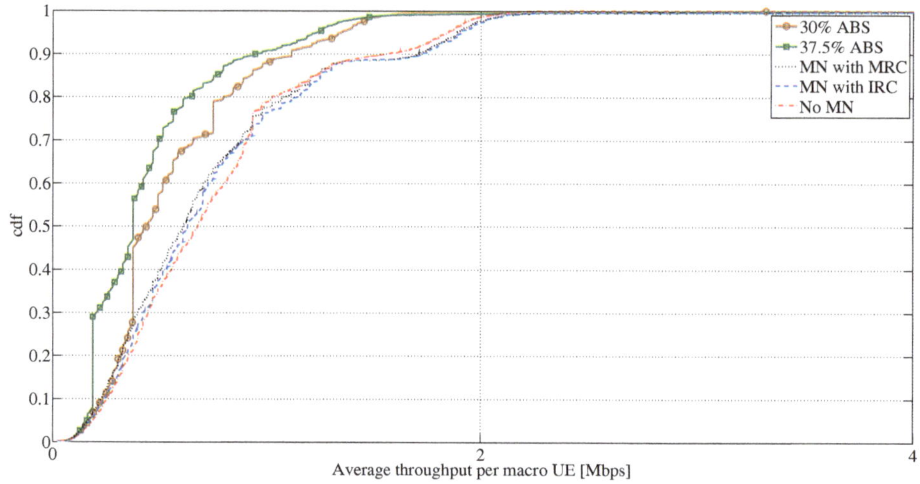

Figure 14 The cdf of macro UE throughput, 25 VUEs per vehicle, VPL at 30 dB. The solid green line with square markers is the average throughput per VUE when the number of ABSs is 37.5% at the macro sectors. The solid brown line with circle markers is the average throughput per VUE when the number of ABSs is 30% at the macro sectors. The dotted black line is the average throughput per VUE when MRC is used at backhaul links of MNs. The dash-dot red line is the average throughput per VUE when no MNs are used. The dashed blue line is the average throughput per VUE when IRC is used at backhaul links of MNs.

this is at a cost of slightly lowering the throughput of the macro UEs.

7.2 Performance analysis for micro UEs

We configure ABSs for the micro BSs to protect the access links of MNs from the interference caused by the micro BS sectors. Figures 19, 20, 21, and 22 plot the throughput cdfs of the micro UEs. Compared with the baseline case, i.e., VUEs served directly via macro BSs, there is a slight degradation in the micro UE throughput in the presence of MNs. This is because when using ABSs, some subframes of the micro BSs are not employed. When the VPL is 15 dB (Figures 19 and 20), the ICIC schemes that offer higher throughput at the backhaul links of MNs have a

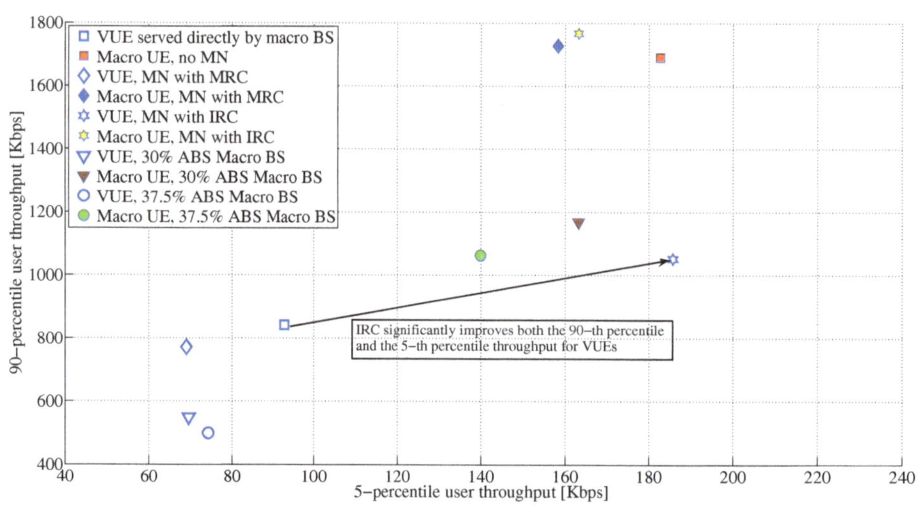

Figure 15 5th percentile vs 90th percentile user throughput for macro UEs and VUEs, uniformly distributed VUEs, VPL at 15 dB. The hollow square marker is the VUE throughput when they are served directly by the macro BS. The solid red square marker is the macro UE throughput when there is no MNs in the system. The hollow diamond marker is the VUE throughput when MRC is used at the backhaul links of MNs. The solid diamond blue marker is the macro UE throughput when MRC is used at the backhaul links of MNs. The hollow star marker is the VUE throughput when IRC is used at the backhaul links of MNs. The solid star yellow marker is the macro UE throughput when IRC is used at the backhaul links of MNs. The hollow triangle marker is the VUE throughput when the number of ABSs is 30% at the macro BS. The solid triangle brown marker is the macro UE throughput when the number of ABSs is 30% at the macro BS. The hollow circle marker is the VUE throughput when the number of ABSs is 37.5% at the macro BS. The solid circle green marker is the macro UE throughput when the number of ABSs is 37.5% at the macro BS.

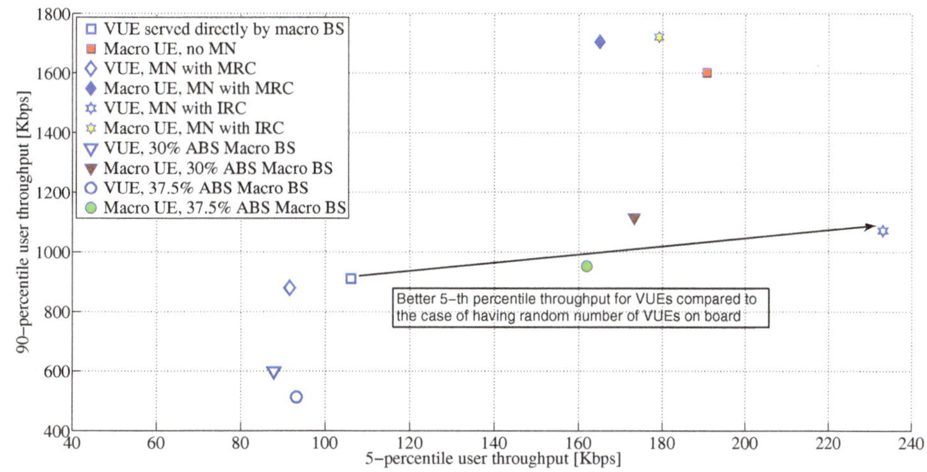

Figure 16 5th percentile vs 90th percentile user throughput for macro UEs and VUEs, 25 VUEs per vehicle, VPL at 15 dB. The hollow square marker is the VUE throughput when they are served directly by the macro BS. The solid red square marker is the macro UE throughput when there is no MNs in the system. The hollow diamond marker is the VUE throughput when MRC is used at the backhaul links of MNs. The solid diamond blue marker is the macro UE throughput when MRC is used at the backhaul links of MNs. The hollow star marker is the VUE throughput when IRC is used at the backhaul links of MNs. The solid star yellow marker is the macro UE throughput when IRC is used at the backhaul links of MNs. The hollow triangle marker is the VUE throughput when the number of ABSs is 30% at the macro BS. The solid triangle brown marker is the macro UE throughput when the number of ABSs is 30% at the macro BS. The hollow circle marker is the VUE throughput when the number of ABSs is 37.5% at the macro BS. The solid circle green marker is the macro UE throughput when the number of ABSs is 37.5% at the macro BS.

greater impact on the throughput of micro UEs. This is because more ABSs are needed for the micro BSs to protect the access links of MNs to accommodate the traffic from their backhaul links. The impact, however, is not very obvious because of the relatively large bandwidth of the micro cells.

When the VPL is at 30 dB (Figures 21 and 22), compared with the baseline case, when the MNs are introduced to the system, there is still a slight degradation in micro UE throughput but not as obvious as when the VPL is at 15 dB. There are two reasons for this. First, the interference power from micro BSs to the access links of MNs can be

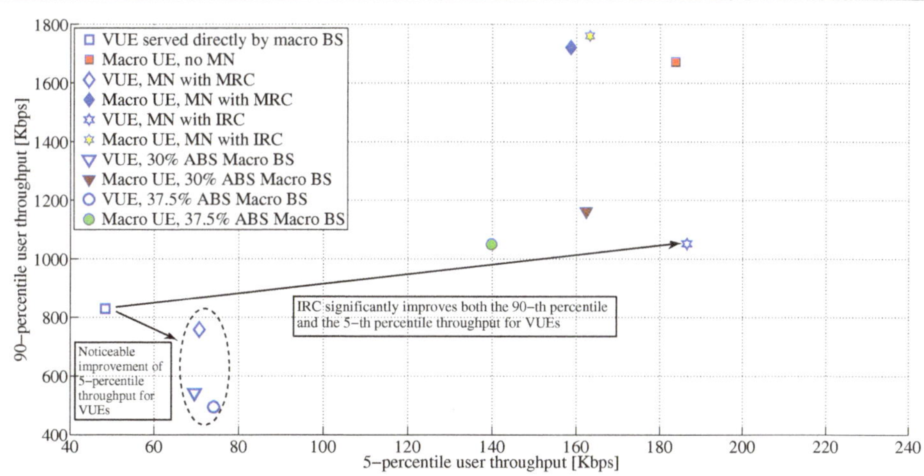

Figure 17 5th percentile vs 90th percentile user throughput for macro UEs and VUEs, uniformly distributed VUEs, VPL at 30 dB. The hollow square marker is the VUE throughput when they are served directly by the macro BS. The solid red square marker is the macro UE throughput when there is no MNs in the system. The hollow diamond marker is the VUE throughput when MRC is used at the backhaul links of MNs. The solid diamond blue marker is the macro UE throughput when MRC is used at the backhaul links of MNs. The hollow star marker is the VUE throughput when IRC is used at the backhaul links of MNs. The solid star yellow marker is the macro UE throughput when IRC is used at the backhaul links of MNs. The hollow triangle marker is the VUE throughput when the number of ABSs is 30% at the macro BS. The solid triangle brown marker is the macro UE throughput when the number of ABSs is 30% at the macro BS. The hollow circle marker is the VUE throughput when the number of ABSs is 37.5% at the macro BS. The solid circle green marker is the macro UE throughput when the number of ABSs is 37.5% at the macro BS.

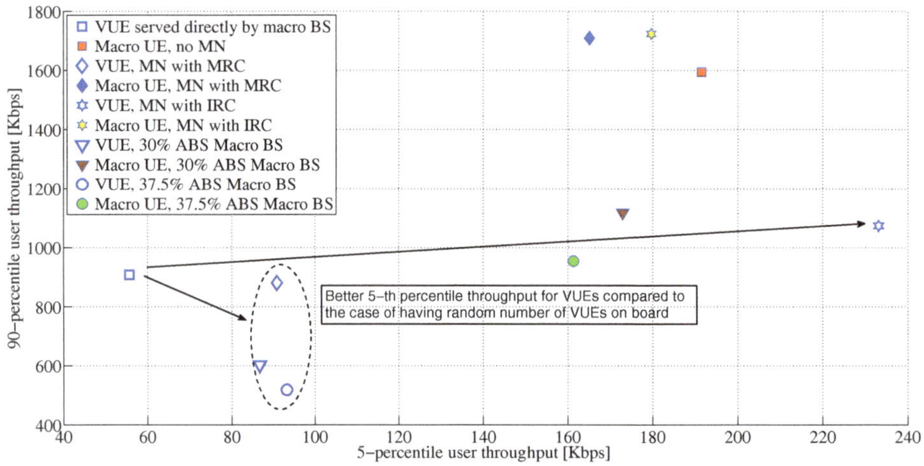

Figure 18 5th percentile vs 90th percentile user throughput for macro UEs and VUEs, 25 VUEs per vehicle, VPL at 30 dB. The hollow square marker is the VUE throughput when they are served directly by the macro BS. The solid red square marker is the macro UE throughput when there is no MNs in the system. The hollow diamond marker is the VUE throughput when MRC is used at the backhaul links of MNs. The solid diamond blue marker is the macro UE throughput when MRC is used at the backhaul links of MNs. The hollow star marker is the VUE throughput when IRC is used at the backhaul links of MNs. The solid star yellow marker is the macro UE throughput when IRC is used at the backhaul links of MNs. The hollow triangle marker is the VUE throughput when the number of ABSs is 30% at the macro BS. The solid triangle brown marker is the macro UE throughput when the number of ABSs is 30% at the macro BS. The hollow circle marker is the VUE throughput when the number of ABSs is 37.5% at the macro BS. The solid circle green marker is the macro UE throughput when the number of ABSs is 37.5% at the macro BS.

significantly attenuated by the vehicles. Therefore, fewer ABSs are required for the micro BSs to protect the access links of MNs. Second, at high VPL, the interference from the access links of MNs to nearby micro UEs (see Figure 2) can also be attenuated by the vehicles.

Moreover, at 30 dB VPL, we observe that the considered ICIC schemes used for the backhaul links of MNs have no noticeable differences with regard to the impact on the throughput of micro UEs. This is because: 1) the vehicle can protect the access link of a MN, because the interference can be significantly attenuated by the VPL; 2) we always consider configuring the ABS patterns for the worst MN in a given area (see Section 6), and there is no noticeable difference in the required ABSs between the considered ICIC schemes. MNs offer very good link qualities to their VUEs because of the relatively short

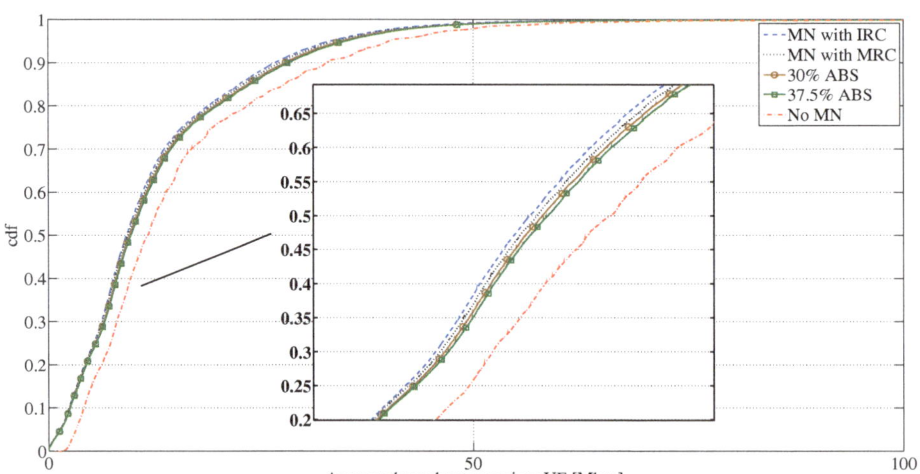

Figure 19 The cdf of micro UE throughput, uniformly distributed VUEs, VPL at 15 dB. The solid green line with square markers is the average throughput per VUE when the number of ABSs is 37.5% at the macro sectors. The solid brown line with circle markers is the average throughput per VUE when the number of ABSs is 30% at the macro sectors. The dotted black line is the average throughput per VUE when MRC is used at backhaul links of MNs. The dash-dot red line is the average throughput per VUE when no MNs are used. The dashed blue line is the average throughput per VnUE when IRC is used at backhaul links of MNs.

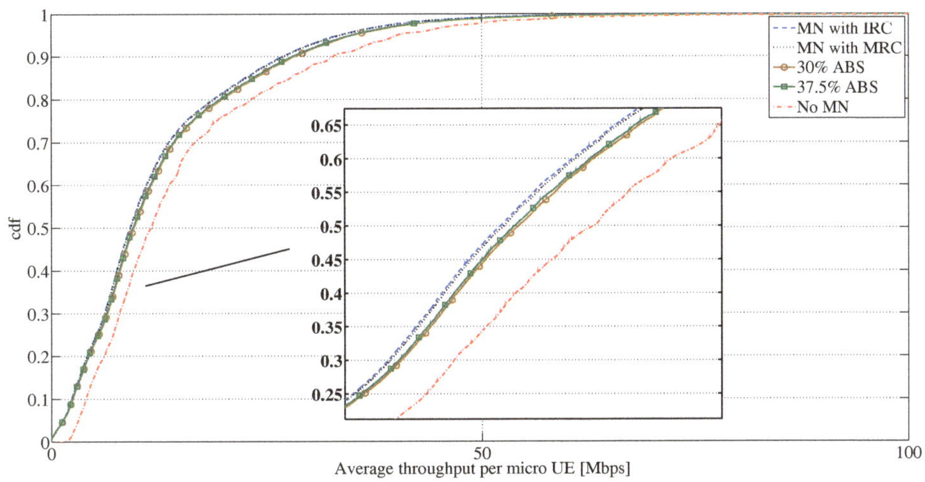

Figure 20 The cdf of micro UE throughput, 25 VUEs per vehicle, VPL at 15 dB. The solid green line with square markers is the average throughput per VUE when the number of ABSs is 37.5% at the macro sectors. The solid brown line with circle markers is the average throughput per VUE when the number of ABSs is 30% at the macro sectors. The dotted black line is the average throughput per VUE when MRC is used at backhaul links of MNs. The dash-dot red line is the average throughput per VUE when no MNs are used. The dashed blue line is the average throughput per VUE when IRC is used at backhaul links of MNs.

distance between their indoor antennas and their VUEs. Because all the micro BSs are muted at the same time, for any of the considered macro ICIC schemes, muting one sub-frame per ABSs duty cycle is sufficient for the access links of MNs to accommodate the traffic from their backhaul links. To better understand the argument above, in Table 2, we list the average ABS ratios for the different schemes. The average ABS ratio is calculated as the total number of ABSs averaged over the total number of sub-frames in all the drops. From Table 2, we observe that when the VPL is low, more ABSs are needed from the micro cells, and the greater the backhaul throughput, the more ABSs are needed. However, when the VPL is 30 dB, only the minimum number of ABSs, i.e., one ABS per ABS duty cycle, are required for the micro cells. Unlike the macro cells, the number of VUEs on board has no observable impact on the micro UE performance.

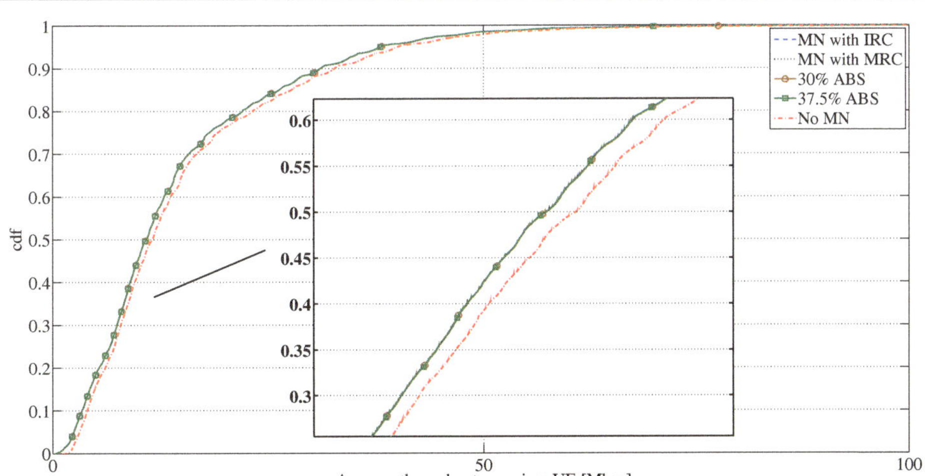

Figure 21 The cdf of micro UE throughput, uniformly distributed VUEs, VPL at 30 dB. The solid green line with square markers is the average throughput per VUE when the number of ABSs is 37.5% at the macro sectors. The solid brown line with circle markers is the average throughput per VUE when the number of ABSs is 30% at the macro sectors. The dotted black line is the average throughput per VUE when MRC is used at backhaul links of MNs. The dash-dot red line is the average throughput per VUE when no MNs are used. The dashed blue line is the average throughput per VUE when IRC is used at backhaul links of MNs.

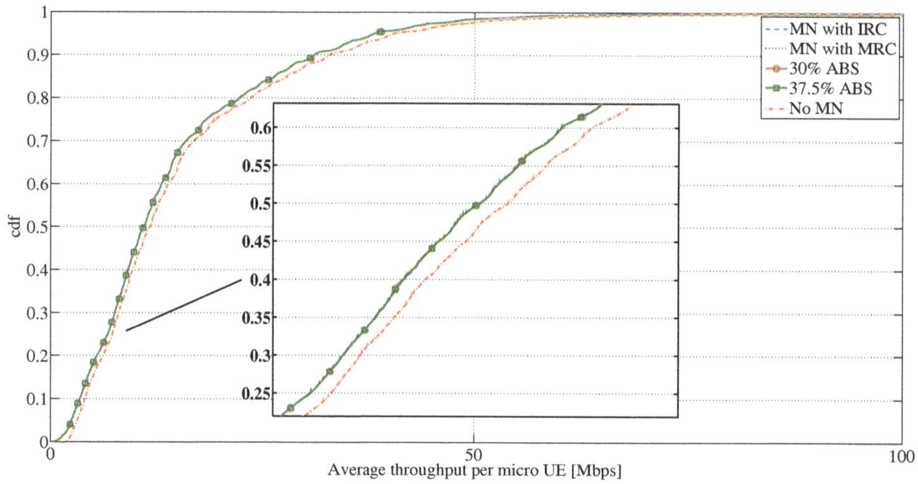

Figure 22 The cdf of micro UE throughput, 25 VUEs per vehicle, VPL at 30 dB. The solid green line with square markers is the average throughput per VUE when the number of ABSs is 37.5% at the macro sectors. The solid brown line with circle markers is the average throughput per VUE when the number of ABSs is 30% at the macro sectors. The dotted black line is the average throughput per VUE when MRC is used at backhaul links of MNs. The dash-dot red line is the average throughput per VUE when no MNs are used. The dashed blue line is the average throughput per VUE when IRC is used at backhaul links of MNs.

8 Discussion

8.1 Complexity of the considered schemes and potential impact on the current LTE-A systems

In general, when introducing MNs to current mobile communication systems, there are foreseeable impacts on both the air interference and system architecture [5,8]. Existing mobile communication systems are expected to support MNs after certain modifications. In this section, we briefly discuss the complexities of using ABSs together with scheduling restrictions for the macro sectors, using multi-antenna receivers for the MNs to enhance the backhaul link receptions, and the application of ABSs for the micro cells to protect the access links of MNs. To facilitate our discussion, we use the 3GPP Rel-12 LTE-A system as our baseline system.

As we can see from the discussion in Section 4.2, the proportional fairness based scheduling restriction is very straightforward to implement, because it only involves ordering the UEs according to their channel quality and average throughput over different protected and non-protected sub-frames. If there are n devices to be scheduled, depending on the sorting algorithm employed, the complexity of the scheduling restriction is between $O(n \log n)$ and $O(n^2)$ ([61] Ch. 2).

However, for this scheme to work, the ABS patterns need to be exchanged between different cells, and the average throughput of a receiver over protected and non-protected sub-frames needs to be estimated. Because the current LTE-A system already supports using ABSs, the configuration of ABSs and exchange of ABS patterns between macro sectors will not introduce additional complexities to the current LTE-A system. To apply the scheduling restrictions, based on the discussion in Section 4, each macro sector needs to identify whether a receiver must be protected from one or two interferers or no protection is needed. According to Equation 9 and Equation 12, the receiver needs to feed back its average throughput when it is in both the protected and non-protected sub-frames. This was a major challenge in the 3GPP Rel-10 and earlier LTE systems, because only one CSI feedback process was allowed. There were no good mechanisms for a receiver to perform accurate CSI measurement over protected sub-frames, especially when different combinations of muting patterns for neighboring interfering cells needed to be considered. However, in the 3GPP Rel-12 LTE-A systems, to support the use of coordinated multi-point (CoMP) transmission, a set of zero-power CSI reference symbols (CSI-RS) is standardized. The receiver can use the zero-power CSI-RS to perform accurate CSI measurement when certain neighboring interfering cells are muted. Several CSI feedback processes can be initiated to report different channel qualities when different muting patterns for neighboring interfering cells are applied. Details about the configuration of

Table 2 Average ABS ratio configured by micro cells

	Backhaul schemes			
Number of VUEs and VPL	IRC	MRC	30% ABS	37% ABS
Random VUEs, 15 dB VPL	0.0920	0.0813	0.0740	0.0663
25 VUEs, 15 dB VPL	0.0884	0.0848	0.0671	0.0697
Random VUEs, 30 dB VPL	0.0250	0.0250	0.0250	0.0250
25 VUEs, 30 dB VPL	0.0250	0.0250	0.0250	0.0250

zero-power CSI-RS and the feedback mechanism can be found in ([62] Ch. 15). Therefore, no new feedback mechanisms need to be standardized to perform the scheduling restriction algorithm considered in this study. We need to emphasize that from Equation 9 and Equation 12, we see that only the average throughput is required to apply the scheduling restrictions. Therefore, a receiver only needs to feed back the CSI measurement or the estimated average throughput for different muting patterns of neighboring interfering cells once per every several sub-frames, e.g., one or several ABS cycles. Therefore, only limited feedback overhead is expected. Hence, the tools offered by the current LTE-A systems can adequately support using ABSs together with the scheduling restrictions considered in our study.

As discussed in Section 5, for using multi-antenna receivers for MNs to enhance the backhaul links, two schemes, i.e., the MRC and IRC receivers, are considered in this study. Because these are schemes applied at the receiver ends, no CSI at the transmitters (CSIT) is required. Therefore, there are no impacts on the network side. For the MRC receivers, it is only necessary to estimate the desired channel coefficient for each of the antennas of the receiver. In our study, we assumed perfect channel knowledge, both for the desired and interference signals. In practical LTE-A systems, the channel estimation is usually achieved through the cell specific reference signal (CRS), which is orthogonal between neighboring cells ([62] Ch. 10). Moreover, joint channel estimation and signal detection can be used to achieve fairly good performance, e.g., using linear-minimum-mean-square-error (LMMSE) estimation [63,64]. The complexity of the channel estimation algorithm varies. If we consider N_c as the number of subcarriers in the OFDMA system, the LMMSE channel estimator has a complexity of $O\left(N_c^3\right)$ [64], but low-complexity channel estimation algorithms with comparable performance to the LMMSE algorithm can reduce the complexity to $O\left(N_c \log N_c\right)$ [64]. Depending on the carrier frequency, the density of CRS in the current LTE-A system supports accurate channel estimation for a user velocity up to 500 km/h ([16] Ch. 1). Therefore, for the speed of the MNs considered in our study, the accuracy of channel estimation is not a major concern.

However, when using IRC receivers, not only the desired channel coefficients at each of the receiving antennas need to be estimated but also a knowledge of the interference channels at the receiver is required. In addition to the complexity of channel estimation, the additional complexity of $O\left(N_c^3\right)$ is introduced by the IRC receivers because of the estimation and inversion of the channel covariance matrix for the interference signals [65]. In practice, as discussed above, the desired channel coefficients can be easily estimated. Various methods can be applied to obtain fairly good estimation of the channel information of the interference signals. To understand the feasibility and performance in using IRC receivers in LTE-A systems, in [66,67] simulations were performed. The results in [66,67] show that by either using data signal-based interference estimation or demodulation reference signal (DMRS)[j]-based interference estimation, the channel of the interfering signaling can be well estimated to enable the IRC receivers to be used in the LTE-A system, and significant cell-edge throughput improvement is observed in the use of IRC receivers. Therefore, the current LTE-A systems have good support for the multi-antenna receivers considered in our study.

Certain changes are required on the current LTE-A system to support the use of ABSs for the micro cells to protect the access links of the MNs. As discussed in Section 6, the micro cells adjust the number of ABSs based on the feedback of MNs. The feedback of the MNs cannot be directly made available at the micro cells, because there are no direct connections between the MNs and the micro cells. Therefore, it is more likely that the network needs to collect the feedback from the MNs, and then configure the ABS patterns at the micro cells accordingly. As discussed in Section 4.1, because the update of ABSs usually occurs in minutes, a slow feedback link between each of the MNs and the network is sufficient to convey this information.

8.2 Limitations in the current LTE-A systems

As mentioned in Section 3, in this study, we align most of our assumptions with the current FDD LTE-A systems. To perform link adaptations at the BSs, the UEs need to feed back their channel quality indication (CQI) to the BSs.[k] In an FDD LTE-A system, a UE can be set up to feed back CQI every 2 ms. However, because of system load, scheduling and frame structures in the current FDD LTE-A systems, a feedback delay between 4 and 10 ms can be expected ([62] Ch. 11, 13). The feedback delay is not a significant problem for low speed UEs but may cause problems at vehicular speeds. In [68], it was argued that the sensitivity of the coding and modulation schemes can be adjusted, and an upper limit of 0.15 for the normalized mean square error of the CQI can be tolerated. In [68], it was further demonstrated that the use of a Kalman-filter-based channel predictor could improve the reliability of the CQI feedback. From Figure 3, we see that most of the UEs, as well as the backhaul links of the MNs, have a fairly good SNR,[l] and therefore, according to the study in [68], fairly good performance of link adaptation can be expected at the velocity of the MNs considered in this study, i.e., 50 km/h. However, this might be a problem for higher speeds. On the one hand, in the future 5G systems, if the length of the sub-frames can be shortened, and the feedback delays can be further reduced, the channel predictors should be able to support much higher speed. On the other hand, because MNs are less constrained by power, antenna element size, and transceiver complexity, it is less costly to integrate more sophisticated multi-antenna schemes and advanced signal processing algorithms into MNs compared with regular UEs. As demonstrated in [41,42], by using the so-called predictor antennas on top of a vehicle, reliable CSIT can be obtained at the transmitter side to support various closed-loop MIMO schemes. This is another big advantage in using MNs to serve VUE devices.

Our results in Section 7 show that the design of the current LTE-A system limits the benefits of using MNs. First, as discussed in Section 7.1, the same coding and modulation schemes are required for the PRBs allocated to the same user at one sub-frame. This limits the gains of frequency adaptive transmission and has a greater impact on the backhaul links of MNs than regular UEs, because the number of PRBs allocated to the backhaul links of an MN is proportional to the number of VUEs it serves, i.e., per-PRB-based link adaptation would be beneficial [69]. Second, even if the SINR at the backhaul links are improved by using various ICIC schemes (see Figure 6), the throughput does not necessarily increase. Certain macro sub-frames need to be muted when using ABS to mitigate interference, therefore improving the SINR statistics, but the total throughput suffers from the muting of the macro sub-frames. Furthermore, in medium SINR regions, the throughput saturates because of the coding and modulation schemes used in the LTE-A system. Currently, only up to 64-QAM is used as the highest modulation order. In the current LTE-A systems, if a single antenna is used at the BS, the throughput saturates at 30 dB SINR ([70] Ch. 5). In Figure 6, we observe that by using MRC, around 30% of the PRBs of the backhaul links of the MNs have more than 30 dB SINR, and if IRC is used over 80% of the PRBs of the backhaul links of the MNs have more than 30 dB SINR. Therefore, if higher order modulation can be used, e.g., 256-QAM ([70] Ch. 5), the throughput of the backhaul links can be further improved. To demonstrate these two points, in Figure 23, we plot the achievable throughput for the VUEs when the rate at backhaul links is calculated on a per PRB basis using the Shannon capacity formula at VPL 15 dB, and MRC receivers are used at the backhaul links of MNs. Compared with when applying LTE-A scheduling restrictions, we observe that the improvement is obvious. Thus, the benefits of MNs can be better exploited in future broadband wireless networks with further enhanced capabilities.

8.3 Further works

Another concern in deploying MNs is that unwanted UEs might be attached to the nearby MNs. This is not a major problem when the public transportation vehicle is moving at a regular speed, because the handover criteria are not easily satisfied unless a UE is moving parallel to the public transportation vehicle at similar speed. However, in situations in which public transportation vehicles move slowly or stop, e.g., in traffic congestion during rush hours or picking up passengers at bus stops, this may be a big problem. Therefore, solutions that optimize the user admission and handover criteria are also of great importance in future studies using MNs. Similar ideas from the research of femto cells can be borrowed [71,72], e.g., an MN only admits the users that have similar mobility characteristic

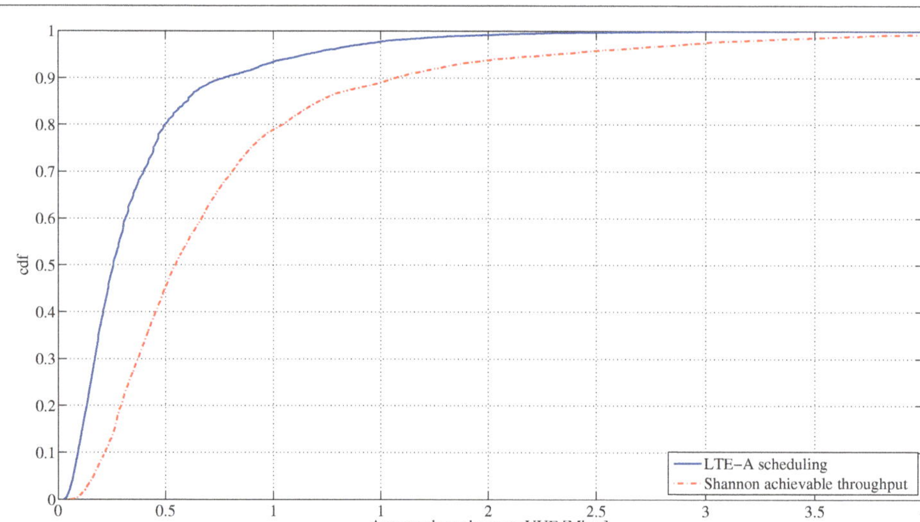

Figure 23 The cdf of VUE Shannon capacity based throughput on a per PRB basis, uniformly distributed VUEs, VPL at 15dB. The solid blue line is the cdf of VUE throughput when we considered the LTE-A scheduling constraints. The dash-dot red line is the cdf of VUE achievable throughput calculated on a per PRB basis by using the Shannon capacity.

to itself, or it sets a relatively higher handover margin to limit its coverage area inside the vehicle as much as possible. In [12], we provided a framework to optimize the handover parameters for the use of MRN. Nevertheless, this issue requires more thorough research.

In this study, we consider the MNs are moving at a constant speed and only serve the users inside the vehicles. However, in scenarios in which MNs have relatively low mobility and high density, their use can be extended in several different ways. For example, during traffic congestion in rush hours, some public transportation vehicles may become stuck in bad coverage areas for quite a long time. Therefore, we may consider enabling multi-hop communication between the backhaul links of MNs. In this way, the capacity of the mobile network can be better distributed across the service area of one or several BSs, and the VUE experience may be significantly improved. Certainly, there are several challenges to support multi-hop communication between the MNs. The future 5G system offers a good opportunity for this, especially when new low latency and high capacity air interfaces are developed. Moreover, this is very similar to a typical MANET scenario, and therefore power control and interference management schemes developed for MANET and D2D communication can be applied to facilitate the multi-hop communication. Furthermore, in such scenarios, we may also consider using the MNs to serve the UEs that are outside the public transportation vehicles. Hence, more users can benefit from the deployment of MNs.

Nowadays, there are commercially available solutions that use a cellular network as a backhaul and use WiFi as the access technology to provide Internet services to VUEs [7,73]. The benefits and limitations of using WiFi to serve VUEs are briefly discussed in [7]. As this solution is totally based on using existing network elements, it can be deployed very fast. Moreover, advantages, such as eliminating the VPL can also be realized. However, as WiFi-based solutions are usually offered by third party companies, rather than mobile operators, and the limitations are also very obvious. As pointed out in Section 3, if a WiFi hot spot is treated simply as a regular UE that has a high data-rate demand, it is not certain that the fairness between VUEs and other UEs served by the same BS can be maintained. Moreover, because WiFi is operating in the open industrial, scientific and medical (ISM) radio bands, the interference management and channel selections for the WiFi access link are achieved in a distributed self-organized manner [74,75]. Therefore, if we further exploit the potential of using WiFi as the access technology to serve VUEs, better interference management schemes need to be developed for the WiFi access links, especially in a densely deployed urban scenario where WiFi hot spots are randomly and densely deployed. Potentially a tightly operator controlled WiFi solution is required in this case.

As pointed out in [76,77], channel modeling is very important in the study of 5G communication systems. The EU 5G project METIS is also developing new channel models that can better serve the needs of 5G system simulations, e.g., the METIS channel models are expected to provide good spatial consistency, which is not well supported in the currently available channel models. Accurate spatial consistency is needed to evaluate the performance of spatial techniques, e.g., multi-user MIMO, where the performance may be exaggerated because of non-independent fading. Before the METIS channel models are finished, the ITU-R Uma and Umi models are suggested by METIS [14] to model small scale fading. We expect limited impacts on the results presented in this study from the new channel models that are under development in METIS. This is because the schemes that we consider in this study do not rely on the use of spatial techniques. Moreover, because of the size of public transportation vehicles, low correlation between antenna elements can be achieved. Around 18 cm distance between antenna elements is needed at 800 MHz to achieve low enough correlations, and cross polarized placements can be used to further lower the correlation between two antenna elements. A single antenna is considered at each BS sector, and adequate separation is assumed between the antennas of each of the sectors. Therefore, as long as enough separation between the receiver antennas can be achieved, we can safely assume independent fading between each of the links. Nevertheless, when extending the study of MNs in future research, especially when various spatial MIMO techniques are considered in both macro and micro cells, using the newly developed METIS channel models or other channel models that provide support for good spatial consistency, should be considered.

The cost of deploying MNs on public transportation vehicles can be recouped by bringing in more business opportunities for both operators and service providers. For example, the MNs can push personalized information to travelers, e.g., weather or traffic information, both to improve customer loyalty and potentially generate more revenue for the operators and service providers. Using MNs can also bring several benefits for the network side and is an indispensable enabler of content caching at the BSs and context aware communication [78-80]. For example, at different times of the day, the MNs can cache commonly accessed content according to the previous habits of the commuters. This not only reduces the latency but also lessens the load on the macro cells.

9 Conclusions

In general, MNs can improve the throughput for the VUEs without significantly affecting the performance of

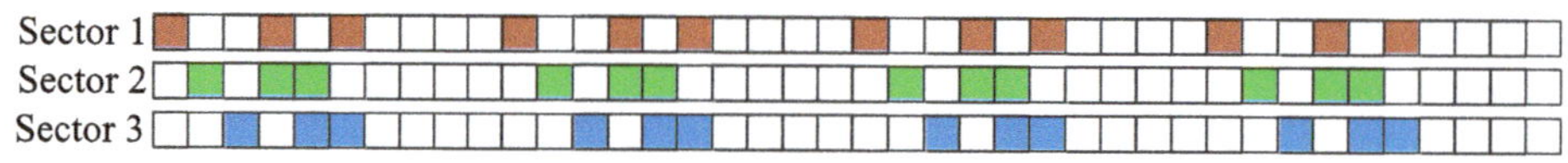

Figure 24 Frames with ABS patterns for three macro sectors in the case of 30% ABS. Colored subframes are ABSs.

outdoor UEs. From the results in Section 7, we see that the backhaul links are the bottlenecks to further improving the performance of the MNs. Compared with noise-limited scenarios and scenarios with limited interference [7,11,12,28], in a densely deployed urban scenario, interference is the biggest factor affecting the performance of MNs, which is why the IRC scheme for the backhaul links of MNs gives the most gain. Compared with regular UEs, public transportation vehicles are less constrained by power, antenna size, and transceiver complexity. Even though, in this study, we only considered using multi-antennas at the receiver end, more advanced signal processing and multi-antenna schemes, e.g., using predictor antennas [41,42], can be realized for MNs to further improve their performance. Moreover, when lacking the CSI from the interferers, ABSs can be combined with the MRC scheme. In this way, potentially fewer ABSs can be used, and better performance can be expected for both the outdoor UEs and the VUEs. Nevertheless, the use of MNs is a good solution in the next-generation system for improving the user experience on VUE.

9.1 Table of notations

A table of notations is provided in Table 3.

Endnotes

[a]5GNOW stands for 5th *G*eneration *N*on-Orthogonal *W*aveforms for *A*synchronous *S*ignaling.

[b]VPL is the outdoor-to-indoor signal attenuation caused when the signal penetrates through the vehicle.

[c]In order to have a concrete and practical description of the challenges in various scenarios faced by future mobile communication system, 12 different test cases have been selected by METIS.

[d]Coordination between several macro and micro cells are needed in order to configure ABSs patterns.

[e]A dominant eigenvector is the eigenvector corresponding to the eigenvalue of the biggest absolute value.

[f]The initial value can be either selected based on long-term observation or simply set to the lowest value.

[g]Software implementation of the METIS pathloss and shadowing is available at https://www.metis2020.com/documents/simulations/, last accessed on 2015-02-01.

[h]Software implementation of the ITU-R channel model is available at https://www.itu.int/oth/R0A06000022/en, last accessed on 2015-02-01.

[i]In the current LTE-A system, all the PRBs that are allocated to the same UE are required to use the same coding and modulation scheme ([16] Ch. 9 and 10).

[j]Unlike CRS, DMRS is UE specific and standardized for more flexible MIMO operations in LTE-A systems ([62] Ch. 10).

[k]Link adaptation only requires the amplitude information of the channel, and no feedback of phase information is needed.

[l]The channel prediction is based on pilot signals from each cell, which is usually orthogonal to each other among neighboring cells. The pilot symbols may be interfered by data symbols, but the system can boost the power of the pilot symbols for better channel estimation and prediction ([62] Ch. 11, 13).

Appendix

The detailed ABS patterns used for the macro BS sectors are plotted on Figures 24 and 25. Please note that these are only representative patterns considered in the simulations. In a practical system, some restrictions may apply to the positions of ABSs [22]. We remark that the choice of 30% ABSs (12 ABSs out of 40 subframes) is based on the SINR distribution when no ICIC scheme is considered (see Figure 6) and the constraint that the number of ABSs needs to be an integer. For the 37.5% ABSs case, compared to the 30% ABSs case, 3 more ABSs are added to have stronger protection of both type 2 and type 3 UEs.

Regarding the ABS configurations for the micro cells, we initialize the first subframe as ABS for all the micro cells. Then, we follow Algorithm 1. If any of the MNs report insufficient number of ABSs, one more arbitrary subframe is added as ABS for more protection. We iterate this process until the average throughput of the access link of the

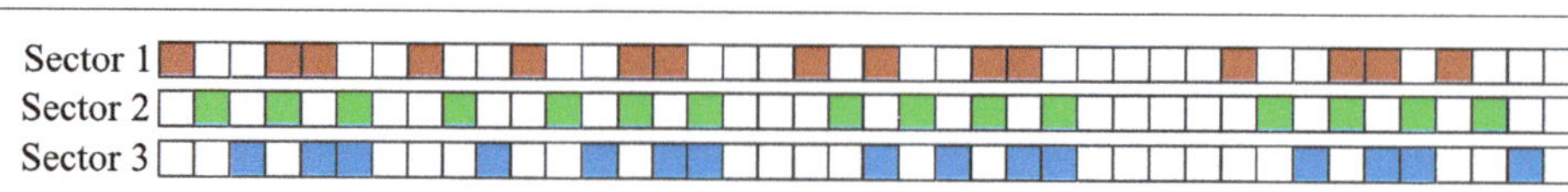

Figure 25 Frames with ABS patterns for three macro sectors in the case of 37.5% ABS. Colored subframes are ABSs.

Table 3 Table of notations

Symbol	Meaning
$b(l, m, u)$	The scheduling decision for the uth user in mth PRB in the lth subframe
$b_p(n)$	Total number of PRBs scheduled for device n over $\mathcal{R}_p$
$b_{np}(n)$	Total number of PRBs scheduled for device n over $\mathcal{R}_{np}$
$d(n)$	A function that measures channel quality difference for the nth device over the protected and non-protected PRBs
i_v	The number of active VUEs on the vth MN
j	The index of an interferer
k	The time index of received signal
l	The index of a subframe
m	The index of a PRB in a given subframe
n	The index of an arbitrary device, i.e., either a UE or an MN, connecting to a macro cell
$n^*(l, m)$	The device is selected to be scheduled at the mth PRB in the lth subframe in the macro cell
p	The number of antennas equipped at the backhaul link of an MN
q_{n_1}	An arbitrary device served by macro BS sector 1
$r(l, m, u)$	The throughput of the uth macro UE at the mth PRB in the lth subframe
$r_{0,i}(v)$	The throughput of the access link of the vth MN when the ith subframe in the micro cells is configured as ABS
$r_{1,i}(v)$	Throughput of the access link of the vth MN when the ith subframe in the micro cells is not configured as ABS
$r_p(n)$	The average rate of device n over $\mathcal{R}_p$
$r_{np}(n)$	The average rate of device n over $\mathcal{R}_{np}$
$s(n)$	The average throughput of device n, which can be either a macro UE or the backhaul link of an MN
$s(u)$	The average throughput of the uth macro UE
$s(v)$	The average throughput of the backhaul link of the vth MN
u	The index of a macro UE
$u^*(l, m)$	The macro UE that is selected to be scheduled at the mth PRB in the lth subframe
x_k	The desired signal transmitted at time k
$x_{j,k}$	The interfering signal from the jth interferer at time k
n_1	The index of a device served by macro BS sector 1
v	The index of an MN
B_p	The size of $\mathcal{R}_p$
B_{np}	The size of $\mathcal{R}_{np}$
$G(\nu)$	$G(\nu) = N\frac{\nu}{(1+\nu)}$ for $\nu \geq 0$, a function defined to facilitate the solution of problem (7)
K	A parameter used in the solution of (11), where $K \in \{0, 1, \ldots, N-1\}$ and $G(d(K+1)) \leq K \leq G(d(K))$
L	The number of total subframes in which the average throughput of a device in the macro cell is calculated
M	The total number of PRBs in each subframe
N	The total number of devices, i.e., $N = N_u + N_m$, connect to a macro cell
N_1	The number of devices served by macro BS sector 1

Table 3 Table of notations *(Continued)*

Symbol	Meaning
N_c	The number of subcarriers in an OFDMA system
N_L	The number of interferers experienced at the backhaul link of an MN
N_m	The number of MNs in a macro cell
N_u	The number of macro UEs in a macro cell
N_{sf}	The number of total subframes that average throughput target of the backhaul link of the MNs are calculated in the micro cells
$\bar{R}(v)$	The average throughput target of the backhaul link of the vth MN during N_{sf} subframes
α	A parameter used in the solution of (11), where $\alpha = \lambda_p - K$
γ_k	The instantaneous output SINR at the MN backhaul receiver at time k
ε	An averaging constant that is required in updating the average throughput of a given device
λ_p	A parameter used in the solution of (11), where $\lambda_p = \max(G(d(K+1)), K)$
λ_{np}	A parameter used in the solution of (11), where $\lambda_{np} = N - \lambda_p$
σ_n^2	The noise power at each antenna
$\mu(l, u)$	The average throughput of user u
$\mathcal{I}_i$	$\mathcal{I}_i \in \{0, 1\}$, $i = 1, 2, 3 \ldots N_{sf}$ indicate the status of the ABSs at the micro cells
$\mathcal{M}$	The sets denotes all the macro UEs in a cell
$\mathcal{R}_p$	The set of protected PRBs in a macro cell
$\mathcal{R}_{np}$	The set of non-protected PRBs in a macro cell
$\mathcal{S}_1$	The set of users only scheduled in R_p
$\mathcal{S}_2$	The set of users only scheduled in R_{np}
$\mathcal{S}_3$	The set of users can be scheduled both in $\mathcal{R}_p$and $\mathcal{R}_{np}$
$\mathcal{U}$	To denote uniform distribution
$\mathcal{V}$	The sets denotes all the MNs in a cell
$\mathbf{h}_k$	The channel vector of the desired signals at each reviver antenna of an MN at time k
$\mathbf{h}_{j,k}$	The channel vector of the jth interfering signal observed at each reviver antenna of an MN at time k
$\mathbf{n}_k$	The vector denotes the thermal noise at each receiver antenna of an MN at time k
$\mathbf{w}_k$	The vector that contains the weight applied at each receiver antenna of an MN at time k
$\mathbf{y}_k$	The vector that contains the received signal at each receiver antenna of an MN at time k
$\mathbf{z}_k$	The combined signal at the output of an MN backhaul receiver at time k
$\mathbf{R}$	The noise covariance matrix of the reviver antennas of an MN
$I_{p\times p}$	The identity matrix of size p
$(\cdot)^H$	To denote the complex conjugate transpose of a matrix
$\mathrm{EIG}_{max}(\cdot)$	To denote the dominant eigenvector of a matrix

worst MN in that area is satisfied. The ABS configurations is reset at the beginning of each drop.

Competing interests
The authors declare that they have no competing interests.

Acknowledgments
Part of this work has been performed in the framework of the FP7 project ICT-317669 METIS, which is partly funded by the European Union. The authors would like to acknowledge the contributions of their colleagues in METIS, although the views expressed are those of the authors and do not necessarily represent the project. This work is also supported by the Swedish Research Council VR under the project 621-2009- 4555 Dynamic Multipoint Wireless Transmission, and in part by United States National Science Foundation grant CNS-1406968.

Author details
[1]Department of Signals and Systems, Chalmers University of Technology, EDIT-huset, Hörsalsvägen 9-11 Gothenburg, SE-41296, Sweden. [2]Electrical and Computer Engineering, Florida International University, 10555 W Flagler Street, EC 3142, Miami, FL 33174-1630, USA.

References
1. I Hwang, B Song, SS Soliman, A holistic view on hyper-dense heterogeneous and small cell networks. IEEE Commun. Mag. **51**, 20–27 (2013). doi:10.1109/MCOM.2013.6525591
2. IS Gaspar, G Wunder, 5G Cellular communications scenarios and system requirements. Technical report. [Online, last accessed 2014-10-01] https://www.5gnow.eu
3. JG Andrews, S Buzzi, W Choi, S Hanly, A Lozano, ACK Soong, JC Zhang, What will 5G be? IEEE J. Sel. Areas Commun., Special Issue on 5G Commun. Syst. **32**(6), 1065–1081 (2014). doi:10.1109/JSAC.2014.2328098
4. P Popovski, V Braun, H-P Mayer, P Fertl, DG-S Zhe Ren, E Ström, T Svensson, H Taoka, P Agyapong, A Benjebbour, G Zimmermann, J Meinilä, J Ylitalo, T Jämsä, P Kyösti, K Dimou, M Fallgren, Y Selen, B Timus, H Tullberg, M Schellmann, Y Wu, M Schubert, DH Kang, JI Markendahl, C Beckman, M Uusitalo, O Yilmaz, C Wijting, Z Li, et al., ICT-317669-METIS/ D1.1 Scenarios, requirements and KPIs for 5G mobile and wireless system. Technical report. [Online, last accessed 2014-10-01] https://www.metis2020.com
5. 3GPP TR 36.836, Technical specification group radio access network: mobile relay for evolved universal terrestrial radio access (E-UTRA). Technical report. [Online, last accessed 2014-10-01] http://www.3gpp.org
6. W Phan, K Horneman, L Yu, J Vihriala, in *Proc. IEEE Veh. Netw. Conf. (VNC)*. Providing enhanced cellular coverage in public transportation with smart relay systems (Jersey City, 2010)
7. Y Sui, J Vihrälä, A Papadogiannis, M Sternad, T Svensson, Moving cells: a promising solution to boost performance for vehicular users. IEEE Commun. Mag. **51**, 62–68 (2013). doi:10.1109/MCOM.2013.6525596
8. D Aronsson, H Bonneville, B Fan, M Färber, J Gora, M Khanfouci, G Liebl, TM de Moraes, A Nicolov, S Redana, A Saadani, E Seidel, M Sternad, T Svensson, J Vihriälä, D3.4 relay configurations. Technical report (Artist4G deliverable, Jul. 2011). [Online, last accessed 2014-10-01] https://ict-artist4g.eu/projet/deliverables
9. E Tanghe, W Joseph, L Verloock, L Martens, Evaluation of vehicle penetration loss at wireless communication frequencies. IEEE Trans. Veh. Technol. **57**(4), 2036–2041 (2008). doi:10.1109/TVT.2007.912164
10. Huawei Technologies: Whitepaper on spectrum. Technical report (Feb. 2013). [Online, last accessed 2014-10-01] http://www.huawei.com
11. Y Sui, A Papadogiannis, T Svensson, in *Proc. IEEE Veh. Tech. Conf. (VTC)*. The potential of moving relays - a performance analysis (Yokohama, 2012)
12. Y Sui, Z Ren, W Sun, T Svensson, P Fertl, in *Proc. IEEE Int. Conf. on Connected Veh. and Expo (ICCVE)*. Performance study of fixed and moving relays for vehicular users with multi-cell handover under co-channel interference (Las Vegas, 2013)
13. A Damnjanovic, J Montojo, Y Wei, T Ji, T Luo, M Vajapeyam, T Yoo, O Song, D Malladi, A survey on 3GPP heterogeneous networks. IEEE Trans. Wireless Commun. **18**(3), 10–21 (2011). doi:10.1109/MWC.2011.5876496
14. P Agyapong, V Braun, M Fallgren, A Gouraud, M Hessler, S Jeux, A Klein, J Lianghai, D Martin-Sacristan, M Maternia, M Moisio, JF Monserrat, K Pawlak, H Tullberg, A Weber, ICT-317669-METIS/D6.1 Simulation guidelines. Technical report (2013). [Online, last accessed 2014-10-01] https://www.metis2020.com
15. HS Dhillon, RK Ganti, F Baccelli, JG Andrews, Modeling and analysis of K-tier downlink heterogeneous cellular networks. IEEE J. Sel. Areas Commun. **30**(3), 550–560 (2012). doi:10.1109/JSAC.2012.120405
16. S Sesia, I Toufik, M Baker, *LTE - The UMTS Long Term Evolution: From Theory to Practice*, 2nd edition. (John Wiley & Sons Ltd, West Sussex, UK, 2011)
17. P Ökvist, A Simonsson, in *Proc. IEEE Veh. Tech. Conf. (VTC)*. LTE HetNet Trial - Range expansion including micro/pico indoor coverage survey (Quebec City, 2012)
18. M Shirakabe, A Morimoto, N Miki, in *Proc. Int. Symp. on Wireless Commun. Syst. (ISWCS)*. Performance evaluation of inter-cell interference coordination and cell range expansion in heterogeneous networks for LTE-Advanced downlink (Aachen, 2011)
19. P Bhat, S Nagata, L Campoy, I Berberana, T Derham, G Liu, X Shen, P Zong, J Yang, LTE-Advanced: an operator perspective. IEEE Commun. Mag. **50**, 104–114 (2012). doi:10.1109/MCOM.2012.6146489
20. TQS Quek, G de la Roche, I Guvenc, M Kountouris (eds.), *Small Cell Networks: Deployment, PHY Techniques, and Resource Management* (Cambridge University Press, Cambridge, UK, 2013)
21. A Anpalagan, M Bennis, R Vannithamby (eds.), *Time and Frequency Domain e-ICIC with Carrier Aggregation in HetNets* (Cambridge University Press, Cambridge, UK, 2014)
22. ST-Ericsson Ericsson, R1-105335: Details of almost blank subframes. Technical report (Oct. 2010). [Online, last accessed 2014-10-01] http://www.3gpp.org
23. S Deb, P Monogioudis, J Miernik, JP Seymour, Algorithms for enhanced inter-cell interference coordination (eICIC) in LTE HetNets. IEEE/ACM Trans. Netw. **22**, 137–150 (2014). doi:10.1109/TNET.2013.2246820
24. NTT DOCOMO, R1-103264: Performance of eICIC with control channel coverage limitation. Technical report (May 2010). [Online, last accessed 2014-10-01] http://www.3gpp.org
25. 3GPP TR 36.866, Study on network-assisted interference cancellation and suppression (NAIC) for LTE. Technical report (2014). [Online, last accessed 2014-10-01] http://www.3gpp.org
26. IEEE 802.16 Broadband Wireless Access Working Group: IEEE C802.16j-07/087r2 mobile relay station operation. Technical report (2007). [Online, last accessed 2014-10-01] http://www.ieee802.org/16/
27. P Mach, Z Becvar, in *Proc. Int. Conf. on Netw. (ICN)*. Path selection in WiMAX networks with mobile relay stations (Toronto, 2011)
28. Y Sui, A Papadogiannis, W Yang, T Svensson, in *Proc. IEEE Global Commun. Conf. (GLOBECOM) Workshops*. Performance comparison of fixed and moving relays under co-channel interference (Anaheim, 2012)
29. Y Sui, A Papadogiannis, W Yang, T Svensson, in *Proc. IEEE Veh. Tech. Conf. (VTC)*. The energy efficiency potential of moving and fixed relays for vehicular users (Las Vegas, 2013)
30. I Chlamtaca, M Contib, JJ-N Liuc, Mobile ad hoc networking: imperatives and challenges. Ad Hoc Netw. **1**, 13–64 (2003). doi:10.1016/S1570-8705(03)00013-1
31. MN Tehrani, M Uysal, H Yanikomeroglu, Device-to-device communication in 5G cellular networks: challenges, solutions, and future directions. IEEE Commun. Mag. **52**, 86–92 (2014). doi:10.1109/MCOM.2014.6815897
32. L Lei, Z Zhong, C Lin, X Shen, Operator controlled device-to-device communications in LTE-advanced networks. IEEE Trans. Wireless Commun. **19**, 96–104 (2012). doi:10.1109/MWC.2012.6231164
33. X Lin, JG Andrews, A Ghosh, R Ratasuk, An overview of 3GPP device-to-device proximity services. IEEE Commun. Mag., 40–48 (2014). doi:10.1109/MCOM.2014.6807945
34. Z Ren, S Stanczak, P Fertl, F Penna, in *Proc. IEEE Int. Conf. on Commun. (ICC)*. Energy-aware activation of nomadic relays for performance enhancement in cellular networks (Sydney, 2014). doi:10.1109/ICC.2014.6883765
35. H Min, J Lee, S Park, D Hong, Capacity enhancement using an interference limited area for device-to-device uplink underlaying cellular networks. IEEE Trans. Wireless Commun. **10**, 3995–4000 (2011). doi:10.1109/TWC.2011.100611.101684
36. G Fodor, E Dahlman, G Mildh, S Parkvall, N Reider, G Miklós, Z Turányi, Design aspects of network assisted device-to-device communications.

IEEE Commun. Mag. **50**, 170–177 (2012). doi:10.1109/MCOM.2012.6163598

37. A Papadogiannis, Y Sui, T Svensson, in *Proc. IEEE Veh. Tech. Conf. (VTC)*. The potential of a hybrid fixed/user relay architecture-a performance analysis (Québec City, 2012). doi:10.1109/VTCFall.2012.6399279
38. NL Pradhan, T Saadawi, Power control algorithms for mobile ad hoc networks. J. Adv. Res. **2**, 199–206 (2011). doi:10.1016/j.jare.2011.04.009
39. P Janis, V Koivunen, C Ribeiro, J Korhonen, K Doppler, K Hug, in *Proc. IEEE Veh. Tech. Conf. (VTC)*. Interference-aware resource allocation for device-to-device radio underlaying cellular networks (Barcelona, 2009). doi:10.1109/VETECS.2009.5073611
40. C-H Yu, K Doppler, CB Ribeiro, O Tirkkonen, Resource sharing optimization for device-to-device communication underlaying cellular networks. IEEE Trans. Wireless Commun. **10**, 2752–2763 (2011). doi:10.1109/TWC.2011.060811.102120
41. M Sternad, M Grieger, R Apelfrojd, T Svensson, D Aronsson, AB Martinez, in *Proc. IEEE Wireless Commun. and Netw. Conf. (WCNC) Workshop*. Using 'predictor antennas' for long-range prediction of fast fading for moving relays (Paris, 2012)
42. D-T Phan-Huy, M Sternad, T Svensson, in *Proc. IEEE Int. Conf. on Connected Veh. and Expo (ICCVE)*. Adaptive large MISO downlink with predictor antennas array for very fast moving vehicles (Las Vegas, 2013)
43. W Li, C Zhang, X Duan, S Jia, Y Liu, L Zhang, in *Proc. IEEE Veh. Tech. Conf. (VTC)*. Performance evaluation and analysis on group mobility of mobile relay for LTE-Advanced system (Quebec City, 2012)
44. P Popovski, V Braun, G Mange, P Fertl, D Gozalvez-Serrano, N Bayer, H Droste, A Roos, G Zimmermann, M Fallgren, A Höglund, H Tullberg, S Jeux, Ö Bulakci, J Eichinger, Z Li, P Marsch, MB Krystian Pawlak, JF Monserrat, ICT-317669-METIS/D6.2 Initial report on horizontal topics, first results and 5G system concept. Technical report (2014). [Online, last accessed 2014-10-01] https://www.metis2020.com
45. P Popovski, G Mange, P Fertl, D GozálvezSerrano, H Droste, N Bayer, A Roos, T Rosowski, G Zimmermann, P Agyapong, M Fallgren, N He, A Höglund, J Söder, H Tullberg, S Jeux, Ö Bulakci, J Eichinger, M Schellmann, J Lianghai, A Rauch, A Klein, M Stamatelatos, Z Li, M Moisio, M Maternia, E Lähetkangas, K Pawlak, JF Monserrat, D Martín-Sacristán, ICT-317669-METIS/D6.3 Intermediate system evaluation results. Technical report (2014). [Online, last accessed 2014-10-01] https://www.metis2020.com
46. F Kelly, A Maulloo, D Tan, Rate control for communication networks: shadow prices, proportional fairness and stability. Oper. Res. Soc. **49**, 237–252 (1998). doi:10.1057/palgrave.jors.2600523
47. MR Jeong, N Miki, A simple scheduling restriction scheme for interference coordinated networks. IEICE Trans. Commun. **E96-B**(6), 1306–1317 (2013)
48. G Caire, R Muller, R Knopp, Hard fairness versus proportional fairness in wireless communications: the single-cell case. IEEE Trans. Inf. Theory. **53**, 1366–1385 (2007). doi:10.1109/TIT.2007.892790
49. P Mogensen, W Na, IZ Kovacs, F Frederiksen, A Pokhariyal, KI Pedersen, T Kolding, K Hugl, M Kuusela, in *Proc. IEEE Veh. Tech. Conf. (VTC)*. LTE capacity compared to the Shannon bound (Dublin, 2007). doi:10.1109/VETECS.2007.260
50. G Piro, LA Grieco, G Boggia, F Capozzi, P Camarda, Simulating LTE cellular systems: an open source framework. IEEE Trans. Veh. Technol. **60**, 498–513 (2011). doi:10.1109/TVT.2010.2091660
51. M Cierny, H Wang, RR Wichman, Z Ding, C Wijting, On number of almost blank subframes in heterogeneous cellular networks. IEEE Trans. Wireless Commun. **12**, 5061–5073 (2013). doi:10.1109/TWC.2013.090513.121756
52. D Lopez-Perez, X Chu, in *Proc. Int. Conf. on Comput. Commun. and Netw (ICCCN)*. Inter-cell interference coordination for expanded region picocells in heterogeneous networks (Maui, 2011)
53. Y Peng, F Qin, in *Proc. IEEE Int. Conf. on Commun. (ICC)*. Exploring Het-Net in LTE-Advanced system, interference mitigation and performance improvement in macro-pico scenario (Kyoto, 2011)
54. AF Molisch, *Wireless Communications*, 2nd edition. (John Wiley & Sons Ltd, West Sussex, UK, 2010)
55. DG Brennan, Linear diversity combining techniques. Proc. IEEE. **47**, 1075–1102 (1959). doi:10.1109/JPROC.2002.808163
56. J Zhang, J Olivier, A Sayeed, BV Veen, in *Proc. IEEE Veh. Tech. Conf. (VTC)*. Low complexity MIMO receiver via maximum SINR interference cancellation (Birmingham, 2002)
57. P Marsch, GP Fettweis (eds.), *Coordinated Multi-Point in Mobile Communications From Theory to Practice* (Cambridge University Press, Cambridge, UK, 2011)
58. A Schrijver, *Theory of Linear and Integer Programming*. (John Wiley & Sons, West Sussex, England, 1998)
59. M Simsek, M Bennis, A Czylwiki, in *Proc. IEEE Global Commun. Conf. (GLOBECOM)*. Dynamic inter-cell interference coordination in HetNets: a reinforcement learning approach (Anaheim, 2012)
60. International Telecommunication Union, Guidelines for evaluation of radio interface technologies for IMT-Advanced. Technical report (2009). [Online, last accessed 2014-10-01] http://www.itu.int/
61. P Gupta, V Agarwal, M Varshney, *Design and Analysis of Algorithms*, Second Edition. (PHI Learning Pvt. Ltd., New Delhi, India, 2012
62. E Dahlman, S Parkvall, J Skold, *4G: LTE/LTE-Advanced for Mobile Broadband*, Second Edition. Academic Press, Waltham, MA, USA, 2014)
63. H Schoeneich, PA Hoeher, Iterative pilot-layer aided channel estimation with emphasis on interleave-division multiple access systems. EURASIP J. Appl. Signal Process. **2006**, 1–15 (2006). doi:10.1155/ASP/2006/81729
64. N Geng, X Yuan, L Ping, Dual-diagonal LMMSE channel estimation for OFDM systems. IEEE Trans. Signal Process. **60**, 4734–4746 (2012). doi:10.1109/TSP.2012.2202112
65. Z YAN, G He, J Ma, An enhanced IRC algorithm for LTE downlink receiver in multi-cell environment. Chin. J. Electron. **23**(2), 377–381 (2014)
66. Y Ohwatari, N Miki, T Asai, T Abe, H Taoka, in *Proc. IEEE Veh. Tech. Conf. (VTC)*. Performance of advanced receiver employing interference rejection combining to suppress inter-cell interference in LTE-Advanced downlink (San Francisco, 2011). doi:10.1109/VETECF.2011.6093196
67. Renesas Mobile Europe Ltd.: 3GPP R1-111562 Interference aware receiver modeling at system level. Technical report (May 2011). [Online, last accessed 2015-01-26] http://www.3gpp.org
68. M Sternad, S Falahati, T Svensson, D Aronsson, in *Proc. 14th European Information Society Technologies (IST) Mobile and Wireless Communications Summit*. Adaptive TDMA/OFDMA for wide-area coverage and vehicular velocities (Dresden, 2005)
69. T Svensson, ChalmersUniversityofTechnology Dept. of Signals and System Goteborg, S Falahati, M Sternad, in *Paper presented at the IEEE 63rd vehicular technology conference*. Coding and resource scheduling in packet oriented adaptive TDMA/OFDMA systems (IEEE Melbourne, Vic., May 2006), p. 7–10
70. 3GPP TR 36.872, Small cell enhancements for E-UTRA and E-UTRAN - physical layer aspects. Technical report (2013). [Online, last accessed 2014-10-01] http://www.3gpp.org
71. V Chandrasekhar, JG Andrews, A Gatherer, Femtocell networks: a survey. IEEE Commun. Mag. **46**, 59–67 (2008). doi:10.1109/MCOM.2008.4623708
72. LB Le, E Hossain, D Niyato, DI Kim, in *Proc. IEEE Int. Conf. on Commun. (ICC)*. Mobility-aware admission control with QoS guarantees in OFDMA femtocell networks, (2013). doi:10.1109/ICC.2013.6654857
73. J Hare, L Hartung, S Banerjee, in *Proc. Int. Conf. on Mobile Systems, Applications, and Services (MobiSys)*. Beyond deployments and testbeds: experiences with public usage on vehicular WiFi hotspots (Low Wood Bay, 2012). doi:10.1145/2307636.2307673
74. DJ Leith, P Clifford, in *Proc. Int. Symposium on Modeling and Optimization in Mobile, Ad Hoc, and Wireless Networks (WiOpt)*. A self-managed distributed channel selection algorithm for WLANS (Boston, 2006). doi:10.1109/WIOPT.2006.1666484
75. VP Mhatre, K Papagiannaki, F Baccelli, in *IEEE Conf. on Computer Commun. (INFOCOM)*. Interference mitigation through power control in high density 802.11 WLANs (Anchorage, 2007)
76. V Nurmela, A Karttunen, A Roivainen, L Raschkowski, T Imai, J Järveläinen, J Medbo, J Vihriälä, J Meinilä, J Kyröläinen, K Haneda, V Hovinen, J Ylitalo, N Omaki, V-M Kolmonen, K Kusume, P Kyösti, T Jämsä, ICT-317669-METIS/D1.2 Initial channel models based on measurements. Technical report (2014). [Online, last accessed 2014-10-01] https://www.metis2020.com
77. J Medbo, K Börner, K Haneda, V Hovinen, T Imai, J Järvelainen, T Jämsä, A Karttunen, K Kusume, J Kyröläinen, P Kyösti, J Meinilä, V Nurmela, L Raschkowski, A Roivainen, J Ylitalo, in *Proc. European Conf. on Networks and Commun. (EuCNC)*. Channel modelling for the fifth generation mobile communications (Bologna, 2014)
78. A Chen, in *Location- and Context-Awareness*, ed. by T Strang, C Linnhoff-Popien. Context-aware collaborative filtering system: predicting the user's preference in the ubiquitous computing environment (Springer Berlin, Germany, 2005), pp. 244–253

79. S Barmpounakis, A Kaloxylos, P Spapis, N Alonistioti, in *Proc. Int. Conf. on Advanced Commun. and Computation (INFOCOMP)*. COmpAsS: A context-aware, user-oriented RAT selection mechanism in heterogeneous wireless networks (Paris, 2014)
80. P Lungaro, Z Segall, J Zander, in *Proc. IEEE Wireless Commun. and Netw Conf. (WCNC)*. ContextShift: a model for efficient delivery of content in mobile networks (Sydney, 2010)

A model for virtual radio resource management in virtual RANs

Sina Khatibi[1,2*] and Luis M Correia[1,2]

Abstract

The combination of Network Function Virtualisation (NFV) and cloud-based radio access network (C-RAN) is a candidate approach for the next generation of mobile networks. In this paper, the novel concept of virtual radio resources, which completes the virtual RAN paradigm, is proposed. The key idea is to aggregate (and manage) all the physical radio resources, to create virtual wireless links, and to offer Capacity-as-a-Service. Due to the isolation among instances, network element abstraction, and a multi-radio access techniques (RAT) structure, the virtualisation approach leads to relatively more efficient and flexible RANs than former ones. Virtual network operators (VNOs) ask for wireless connectivity in the form of capacity per service, hence, not dealing with physical radio resources at all. A model for the management of virtual radio resources is proposed, which can even support the shortage of resources. A practical heterogeneous cellular network is considered as a case study, and results are presented, showing how the virtual radio resource management allocates capacity to services of different VNOs, with different service-level agreements (SLAs) and priority when the overall network capacity reduces down to 45% of the initial one.

Keywords: Virtualisation of radio resources; Virtual radio resource management; Radio access networks; Network Function Virtualisation

1 Introduction

Future mobile networks will have to face the rapid growth of mobile data demand [1]. The candidate approach is to use small cell networks with a dense deployment of base stations (BSs); however, traffic varies drastically, both geographically and temporally [2], which creates constraints that are not solved by this approach. The dimensioning of radio access networks (RANs) for busy hours (i.e. the current approach), guarantees the desired performance during that interval, yet it leads to an inefficient resource usage for the remainder of the time, with relatively high capital and operational expenditure (CAPEX and OPEX) costs.

A solution for this matter lays in the ability of adapting RAN during runtime, based on network changes and traffic demands. Hence, flexibility [3] and cost reduction [4] in RANs became the motivation for their implementation in cloud data centres, in order to achieve centralised processing, collaborative radio, real-time cloud computing [5], and clean RAN systems [6], also known as cloud-based RAN (C-RAN).

Recent studies are focused on the extension of RANs using Network Function Virtualisation (NFV) [7] to add multi-tenancy support, enabling that multiple virtual network operators (VNOs) can be served over the same infrastructure. The concept of a virtualised eNodeB is introduced in [8], by adding an entity, called 'hypervisor', on the top of physical resources, which allocates these resources among various virtual instances. Using the concept of RAN sharing, the air interface resources (i.e. the LTE spectrum) are dynamically divided among various virtual eNodeBs by the hypervisor. In [9], an adaptive allocation of virtual radio resources in heterogeneous networks is analysed, sharing spectrum among VNOs. In [10], the advantage of a virtualised LTE system is investigated by an analytical model for FTP (File Transfer Protocol) transmission. The concept of joint NFV and C-RAN is discussed in [11,12]. This solution, which is called virtual RAN (V-RAN), provides operators with RAN-as-a-Service (RANaaS).

In this paper, the concept of virtualisation of radio resources to achieve virtual wireless links and to have

* Correspondence: sina.khatibi@inov.pt
[1]Department of Electrical and Computer Engineering, Instituto Superior Técnico (IST), University of Lisbon, Av. Rovisco Pais, Lisbon 1049-001, Portugal
[2]INOV-INESC, Rua Alves Redol Lisbon, 1000-029, Portugal

end-to-end virtual networks [13], by aggregating all the physical resources from different radio access techniques (RATs), in order to offer VNOs with a more efficient wireless connectivity is proposed. In this novel methodology, VNOs ask for wireless capacity from a set of physical network providers to serve their subscribers, and they do not have to deal with the physical infrastructure at all. The RAN provider (RANP), owning the physical infrastructure, is capable of offering Capacity-as-a-Service to VNOs. The advantages of RAN virtualisation compared to RAN sharing (where each operator is allocated a portion of spectrum) comes from network element abstraction, isolation among virtual instances, and the ability to support multi-RATs.

A differentiation of these two concepts can be addressed using the analogy presented in [14], where a process on an operating system (OS) is presented as the equivalent of a session in a network. As depicted in Figure 1, the V-RAN and the virtual machine (VM) can be claimed to be a realisation of the corresponding concepts; likewise, RAN sharing is the equivalent of multi-tasking in OSs. In the virtualisation solutions, there is always a virtual manager, such as VMware, offering isolation and abstraction to the upper levels. The offered isolation makes it possible to have multiple instances with different configurations running over the same physical infrastructure, and it relatively reduces the system downtime. The ease of use is the result of the offered abstraction, since virtual instances do not have to deal with physical resources and their complexity.

The novelty of this paper, besides the presented concept of virtualisation of radio resources, is the proposal of an analytical model for virtual radio resource management (VRRM). For a network with multiple RATs, such as GSM, UMTS, and LTE, the model is capable of estimating the overall network capacity based on a given number of the available radio resource units (RRUs) from each RAT. It also shares the available capacity among the different services of the VNOs, the allocation being based on the VNOs' service level agreements (SLAs), in which VNOs may be guaranteed with a minimum as well as a maximum capacity per service, or simply served in a best-effort approach. The presented VRRM model satisfies SLAs when there are enough RRUs and minimises SLA violations in resource shortage cases; in both cases, fairness of resource allocation is considered. In addition to the proposition of the novel VRRM model, an architecture for a V-RAN based on a C-RAN infrastructure, with its required modifications to support virtualisation of radio resources, is briefly addressed.

The rest of this paper is organised as follows. Section 2 presents the V-RAN architecture, Section 3 is about the modelling of the problem of management of virtual radio resources. Section 4 describes the details of the scenario, based on which the proposed model is evaluated; numeric results are being discussed in Section 5. Finally, conclusions are presented in the final section of this paper.

2 Virtual radio access network architecture

In this section, the architecture for a V-RAN using virtualisation of radio resources is discussed. It is based on a C-RAN, with modifications to support Capacity-as-a-Service, as depicted in Figure 2. The key architectural elements are as follows:

- VNOs: network operators that do not own a RAN infrastructure. They ask the virtualisation platform for wireless connectivity in terms of capacity to carry various services traffic with various quality-of-service (QoS) requirements to/from their subscribers.
- Backhaul transport network: a low latency optical transport network, which connects the operators' cores to the physical infrastructure of a RAN.
- Virtualisation platform: the key difference between a C-RAN and a V-RAN. On the one hand, it is in charge of abstracting the physical infrastructure for the VNOs, while on the other hand, it handles the request of VNOs through the available physical

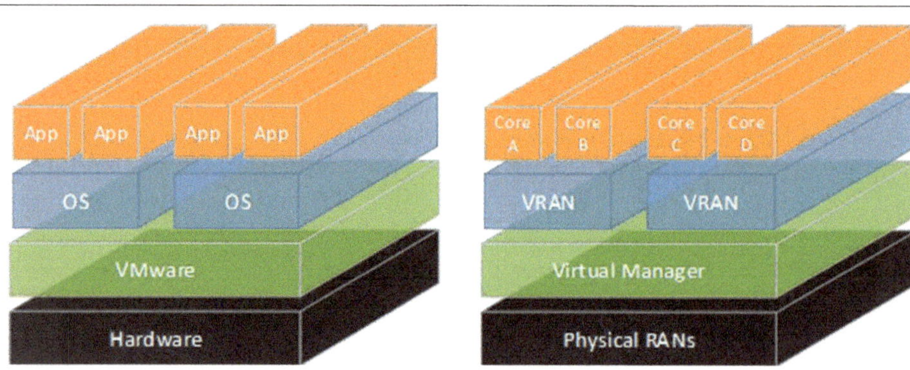

Figure 1 Comparison between V-RAN and VM.

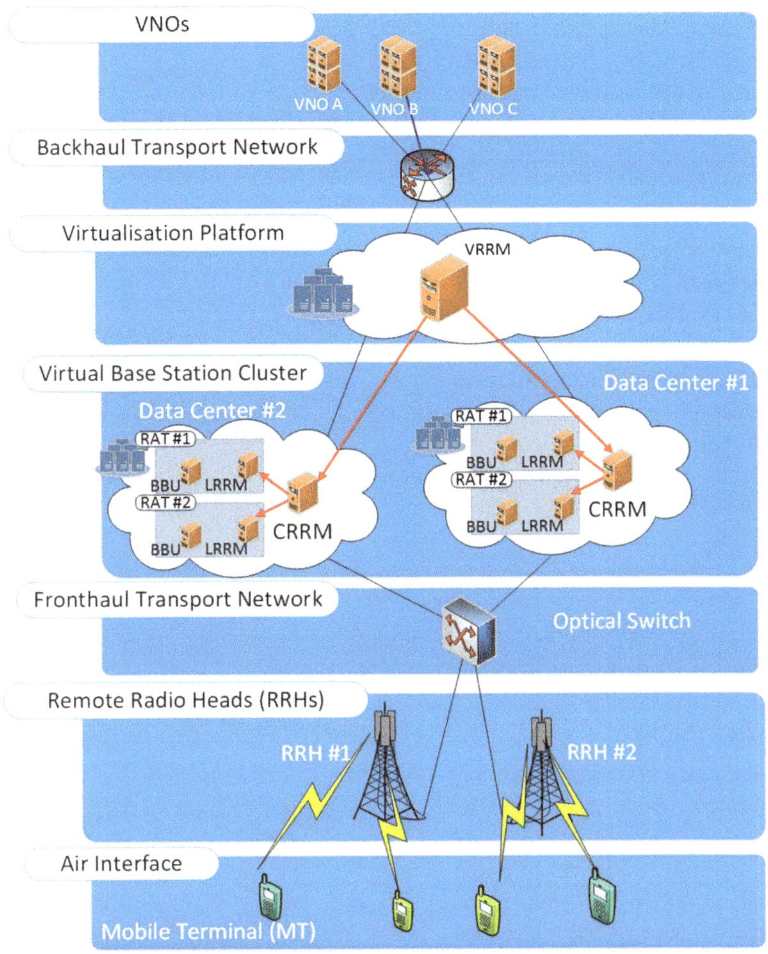

Figure 2 **Architecture of V-RAN.**

resources. The most important functionality of the virtualisation platform is the VRRM, the highest manager, which is in charge of translating VNO requirements and SLAs through sets of polices onto the lower levels. It optimises the usage of virtual radio resources without dealing with the management of physical resources. Nevertheless, reports and monitoring information (e.g. estimated remained capacity) received from lower levels enable it to improve policies.

- BBU (baseband units) pools' data centre: a set of VMs used for baseband processing of traffic among user terminals and network cores.
- Fronthaul transport network: it transmits digitised radio signals between BBU pools and remote radio heads (RRH), using Common Public Radio Interface (CPRI) with high data rates over optical fibres. The optical equipment needs to have the lowest delay possible, since the maximum round-trip delay must be below 150 μs (i.e. a maximum of 15 km of BBU-RRH distance) [15]. The optical switch is a non-smart manageable switch, enabling the scaling or the migration of BBU pools among multiple data centres.
- RRHs: the transceivers in charge of exchanging data and control traffic to/from mobile terminals (MTs) through the air interface, supporting multiple RATs.

By comparing C-RANs with current mobile networks, it can be seen that the eNodeB has been divided into RRHs, fibre optics, and BBU pools. The virtualisation platform, which offers isolation, element abstraction, and multi-tenancy, does not exist in current networks. The changes in the architecture and the dedicated hardware replacement by VMs in data centres provide high flexibility, resource efficiency, and cost reduction.

3 Modelling of virtual radio resource management

The management hierarchy of virtual radio resources is also shown in Figure 2, consisting of VRRM on the top of the usual radio resource management entities in heterogeneous access networks [16], common RRM

(CRRM), and local RRM (LRRM). The VRRM estimates the total network data rate, then, having the available estimated resources, it allocates capacity to the different services of each VNO so that minimum and maximum guaranteed capacities are met. This section presents the analytical modelling, considering the estimation of resources, and their allocation without and with violation of SLAs.

3.1 Estimation of available resources

In general, the data rate of an RRU assigned to an MT varies between zero and the maximum data rate based on various parameters, e.g. RAT, modulation, and coding schemes. Therefore, it can be given as a function of channel quality, i.e. signal to noise ratio (SNR), as follows:

$$R_{b_{\mathrm{RAT}_i}[\mathrm{Mbps}]}(\rho_{\mathrm{in}}) \in \left[0, R^{\max}_{b_{\mathrm{RAT}_i}[\mathrm{Mbps}]}\right] \quad (1)$$

where

- $R_{b\mathrm{RAT}_i}$ is the data rate of a single RRU of the ith RAT,
- ρ_{in} is the input SNR, and
- $R^{\max}_{b\mathrm{RAT}_i}$ is the maximum data rate of a single RRU of the ith RAT.

In [17], a heterogeneous cellular network is modelled as a K-tier network, where each tier models the BSs of a particular class. It is assumed that the BSs in a given tier are spatially distributed as a Poisson point process (PPP) with a given density and transmission power. The received power is assumed to be exponentially distributed (i.e. Rayleigh fading is assumed for the signal magnitude). It is shown that the cumulative distributed function (CDF) of the input SNR for an interference limited network, where MTs are connected to the BS with the strongest signal, can be written as follows:

$$P_{\rho_{[\mathrm{dB}]}}(\rho_{\mathrm{in}}) = 1 - e^{-\frac{0.2}{\alpha_p}\ln(10)\rho_{\mathrm{in}\,[\mathrm{dB}]}} \quad (2)$$

where

- α_p is the path loss exponent (which values are larger or equal to 2).

Based on real logs, the data rates of different access technologies, as a function of input SINR and vice versa, have been presented in [18]. In the next step, for the sake of simplicity, these functions have been approximated by an equivalent polynomial of degree 5; hence, the SINR can be written as a function of data rate as follows:

$$\rho_{\mathrm{in}\,[\mathrm{dB}]}(R_{b\mathrm{RAT}_i}) = \sum_{k=0}^{5} a_{i\left[\frac{\mathrm{dB}}{\mathrm{Mbps}^k}\right]} R^k_{b\mathrm{RAT}_i[\mathrm{Mbps}]} \quad (3)$$

where

- a_k are coefficients of a polynomial approximation of SINR as a function of data rate in each access technology listed in [13].

By substituting this polynomial in (2) and adding the boundary conditions addressed in (1), the CDF of a single RRU of RAT_i is:

$$P_{\mathrm{Rb}}(R_{b\mathrm{RAT}_i[\mathrm{Mbps}]}) = \frac{e^{-\frac{0.2}{\alpha_p}\ln(10)a_0} - e^{-\frac{0.2}{\alpha_p}\ln(10)\sum_{k=0}^{5} a_k (R_{b\mathrm{RAT}_i})^k}}{e^{-\frac{0.2}{\alpha_p}\ln(10)a_0} - e^{-\frac{0.2}{\alpha_p}\ln(10)\sum_{k=0}^{5} a_k \left(R^{\max}_{b\mathrm{RAT}_i}\right)^k}} \quad (4)$$

In the next step, the overall capacity of a RAT is estimated as follows:

$$R^{\mathrm{RAT}_i}_{b_{\mathrm{tot}}} = \sum_{n=1}^{N^{\mathrm{RAT}_i}_{\mathrm{RRU}}} R^{\mathrm{RAT}_i}_{b_n} \quad (5)$$

where

- $N^{\mathrm{RAT}_i}_{\mathrm{RRU}}$ is the number of RRUs in the ith RAT,
- $R^{\mathrm{RAT}_i}_{b_{\mathrm{tot}}}$ is the data rate from a ith RAT pool, and
- $R^{\mathrm{RAT}_i}_{b_n}$ is the data rate from the nth RRU of the ith RAT.

Based on [19], the probability density function (PDF) of each RAT, assuming the RRUs are independent, is equal to the convolution of all the PDFs of that RAT's RRU. From (4), the PDF of a single RRU is calculated (and then numerically sampled with a step of 10 kbps). To compute the total data rate PDF of each RAT, the PDF of the entire RRUs is convolved.

The resource pools of RATs can be aggregated under the supervision of the CRRM, and the total data rate from all RATs is the summation of the total data rate from each of them:

$$R^{\mathrm{CRRM}}_{b\,[\mathrm{Mbps}]} = \sum_{i=1}^{N_{\mathrm{RAT}}} R^{\mathrm{RAT}_i}_{b_{\mathrm{tot}}[\mathrm{Mbps}]} \quad (6)$$

The PDF of the total network data rate is computed by convolving all the RATs' PDFs. By obtaining the total network CDF and PDF, an estimation of available network capacity is in hand to be used in the allocation procedure, as described in the next subsections.

3.2 Allocation of resources

After estimating the total network capacity, the VRRM has to allocate it to the various services of the networks. The key objective in the allocation of resources is to maximise the usage efficiency in addition to meeting the constraint set. The algorithms of resource allocation have also to consider the priority of the different services of different VNOs based on their SLAs. For instance, conversation (e.g. VoIP) and streaming (e.g. video streaming) service classes are delay sensitive, but they have almost constant data rates. The allocation to these services of data rates higher than contracted capacities do not increase quality of service (QoS), in contrast to interactive (e.g. FTP) and background (e.g. email) service classes, where the increase of data rates can indeed improve a user's quality of experience (QoE); hence, operators offering the former services are not interested in allocating higher data rates. Based on the service set and requirements, VNOs may have different SLAs, but these SLAs can generally be categorised into three types of contract:

- Guaranteed bit rate (GB), in which the VNO is guaranteed a minimum as well as a maximum level of data rates, regardless of the network status. In other words, the total satisfaction of the VNO is achieved when the maximum guaranteed data rate is allocated to it. The upper boundary in this type of SLA enables VNOs to have full control on their networks. For instance, a VNO offering VoIP to its subscribers may foresee to offer this service to only 30% up 50% of its subscribers simultaneously. The VNO can put this policy into practice by choosing a guaranteed SLA for its VoIP service. It is expected that subscribers always experience a good QoS in return of relatively more expensive services.
- Best effort with minimum guaranteed (BG), where the VNO is guaranteed with a minimum level of service, but the request for higher data rates than the guaranteed one is served in a best-effort manner. In this case, although VNOs do not invest as much as former ones, they can still guarantee the minimum QoS to their subscribers. From the subscribers' viewpoint, the acceptable service (not as good as the other ones) is offered with a relatively lower cost.
- Best effort (BE), in which the VNO is served in a pure best-effort manner. Operators, and consequently their subscribers, in return, may suffer from low QoS and resource starvation during busy hours.

Hence, the objective function is formed as the total weighted network data rate:

$$f_{\mathbf{R_b}}^{\text{cell}}(\mathbf{R_b}) = \sum_{i=1}^{N_{\text{VNO}}}\sum_{j=1}^{N_{\text{srv}}} W_{ji}^{\text{Srv}} R_{b_{ji}\,[\text{Mbps}]}^{\text{Srv}} \tag{7}$$

where

- $R_{b_{ji}}^{\text{Srv}}$ is the serving data rate for service j of VNO i,
- $\mathrm{R_b}$ is the vector of serving data rates,
- N_{VNO} is the number of served VNOs,
- N_{srv} is the number of services for each VNO, and
- W_{ji}^{Srv} is the weight of serving unit of data rate for the jth service of the ith VNO, where $W_{ji}^{\text{Srv}} \in [0, 1]$.

The weights in (7) are used to prioritise the allocation of data rates to different services of different VNOs. Services with the higher weights are served with the higher data rates. The choice of these weights is based on the SLAs between the VNOs and VRRM.

There are also constraints in the allocation of data rates that should not be violated. The fundamental constraint is the total network capacity, estimated in the previous subsection. The summation of the entire assigned data rate to all services should not be greater than the total estimated capacity of the network:

$$\sum_{i=1}^{N_{\text{VNO}}}\sum_{j=1}^{N_{\text{srv}}} R_{b_{ji}\,[\text{Mbps}]}^{\text{Srv}} \leq R_{b\,[\text{Mbps}]}^{\text{CRRM}} \tag{8}$$

where

- R_b^{CRRM} is the estimated total data rate that can be provided by CRRM from various RATs.

However, the optimisation of this objective function in the current situation may not lead to a desirable situation: services with the highest serving weight receive almost all the resources, while the other services are allocated by the minimum possible data rate; this way of resource allocation is neither fair nor desirable. In contrast, the ideal case is when the normalised data rate (i.e. the data rate divided by the serving weight) of all services, and consequently the normalised average, has the same value. This can be expressed as follows:

$$\frac{R_{b_{ji}\,[\text{Mbps}]}^{\text{Srv}}}{W_{ji}^{\text{Srv}}} - \frac{1}{N_{\text{VNO}} N_{\text{srv}}} \sum_{i=1}^{N_{\text{VNO}}}\sum_{j=1}^{N_{\text{srv}}} \frac{R_{b_{ji}\,[\text{Mbps}]}^{\text{Srv}}}{W_{ji}^{\text{Usg}}} = 0 \tag{9}$$

Nevertheless, resource efficiency and fair allocation are two contradict goals. For instance, if one assumes a network with a 100-Mbps capacity to serve two services with serving weights of 0.8 and 0.2, by considering only (7), all the network capacity has to be allocated to the first service (the one with a serving weight of 0.8), while a fair allocation is achieved when the first service receives 80 Mbps and the other 20 Mbps. As expected, the increment of the data rate in one of them leads to the decrement in the other; hence, instead of having the fairest allocation possible (i.e. the deviation of all normalised data rates from the normalised average is zero), the minimisation of the total deviation from the normalised average is used:

$$f_{\mathbf{R_b}}^{fr} = \min_{R_{b_{ji}}^{\mathrm{Srv}}} \left\{ \sum_{i=1}^{N_{\mathrm{VNO}}} \sum_{j=1}^{N_{\mathrm{srv}}} R_{b_{ji}[\mathrm{Mbps}]}^{D} \right\} \tag{10}$$

where

- $f_{\mathbf{R_b}}^{fr}$ is the fairness objective function and
- $R_{b_{ji}}^{D}$ is the deviation from the normalised average for service j of VNO i, given by the following:

$$R_{b_{ji}[\mathrm{Mbps}]}^{D} = \left| \frac{R_{b_{ji}\,[\mathrm{Mbps}]}^{\mathrm{Srv}}}{W_{ji}^{\mathrm{Srv}}} - \frac{1}{N_{\mathrm{VNO}} N_{\mathrm{srv}}} \sum_{i=1}^{N_{\mathrm{VNO}}} \sum_{j=1}^{N_{\mathrm{srv}}} \frac{R_{b_{ji}\,[\mathrm{Mbps}]}^{\mathrm{Srv}}}{W_{ji}^{\mathrm{Srv}}} \right| \tag{11}$$

In order to convert the problem into a linear form, (11) can be written as follows:

$$-R_{b_{ji}[\mathrm{Mbps}]}^{f} \leq \frac{R_{b_{ji}\,[\mathrm{Mbps}]}^{\mathrm{Srv}}}{W_{ji}^{\mathrm{Srv}}} - \frac{1}{N_{\mathrm{VNO}} N_{\mathrm{srv}}} \sum_{i=1}^{N_{\mathrm{VNO}}} \sum_{j=1}^{N_{\mathrm{srv}}} \frac{R_{b_{ji}\,[\mathrm{Mbps}]}^{\mathrm{Srv}}}{W_{ji}^{\mathrm{Srv}}} \leq R_{b_{ji}[\mathrm{Mbps}]}^{f} \tag{12}$$

where

- $R_{b_{ji}}^{f}$ is the boundary for deviation data rate from the normalised average for service j of VNO i.

According to (12), $R_{b_{ji}}^{f}$ is always larger or equal to $R_{b_{ji}}^{D}$, and minimising the former leads to the minimisation of the latter. Therefore, (10) reformulated into a form originated from [20] as follows:

$$f_{\mathbf{R_b}}^{fr} = \min_{R_{b_{ji}}^{\mathrm{Srv}},\, R_{b_{ji}}^{f}} \left\{ \sum_{i=1}^{N_{\mathrm{VNO}}} \sum_{j=1}^{N_{\mathrm{srv}}} R_{b_{ji}[\mathrm{Mbps}]}^{f} \right\}$$

$$s.t. \begin{cases} \dfrac{R_{b_{ji}\,[\mathrm{Mbps}]}^{\mathrm{Srv}}}{W_{ji}^{\mathrm{Srv}}} - \dfrac{1}{N_{\mathrm{VNO}} N_{\mathrm{srv}}} \displaystyle\sum_{i=1}^{N_{\mathrm{VNO}}} \sum_{j=1}^{N_{\mathrm{srv}}} \dfrac{R_{b_{ji}\,[\mathrm{Mbps}]}^{\mathrm{Srv}}}{W_{ji}^{\mathrm{Srv}}} \leq R_{b_{ji}[\mathrm{Mbps}]}^{f} \\ \dfrac{-R_{b_{ji}\,[\mathrm{Mbps}]}^{\mathrm{Srv}}}{W_{ji}^{\mathrm{Srv}}} + \dfrac{1}{N_{\mathrm{VNO}} N_{\mathrm{srv}}} \displaystyle\sum_{i=1}^{N_{\mathrm{VNO}}} \sum_{j=1}^{N_{\mathrm{srv}}} \dfrac{R_{b_{ji}\,[\mathrm{Mbps}]}^{\mathrm{Srv}}}{W_{ji}^{\mathrm{Srv}}} \leq R_{b_{ji}[\mathrm{Mbps}]}^{f} \end{cases} \tag{13}$$

It is worthwhile noting that the fairness for services with minimum guaranteed data rates applies only to the amount exceeded over the minimum guaranteed level. As the network capacity increases, the summation in (7) increases as well; therefore, in order to combine it with (13), the fairness intermediate variable, R_{ji}^{f}, has to adapt to the network's capacity:

$$f_{\mathbf{R_b}}^{v}(\mathbf{R_b}) = f_{\mathbf{R_b}}^{\mathrm{cell}}\left(R_{b_{ji}}^{\mathrm{cell}}\right) - W^{f} f_{\mathbf{R_b}}^{f}\left(\mathbf{R_b^f}\right) \tag{14}$$

where

- $W^{f} \in [0, 1]$ is the fairness weight in the objective function, indicating how much weight should be put on the fair allocation and
- $\mathbf{R_b^f}$ is the vector of intermediate fairness variable:

$$\mathbf{R_b^f} = \left\{ R_{b_{ji}}^{f} \mid j = 1, 2, \ldots, N_{\mathrm{srv}} \text{ and } i = 1, 2, \ldots, N_{\mathrm{VNO}} \right\} \tag{15}$$

- $f_{\mathbf{R_b}}^{f}$ is the fairness function:

$$f_{\mathbf{R_b}}^{fr}\left(\mathbf{R_b^f}\right) = \sum_{i=1}^{N_{\mathrm{VNO}}} \sum_{j=1}^{N_{\mathrm{srv}}} \left(\frac{R_{b\,[\mathrm{Mbps}]}^{\mathrm{CRRM}}}{R_{b[\mathrm{Mbps}]}^{\mathrm{min}}} R_{b_{ji}[\mathrm{Mbps}]}^{f} \right) \tag{16}$$

where

- $\overline{R_{b}^{\mathrm{min}}}$ is the minimum average data rate among all the network services (i.e. VoIP).

The division of the network capacity by the minimum average data rate of services gives the maximum possible number of users in the network with a given network capacity and service set. By multiplying the fairness variable by the maximum number of users, the balance of these two objectives (i.e. network throughput and fairness) can be kept.

In addition, there are more constraints for VRRM to allocate data rates to various services, which should not be violated. The very fundamental constraint is the total network capacity estimated in the previous section. The summation of the entire assigned data rates to all services should always be smaller than the total estimated capacity of the network:

$$\sum_{i=1}^{N_{\mathrm{VNO}}} \sum_{j=1}^{N_{\mathrm{srv}}} R_{b_{ji}\,[\mathrm{Mbps}]}^{\mathrm{Srv}} \leq R_{b\,[\mathrm{Mbps}]}^{\mathrm{CRRM}} \tag{17}$$

The offered data rate to the guaranteed and the best effort with minimum guaranteed services imposes the next constraints. The allocated data rates related to these services have to be higher than the minimum guaranteed level (for guaranteed and best effort with minimum guaranteed) and lower than the maximum guaranteed one (for guaranteed services only):

$$R^{\text{Min}}_{b_{ji}[\text{Mbps}]} \leq R^{\text{Srv}}_{b_{ji}\,[\text{Mbps}]} \leq R^{\text{Max}}_{b_{ji}[\text{Mbps}]} \tag{18}$$

where

- $R^{\text{Min}}_{b_{ji}}$ is the minimum guaranteed data rate of service j of VNO i and
- $R^{\text{Max}}_{b_{ji}}$ is the maximum guaranteed data rate of service j of VNO i.

3.3 Resource allocation with violation

However, in the allocation process, there are situations where the resources are not enough to meet all guaranteed capacity and the allocation optimisation is no longer feasible. Data centre migration is a practical example of this case. A simple approach in these cases is to relax the constraints by the introduction of violation (also known as slack) variables. In case of VRRM, the relaxed constraint is as follows:

$$\begin{aligned} &R^{\text{Min}}_{b_{ji}[\text{Mbps}]} \leq R^{\text{Srv}}_{b_{ji}\,[\text{Mbps}]} + \Delta R^{v}_{b_{ji}\,[\text{Mbps}]} \\ &\Delta R^{v}_{b_{ji}\,[\text{Mbps}]} \geq 0 \end{aligned} \tag{19}$$

where

- $\Delta R^{v}_{b_{ji}}$ is the violation variable for the minimum guaranteed data rate of service j of VNO i.

By introducing the violation parameter, the former infeasible optimisation problem turns into a feasible one. The optimal solution maximises the objective function and minimises the weighted average constraint violations. The weighted average constraint violation is defined as follows:

$$\Delta \bar{R}^{v}_{b}[\text{Mbps}] = \frac{1}{N_{\text{VNO}} N_{\text{srv}}} \sum_{i=1}^{N_{\text{VNO}}} \sum_{j=1}^{N_{\text{srv}}} W^{v}_{ji} \Delta R^{v}_{b_{ji}\,[\text{Mbps}]} \tag{20}$$

where

- $\Delta \bar{R}^{v}_{b}$ is the average constraint violation and
- W^{v}_{ji} is the weight of violating minimum guaranteed data rate of service j of VNO i, where $W^{v}_{ji} \in [0, 1]$.

The objective function presented in (14) has also to be changed. The new objective function, the relaxed one, has to contain the minimisation of violations in addition to the maximisation of former objectives. Although the average constraint violation has a direct relation with the allocated data rate to services, where the increment in one leads to the decrement of the other, it does not have the same relation with fairness. It can be claimed that the maximisation of fairness and minimisation of constraint violations are independent. Therefore, the final objective function considering both issues has to consider the same approach for minimisation of the violations as well as fairness. In other words, the fairness variable is weighted as it is presented in (17) to compensate the summation of weighted data rate of various services. The derivation from fair allocation, which is desired to be as minimum as possible, leads to a relatively higher weight in the objective function and may confiscate the constraint violation strategies. Therefore, the average constraint violation also has to be placed in the objective function in a similar way:

$$f^{v}_{\mathbf{R_b}}(\mathbf{R_b}) = f^{\text{cell}}_{\mathbf{R_b}}(\mathbf{R^{cell}_b}) - f^{vi}_{R^{v}_{b}}(\Delta \bar{R}^{v}_{b}) - W^{f} f^{f}_{R_b}\left(\mathbf{R^{f}_b}\right) \tag{21}$$

where $f^{vi}_{R^{v}_{b}}$ is the constraint violation function:

$$f^{vi}_{R^{v}_{b}}(\Delta \bar{R}^{v}_{b}) = \frac{R^{\text{CRRM}}_{b\,[\text{Mbps}]}}{R^{\text{min}}_{b[\text{Mbps}]}} \Delta \bar{R}^{v}_{b}[\text{Mbps}] \tag{22}$$

However, the definition of fairness in a congestion situation is not the same. The fairness objective in the normal case is to have the same normalised data rate for all services. As a reminder, when the network faces congestion, there are not enough resources to serve all services with the minimum acceptable data rates. Therefore, some of best-effort services are not allocated any capacity at all, and some violation is also introduced in the guaranteed data rates. In this case, fairness is to make sure that the weighted violation of all services is the same. The ideal fairness with this approach is as follows:

$$W^{v}_{ji} \Delta R^{v}_{b_{ji}\,[\text{Mbps}]} - \frac{1}{N_{\text{VNO}} N_{\text{srv}}} \sum_{i=1}^{N_{\text{VNO}}} \sum_{j=1}^{N_{\text{srv}}} W^{v}_{ji} \Delta R^{v}_{b_{ji}\,[\text{Mbps}]} = 0 \tag{23}$$

The violation data rates for the best-effort services are always zero; consequently, (13) is changed to the following:

$$\begin{aligned} &f^{fr}_{\mathbf{R_b}} = \min_{R^{\text{Srv}}_{b_{ji}},\, R^{f}_{b_{ji}}} \left\{ \sum_{i=1}^{N_{\text{VNO}}} \sum_{j=1}^{N_{\text{srv}}} R^{f}_{b_{ji}[\text{Mbps}]} \right\} \\ &s.t. \begin{cases} W^{v}_{ji} \Delta R^{v}_{b_{ji}\,[\text{Mbps}]} - \dfrac{1}{N_{\text{VNO}} N_{\text{srv}}} \displaystyle\sum_{i=1}^{N_{\text{VNO}}} \sum_{j=1}^{N_{\text{srv}}} W^{v}_{ji} \Delta R^{v}_{b_{ji}\,[\text{Mbps}]} \leq R^{f}_{b_{ji}[\text{Mbps}]} \\ -W^{v}_{ji} \Delta R^{v}_{b_{ji}\,[\text{Mbps}]} + \dfrac{1}{N_{\text{VNO}} N_{\text{srv}}} \displaystyle\sum_{i=1}^{N_{\text{VNO}}} \sum_{j=1}^{N_{\text{srv}}} W^{v}_{ji} \Delta R^{v}_{b_{ji}\,[\text{Mbps}]} \leq R^{f}_{b_{ji}[\text{Mbps}]} \end{cases} \end{aligned} \tag{24}$$

The management of virtual radio resources is a complex optimisation problem since the network status and constraints vary in time. Among various possible techniques and approaches for solving this problem, partial

VRRM seems to be the simplest one. In this approach, the main optimisation problem is broken into multiple sub-problems. In other words, the time axis is divided into decision windows, and VRRM maximises the objective function in each of these intervals, independently. However, it is worth noting that decisions in each interval affect directly the network state, and the outcome at a certain interval depends on the decisions and states in previous intervals; the optimal solution has to take this dependency into consideration. As a consequence, the output of partial VRRM may only be a local minimum and not the global one. Nevertheless, partial VRRM is a simple solution, which can be used as the starting step and reference point.

Figure 3 illustrates a decision window of VRRM, CRRM, and LRRMs. The VRRM decision window contains multiple CRRM ones, during which CRRM applies the decided policy set. In the next decision window of VRRM, after multiple network stages, the VRRM updates the network situation and makes the new decision for the next time interval.

The aforementioned optimisation problem is solved by MATLAB Linear Programming (LP) problem solver (i.e. linprog function) [21]. The method used in this function is the interior-point LP [22], which is a variant of Mehrotra's predictor-corrector algorithm [23], a primal-dual interior-point method. The termination tolerance on the function is chosen to be 10^{-8}.

4 Scenario

A number of scenarios are chosen to evaluate the performance of the proposed model. The key parameters of these scenarios are cell layout, the RATs' configuration, the VNOs, and the service set.

The RRHs are capable of supporting multiple RATs, which are OFDMA (based on LTE-Advance), CDMA (based on UMTS), and FDMA/TDMA (based on GSM), and their flexibility enables various cell layout for these RATs. The considered layout, illustrated in Figure 4, offers full coverage using TDMA cells with the radius of 1.6 km, CDMA cells with 1.2 km, and OFDMA cells with 0.4 km. It is assumed that the coverage area is divided into serving areas, over which a VRRM is operating. Dividing the coverage area to different serving areas makes it possible to consider different policies for different regions (e.g. for residential or commercial regions). In these scenarios, the serving area for each VRRM is considered to be as big as the TDMA cell; hence, each serving area is covered by 1 TDMA cell, approximately 1.7 CDMA cells, and 16 OFDMA ones.

The details of each RAT configuration, such as the number of cells and the number of RRU per RAT, are presented in Table 1. For the CDMA cells, in which the serving area covers an area equivalent to area of 1.7 cells, it is assumed that the radio resources are distributed uniformly and the available resources for this RAT are 1.7 times the resources of a single cell. Moreover, variations of the reference scenario are considered, in which the serving area is covered with a lower number of OFDMA cells temporarily. A lower cell number leads to a lower network capacity; hence, network capacity and VRRM performance are compared in these scenarios. The minimum number of OFDMA cells is chosen to be 5, an extreme case where the network capacity is reduced to 45% of the reference scenario's capacity.

Furthermore, 3 VNOs, each one with 300 subscribers, are assumed to operate in this area, and the average required data rate for each of them is 6.375 Mbps [24]. Hence, the contracted data rate for each of these operators is 1,912.5 Mbps. It is worth noting that the choice of the average data rate is just used to consider realistic boundaries for the guaranteed data rates. Although they have the same number of subscribers and contracted data rate, they are different SLAs as follows:

- VNO GB, the allocated data rates for services are guaranteed to be in a range [50, 100]% of the service data rate.
- VNO BG has best effort with a minimum of 25% of service data rate guaranteed SLA.
- Services of VNO BE are served all in a best-effort manner.

All of these VNOs offer the same set of services to their subscribers. These services and their volume share

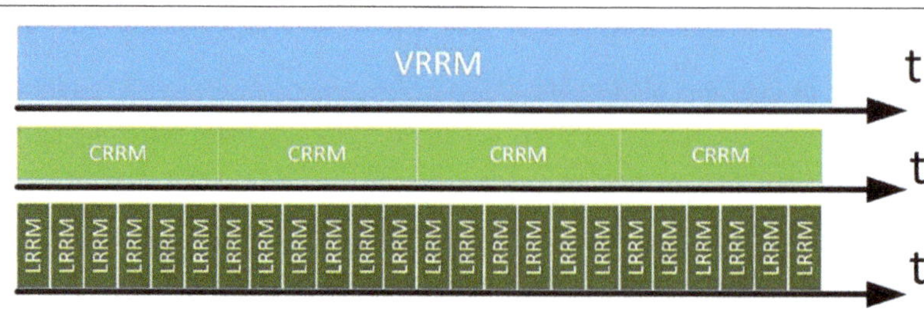

Figure 3 Decision window of VRRM and CRRM.

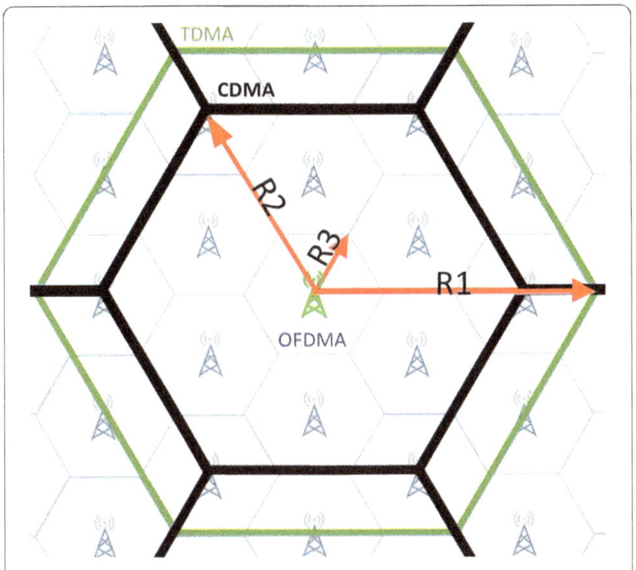

Figure 4 Network cell layout (R1 = 1.6 km, R2 = 1.2 km, R3 = 0.4 km).

of an operator traffic are listed in Table 2 which are adopted from [25,26].

Finally, the serving and the violation weights of the services are based on general service classes: conversational (0.4), streaming (0.3), interactive (0.2), and best effort (0.05); in order not to compromise the objective function for having a higher fairness, the fairness weight, W^f, is heuristically chosen to be equal to the lowest serving weight (0.05).

5 Analysis of results

Results for the reference scenario and its variation were obtained, being presented and analysed from three main perspectives: the total network capacity and the capacity of VNOs, the allocated data rate to each service of a VNO in the reference scenario, and finally the allocated data rate to each service class in VNO GB.

5.1 Total network and VNO capacity

The total network capacity of network is achieved by obtaining the PDF of different RATs, as presented in (4). One compares the concept of virtualisation of radio resources and RAN sharing by considering the CDFs of the total network. Since all three VNOs have the same traffic demand, RRUs are divided into three equal parts in RAN sharing, whereas in the V-RAN approach all RRUs are aggregated. Taking the data from Table 1, the CDF of the total network for RAN sharing and the V-RAN approach using (6) is obtained (Figure 5).

For the sake of simplicity, the RAN sharing CDF using one third of resources is multiplied by three. However, it should be reminded that the total capacity of the network using all the aggregation can be achieved by convolving the PDF of each spectrum slice and not simply summing them. It can be seen that for 50% of the time, the total V-RAN network capacity is 1,800 Mbps, where RAN sharing offers less than 1,782 Mbps. The highest difference can be seen where the CDF is equal to 0.1, in which case, the relative data rate for the V-RAN is 1 725 Mbps, while RAN sharing offers only 1,656 Mbps.

Figure 6 illustrates the total network capacity when different number of cells is used to cover the serving area. The total network capacity with the 16 cells (i.e. the reference scenario) is 1,800 Mbps. It reduces to 48.5% of its initial value (i.e. 872.4 Mbps) when the full

Table 1 Different RAT cell radius

RAT	Number cells	System	$N_{RRU}^{RAT_i}$:	Total RRUs
OFDMA	16	LTE	500	8,000
CDMA	1.7	UMTS	45	80
TDMA	1	GSM	75	75

Table 2 Network traffic mixture

Service		Volume [%]	W_{ji}^{Srv}	W_{ji}^{v}
VoIP (VoI)		1.0	0.04	0.36
Music streaming (MuS)		3.0	0.03	0.27
File sharing (FTP)		3.5	0.02	0.18
Web browsing (WWW)		11.9	0.02	0.18
Social networking (SoN)		14.4	0.02	0.18
Email (Ema)		1.0	0.005	0.045
M2M	Smart metres (MMM)	1.475	0.005	0.045
	e-Health (MME)	1.475	0.02	0.18
	ITS (MMI)	1.475	0.04	0.36
	Surveillance (MMS)	1.475	0.03	0.27
Mobile video	Video calling (ViC)	2.75	0.04	0.36
	Video streaming (VoS)	56.95	0.03	0.27

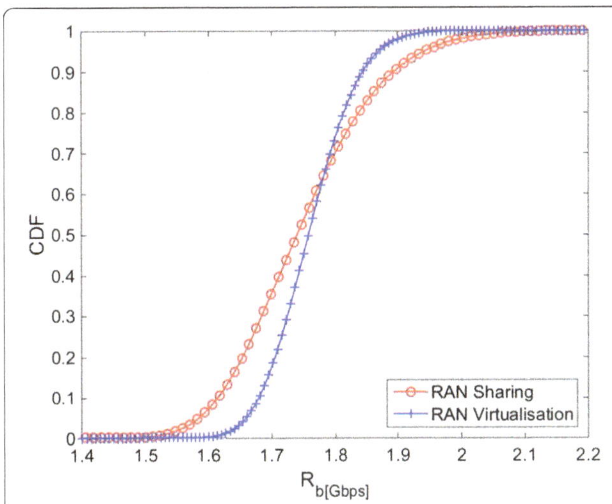

Figure 5 CDF of network capacity for V-RAN and RAN sharing.

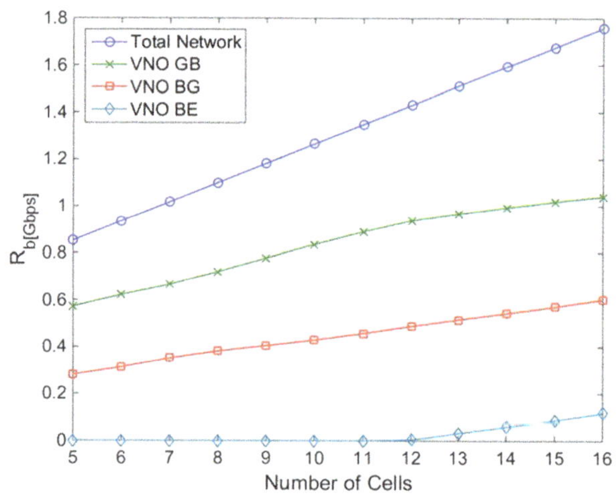

Figure 6 The total network and VNO capacity.

Table 3 Allocated data rate to services when all the cells are available

Services	$R_{b_{ji}}^{Srv}$ [Mbps]		
	VNO GE	VNO BG	VNO BE
VoIP	19.12	21.47	16.69
Music streaming	40.30	25.96	11.62
File sharing	41.21	24.48	7.74
Web browsing	121.54	64.64	7.74
Social networking	145.44	93.33	7.74
Email	11.50	6.72	1.94
M2M-SM	15.08	8.51	1.94
M2M-eH	20.89	14.32	7.74
M2M-ITS	26.30	23.26	16.69
M2M-SV	24.77	18.19	11.62
Video streaming	556.24	283.93	11.62
Video call	42.83	29.82	16.69

coverage is obtained by only five OFDMA BSs. According to the scenario definition, the total guaranteed data rate is 1,434.37 Mbps, which means that there is enough capacity to serve the guaranteed data rate plus the best-effort services. The layout with 12 cells is the marginal point where the network capacity and the total guaranteed data rate are almost equal; the use of only five cells provides a very low capacity.

Considering the allocated data rate to VNOs, as expected, all the capacity allocated to the VNOs decreases by reducing the number of cells. Capacity reduction has a higher impact on VNO BE (the best-effort operator) comparing to VNO's GE and BG, since the network tries to meet these latter VNOs' guaranteed capacity before serving the best-effort one. When there are 12 cells, VNO BE gets almost no data rate, but the other two VNOs still have a relatively acceptable data rate. In this situation, the total network capacity is still higher than the total guaranteed capacity.

The network capacity shrinks to 1,378.15 Mbps when another cell is reduced, i.e. 11 cells, which is lower than the guaranteed data rate. The violation is inevitable for the cell layout with less than 12 cells. In these situations, the main objective function becomes infeasible and VRRM switches to the objective function with violation, presented in (21). While no capacity is allocated to VNO BE, the other two VNOs share the violation between them. Since the model tries to minimise the weighted average violation, it can be seen that VNO GE always receive a relatively larger portion of the network capacity, since it has a higher guarantee rate.

5.2 Data rate allocation in service level

At the service level, Table 3 presents the allocation of data rates to the services of all three VNOs for the reference case with 16 cells; in these conditions, the VRRM is able to allocate the capacity to all services without violating any constraints. As expected, the highest data rate is allocated to video streaming of VNO GB, since it has the highest guaranteed data rate. The lowest data rates are given to Email and M2M Smart Meter services, since they are background ones with the lowest serving weight. The best demonstration of prioritising the services based on their serving weights can be seen in VNO BE, where there is no minimum guaranteed data rate for the services. The highest capacities belong to VoIP, M2M-ITS, and video calls, which are services from the conversational class with the highest data rates; since these services have the same serving weight, they receive the same capacity. Music, M2M-SV, and video streaming are in the second group, i.e. streaming. Services of the interactive class, i.e. FTP, web browsing, social networking, and M2M-eH, received all 7.74 Mbps. The effect of fairness is very well demonstrated in services of VNO BE; although the services have different serving weights, they are served relatively well based on their serving weight. In addition, services with the same serving weights are allocated the same capacity. For the other two VNOs, the services have different guaranteed capacities, and the fairness effect is not as obvious as in VNO BE.

It is worth noting that Table 3 is also showing an interesting difference of best effort with minimum guaranteed services and guaranteed ones. Guaranteed services are bounded by the maximum capacity, and the allocated capacity cannot go higher than this boundary, while best effort with minimum guaranteed service does not have this limitation. Considering VoIP in VNO GB and VNO BG, it can be seen that the latter is allocated with a higher capacity since this service of VNO GB is assigned with the maximum capacity; for VNO BE, VoIP shows the effect of fairness among services.

5.3 Allocated data rate to each service class in VNO GB

Finally, to study how different services are affected by the changes of the total network capacity from the VNO GB, four service classes are considered, with different serving and violation weights. Figure 7 illustrates the percentage of violation for each of these services by shutting down cells.

Obviously, there is no violation of guaranteed capacity as long as there are more than 12 cells. However, VRRM has no other choice than start violating the level of data rate guaranteed when there are fewer cells. As a matter of fact, the violations have to start by the service with the lowest violation weight. According to the weights presented in Table 2, background services are the first candidate for violation, since they have the lowest serving and violation weights. When there are only 11 cells, the background traffic violation reaches 100%, which means that it is not allocated any capacity at all. Since these services (i.e. Email and M2M-SV) are low-volume services, even the total violation of their capacity cannot cover the shortage of network capacity; therefore, interactive services, the ones with the second lowest violation weights, have also to be subject of capacity violations. Finally, when there are only five cells (the worst network situation considered in this paper), background and interactive services have to be totally shut down (i.e. 100% violation), video streaming suffering a violating up to 8%. Nevertheless, conversational services, the ones with the highest weights, are served without any violations.

In conclusion, it can be seen that by aggregating all the radio resources, their efficiency use increases. In addition, the effect of having different SLAs (i.e. best effort or guaranteed) and priority (i.e. serving weights) is demonstrated by means of these results.

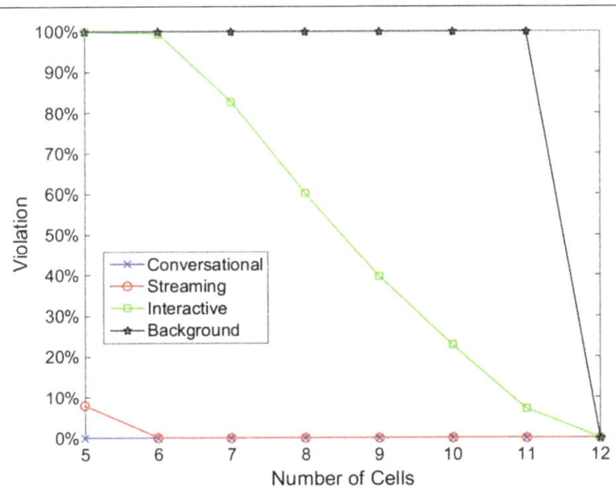

Figure 7 Percentage of violation of services of VNO GB.

6 Conclusions

This paper presents the concept of virtualisation of radio resources as the final step towards an end-to-end virtual network by realising a virtual wireless link. It is suggested to aggregate all the physical radio resources and to have a central management to offer VNO's Capacity-as-a-Service. In this solution, VNOs no longer have to deal with the management of physical infrastructure, but rather to ask for capacity in order to serve their subscribers. Using the proposed technique, not only heterogeneous access networks are shared among multiple VNOs but also the ease of use and VNO specific configuration are achieved as the result of network element abstraction and isolation.

In addition, a model for the management of virtual radio resources with the capability of supporting various situations is proposed. The model takes a number of available RRUs in different RATs as the input and maps them onto the total network capacity. Having an estimation of the available capacity, the model formulates the allocation problem into a linear optimisation problem. The objective function in this problem is to maximise the weighted throughput of network and fairness among the services of the VNOs. The suggested model also tries to meet all guaranteed service levels while offering fairness. When the model fails to find any feasible solution to serve all guaranteed data rates, due to the shortage of resources, it introduces violations to guaranteed data rates. This approach changes the former infeasible solution to a feasible one, and the model aims at minimising the summation of weighted violations. This way, the services with less importance are facing the violation of guaranteed data rates while the more important ones are served properly.

The proposed model is evaluated in a practical set of scenarios, and numeric results are obtained for them. The results indicate that to cover the serving area with the mentioned number of VNOs, subscribers, and SLAs, at least 11 OFDMA cells are required. The reference scenario assumes 16 cells, where the total network capacity is estimated as 1,800 Mbps. By reducing the number of cells to five, the total network capacity shrinks to almost half of its initial value (i.e. 48% of reference case). The changes in network capacity mostly influence the VNO with best-effort services, while the other types of VNOs suffer relatively less from the reduction of resources. Among guaranteed services, the violation starts from the service(s) with the lowest violation weight, which in our case study are background ones. The numeric results justify that the model is able to prioritise the service according to their serving and violation weights. In the worst case studied in this paper, background and interactive services are totally shutdown while conversational ones experience no violation, as expected.

Competing interests
The authors declare that they have no competing interests.

Acknowledgements
The research leading to these results was partially funded by the European Union's Seventh Framework Programme Mobile Cloud Networking project (FP7-ICT-318109).

References
1. Cisco Systems, *Global Mobile Data Traffic Forecast Update, 2012 - 2017* (Visual Network Index White Paper, Cisco Systems, Palo Alto, CA, USA, 2013). http://www.cisco.com/c/en/us/solutions/collateral/service-provider/visual-networking-index-vni/white_paper_c11-520862.pdf
2. H Guan, T Kolding, P Merz, Discovery of cloud-RAN, in *Proc. of NSN cloud-RAN workshop, Beijing, China*, 2010
3. D Sabella, P Rost, S Yingli, E Pateromichelakis, U Salim, P Guitton-Ouhamou, M Di Girolamo, G Giuliani, RAN as a service: challenges of designing a flexible RAN architecture in a cloud-based heterogeneous mobile network, in *Proc. of FuNeMS'2013 - 22nd future networks and mobile summit, Lisbon, Portugal*, 2013
4. R Yrjo, D Rushil, Cloud computing in mobile networks - case MVNO, in *Proc. of ICIN'11 - 15th international conference on intelligence in next generation networks, Berlin, Germany*, 2011
5. A Klein, C Mannweiler, J Schneider, HD Schotten, Access schemes for mobile cloud computing, in *Proc. of MDM'10 - 11th international conference on mobile data management, Kansas City, MI, USA*, 2010
6. C Mobile, *C-RAN - road towards green radio access network* (China, White Paper, China Mobile Research Institute, Shanghai, 2011) (http://labs.chinamobile.com/cran/wp-content/uploads/CRAN_white_paper_v2_5_EN.pdf)
7. M Chiosi, D Clarke, P Willis, A Reid, J Feger, M Bugenhagen, W Khan, M Fargano, C Cui, H Deng, J Benitez, U Michel, H Damker, K Ogaki, T Matsuzaki, *Network function virtualisation: an introduction, benefits, enabler, challenges, and call for action* (France, White Paper, European Telecommunications Standards Institute, Sophia-Antipolis, 2012) (http://portal.etsi.org/NFV/NFV_White_Paper.pdf)
8. Y Zaki, Z Liang, C Goerg, A Timm-Giel, LTE wireless virtualization and spectrum management, in *Proc. of WMNC'2010 - 3rd joint IFIP wireless and mobile networking conference, Budapest, Hungary*, 2010
9. L Caeiro, FD Cardoso, LM Correia, Adaptive allocation of virtual radio resources over heterogeneous wireless networks, in *Proc. of EW'2012 - 18th European wireless conference, Poznan, Poland*, 2012
10. Z Liang, L Ming, Y Zaki, A Timm-Giel, C Gorg, LTE virtualization: from theoretical gain to practical solution, in *Proc. of 23rd International Teletraffic Congress, San Francisco, CA, USA*, 2011
11. X Costa-Perez, J Swetina, G Tao, R Mahindra, S Rangarajan, Radio access network virtualization for future mobile carrier networks. IEEE Commun Mag **51**(7), 27–35 (2013)
12. T Metsch, P Gray (eds.), *"Infrastructure management foundations - specifications & design for mobile cloud framework"*, 2013. Deliverable D3.1, mobile cloud networking project, (http://www.mobile-cloud-networking.eu)
13. S Khatibi, LM Correia, Modelling of virtual radio resource management for cellular heterogeneous access networks, in *Proc. of PIMRC'14 - IEEE 25th annual international symposium on personal, indoor, and mobile radio communications, Washington, DC, USA*, 2014
14. A Khan, A Zugenmaier, D Jurca, W Kellerer, Network virtualization: a hypervisor for the Internet? IEEE Commun Mag **50**(1), 136–143 (2012)
15. B Haberland, F Derakhshan, H Grob-Lipski, R Klotsche, W Rehm, P Schefczik, M Soellner, Radio base stations in the cloud. Bell Labs Tech J **18**(1), 129–152 (2013)
16. J Pérez-Romero, X Gelabert, O Sallent, Radio resource management for heterogeneous wireless access, in *Heterogeneous wireless access networks*, ed. by E Hossain (USA, Springer, New York, NY, 2009)
17. HS Dhillon, RK Ganti, F Baccelli, JG Andrews, Modelling and analysis of K-tier downlink heterogeneous cellular networks. IEEE J Sele Areas in Comm **30**(3), 550–560 (2012)
18. Jacinto, N. M. d. S., "Performance gains evaluation from UMTS/HSPA+ to LTE at the radio network level", Master of Science, Department of Electrical and Computer Engineering, Instituto Superior Técnico, University of Lisboa, Lisbon, Portugal, 2009 (http://grow.inov.pt/wp-content/uploads/2014/01/2009_NunoJacinto.pdf).
19. A Papoulis, SU Pillai, *Probability, random variables, and stochastic processes* (McGraw-Hill, New York, NY, USA, 2002)
20. M Buehrer, RM Buehrer, *Code division multiple access (CDMA)* (Morgan & Claypool Publishers, San Rafael, CA, USA, 2006)
21. MATLAB and Statistics Toolbox Release 2012b, The MathWorks, Inc., Natick, Massachusetts, United States, http://www.mathworks.com, Accessed Feb, 2014.
22. Y Zhang, *Solving large-scale linear programs by interior-point methods under the MATLAB environment, Technical Report TR96-01* (University of Maryland, Baltimore, MD, USA, 1995)
23. S Mehrotra, On the implementation of a primal-dual interior point method. SIAM J Optim **2**, 575–601 (1992)
24. C Systems, *The Zettabyte era - trends and analysis* (Visual Network Index White Paper, Cisco Systems, Palo Alto, CA, USA, 2013) (http://www.cisco.com/c/en/us/solutions/collateral/service-provider/visual-networking-index-vni/VNI_Hyperconnectivity_WP.pdf)
25. C Systems, CVN Index, *Global mobile data traffic forecast update, 2011-2016, from visual network index (VNI) white paper* (Cisco Systems, California, USA, 2012)
26. Scalable and adaptive Internet solutions (SAIL) project, (http://www.sail-project.eu/), 2015.

Permissions

All chapters in this book were first published in EURASIP-JWCN, by Springer; hereby published with permission under the Creative Commons Attribution License or equivalent. Every chapter published in this book has been scrutinized by our experts. Their significance has been extensively debated. The topics covered herein carry significant findings which will fuel the growth of the discipline. They may even be implemented as practical applications or may be referred to as a beginning point for another development.

The contributors of this book come from diverse backgrounds, making this book a truly international effort. This book will bring forth new frontiers with its revolutionizing research information and detailed analysis of the nascent developments around the world.

We would like to thank all the contributing authors for lending their expertise to make the book truly unique. They have played a crucial role in the development of this book. Without their invaluable contributions this book wouldn't have been possible. They have made vital efforts to compile up to date information on the varied aspects of this subject to make this book a valuable addition to the collection of many professionals and students.

This book was conceptualized with the vision of imparting up-to-date information and advanced data in this field. To ensure the same, a matchless editorial board was set up. Every individual on the board went through rigorous rounds of assessment to prove their worth. After which they invested a large part of their time researching and compiling the most relevant data for our readers.

The editorial board has been involved in producing this book since its inception. They have spent rigorous hours researching and exploring the diverse topics which have resulted in the successful publishing of this book. They have passed on their knowledge of decades through this book. To expedite this challenging task, the publisher supported the team at every step. A small team of assistant editors was also appointed to further simplify the editing procedure and attain best results for the readers.

Apart from the editorial board, the designing team has also invested a significant amount of their time in understanding the subject and creating the most relevant covers. They scrutinized every image to scout for the most suitable representation of the subject and create an appropriate cover for the book.

The publishing team has been an ardent support to the editorial, designing and production team. Their endless efforts to recruit the best for this project, has resulted in the accomplishment of this book. They are a veteran in the field of academics and their pool of knowledge is as vast as their experience in printing. Their expertise and guidance has proved useful at every step. Their uncompromising quality standards have made this book an exceptional effort. Their encouragement from time to time has been an inspiration for everyone.

The publisher and the editorial board hope that this book will prove to be a valuable piece of knowledge for researchers, students, practitioners and scholars across the globe.

List of Contributors

Ao Zhou
State Key Laboratory of Networking and Switching Technology, Beijing University of Posts and Telecommunications, Beijing 100876, China

Jinglin Li
State Key Laboratory of Networking and Switching Technology, Beijing University of Posts and Telecommunications, Beijing 100876, China

Qibo Sun
State Key Laboratory of Networking and Switching Technology, Beijing University of Posts and Telecommunications, Beijing 100876, China

Cunqun Fan
National Satellite Meteorological Centre, China Meteorological Administration, Beijing 100876, China

Tao Lei
State Key Laboratory of Networking and Switching Technology, Beijing University of Posts and Telecommunications, Beijing 100876, China

Fangchun Yang
State Key Laboratory of Networking and Switching Technology, Beijing University of Posts and Telecommunications, Beijing 100876, China

Ali Sercan Kesten
Division of Civil Engineering, Department of Engineering, Işık University, Istanbul, Turkey

Murat Ergün
Transportation Engineering Study Group, Department of Civil Engineering, Istanbul Technical University, Istanbul, Turkey

Hassan Halabian
Department of Systems and Computer Engineering, Carleton University, 1125 Colonel By Dr, Ottawa K1S 5B6, Canada

Perumalraja Rengaraju
Department of Information Technology, Velammal College of Engineering and Technology, Viraganoor, Madurai 625 009, India

Chung-Horng Lung
Department of Systems and Computer Engineering, Carleton University, 1125 Colonel By Dr, Ottawa K1S 5B6, Canada

Ioannis Lambadaris
Department of Systems and Computer Engineering, Carleton University, 1125 Colonel By Dr, Ottawa K1S 5B6, Canada

Tao Jing
School of Electronics and Information Engineering, Beijing Jiaotong University, No.3 Shangyuancun, 100044 Beijing, China

Xuewei Cui
School of Electronics and Information Engineering, Beijing Jiaotong University, No.3 Shangyuancun, 100044 Beijing, China

Wei Cheng
Department of Computer Science, Virginia Commonwealth University, 401 West Main Street, 23284 Richmond, VA, USA

Shixiang Zhu
School of Electronics and Information Engineering, Beijing Jiaotong University, No.3 Shangyuancun, 100044 Beijing, China

Yan Huo
School of Electronics and Information Engineering, Beijing Jiaotong University, No.3 Shangyuancun, 100044 Beijing, China

Xiuzhen Cheng
Department of Computer Science, The George Washington University, 801 22nd Street NW, 20052 Washington DC, USA

Wei Liu
University Ghent - iMinds, Gaston Crommenlaan 8, 9050 Ghent, Belgium

Mikolaj Chwalisz
Telecommunication Networks Group (TKN)/TU Berlin, Sekretariat FT 5, Einsteinufer 25, 10587 Berlin, Germany. 3imec, Kapeldreef 75, Leuven, 3001, Belgium, 3000 Leuven, Belgium

Carolina Fortuna
University Ghent - iMinds, Gaston Crommenlaan 8, 9050 Ghent, Belgium

Eli De Poorter
University Ghent - iMinds, Gaston Crommenlaan 8, 9050 Ghent, Belgium

Jan Hauer
Telecommunication Networks Group (TKN)/TU Berlin, Sekretariat FT 5, Einsteinufer 25, 10587 Berlin, Germany

Daan Pareit
University Ghent - iMinds, Gaston Crommenlaan 8, 9050 Ghent, Belgium

Lieven Hollevoet
imec, Kapeldreef 75, Leuven, 3001, Belgium, 3000 Leuven, Belgium

Ingrid Moerman
University Ghent - iMinds, Gaston Crommenlaan 8, 9050 Ghent, Belgium

Jiri Blumenstein
Department of Radio Electronics, Brno University of Technology, Technicka 12, 612 00, Brno, Czech Republic

Ales Proke
Department of Radio Electronics, Brno University of Technology, Technicka 12, 612 00, Brno, Czech Republic

Tomas Mikulasek
Department of Radio Electronics, Brno University of Technology, Technicka 12, 612 00, Brno, Czech Republic

Roman Marsalek
Department of Radio Electronics, Brno University of Technology, Technicka 12, 612 00, Brno, Czech Republic

Thomas Zemen
AIT Austrian Institute of Technology GmbH, Donau-City-Straße 1, 1220 Vienna, Austria

Christoph Mecklenbräuker
Department of Radio Electronics, Brno University of Technology, Technicka 12, 612 00, Brno, Czech Republic
Institute of Telecommunications, Vienna University of Technology, Gußhausstraße 25/E 389, 1040 Vienna, Austria

Yasamine Zrelli
Université Européenne de Bretagne, 5 Boulevard Laënnec, 35000 Rennes, France
Université de Brest; CNRS, UMR 6285 Lab-STICC, 6 avenue Victor Le Gorgeu, 29238 Brest, France

Roland Gautier
Université Européenne de Bretagne, 5 Boulevard Laënnec, 35000 Rennes, France
Université de Brest; CNRS, UMR 6285 Lab-STICC, 6 avenue Victor Le Gorgeu, 29238 Brest, France

Eric Rannou
Université Européenne de Bretagne, 5 Boulevard Laënnec, 35000 Rennes, France
Université de Brest; CNRS UMR 6205, Laboratoire de Mathématiques Bretagne Atlantique, 6 avenue Victor Le Gorgeu, 29238 Brest, France

MélanieMarazin
Université Européenne de Bretagne, 5 Boulevard Laënnec, 35000 Rennes, France
Université de Brest; CNRS, UMR 6285 Lab-STICC, 6 avenue Victor Le Gorgeu, 29238 Brest, France

Emanuel Radoi
Université Européenne de Bretagne, 5 Boulevard Laënnec, 35000 Rennes, France
Université de Brest; CNRS, UMR 6285 Lab-STICC, 6 avenue Victor Le Gorgeu, 29238 Brest, France

Mohammed Ridouani
Computer Sciences Department, GREENTIC/UH2C/EST, Casablanca, Morocco

Aawatif Hayar
Computer Sciences Department, GREENTIC/UH2C/ENSEM, Casablanca, Morocco

Abdelkrim Haqiq
Department of Mathematics and Computer, GREENTIC/UH1/FST, Settat, Morocco

Yang Yang
State Key Laboratory of Networking and Switching Technology, Beijing University of Posts and Telecommunications, No. 10 Xi Tu Cheng Road, Beijing 100876, China

Qian Liu
State Key Laboratory of Networking and Switching Technology, Beijing University of Posts and Telecommunications, No. 10 Xi Tu Cheng Road, Beijing 100876, China

Zhipeng Gao
State Key Laboratory of Networking and Switching Technology, Beijing University of Posts and Telecommunications, No. 10 Xi Tu Cheng Road, Beijing 100876, China

Xuesong Qiu
State Key Laboratory of Networking and Switching Technology, Beijing University of Posts and Telecommunications, No. 10 Xi Tu Cheng Road, Beijing 100876, China

Lanlan Rui
State Key Laboratory of Networking and Switching Technology, Beijing University of Posts and Telecommunications, No. 10 Xi Tu Cheng Road, Beijing 100876, China

Xin Li
China Electronic Systems Engineering Corporation, Xiaotun Road, Beijing 100039, China

Chamara N Devanarayana
Department of Electrical and Computer Engineering, University of Manitoba, E1-551 EITC, 75A Chancellors Cir., R3T 5V6 Winnipeg, MB, Canada

Attahiru S Alfa
Department of Electrical and Computer Engineering, University of Manitoba, E3-504A EITC, 75A Chancellors Cir., R3T 5V6 Winnipeg, MB, Canada

Erik Ortiz Guerra
Central University of Las Villas (UCLV), Camajuaní Highway Km 5 1/2, Santa Clara, Cuba

Vitalio Alfonso Reguera
Central University of Las Villas (UCLV), Camajuaní Highway Km 5 1/2, Santa Clara, Cuba

Richard D Souza
Federal University of Technology, Paraná (UTFPR), Avenida Sete de Setembro, Curitiba, Brazil

Evelio G Fernández
Federal University of Paraná (UFPR), Rua XV de Novembro, 1299 - Centro, Curitiba, Brazil

Marcelo E Pellenz
Pontifical Catholic University of Paraná (PUC), Rua Imaculada Conceicao, 1155 Curitiba, Brazil

Dawei Li
The Key Laboratory of All Optical Network and Advanced Telecommunication Network of Ministry of Education, No.3 Shangyuancun, Haidian District, Beijing, China, 100044
The Institute of Lightwave Technology, Beijing Jiaotong University, No.3 Shangyuancun, Haidian District, Beijing, China, 100044

Junhong Wang
The Key Laboratory of All Optical Network and Advanced Telecommunication Network of Ministry of Education, No.3 Shangyuancun, Haidian District, Beijing, China, 100044
The Institute of Lightwave Technology, Beijing Jiaotong University, No.3 Shangyuancun, Haidian District, Beijing, China, 100044

Meie Chen
The Key Laboratory of All Optical Network and Advanced Telecommunication Network of Ministry of Education, No.3 Shangyuancun, Haidian District, Beijing, China, 100044
The Institute of Lightwave Technology, Beijing Jiaotong University, No.3 Shangyuancun, Haidian District, Beijing, China, 100044

Zhan Zhang
The Key Laboratory of All Optical Network and Advanced Telecommunication Network of Ministry of Education, No.3 Shangyuancun, Haidian District, Beijing, China, 100044
The Institute of Lightwave Technology, Beijing Jiaotong University, No.3 Shangyuancun, Haidian District, Beijing, China, 100044

Zheng Li
The Key Laboratory of All Optical Network and Advanced Telecommunication Network of Ministry of Education, No.3 Shangyuancun, Haidian District, Beijing, China, 100044
The Institute of Lightwave Technology, Beijing Jiaotong University, No.3 Shangyuancun, Haidian District, Beijing, China, 100044

Errong Pei
Engineering Research Center of Communication Network Testing Technologies, Chongqing University of Posts and Telecommunications, No. 2, Chongwen Rd, Nan'an Dist, Chongqing 400065, China

Haozhe Han
Engineering Research Center of Communication Network Testing Technologies, Chongqing University of Posts and Telecommunications, No. 2, Chongwen Rd, Nan'an Dist, Chongqing 400065, China

Zehua Sun
Engineering Research Center of Communication Network Testing Technologies, Chongqing University of Posts and Telecommunications, No. 2, Chongwen Rd, Nan'an Dist, Chongqing 400065, China

Bin Shen
Engineering Research Center of Communication Network Testing Technologies, Chongqing University of Posts and Telecommunications, No. 2, Chongwen Rd, Nan'an Dist, Chongqing 400065, China

Tianqi Zhang
Chongqing Key Laboratory of Signal and Information Processing, Chongqing University of
Posts and Telecommunications (CQUPT), No. 2, Chongwen Rd, Nan'an Dist, Chongqing 400065, China

Juan Fang
NYU WIRELESS, Department of Electrical and Computer Engineering, Polytechnic School of Engineering of New York University, 6 MetroTech Center, Brooklyn 11201, USA

I-Tai Lu
NYU WIRELESS, Department of Electrical and Computer Engineering, Polytechnic School of Engineering of New York University, 6 MetroTech Center, Brooklyn 11201, USA

Zhihui Liu
Key Lab of Universal Wireless Communications, Ministry of Education, Beijing University of Posts and Telecommunications, Xitucheng Road 10, Haidian District, 100876 Beijing, China

Wenjun Xu
Key Lab of Universal Wireless Communications, Ministry of Education, Beijing University of Posts and Telecommunications, Xitucheng Road 10, Haidian District, 100876 Beijing, China

Shengyu Li
Key Lab of Universal Wireless Communications, Ministry of Education, Beijing University of Posts and Telecommunications, Xitucheng Road 10, Haidian District, 100876 Beijing, China

Jiaru Lin
Key Lab of Universal Wireless Communications, Ministry of Education, Beijing University of Posts and Telecommunications, Xitucheng Road 10, Haidian District, 100876 Beijing, China

Pavel Kukolev
Department of Radio Electronics, Brno University of Technology, Technicka 12, 61600 Brno, Czech Republic

Aniruddha Chandra
Department of Radio Electronics, Brno University of Technology, Technicka 12, 61600 Brno, Czech Republic

Tomáš Mikulášek
Department of Radio Electronics, Brno University of Technology, Technicka 12, 61600 Brno, Czech Republic

Aleš Prokeš
Department of Radio Electronics, Brno University of Technology, Technicka 12, 61600 Brno, Czech Republic

Thomas Zemen
AIT Austrian Institute of Technology GmbH, 1220 Vienna, Austria
Telecommunications Research Center Vienna (FTW), Donau-City-Straße 1, 1220 Vienna, Austria

Christoph F Mecklenbräuker
Institute of Telecommunications, Vienna University of Technology, Gusshausstraße 25, 1040 Vienna, Austria

Changho Yun
Ocean System Engineering Research Division, KRISO, 32, Yuseong-daero 1312 beon-gil, Yuseong-gu, Daejeon 305-343, Republic of Korea

Yong-kon Lim
Ocean System Engineering Research Division, KRISO, 32, Yuseong-daero 1312 beon-gil, Yuseong-gu, Daejeon 305-343, Republic of Korea

Yuzhong Chen
College of Mathematics and Computer Science, Fuzhou University, No. 2 Xue Yuan Road, 350108 Fuzhou, China
Fujian Key Laboratory of Network Computing and Intelligent Information Processing, No. 2 Xue Yuan Road, 350108 Fuzhou, China

Mingyue Fang
College of Mathematics and Computer Science, Fuzhou University, No. 2 Xue Yuan Road, 350108 Fuzhou, China
Fujian Key Laboratory of Network Computing and Intelligent Information Processing, No. 2 Xue Yuan Road, 350108 Fuzhou, China

Song Shi
College of Mathematics and Computer Science, Fuzhou University, No. 2 Xue Yuan Road, 350108 Fuzhou, China
Fujian Key Laboratory of Network Computing and Intelligent Information Processing, No. 2 Xue Yuan Road, 350108 Fuzhou, China

Wenzhong Guo
College of Mathematics and Computer Science, Fuzhou University, No. 2 Xue Yuan Road, 350108 Fuzhou, China
Fujian Key Laboratory of Network Computing and Intelligent Information Processing, No. 2 Xue Yuan Road, 350108 Fuzhou, China

Xianghan Zheng
College of Mathematics and Computer Science, Fuzhou University, No. 2 Xue Yuan Road, 350108 Fuzhou, China
Fujian Key Laboratory of Network Computing and Intelligent Information Processing, No. 2 Xue Yuan Road, 350108 Fuzhou, China

Yutao Sui
Department of Signals and Systems, Chalmers University of Technology, EDIT-huset, Hörsalsvägen 9-11 Gothenburg, SE-41296, Sweden

Ismail Guvenc
Electrical and Computer Engineering, Florida International University, 10555 W Flagler Street, EC 3142, Miami, FL 33174-1630, USA

Tommy Svensson
Department of Signals and Systems, Chalmers University of Technology, EDIT-huset, Hörsalsvägen 9-11 Gothenburg, SE-41296, Sweden

Sina Khatibi
Department of Electrical and Computer Engineering, Instituto Superior Técnico (IST), University of Lisbon, Av. Rovisco Pais, Lisbon 1049-001, Portugal
INOV-INESC, Rua Alves Redol Lisbon, 1000-029, Portugal

Luis M Correia
Department of Electrical and Computer Engineering, Instituto Superior Técnico (IST), University of Lisbon, Av. Rovisco Pais, Lisbon 1049-001, Portugal
INOV-INESC, Rua Alves Redol Lisbon, 1000-029, Portugal